AF521458

R. Gupta's®

POPULAR MASTER GUIDE

M.Sc. BIOTECHNOLOGY & BIOCHEMISTRY

Entrance Examination

Dr. B.B. Singh
Ph.D., FISEP, FSLW

2026
EDITION

RAMESH PUBLISHING HOUSE, NEW DELHI

Published by
O.P. Gupta *for* Ramesh Publishing House

Admin. Office
12-H, New Daryaganj Road, Opp. Officers' Mess,
New Delhi-110002 ✆ 23275224, 23245124

E-mail: info@rameshpublishinghouse.com
For Online Shopping: www.rameshpublishinghouse.com

Showroom
- Balaji Market, Nai Sarak, Delhi-6 ✆ 23253720, 23282525
- 4457, Nai Sarak, Delhi-6, ✆ 23918938

© *Reserved with the Publisher*

No Part of this book may be reproduced or transmitted in any form or by any means, electronic or mechanical including photocopying, recording or by any transformation storage and retrieval system without written permission from the Publisher.

Indemnification Clause: This book is being sold/distributed subject to the exclusive condition that neither the author nor the publishers, individually or collectively, shall be responsible to indemnify the buyer/user/possessor of this book beyond the selling price of this book for any reason under any circumstances. If you do not agree to it, please do not buy/accept/use/possess this book.

Book Code: R-1859

ISBN: 978-93-5012-896-1

Price: ₹ 360

Printed at: Deepak Offset, Delhi

CONTENTS

Biotechnology and Biochemistry

●●●

Previous Years' Paper

GRADUATE APTITUDE TEST–BIOTECHNOLOGY (GAT-B)

ENTRANCE EXAM-2024

(For M.Sc. Biotechnology and Allied Areas; M.Tech. Biotechnology and Allied Areas; M.Sc. Agricultural Biotechnology and M.V.Sc. Animal Biotechnology)

(Exam held on 20 April, 2024)

SECTION-A

1. Assume that a narrow tunnel is dug between two diametrically opposite points on the earth's surface. If a particle is released in this tunnel, it will execute a simple harmonic motion. What will be the time period of SHM of this particle?

A. $\frac{1}{2\pi}\sqrt{\frac{R^3}{GM}}$ B. $\frac{1}{2\pi}\sqrt{\frac{GM}{R^3}}$

C. $2\pi\sqrt{\frac{R^3}{GM}}$ D. $2\pi R\sqrt{\frac{1}{GM}}$

2. If a body is performing uniform circular motion with velocity v and radius R, then identify the true statements from the following:

(*a*) Its velocity v is constant.

(*b*) Acceleration is always directed towards the centre and its magnitude is $a = v^2/R$.

(*c*) Angular momentum is constant in magnitude but its direction keeps changing.

(*d*) Angular velocity of the body $\omega = v/R$.

Choose the most appropriate answer from the options given below.

A. (*a*) and (*c*) only

B. (*b*) and (*d*) only

C. (*a*), (*b*) and (*d*) only

D. (*a*) and (*d*) only

3. A proton and a deuteron moving with equal kinetic energy enter perpendicularly into a magnetic field. What will be the ratio of radii of the circular path of the proton to that of the deuteron?

A. 1 B. 2

C. $\frac{1}{2}$ D. $\frac{1}{\sqrt{2}}$

4. A big oil droplet of radius 10 cm is broken into a thousand equal droplets. What will be the gain in surface energy? (Surface tension of the oil is 0.1 Nm^{-1})

A. 5 J B. 10 J

C. 0.11 J D. 0.25 J

5. Two parallel rail tracks run east-west. Train P moves in east direction with a speed of 36 kmh^{-1} and train Q moves with a speed of 72 kmh^{-1} in west direction. What is the velocity of Q with respect to P?

A. 30 m/s from east to west

B. 30 m/s from west to east

C. 36 m/s from west to east

D. 10 m/s from east to west

1. C **2.** B **3.** D **4.** C **5.** A

6. Identify the statement which in NOT true for a 'conservative force'.

A. The work done by the conservative force depends only on the end points.
B. The work done by a conservative force in a closed path is zero.
C. Spring force and frictional force are conservative.
D. The total mechanical energy of a system is conserved if forces doing work on it are conservative.

7. A boy sitting on a surface inside a satellite moving around the earth feels weightless because:

A. the earth does not attract the object in a satellite
B. the reaction on the person balances the gravitational force
C. a person sitting in the satellite is not accelerated
D. the normal force (reaction) is zero

8. Two cylindrical vessels of equal cross-sectional area A contain water up to height h_1 and h_2. The vessels are interconnected so that the levels in them become equal. What is the work done during this process if ρ is the density of water?

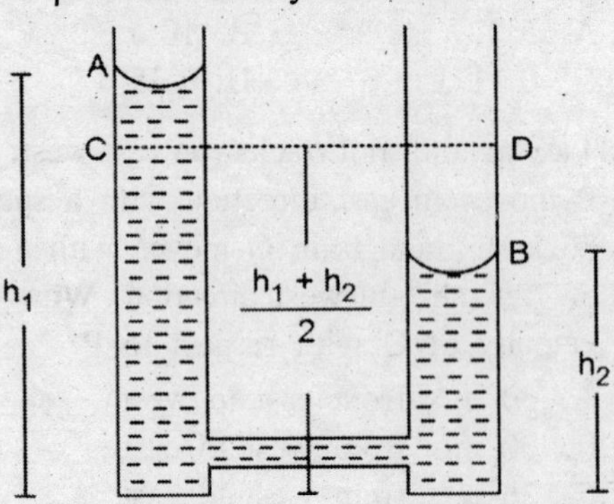

A. $\rho.A.(h_1 - h_2)$ B. $\rho.A.(h_1 - h_2)/2$
C. $\rho.A.(h_1 - h_2)^2.g$ D. $\rho.A.[(h_1 - h_2)/2]^2.g$

9. A bullet of mass 20 g, moving at 50 m/s penetrates 20 cm into a wooden block. What is the magnitude of the force exerted on the wooden block?

A. 625 N B. 225 N
C. 125 N D. 725 N

10. Identify which of the following statements regarding significant figures are correct:

(*a*) 6.405 has four significant figures
(*b*) 12300 has five significant figures
(*c*) 0.00421 has five significant figures
(*d*) 4.500 has four significant figures

Choose the most appropriate answer from the options given below.

A. (*a*), (*b*) and (*c*) only
B. (*a*) and (*d*) only
C. (*c*) and (*d*) only
D. (*b*) and (*d*) only

11. The cross product of vector $\vec{A}$ and vector $\vec{B}$ has a magnitude of 50 unit, where vector $\vec{A}$ has a magnitude of 10. The angle between vector $\vec{A}$ and $\vec{B}$ is 60 degrees. What is the magnitude of vector $\vec{B}$?

A. $\frac{5}{\sqrt{2}}$ B. $\frac{10}{\sqrt{2}}$
C. $\frac{10}{\sqrt{3}}$ D. $\frac{5}{\sqrt{3}}$

12. A resistor R dissipates power P when connected to a generator. If another resistor Q is put in series with R, the power dissipated by R will:

A. Increase
B. Decrease
C. Remain the same
D. Increase or decrease depending on the values of R and Q

6. C **7.** D **8.** D **9.** C **10.** B **11.** C **12.** B

13. The electric charge on a body is always an integral multiple of '*e*' where '*e*' is the charge that an electron or proton carries. This concept is known as:

A. Additivity of charges
B. Quantization of charges
C. Conservation of charges
D. Principle of superposition

14. Match items in List-I with items in List-II.

List-I (Type of thermodynamic process)	**List-II (Work done)**
(*a*) Isothermal	I. Zero
(*b*) Adiabatic	II. $\mu R\ (T_2 - T_1)$
(*c*) Isochoric	III. $\mu RT \ln V_2/V_1$
(*d*) Isobaric	IV. $\mu R\ (T_1 - T_2)/(\gamma - 1)$

Choose the correct answer from the options given below:

	(*a*)	(*b*)	(*c*)	(*d*)
A.	III	IV	I	II
B.	IV	III	I	II
C.	III	IV	II	I
D.	III	I	IV	II

15. Two parallel plate capacitors each of 15 μF capacity are connected in series. The space between the plates of one capacitor is filled with a dielectric material of dielectric constant K = 2. The equivalent capacitance of the system will be:

A. 45 μF B. 30 μF
C. 10 μF D. 15 μF

16. Many enzymes catalyze both forward and reverse reactions. Which one of the following statement is NOT correct?

A. An equilibrium is established after some time.
B. It is possible to control the directions of the reaction by suitably removing the formed product.
C. These reactions are both temperature and concentration dependent.
D. The forward and reverse reactions proceed via different activation complexes.

17. Sanger reaction (Sequencing) is an example of:

A. electrophilic substitution
B. hydrolysis
C. esterification
D. nucleophilic substitution

18. Which one of the following is an aromatic compound?

(*a*) (*b*) (*c*) (*d*)

A. (*a*) B. (*b*)
C. (*c*) D. (*d*)

19. Given below are two statements.

Statement I: Precision refers to the closeness of various measurements for the same quantity.

Statement II: Accuracy is the agreement of the obtained value with the known or true value of the quantity.

In light of the above statements, choose the correct answer from the options given below:

A. Both Statement I and Statement II are correct
B. Both Statement I and Statement II are NOT correct
C. Statement I is correct, but Statement II is not correct
D. Statement I is not correct, but Statement II is correct

20. Which one of the following transition metals is present in Vitamin B12?

A. Mn B. Co
C. Zn D. Cu

13. B **14.** A **15.** C **16.** D **17.** D **18.** A **19.** A **20.** B

21. Which one of the following drugs contains β-lactam structure?

A. Penicillin B. Sulphanilamide
C. Erythromycin D. Chloramphenicol

22. Which law of thermodynamics states that "energy of an isolated system is constant"?

A. First B. Second
C. Third D. Zeroth

23. According to the molecular orbital theory, which of the following molecules should exhibit paramagnetism?

A. O_2 B. N_2
C. F_2 D. C_2

24. It is possible to separate *o*-nitrophenol and *p*-nitrophenol using steam distillation because *o*-nitrophenol has:

A. Van der Waals force
B. Steric hindrance
C. Intermolecular H-bonding
D. Intramolecular H-bonding

25. Natural rubber is a polymer of .... while synthetic rubber neoprene is formed by polymerization of ....

A. 1,3-butadiene; acrylonitrile
B. 2-chloro-1,3-butadiene; 1,3-butadiene
C. 2-methyl-1,3-butadiene;2-chloro-1, 3-butadiene
D. Acrylonitrile; 2-methyl-1,3-butadiene

26. Arrange the following in decreasing order of their acidic strength:

(*a*) CH_3COOH (*b*) $ClCH_2COOH$
(*c*) $Cl_2CHCOOH$ (*d*) $Cl_3CHCOOH$
(*e*) F_3CCOOH

Choose the correct answer from the options given below:

A. (*a*) > (*b*) > (*c*) > (*d*) > (*e*)
B. (*e*) > (*d*) > (*c*) > (*b*) > (*a*)
C. (*a*) > (*e*) > (*d*) > (*c*) > (*b*)
D. (*b*) > (*c*) > (*d*) > (*a*) > (*e*)

27. Which one of the following modifications is NOT a natural N-terminal modification of proteins?

A. Acetylation B. Benzylation
C. Myristoylation D. Sumoylation

28. Match the items in List-I with items in List-II.

List-I	List-II
(*a*) Enzymes	I. Amino acid
(*b*) Glucose	II. Biocatalysts
(*c*) Lactose	III. Aldohexose
(*d*) Methionine	IV. Disaccharide

Choose the correct answer from the options given below:

	(*a*)	(*b*)	(*c*)	(*d*)
A.	I	II	III	IV
B.	II	III	IV	I
C.	II	IV	III	I
D.	I	III	IV	II

29. A nucleic acid chain comprises of:

(*a*) Phosphate group
(*b*) Nitrogen base
(*c*) Pentose sugar
(*d*) Thiol group
(*e*) β (1-4) linkage

Choose the correct answer from the options given below:

A. (*b*) and (*c*) only
B. (*b*), (*d*) and (*e*) only
C. (*a*), (*d*) and (*e*) only
D. (*a*), (*b*) and (*c*) only

30. Which of the following elements readily react with oxygen to form their oxides?

A. Au and Pt B. Ne and Ar
C. Al and Cu D. Cu and Pt

21. A **22.** A **23.** A **24.** D **25.** C **26.** B **27.** B **28.** B **29.** D **30.** C

31. Which one of the following is caused by point mutation?
A. Turner's syndrome
B. Down's syndrome
C. Sickle cell anemia
D. Klinefelter's syndrome

32. Which is NOT the function of placenta?
A. Supply of oxygen and nutrients to the embryo
B. Removal of excretory waste products produced by embryo
C. Production of hCG and HPL
D. Supply all types of antibodies to the embryo

33. Which statement is true with respect to colostrum?
A. It is a yellowish fluid secreted by the mother during later days of lactation.
B. Colostrum provides passive immunity to the infant.
C. Colostrum is rich in carbohydrates and has no antibodies.
D. Colostrum provides active immunity to the infant.

34. *Bacillus thurinigiensis CryA* controls certain caterpillar pests by:
A. turning toxic in the acidic pH of their gut medium
B. turning toxic in the alkaline medium of their gut
C. repelling them from the crops
D. inducing satiation in them

35. Which of the following options correctly match the name of gene and its function in cloning vector pBR322?
A. Cla I - Acts as selectable marker to identify non-transformants
B. amp^r - Codes for plasmid amplifying enzymes
C. rop - Codes proteins required for plasmid replication
D. ori - Controls plasmid size

36. Which one of the following statements about AIDS caused by HIV is correct?
A. The time lag between the HIV infection and AIDS manifestation varies from 2-3 weeks.
B. After entering the body, HIV enter B-lymphocytes.
C. The AIDS-affected individuals are more susceptible to Tuberculosis.
D. HIV infection depletes only the CD8 lymphocytes in the body.

37. The two heavy chains of a human antibody are linked to each other by:
A. hydrogen bond
B. glycosidic bond
C. phosphodiester bond
D. disulfide bond

38. Aldosterone regulates the water and electrolyte balance in human body by:
A. Stimulating the H_2O and Na^+ re-absorption, while K^+ and PO_4^{3-} excretion
B. Stimulating the Na^+ and K^+ reabsorption, while H_2O and PO_4^{3-} excretion
C. Stimulating the H_2O reabsorption and Na^+ excretion
D. Stimulating the Na^+ and PO_4^{3-} reabsorption, while K^+ excretion

39. At G_2/M checkpoint the cell cycle will arrest if:
A. The cell has not achieved an adequate size
B. The spindle fibre formation has not occurred
C. The DNA replication or repair of DNA damage has not been completed
D. The attachment of the spindle fibres to the kinetochore of centromeres is not adequate

31. C	**32.** D	**33.** B	**34.** B	**35.** C	**36.** C	**37.** D	**38.** A	**39.** C

40. Choose the option that correctly matches for an immunosuppressant and its origin:

A. Cholesterol - Palm Oil

B. Cyclosporin A - *Trichoderma polysporum*

C. Streptokinase - *Streptococcus*

D. Botulinum toxin - *Clostridium botulinum*

41. Which statement is true with respect to meiosis?

A. Meiosis involves two sequential cycles of nuclear and cell division but only a single cycle of DNA replication.

B. Meiosis involves one cycle of nuclear and cell division but two cycles of DNA replication.

C. Four diploid cells are formed at the end of meiosis.

D. Two haploid cells are formed at the end of meiosis.

42. The principle driving force behind movement of water in plants is known as:

A. Ionic potential

B. Membrane potential

C. Soil temperature

D. Water potential

43. Which one of the following categories of methods CANNOT be used for animal virus detection?

A. Serology

B. Nucleic acid hybridization

C. Hematology

D. Hemagglutination

44. Respiratory pathway is considered as a .... pathway.

A. Catabolic

B. Anabolic

C. Amphibolic

D. Fermentative

45. Match the items of List-I with the items in List-II.

List-I	List-II
(*a*) Diabetes Insipidus	I. Dysregulation of glucagon
(*b*) Exophthalmic Goiter	II. Water loss and dehydration
(*c*) Acromegaly	III. Grave's disease
(*d*) Hyperglycemia	IV. Disfigurement of face

Choose the correct answer from the options given below:

	(*a*)	(*b*)	(*c*)	(*d*)
A.	I	III	IV	II
B.	IV	III	II	I
C.	II	I	III	IV
D.	II	III	IV	I

46. An equilateral triangle of side 6 cm has its corners cut-off to form a regular hexagon. The area of regular hexagon is:

A. $2\sqrt{3}$ cm^2 B. $3\sqrt{2}$ cm^2

C. $6\sqrt{3}$ cm^2 D. $3\sqrt{6}$ cm^2

47. A train passes a standing man in 6 seconds and 210 m long platform in 16 seconds. The length and speed of the train, respectively, is:

A. 126 m, 21 m/s

B. 120 m, 20 m/s

C. 110 m, 20 m/s

D. 63 m, 21 m/s

48. In an election contested by two candidates, one candidate got 30% of total votes and lost by 500 votes. The total number of votes polled is:

A. 1350

B. 1450

C. 1150

D. 1250

40. B **41.** A **42.** D **43.** C **44.** C **45.** D **46.** C **47.** A **48.** D

49. $9.6 \times 3.6 \div 7.2 + 10.8$ of $1/18 - 1/10 = ?$

A. 15.56 B. 10.56
C. 5.3 D. 15.36

50. The traffic lights at three different road crossings change after every 48 s, 72 s and 108 s, respectively. If they all change simultaneously at 8:20:00 h, when will they change again simultaneously?

A. 8:27:12 h
B. 8:25:10 h
C. 8:26:12 h
D. 8:24:10 h

51. The sum of four consecutive even numbers is 107 more than the sum of three consecutive odd numbers. If the sum of smallest odd number and the smallest even number is 55, then what is the smallest even number?

A. 36 B. 40
C. 32 D. 38

52. Maximum distance between any two points inside or on cube of side 1 cm is equal to:

A. 1 cm B. $\sqrt{2}$ cm
C. $\sqrt{3}$ cm D. 6 cm

53. Number of natural numbers that can be formed using digits 1, 2, 3, 4, 5, 6, 7 each exactly once so that digits 3, 4 and 5 are always in the middle is equal to:

A. 24 B. 144
C. 5040 D. 720

54. The acute angle between hour and minute hands of a wall clock when the time shown by it is 02:15 is equal to:

A. 30°
B. 26.25°
C. 22.5°
D. 37.5°

55. Number of squares in a chess-board is equal to:

A. 64 B. 81
C. 204 D. 284

56. $x^2 + ax + 1 = 0$ has no real root. Which one of the following is correct?

A. $a \leq 2$
B. $a \geq 2$
C. $-2 \leq a < 2$
D. $-2 < a < 2$

57. There are 30 boys and 60 girls in a class. If the average age of boys is 12 years and average age of girls is 10 years, what is the average age of the whole class?

A. 10.11 years
B. 10.66 years
C. 11.66 years
D. 11.11 years

58. The diagonals of a rhombus are 16 cm and 12 cm. The side of the rhombus would be:

A. 10 cm B. 11 cm
C. 8 cm D. 9 cm

59. For $x > 0$, if variable takes discrete values $x + 4$, $x - 3.5$, $x - 2.5$, $x - 3$, $x - 2$, $x + 0.5$, $x - 0.5$, $x + 5$, then the value of median is:

A. $x - 1.25$
B. $x - 0.5$
C. $x + 0.5$
D. $x + 1.25$

60. The salary of a worker is first increased by 5% and then it is decreased by 5%. What is the change in his salary?

A. Decrease in salary 0.25%
B. Increase in salary 0.50%
C. No change in salary
D. Decrease in salary 0.50%

49. C **50.** A **51.** D **52.** C **53.** B **54.** C **55.** C **56.** D **57.** B **58.** A **59.** A **60.** A

SECTION-B

61. Which one of the following antibody types protects against inhaled and ingested pathogens?

A. IgG B. IgD
C. IgM D. IgA

62. Which one of the following hormones transmit their signal via nuclear receptors?

A. Thyroid hormone
B. Follicle Stimulating hormone
C. Insulin
D. Luteinizing hormone

63. Which one of the following amino acids is coded by a single codon?

A. Valine B. Threonine
C. Tryptophan D. Isoleucine

64. Which one of the following statements is NOT correct?

A. Glucose is stored in animals as glycogen.
B. Glucose is stored in plants as starch.
C. Cellulose is a polymer of only glucose.
D. Hemicellulose is a polymer of only glucose.

65. Match the items in List-I with the items in List-II.

List-I (Organelle)	List-II (Function)
(*a*) Mitochondria	I. Protein processing and transport
(*b*) Nucleolus	II. Protein synthesis
(*c*) Golgi complex	III. Energy production
(*d*) Endoplasmic reticulum	IV. Ribosomal RNA synthesis

Choose the correct answer from the options given below:

	(*a*)	(*b*)	(*c*)	(*d*)
A.	I	III	II	IV
B.	II	III	I	IV
C.	III	II	I	IV
D.	III	IV	I	II

66. In which one of the following compartments of the cell, carbohydrates are added to a protein during glycoprotein synthesis?

A. Mitochondria B. Lysosome
C. Nucleus D. Golgi complex

67. If a single base in the middle of a protein coding open reading frame is replaced with another base, which statement from the following is NOT a likely possibility?

A. It may not make any difference to the protein sequence.
B. It may cause a single amino acid mutation.
C. It may create a premature stop codon.
D. The mRNA will not be recruited for translation.

68. The indigenous vaccine, Covaxin against SARS Coronavirus-2 contains:

A. the mRNA expressing viral spike protein
B. inactivated whole virions
C. the purified viral envelope protein
D. the DNA coding for viral spike protein

69. Antibody diversity is an example of:

A. Gene rearrangement
B. Domain swapping
C. Post-translational modification
D. Proteolytic processing

70. Which of the following immunoglobulins primarily pass through the placenta to provide passive immunity to the fetus?

A. IgM only B. IgM and IgG
C. IgA and IgG D. IgG only

61. D **62.** A **63.** C **64.** D **65.** D **66.** D **67.** D **68.** B **69.** A **70.** D

71. Which of the following pair of mono-saccharides contains epimers of each other?

A. D-Mannose and D-Glucose
B. D-Gluose and D-Glucose
C. D-Arabinose and L-Arabinose
D. D-Glucose and D-Fructose

72. Match the items in List-I with items in List-II.

List-I	List-II
(*a*) Prophase	I. Reformation of the nuclear envelope around daughter chromosomes
(*b*) Metaphase	II. Separation of the two daughter chromosomes
(*c*) Anaphase	III. Condensation of DNA into chromatids
(*d*) Telophase	IV. Chromatids line up along an axis

Choose the correct answer from the options given below:

	(*a*)	(*b*)	(*c*)	(*d*)
A.	III	II	IV	I
B.	III	IV	II	I
C.	IV	III	II	I
D.	II	IV	I	III

73. The specificity in an antibody molecule is provided by the:

A. Light chain variable region
B. Light chain constant region
C. Heavy chain constant region-I
D. Hinge region

74. Calcium alginate based synthetic seeds tend to lose water rapidly and become hard pellet. This problem can be overcome by:

A. Coating the capsule with polyethylene glycol
B. Preserving the seeds in the airtight packaging till sowing
C. Treating the somatic embryos with sterile water for 3 hours before encapsulation
D. Coating the capsules with Elvax 4260

75. Which one of the following statements most appropriately describes the concept of 'Codon Bias'?

A. Some codons for a particular amino acid are used more frequently.
B. There has been an element of human bias for assigning specific codons to an amino acid.
C. There is no codon bias in plants.
D. The usage of codons varies for different proteins in an organism.

76. Which one of the following is NOT true for Quantum dots as fluorescent probes in fluorescence microscopy?

A. They are highly resistant to photo-bleaching.
B. They can generate fluorescence of different emission wavelengths.
C. They are nanocrystals of different sizes.
D. Their flurorescence properties do not depend on the size of the Quantum dots.

77. Francis & Crick proposed the scheme called Central Dogma in 1958. Which of the following processes was NOT covered in this scheme?

A. Replication
B. Transcription
C. Reverse transcription
D. Translation

78. *Beggiatoa*, a bacterium depends on organic carbon, inorganic chemicals and inorganic electron donor for its nutrition. On the basis of its nutritional type, it is classified as:

A. Photoorganoheterograph
B. Chemolithoautotroph
C. Chemolithoheterotroph
D. Chemoorganoheterotroph

71. A **72.** B **73.** A **74.** D **75.** A **76.** D **77.** C **78.** C

79. Which one of the following statements is correct about various microbial culture media?

A. Mannitol salt agar is an enriched and differential media.
B. Selective components in MacConkey (MAC) agar are eosin Y and methylene blue which inhibits the growth of gram positive bacteria.
C. Blood agar is a differential media which is differentiated on the basis of bacterial ability to produce hemolysins.
D. Bile salts and crystal violet present in the EMB agar media inhibits the gram positive bacteria growth and hence helps to differentiate between gram positive and gram negative.

80. The microscope which uses lasers to scan the specimen at a specific depth, illuminates one area at a time and blocks stray light to give an image with excellent contrast and resolution is:

A. Differential Interference Contrast (DIC) Microscope
B. Confocal Microscope
C. Scanning Electron Microscope
D. Phase Contrast Microscope

81. A biochemist is pelleting down the microsomal fraction from a sample using ultracentrifuge at a speed of 20000 rpm. What would be RCF if the diameter of the rotor is 7 cm?

A. 15680 B. 31360
C. 7840 D. 3920

82. In which of the given centrifuge rotors the value of r_{min} (radius minimum), r_{max} and r_{av} have the minimum deviation?

A. Fixed-angle rotor only
B. Vertical rotor only
C. Swing rotor only
D. Fixed-angle and Vertical rotors

83. Match the items in List-I with items in List-II.

List-I	List-II
(*a*) mRNA	I. inhibits gene expression
(*b*) tRNA	II. carries amino acids for translation
(*c*) snRNA	III. provides template for translation
(*d*) siRNA	IV. involved in RNA splicing

Choose the correct answer from the options given below:

	(*a*)	(*b*)	(*c*)	(*d*)
A.	III	II	IV	I
B.	II	III	I	IV
C.	IV	III	II	I
D.	II	IV	I	III

84. Which of the following cell types has the highest surface area to volume ratio?

A. RBC B. Fibroblast
C. Keratinocyte D. Hepatocyte

85. How many grams of Albumin and Aspirin will be required to set a reaction between one millimole of Albumin and 0.5 millimole of Aspirin? Given the molecular weight of Albumin is 67,000 Da and that of Aspirin is 180 Da:

A. 1 g, 1 mg B. 67 g, 90 mg
C. 0.1 g, 70 mg D. 67 μg, 90 mg

86. Of the amino acids listed below, which three amino acids can undergo posttranslational modification?

A. Glycine, Leucine, Tryptophan
B. Serine, Threonine, Tyrosine
C. Cysteine, Glutamine, Proline
D. Glutamic acid, Arginine, Methionine

87. What is the common feature of the following peptides?

GKWLY, YLWKG, WGKLY, WLKGY

A. Same sequence
B. Same amino acid composition
C. Same conformation
D. Same interactome

79. C **80.** B **81.** A **82.** B **83.** A **84.** A **85.** B **86.** B **87.** B

88. What is the final concentration of NaCl upon mixing 10 ml of 10 mM NaCl with 990 ml of 10 mM NaCl?

A. 0.1 mM B. 0.1 M

C. 0.01 M D. 1.0 mM

89. Which of the following molecules is a Peptide Nucleic Acid (PNA)?

A.

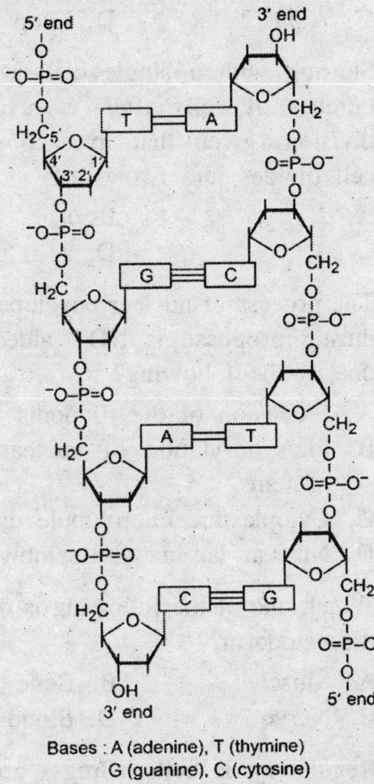

B.

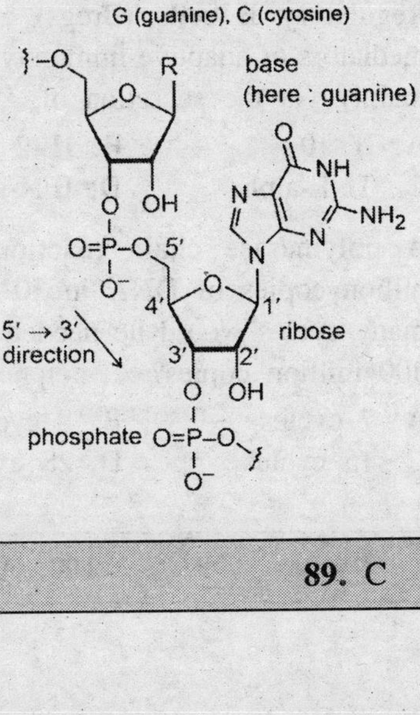

C.

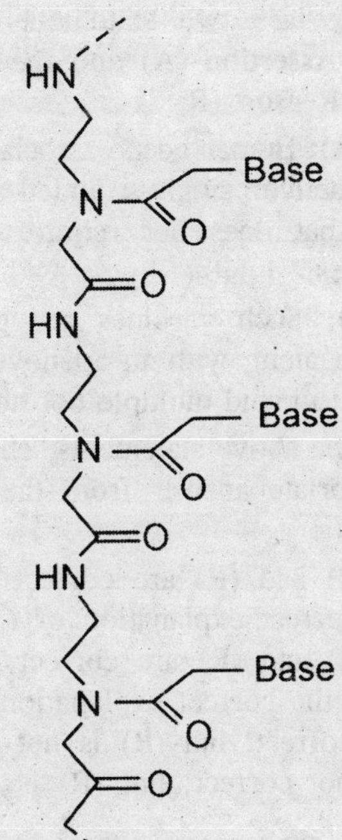

D.

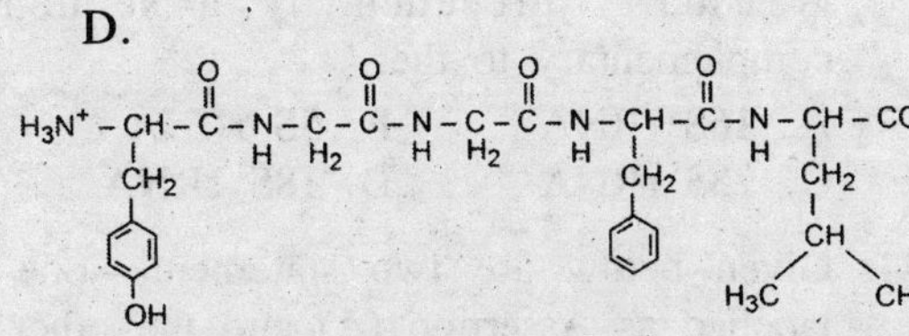

90. Among Wheat, Moong Dal, Rice and Bajra, the one with the highest protein content is:

A. Bajra

B. Wheat

C. Moong dal

D. Rice

91. Which one of the following is the most effective strategy in delivering a gene of interest in non-proliferating terminally differentiated cells?

A. Adeno-associated viral particle

B. Retroviral particle

C. Calcium chloride

D. Lipofectamine

88. C	89. C	90. C	91. A

92. Given below are two statements—one is labelled as Assertion (A) and the other is labelled as Reason (R):

Assertion (A): Human adeno-associated virus is used to deliver single-stranded DNA as a vaccine that does not require multiple booster doses.

Reason (R): Such vaccines are generally administered along with an adenovirus or a herpesvirus to avoid multiple booster doses.

In light of the above statements, choose the most appropriate answer from the options given below.

A. Both (A) and (R) are correct and (R) is the correct explanation of (A).
B. Both (A) and (R) are correct, but (R) is NOT the correct explanation of (A).
C. (A) is correct, but (R) is not correct.
D. (A) is not correct, but (R) is correct.

93. In case of prokaryotes, the start codon is usually preceded by a sequence complementary to the:

A. 16S rRNA B. 5S rRNA
C. 28S rRNA D. 18S rRNA

94. Given below are two statements—one is labelled as Assertion (A) and the other is labelled as Reason (R):

Assertion (A): In the eukaryotic genes, TATA box aids in transcription.

Reason (R): The TATA box facilitates formation of pre-initiation complex for transcription initiation.

In light of the above statements, choose the most appropriate answer from the options given below.

A. Both (A) and (R) are correct and (R) is the correct explanation of (A)
B. Both (A) and (R) are correct, but (R) is NOT the correct explanation of (A)
C. (A) is correct, but (R) is not correct
D. (A) is not correct, but (R) is correct

95. Which one of the following cell types is involved in retaining the tattoo ink?

A. Macrophages B. Melanocytes
C. Keratinocytes D. Lymphocytes

96. Which one of the following does NOT refer to secondary structures in protein?

A. Beta sheet B. Twist
C. Alpha helix D. Loop

97. Starting with a single cell, what will be number of cells after '*n*' cycles of cell division, given that in each cycle every cell divides into two cells?

A. 2^2 B. n^n
C. n^2 D. 2^n

98. The process of nuclear envelope breakdown during prophase is NOT aided by which one of the following?

A. Extension of the filopodia
B. Phosphorylation of nuclear membrane proteins
C. Cytoplasmic microtubule dynamics
D. Nuclear lamina disassembly

99. Which one of the following is derived from the ectoderm?

A. Muscle B. Bone
C. Nerve D. Blood

100. Regulatory B cells (Bregs) are important mediators of adaptive immunity and function mainly via the secretion of:

A. IL-10 B. IL-2
C. TNF-alpha D. IFN-gamma

101. A polymerase chain reaction yields 1.2 billion copies of DNA in 30 cycles. How many cycles would be needed to obtain its 300 million copies?

A. 7 cycles B. 8 cycles
C. 15 cycles D. 28 cycles

92. A **93.** A **94.** A **95.** A **96.** B **97.** D **98.** A **99.** C **100.** A **101.** D

102. Neoschizomers are the restriction endonucleases with:

A. identical recognition site but different cleavage sites
B. different recognition sites but same cleavage site
C. different recognition site but producing same sticky ends
D. identical recognition and cleavage sites

103. Telomerase, an RNA-protein complex adds telomeres at the end of chromosomes. What kind of enzymatic activity does it possess?

A. DNA-dependent DNA polymerase
B. DNA-dependent RNA polymerase
C. RNA-dependent DNA polymerase
D. RNA-dependent RNA polymerase

104. Which of the following is NOT true for the layers of gastrula?

A. The linning of the digestive tract will be formed by the endoderm.
B. The bones will be formed by the mesoderm.
C. The nerves will be formed by the ectoderm.
D. The skin will be formed by the mesoderm.

105. Which of the following statement is NOT correct?

A. Transcription takes place in the nucleus of eukaryotic cells.
B. In prokaryotes mRNA is not capped.
C. Translation in eukaryotes takes place in the nucleus.
D. In prokaryotes, DNA is replicated in the cytoplasm.

106. Rancidity in spoiled foods is mainly due to:

A. Proteolytic enzymes
B. Photosynthetic microbes
C. Saccharolytic microbes
D. Lipolytic microbes

107. The helical content of a protein can be directly determined using:

A. infrared spectrometer
B. fluorescence
C. circular dichroism
D. UV-visible spectrophotometer

108. DNA conformation is left handed in:

A. DNA B B. DNA C
C. DNA Z D. DNA A

109. Which one of the following tissue culture approaches is most appropriate for production of double halploid plants?

A. Protoplast fusion B. Embryo rescue
C. Anther culture D. Meristem culture

110. Which one of the following is NOT an auxin?

A. Indole acetic acid (IAA)
B. Indole butyric acid (IBA)
C. 2,4-dichlorophenoxy acetic acid (2,4-D)
D. 6-Benzylaminopurine (BAP)

111. Which of the following is a heuristic algorithm that works faster than those driven by dynamic programming?

A. Needleman-Wunsch
B. Smith-Waterman
C. BLAST
D. Gradient Descent

112. For an imaginary Martian species with three nucleotides (X, Y and Z), how many 3-letter codons are possible?

A. 64 B. 27
C. 9 D. 4

113. What is the likely number of amino acids in a 11 KDa protein?

A. 90 B. 100
C. 110 D. 120

102. A	**103.** C	**104.** D	**105.** C	**106.** D	**107.** C
108. C	**109.** C	**110.** D	**111.** C	**112.** B	**113.** B

114. For a normal (Gaussian) distribution, decreasing the spread and increasing the height would lead to a:

A. smaller value of standard deviation
B. higher value of standard deviation
C. smaller value of mean
D. higher value of mean

115. Which of the following is a method to conduct phylogeny of protein and DNA sequences?

A. BLAST
B. OMNIBUS
C. Maximum likelihood
D. DAVID

116. The degree of inhibition for an enzyme catalyzed reaction at a particular inhibitor concentration is independent of the intitial substrate concentration. This is:

A. Un-competitive inhibition
B. Non-competitive inhibition
C. Competitive inhibition
D. Mixed inhibition

117. An enzymatic reaction exhibits Michaelis-Menten Kinetics. What will happen if the concentration of enzyme is doubled keeping $[S_0] >> [E]$?

A. Both K_m and V_{max} will remain same
B. Both K_m and V_{max} will increase
C. V_{max} will increase; K_m will remain same
D. K_m will increase; V_{max} will remain same

118. A zero order liquid phase reaction $A \xrightarrow{k} B$, is being carried out in a batch with $k = 10^{-3}$ moles/min. Reactor volume is 100 L. Initial concentration of A is 0.1 moles/L. What is the earliest time at which A is completely exhausted in the system?

A. 100 min
B. 200 min
C. 300 min
D. 40 min

119. If the average diameter of air bubbles in a bioreactor is 2 mm and the gas hold up is 10% then the surface area of gas bubbles per liter of reactor is:

A. 30 cm^2
B. 300 cm^2
C. 3000 cm^2
D. 30000 cm^2

120. A good resolution in ion exchange chromatography is obtained when the two proteins have a:

A. large difference in binding affinity and large dispersion
B. small difference in binding affinity and large dispersion
C. large difference in binding affinity and small dispersion
D. small difference in binding affinity and small dispersion

121. Given below are two statement - one is labelled as Assertion (A) and the other is labelled as Reason (R):

Assertion (A): Bacterial lipopolysaccharide (LPS) on its own does not induce memory B-cell in humans.

Reason (R): LPS does not activate T-cell.

In light of the above statements, choose the most appropriate answer from the options given below:

A. Both (A) and (R) are correct and (R) is the correct explanation of (A)
B. Both (A) and (R) are correct, but (R) is NOT the correct explanation of (A)
C. (A) is correct, but (R) is not correct
D. (A) is not correct, but (R) is correct

122. The allergic immune response is characterized by the increased levels of:

A. IgE
B. IgA
C. IgG
D. IgM

114. A **115.** C **116.** B **117.** C **118.** A **119.** C **120.** C **121.** A **122.** A

123. The presence of antibody in infected patients serum can be detected by:

A. ELISPOT B. PCR
C. Northern blot D. Western blot

124. What will happen to immune cell development if we remove thymus from neonatal mice?

A. B-cell maturation will be impaired
B. Both B- and T-cell maturation will be impaired
C. T-cell maturation will be impaired
D. No effect on B- and T-cell maturation

125. A series of spontaneous point mutations that occur gradually resulting in changes in Influenza virus surface antigens over a time is called:

A. genomic instability
B. antigenic shift
C. antigenic drift
D. chromosome translocation

126. *Myasthenia gravis* is an autoimmune disease where patient makes antibodies for its own:

A. Acetylcholine receptor protein
B. NOD1 protein
C. TLR11 protein
D. RIG-I protein

127. Match the items in List-I with items in List-II.

List-I	List-II
(*a*) Toll like receptor 9	I. Recognition of unmethylated CpG dinucleotide
(*b*) T-helper cells	II. Recognition of antigen with MHC II complex
(*c*) T-cytotoxic cells	III. Recognition of antigen with MHC I complex
(*d*) Plasmacytoid dendritic cells (pDCs)	IV. Type I interferon (IFN) production

Choose the correct answer from the options given below:

	(*a*)	(*b*)	(*c*)	(*d*)
A.	II	III	I	IV
B.	I	III	IV	II
C.	IV	II	III	I
D.	I	II	III	IV

128. What is the length of peptides binding to Major Histocompatibility Complex (MHC) class II molecule?

A. 8-11 amino acids
B. 21-27 amino acids
C. 15-20 amino acids
D. 507 amino acids

129. Tuberculosis (TB) is caused by *Mycobacterium tuberculosis*. The TB vaccine is made using:

A. *Mycobacterium tuberculosis*
B. *Mycobacterium bovis*
C. Tuberculin
D. Mycobacterial DNA

130. CD4 antigen is absent on:

A. B-cells B. T-cells
C. macrophage cells D. gamma-delta T cells

131. Match the items in List-I with items in List-II.

List-I	List-II
(*a*) HIV	I. RNA-dependent RNA polymerase
(*b*) Influenza virus	II. dsDNA virus
(*c*) Hepatitis C virus	III. Segmented RNA genome
(*d*) Pox virus	IV. Reverse Transcriptase

Choose the correct answer from the options given below:

	(*a*)	(*b*)	(*c*)	(*d*)
A.	III	I	IV	II
B.	IV	I	II	III
C.	II	I	III	IV
D.	IV	III	I	II

123. D **124.** C **125.** C **126.** A **127.** D **128.** C **129.** B **130.** A **131.** D

132. Hepatitis B virus genome is:
A. ssDNA
B. dsDNA
C. ssRNA
D. partially dsDNA

133. Herpes simplex virus maintains latency in:
A. Neuronal cells B. Liver cells
C. Epithelial cells D. Kidney cells

134. Pox virus replicates in the:
A. Cytoplasm B. Nucleus
C. Golgi D. Mitochondria

135. Which of the following viruses is a plus-sense single-stranded RNA virus?
A. Dengue virus
B. Rotavirus
C. Adenovirus
D. Influenza virus

136. In N-linked glycoproteins, carbohydrate moiety is attached to which of the following amino acids?
A. Valine B. Asparagine
C. Serine D. Threonine

137. The catalytic triad of Chymotrypsin is composed of:
A. Asp, Ser, His
B. Arg, Ser, His
C. Glu, Thr, Lys
D. Glu, Asp, Tyr

138. A mixture of proteins (A, B, C and D) is separated on a Sephadex G-200 column. The proteins elute in the order of A, B, C and D. Assuming that all proteins are globular and monomeric, the protein with minimum electrophoretic mobility on SDS-PAGE will be:
A. A B. B
C. C D. D

139. Match the items in List-I with items in List-II.

List-I	List-II
(*a*) β-Oxidation	I. Ribulose Bisphosphate Carboxylase
(*b*) Glycolysis	II. Phosphofructo kinase-l
(*c*) Gluconeo-genesis	III. Phosphoenolpyruvate carboxylase
(*d*) Calvin cycle	IV. Thiolase

Choose the correct answer from the options given below:

	(*a*)	(*b*)	(*c*)	(*d*)
A.	III	IV	II	I
B.	II	IV	I	III
C.	IV	II	III	I
D.	III	II	IV	I

140. Lipid Rafts are composed of the:
A. cholesterol and cardiolipin
B. sphingolipid and cardiolipin
C. sphingolipid and cholesterol
D. cholesterol but no sphingolipid

141. Match the items in List-I with items in List-II.

List-I (Pollutant)	List-II (Impact on environment)
(*a*) Carbon monoxide	I. Greenhouse effect
(*b*) Hydrocarbons	II. Photochemical smog
(*c*) Oxides of nitrogen	III. Acid rain
(*d*) Ozone near earth's surface	IV. Impaired plant growth

Choose the correct answer from the options given below:

	(*a*)	(*b*)	(*c*)	(*d*)
A.	II	III	IV	I
B.	III	II	I	IV
C.	II	III	I	IV
D.	I	II	III	IV

132. D **133.** A **134.** A **135.** A **136.** B **137.** A **138.** A **139.** C **140.** C **141.** D

142. The presence of excess nutrients in aquatic system will lead to:

A. Crustacean bloom
B. Algal bloom
C. Coral bloom
D. Lotus bloom

143. Which one of the following is commonly used for converting cellulose raw materials into glucose?

A. *Saccharomyces cerevisiae*
B. *Acinetobacter radioresistens*
C. *Trichoderma viride*
D. *Bacillus amyloliquifaciens*

144. The biocide DDT (a chlorinated hydrocarbon) has a half-life of around:

A. < 1 year
B. 2-15 years
C. 16-30 years
D. > 30 years

145. Which one of the following waste treatment system is devoid of any packing material, and it recycles internal biomass based on gravity?

A. UASB
B. FSSB
C. RBC
D. Trickling filter

146. Which of the following is true for a water sample with a BOD value of more than 50 ppm?

A. The DO content would be less than 6 ppm
B. The water is clean and potable
C. Aquatic life will thrive
D. The COD of the sample is 25 ppm

147. The most widely used method for removing of particulate matter from gas is:

A. Electrostatic precipitation
B. Chemo-osmotic precipitation
C. Magnetostatic precipitation
D. Chemo-electrostatic precipitation

148. The acid involved in ocean acidification is:

A. Carbonic acid
B. Sulphuric acid
C. Phosphoric acid
D. Nitric acid

149. Which of the following continent is the driest one?

A. Africa
B. Antarctica
C. Australia
D. Europe

150. Enhanced CO_2 concentration in environment would lead to .... in plants.

A. increased water uptake and reduced photosynthesis
B. increased photosynthesis and increased water demand
C. decreased photosynthesis and decreased water demand
D. decreased O_2 emission and no change in photosynthesis

151. In plant mycorrhizal fungi association, what is the most appropriate exchange between two organisms or partners?

A. Plant provides carbon to fungi and in return gets minerals
B. Fungi provides protein to plant and in return gets water
C. Plant provides minerals to fungi and in return gets carbon
D. Plant and fungi do not exchange anything

152. Pattern of inheritance of flower colour in *Mirabilis jalapa* is similar to that of:

A. ABO blood group in human beings
B. Flower colour in snapdragon
C. Fur colour in rabbit
D. Skin colour in human beings

142. B **143.** C **144.** B **145.** A **146.** A **147.** A **148.** A **149.** B **150.** B **151.** A **152.** B

153. Given below are two statements:

Statement I: In general, a higher auxin : cytokinin ratio will induce root formation under *in vitro* culture conditions in plants.

Statement II: NAA is a cytokinin and BAP is an auxin.

In light of the above statements, choose the most appropriate answer from the options given below.

A. Both Statement I and Statement II are correct.
B. Both Statement I and Statement II are NOT correct.
C. Statement I is correct but Statement II is not correct.
D. Statement I is not correct but Statement II is correct.

154. Complete the following statement with the correct option:

Agrobacterium-mediated plant transformation in the laboratory:

A. is not influenced by the genotype of the host plant.
B. always leads to integration of a single copy of the T-DNA in the host cell.
C. is faciliated by the use of selection marker genes to allow preferential growth of transformed cells.
D. requires the expression of opine genes for the production of transgenic plants.

155. Which one of the following plant tissue culture techniques can be most effectively used for production of virus-free plants?

A. Protoplast culture
B. Culture of shoot apical meristem
C. Somatic embryogenesis from calli of leaf explants
D. Production of cybrids

156. A suicide plasmid vector lacks the following:

A. antibiotic marker
B. origin of replication
C. multiple cloning sites
D. site for integration

157. Animal gut does NOT possess the enzymes required for digesting:

A. glycogen
B. starch
C. cellulose
D. proteins

158. Foreign DNA can NOT be transferred into a zygote by which one of the following methods?

A. Transduction
B. Microinjection
C. Electroporation
D. Conjugation

159. Leptin receptor is primarily present in the following tissue:

A. Hepatic
B. Muscle
C. Adipose
D. Neuronal

160. Match the items in List-I with items in List-II.

List-I	List-II
(*a*) Beriberi	I. Cobalamin
(*b*) Megaloblastic Anemia	II. Thiamin
(*c*) Scurvy	III. Folic acid
(*d*) Pernicious Anemia	IV. Ascorbic Acid

Choose the correct answer from the options given below:

	(*a*)	(*b*)	(*c*)	(*d*)
A.	II	IV	III	I
B.	III	II	IV	I
C.	I	III	IV	II
D.	II	III	IV	I

153. C **154.** C **155.** B **156.** B **157.** C **158.** D **159.** D **160.** D

EXPLANATORY ANSWERS

1. (C): When a particle is inside the Earth at a distance r from the center, the gravitational force acting on it is $F = -\frac{GMm}{R^3}r$, where M is the mass of the Earth and R is its radius. This force is proportional to the displacement r and directed towards the center, satisfying the conditions for simple harmonic motion (SHM). The angular frequency ω of the SHM is given by $\omega = \sqrt{\frac{GM}{R^3}}$.

Therefore, the time period T of the motion is $T = \frac{2\pi}{\omega} = 2\pi\sqrt{\frac{R^3}{GM}}$.

2. (B):

(*a*) **False.** In uniform circular motion, the speed (magnitude of velocity) is constant, but the velocity vector changes direction continuously; thus, the velocity is not constant.

(*b*) **True.** The acceleration in uniform circular motion is the centripetal acceleration, which is always directed towards the center of the circle with magnitude $a = \frac{v^2}{R}$.

(*c*) **False.** In uniform circular motion, the angular momentum vector remains constant in both magnitude and direction if the motion is in a plane and no external torque is acting.

(*d*) **True.** The angular velocity ω is related to linear speed v by $\omega = \frac{v}{R}$.

3. (D): When a charged particle moves perpendicular to a magnetic field, it follows a circular path with radius $r = \frac{mv}{qB}$.

For equal kinetic energies,

$$\frac{1}{2}m_p v_p^2 = \frac{1}{2}m_d v_d^2$$

Since, $m_d = 2m_p$

it follows that

$$m_p v_p^2 = 2m_p v_d^2$$

$$\Rightarrow \quad v_p^2 = 2v_d^2$$

$$\Rightarrow \quad v_p = \sqrt{2}v_d$$

Substituting back into the radius formula:

$$r_p = \frac{m_p v_p}{eB} = \frac{m_p\sqrt{2}v_d}{eB}$$

$$r_d = \frac{2m_p v_d}{eB}$$

Therefore, the ratio of the radii is:

$$\frac{r_p}{r_d} = \frac{m_p\sqrt{2}v_d}{2m_p v_d} = \frac{\sqrt{2}}{2} = \frac{1}{\sqrt{2}}$$

Thus, the ratio of the radii of the proton to the deuteron is $\frac{1}{\sqrt{2}}$.

4. (C): When a large droplet is broken into smaller droplets, the total volume remains constant. The initial droplet has radius R = 0.1 m, so its volume is $V = \frac{4}{3}\pi R^3$.

The volume of one small droplet with radius r is $v = \frac{4}{3}\pi r^3$ and there are 1000 such droplets.

Setting total volumes equal:

$$1000 \times v = V$$

$$1000 \times \frac{4}{3}\pi r^3 = \frac{4}{3}\pi R^3$$

$$1000r^3 = R^3$$

$$r^3 = \frac{R^3}{1000}$$

$$r = \left(\frac{0.1^3}{1000}\right)^{1/3} = 0.01 \text{ m}$$

Initial surface area:

$$A_i = 4\pi R^2 = 4\pi(0.1)^2 = 0.04\pi \text{ m}^2$$

Total final surface area:

$$A_f = 1000 \times 4\pi r^2 = 1000 \times 4\pi(0.01)^2 = 0.4\pi \text{ m}^2$$

Increase in surface area:

$$\Delta A = A_f - A_i = 0.4\pi - 0.04\pi = 0.36\pi \text{ m}^2$$

Gain in surface energy:

$$\Delta E = \text{Surface tension} \times \Delta A = 0.1 \text{ N/m} \times 0.36\pi \text{ m}^2 = 0.036\pi \text{ J} \approx 0.113 \text{ J}$$

Therefore, the gain in surface energy is approximately 0.11 J.

5. **(A):** To find the velocity of train Q relative to train P, we use

$$v_{QP} = v_Q - v_P$$

Assigning east as the positive direction:

- Train P moves east at $v_P = +10$ m/s.
- Train Q moves west at $v_Q = -20$ m/s.

Therefore:

$v_{QP} = -20 \text{ m/s} - (+10 \text{ m/s}) = -30 \text{ m/s}.$

The negative sign indicates that from the frame of train P, train Q is moving westward at 30 m/s, that is, from east to west.

6. **(C):** A conservative force is one for which the work done is path-independent and depends only on the initial and final positions. The work done by a conservative force around a closed path is zero. While the spring force is conservative, the frictional force is non-conservative because it depends on the path and dissipates mechanical energy as heat. Therefore, the statement that both spring force and frictional force are conservative is not true.

7. **(D):** In an orbiting satellite, both the satellite and any objects inside it are in free fall towards the Earth, experiencing the same gravitational acceleration. Because there is no contact force (normal force) acting on the person from the floor of the satellite, the normal force is zero. Weightlessness is felt when the normal force is zero because weight is the normal force exerted by the surface supporting the person. Therefore, the boy feels weightless because the normal force (reaction) is zero.

8. **(D):** To solve the problem, we start by understanding the setup: two cylindrical vessels of equal cross-sectional area are connected, allowing water to flow between them until the water levels in both are equal.

   1. **Final Water Level:** Given that the vessels are interconnected and the areas are the same, the final water level h_f is the average of h_1 and h_2, calculated as:

   $$h_f = \frac{h_1 + h_2}{2}$$

   2. **Change in Water Level:** The water level in the first vessel decreases from h_1 to h_f, so the change in height is $h_1 - h_f$. The water level in the second vessel increases from h_2 to h_f, so the change in height is $h_f - h_2$.
   3. **Change in Potential Energy:** The change in gravitational potential energy (GPE) for water shifting from one vessel to another is given by:

   $$\Delta U = mgh = \rho V gh = \rho A h \Delta h g$$

   For the first vessel:

   $$\Delta U_1 = \rho A(h_1 - h_f)g(h_1 - h_f)/2$$

   For the second vessel:

   $$\Delta U_2 = \rho A(h_f - h_2)g(h_f - h_2)/2$$

   4. **Total Change in GPE:** Summing the changes, considering one vessel gains what the other loses, the net change in energy should consider the effective distance water moved under gravity:

   $$\Delta U_{total} = \rho Ag((h_1 - h_f)^2 + (h_f - h_2)^2)/2$$

5. **Simplification:** The final height $h_f = \frac{h_1 + h_2}{2}$, thus

$$h_1 - h_f = h_f - h_2 = \frac{h_1 - h_2}{2}$$

Substituting these:

$$\Delta U_{total} = \rho A_g \left(\left(\frac{h_1 - h_2}{2}\right)^2 + \left(\frac{h_1 - h_2}{2}\right)^2 \right) / 2$$

$$= \rho A_g \left(2\left(\frac{h_1 - h_2}{2}\right)^2 \right) / 2$$

$$= \rho A_g \left(\frac{(h_1 - h_2)^2}{4} \right)$$

6. **Finding Work Done:** The work done is equal to the change in gravitational potential energy but in the opposite sign since it's energy lost by the system:

$$W = -\Delta U_{total} = -\rho A_g \left(\frac{(h_1 - h_2)^2}{4} \right)$$

Given the answer options, the correct answer would be option (D):

9. **(C):** The bullet comes to rest after penetrating the block, so we can calculate the deceleration using $v^2 = u^2 + 2as$. Since the final velocity $v = 0$ and initial velocity $u = 50$ m/s, and the penetration distance $s = 0.2$ m:

$$0 = (50)^2 + 2a(0.2)a$$

$$= -\frac{(50)^2}{2 \times 0.2} = -6250 \text{ m/s}^2$$

The negative sign indicates deceleration. The force exerted on the block is F = *ma*:
F = 0.02 kg × (-6250 m/s²) = −125 N
The magnitude of the force is 125 N.

10. **(B):** (*a*) and (*d*) only:

(*a*) **Correct.** All non-zero digits and zeros between them are significant; thus, 6.405 has four significant figures.

(*b*) **Incorrect.** Trailing zeros in a whole number without a decimal point are not significant unless specified; thus, 12300 has three significant figures.

(*c*) **Incorrect.** Leading zeros are not significant; thus, 0.00421 has three significant figures.

(*d*) **Correct.** Zeros to the right of a decimal point and after a non-zero digit are significant; thus, 4.500 has four significant figures.

11. **(C):** The magnitude of the cross product is $|A \times B| = |A|\ |B| \sin \theta$.
Given, $|A| = 10$, $|A \times B| = 50$ and $\theta = 60°$:

$$50 = 10 \times |B| \times \sin 60°$$

$$\sin 60° = \frac{\sqrt{3}}{2}$$

$$50 = 10 \times |B| \times \frac{\sqrt{3}}{2}$$

$$|B| = \frac{50 \times 2}{10\sqrt{3}} = \frac{10}{\sqrt{3}}.$$

12. **(B):** Adding resistor Q in series increases the total resistance, reducing the current I. Since power dissipated by R is $P = I^2R$, the decreased current leads to a decrease in power dissipated by R.

13. **(B):** Quantization of charges: Electric charge exists in discrete packets and is always an integral multiple of the elementary charge *e*. This principle is known as the quantization of charge.

14. **(A):**

(*a*) Isothermal — III: $W = \mu RT \ln \frac{V_2}{V_1}$

(*b*) Adiabatic — IV: $W = \frac{\mu r(T_1 - T_2)}{\gamma - 1}$

(*c*) Isochoric — I: W = 0 (no volume change)

(*d*) Isobaric — II: $W = \mu R(T_2 - T_1)$.

15. (C): With the dielectric, one capacitor becomes, 30 μF, the other remains 15 μF. In series:

$$\frac{1}{C_{eq}} = \frac{1}{30} + \frac{1}{15} = \frac{1}{30} + \frac{2}{30} = \frac{3}{30}$$

$$C_{eq} = \frac{30}{3} = 10 \ \mu F.$$

16. (D): The forward and reverse reactions proceed via different activation complexes: Enzymatic reactions involve the same activation complex for both forward and reverse reactions. Therefore, this statement is not correct.

17. (D): Nucleophilic substitution: The Sanger reaction involves the nucleophilic attack of an amino group on the electrophilic carbon of FDNB, displacing a fluoride ion. This is a classic nucleophilic substitution reaction.

18. (A): This structure is aromatic because it features a three-membered ring containing an oxygen atom with a positive charge, known as an oxonium ion. Although not a common example of aromaticity, it is considered aromatic in this context because it can delocalize the lone pair of electrons from the oxygen across the ring, contributing to a stable, conjugated π-electron system. This delocalization satisfies Huckel's rule for aromaticity, which requires the molecule to be cyclic, planar, and to have a continuous π-electron system, specifically with 2 π-electrons, matching the $4n+2$ rule with $n = 0$.

19. (A): Both Statement I and Statement II are correct:

- **Statement I is correct.** Precision refers to the closeness of multiple measurements of the same quantity. It indicates how reproducible the measurements are, regardless of whether they are close to the true value.
- **Statement II is correct.** Accuracy refers to how close a measured value is to the true or accepted value of the quantity. It indicates the correctness of the measurement.

Therefore, both statements are correct.

20. (B): Vitamin B_{12} contains the transition metal Cobalt (Co). Cobalt is centrally located in the corrin ring structure of Vitamin B_{12}, playing a crucial role in its biological functions, including DNA synthesis and metabolism of fatty acids.

21. (A): Penicillin contains a β-lactam ring structure, which is essential for its antibiotic activity. The β-lactam ring interferes with the synthesis of the bacterial cell wall by binding to penicillin-binding proteins, ultimately leading to cell lysis and death of the bacterium.

22. (A): The First Law of Thermodynamics states that the energy of an isolated system is constant. This law embodies the principle of conservation of energy, meaning that energy cannot be created or destroyed but can only be transformed from one form to another.

23. (A): According to molecular orbital theory, O_2 (oxygen) exhibits paramagnetism due to the presence of two unpaired electrons in its π* antibonding molecular orbitals. In contrast, molecules like N_2, F_2, and C_2 have all electrons paired and are therefore diamagnetic.

24. (D): o-Nitrophenol can form intramolecular hydrogen bonds between the hydroxyl group and the nitro group. This intramolecular bonding affects its physical properties, such as boiling point and solubility, allowing it to be separated from p-nitrophenol using techniques like steam distillation. p-Nitrophenol lacks this intramolecular hydrogen bonding, which results in different physical behaviours.

25. **(C):** 2-methyl-1,3-butadiene; 2-chloro-1,3-butadiene:

- Natural rubber is a polymer of 2-methyl-1,3-butadiene (isoprene).
- Synthetic rubber neoprene is formed by the polymerization of 2-chloro-1,3-butadiene (chloroprene).

Therefore, the correct pairing is Option C.

26. **(B):** The acidic strength of carboxylic acids increases with the number of electron-withdrawing groups attached to the carbon adjacent to the carboxyl group. Electron-withdrawing groups stabilize the negative charge on the carboxylate ion, enhancing acidity.

(*e*) **F_3CCOOH (Trifluoroacetic acid):** Three highly electronegative fluorine atoms significantly increase acidity.

(*d*) **$Cl_3CHCOOH$:** Three chlorine atoms also withdraw electron density, though less effectively than fluorine.

(*c*) **$Cl_2CHCOOH$:** Two chlorine atoms provide moderate electron withdrawal.

(*b*) **$ClCH_2COOH$:** One chlorine atom offers minimal electron withdrawal.

(*a*) **CH_3COOH (Acetic acid):** No electron-withdrawing groups, making it the least acidic among the listed.

Hence, the order is (*e*) > (*d*) > (*c*) > (*b*) > (*a*).

27. **(B):** Benzylation is not a natural N-terminal modification of proteins. Natural N-terminal modifications include:

- **Acetylation:** Addition of an acetyl group.
- **Myristoylation:** Addition of a myristoyl group (a type of fatty acid).
- **Sumoylation:** Attachment of a SUMO protein.

28. **(B):**

(*a*) **Enzymes → II. Biocatalysts:** Enzymes function as biocatalysts, speeding up biochemical reactions without being consumed.

(*b*) **Glucose → III. Aldohexose:** Glucose is an aldohexose, a six-carbon sugar with an aldehyde group.

(*c*) **Lactose → IV. Disaccharide:** Lactose is a disaccharide, composed of glucose and galactose units.

(*d*) **Methionine → I. Amino acid:** Methionine is an amino acid, one of the building blocks of proteins.

Therefore, the correct matching is Option B.

29. **(D):** A nucleic acid chain, such as DNA or RNA, is composed of repeating units called nucleotides. Each nucleotide consists of three main components:

- **Phosphate group (*a*):** This forms the backbone of the nucleic acid chain by linking the sugars of adjacent nucleotides through phosphodiester bonds.
- **Nitrogenous base (*b*):** These are adenine (A), thymine (T), cytosine (C), guanine (G) in DNA, and uracil (U) replaces thymine in RNA.
- **Pentose sugar (*c*):** This is either deoxyribose in DNA or ribose in RNA.

Incorrect Components:

- **Thiol group (*d*):** Thiol groups are characteristic of certain amino acids like cysteine and are not a component of nucleic acids.
- **ß (1-4) linkage (*e*):** This type of linkage is typical in polysaccharides like starch and cellulose, not in nucleic acids. Nucleic acids have phosphodiester linkages.

Therefore, the correct components are (*a*), (*b*) and (*c*) only.

30. **(C):** Elements that readily react with oxygen to form their oxides are typically metals that are not noble gases or highly unreactive metals. Let's evaluate each option:

A. **Au and Pt:** Gold (Au) and Platinum (Pt) are noble metals. They are highly resistant to oxidation and do not readily form oxides under normal conditions.

B. **Ne and Ar:** Neon (Ne) and Argon (Ar) are noble gases. They are inert and do not react with oxygen to form oxides.

C. **Al and Cu:** Aluminum (Al) and Copper (Cu) are reactive metals that readily form their respective oxides when exposed to oxygen. For example, Aluminum forms aluminum oxide (Al_2O_3), and Copper forms copper oxide (CuO or Cu_2O).

D. **Cu and Pt:** While Copper (Cu) readily reacts with oxygen, Platinum (Pt) does not.

Thus, Aluminum (Al) and Copper (Cu) are the elements that readily react with oxygen to form their oxides.

31. **(C):** A point mutation refers to a change in a single nucleotide base in the DNA sequence. This can lead to various genetic disorders depending on where the mutation occurs and how it affects protein function.

A. **Turner's syndrome:** This is a chromosomal disorder caused by the complete or partial absence of one of the X chromosomes (45, X karyotype), not a point mutation.

B. **Down's syndrome:** This is caused by trisomy of chromosome 21, a chromosomal abnormality, not a point mutation.

C. **Sickle cell anemia:** This disease is caused by a point mutation in the beta-globin gene of hemoglobin. Specifically, it involves the substitution of valine for glutamic acid at the sixth position of the beta-globin chain, leading to the formation of abnormal hemoglobin (HbS).

D. **Klinefelter's syndrome:** This is a chromosomal disorder in males caused by the presence of an extra X chromosome (47, XXY), not a point mutation.

Therefore, sickle cell anemia is caused by a point mutation.

32. **(D):** The placenta serves several vital functions during pregnancy, including:

A. **Supply of oxygen and nutrients to the embryo:** The placenta facilitates the transfer of oxygen and nutrients from the mother's blood to the developing embryo/fetus.

B. **Removal of excretory waste products produced by the embryo:** It also removes waste products like carbon dioxide and urea from the fetus.

C. **Production of hCG and HPL:** The placenta produces hormones such as Human Chorionic Gonadotropin (hCG) and Human Placental Lactogen (HPL), which are crucial for maintaining pregnancy and fetal development.

D. **Supply all types of antibodies to the embryo:** This statement is NOT true. The placenta primarily transfers IgG antibodies from the mother to the fetus, providing passive immunity. It does not supply all types of antibodies (such as IgA, IgM, etc.) to the embryo.

Thus, Option D is NOT a function of the placenta.

33. **(B):** Colostrum is the first form of milk produced by mammals immediately following delivery of the newborn. Its characteristics include:

A. **It is a yellowish fluid secreted by the mother during later days of lactation:** Incorrect. Colostrum is produced in the initial days post-delivery, not during the later stages of lactation. Later milk is called "mature milk."

B. **Colostrum provides passive immunity to the infant:** Correct. Colostrum is rich in antibodies, particularly Immunoglobulin A (IgA), which provides passive immunity to the newborn, protecting it against infections.

C. **Colostrum is rich in carbohydrates and has no antibodies:** Incorrect. While colostrum does contain carbohydrates, it is especially rich in antibodies.

D. **Colostrum provides active immunity to the infant:** Incorrect. Active immunity involves the production of antibodies by the individual's own immune system, whereas passive immunity is conferred by antibodies from another source (mother to infant via colostrum).

Therefore, Option B is the correct statement.

34. (B): *Bacillus thuringiensis* (Bt) produces crystal proteins known as Cry toxins (e.g., CryA). These toxins are specifically effective against certain insect pests, such as caterpillars.

Mechanism of Action:

1. **Ingestion:** The insect pest consumes plant material treated with Bt toxins.
2. **Activation:** In the alkaline environment of the insect's gut, the Cry proteins are solubilized and then activated by proteolytic enzymes.
3. **Binding and Pore Formation:** The activated toxins bind to specific receptors on the gut epithelial cells, leading to the formation of pores.
4. **Cell Lysis and Death:** Pore formation disrupts the gut cell membranes, causing cell lysis, gut paralysis, and ultimately death of the insect.

Option B correctly describes that the Cry toxins become toxic in the alkaline medium of the insect gut, making it the accurate choice.

35. (C): The cloning vector pBR322 is one of the most widely used plasmids in molecular biology. It contains several important genes and regions:

- **ClaI (A):** This refers to a restriction enzyme site used for cutting the plasmid DNA, not a gene. It does not act as a selectable marker.
- **amp^r (B):** This gene confers ampicillin resistance, serving as a selectable marker to identify transformants (cells that have successfully taken up the plasmid), not plasmid amplifying enzymes.
- **rop (C):** This gene codes for the Rop protein, which is involved in regulating the copy number of the plasmid by affecting its replication. It ensures proper plasmid replication within the host cell.
- **ori (D):** This is the origin of replication, a crucial region that allows the plasmid to be replicated within the host cell. It controls the initiation of replication but does not directly control plasmid size.

Correct Matching:

- rop codes for proteins required for plasmid replication.

Therefore, Option C is correct.

36. (C): Acquired Immunodeficiency Syndrome (AIDS) is caused by the Human Immunodeficiency Virus (HIV), which targets the immune system, specifically CD4+ T lymphocytes. This immunosuppression makes individuals highly susceptible to opportunistic infections.

A. **The time lag between the HIV infection and AIDS manifestation varies from 2-3 weeks:** Incorrect. The progression from HIV infection to AIDS typically takes several years, not weeks.

B. **After entering the body, HIV enter B-lymphocytes:** Incorrect. HIV primarily infects CD4+ T cells, macrophages, and dendritic cells, not B-lymphocytes.

C. **The AIDS-affected individuals are more susceptible to Tuberculosis:** Correct. Due to the compromised immune system, individuals with AIDS are more vulnerable to opportunistic infections like Tuberculosis (TB), which is a leading cause of death among HIV-positive individuals.

D. HIV infection depletes only the CD8 lymphocytes in the body: Incorrect. HIV primarily depletes CD4+ T lymphocytes, not CD8+ lymphocytes. The reduction in CD4+ cells impairs the immune response.

Therefore, Option C is the correct statement.

37. (D):

- **Antibodies Structure:** Human antibodies (immunoglobulins) are Y-shaped proteins composed of two identical heavy chains and two identical light chains. These chains are held together by various types of bonds to maintain the antibody's structure and functionality.
- **Linkage Between Heavy Chains:** The two heavy chains of an antibody are linked to each other primarily by disulphide bonds. These are strong covalent bonds formed between the sulphur atoms of cysteine residues in the heavy chains. Disulphide bonds provide structural stability, ensuring the antibody maintains its correct three-dimensional shape necessary for antigen binding.

Other Bonds in Antibodies:

- **Hydrogen Bonds (Option A):** While hydrogen bonds are present within antibodies, especially in maintaining secondary and tertiary structures, they are not responsible for linking the heavy chains together.
- **Glycosidic Bonds (Option B):** These bonds link sugars to other molecules, such as in nucleic acids or glycoproteins, and are not involved in antibody structure.
- **Phosphodiester Bonds (Option C):** These bonds link nucleotides in DNA and RNA, not proteins like antibodies.

Therefore, the correct linkage between the two heavy chains of a human antibody is through disulphide bonds.

38. (A): Role of Aldosterone: Aldosterone is a steroid hormone produced by the adrenal cortex. It plays a crucial role in regulating electrolyte and water balance in the body, which in turn affects blood pressure and fluid volume.

Mechanism of Action:

- **Sodium (Na^+) Reabsorption:** Aldosterone acts on the cells of the distal tubules and collecting ducts in the kidneys to increase the reabsorption of sodium ions back into the bloodstream. This reabsorption creates an osmotic gradient that facilitates the reabsorption of water (H_2O), thereby increasing blood volume and pressure.
- **Water (H_2O) Reabsorption:** The increased sodium reabsorption indirectly promotes water reabsorption through osmosis, helping to maintain fluid balance.
- **Potassium (K^+) Excretion:** As sodium ions are reabsorbed, potassium ions are secreted into the urine to maintain electrochemical balance.
- **Phosphate (PO_4^{3-}) Excretion:** Aldosterone also promotes the excretion of phosphate ions, aiding in maintaining phosphate balance in the body.

Therefore, Option A is the correct answer.

39. (C): Cell Cycle Checkpoints: The cell cycle includes several checkpoints to ensure the integrity and completeness of cell division processes. The G_2/M checkpoint is critical for ensuring that the cell is ready to enter mitosis (M phase).

G_2/M Checkpoint Function:

- **DNA Replication Verification:** This checkpoint verifies that DNA replication in the S phase has been completed successfully.
- **DNA Damage Repair:** It also ensures that any DNA damage detected is repaired before the cell proceeds to mitosis. This prevents the transmission of damaged or incomplete genetic material to daughter cells.

Consequences of Incomplete Processes:

- **Incomplete DNA Replication:** If DNA replication is incomplete, the cell cycle is arrested at the G_2/M checkpoint to prevent mitosis with unreplicated DNA.
- **Unrepaired DNA Damage:** Similarly, if DNA damage is present and not repaired, the checkpoint halts cell cycle progression to maintain genomic integrity.

Therefore, the cell cycle will arrest at the G_2/M checkpoint if DNA replication is incomplete or if there is unrepaired DNA damage, making Option C the correct answer.

40. (B): Immunosuppressants and Their Origins:

Cyclosporin A: An immunosuppressant used to prevent organ transplant rejection. It is produced by the fungus *Trichoderma polysporum*.

Other Options:

- **Option A:** Cholesterol is a lipid molecule found in animal cell membranes and is not an immunosuppressant. Palm Oil is a plant-derived oil.
- **Option C:** Streptokinase is an enzyme derived from *Streptococcus bacteria*, used as a thrombolytic agent to dissolve blood clots, not as an immunosuppressant.
- **Option D:** Botulinum toxin is produced by *Clostridium botulinum* bacteria and is used for cosmetic and medical purposes, such as reducing muscle spasms, but it is not an immunosuppressant.

Correct Matching:

- **Cyclosporin A** is correctly matched with its origin, *Trichoderma polysporum*, making Option B the correct choice.

Therefore, Option B is the correct answer.

41. (A): Overview of Meiosis:

- **Meiosis I:** Reduction division where homologous chromosomes are separated, resulting in two haploid cells.
- **Meiosis II:** Equational division similar to mitosis where sister chromatids are separated, leading to four haploid cells in total.

DNA Replication:

- **Single Cycle:** DNA replication occurs once during the S phase before meiosis begins. This ensures that each chromosome consists of two sister chromatids.
- **Sequential Cycles:**

Two Divisions: Meiosis comprises two sequential cycles of nuclear (chromosome segregation) and cell division (cytokinesis), but DNA replication does not occur again between Meiosis I and Meiosis II.

Therefore, Option A is the correct answer.

42. (D): Water Movement in Plants:

- **Principle:** The movement of water in plants is driven by differences in water potential.
- **Water Potential (ψ):** A measure of the potential energy of water in a system, influencing the direction water will flow. Water moves from regions of higher water potential to lower water potential.

Components of Water Potential:

- **Solute Potential (ψ_S):** Influenced by the concentration of solutes; more solutes lower water potential.
- **Pressure Potential (Ψ_P):** Physical pressure on water; positive pressure increases water potential.

Driving Forces:

- **Transpiration:** Evaporation of water from leaves creates a negative pressure potential, pulling water upward through the xylem from roots where water potential is higher.
- **Osmosis:** Movement of water across cell membranes from areas of higher to lower solute concentration, contributing to overall water potential gradients.

Therefore, Option D is the correct answer.

43. (C): Methods for Animal Virus Detection:

- **Serology (Option A):** Involves detecting antibodies or antigens related to viruses using techniques like ELISA, Western blotting, or immunofluorescence.
- **Nucleic Acid Hybridization (Option B):** Detects viral DNA or RNA sequences using methods such as PCR, Southern blotting, or in situ hybridization.
- **Hemagglutination (Option D):** Utilizes the ability of certain viruses (e.g., influenza) to agglutinate red blood cells, serving as a detection method.

Hematology (Option C):

- **Definition:** The study of blood, including its components like red and white blood cells, hemoglobin, and platelets.
- **Relevance to Virus Detection:** Hematology focuses on blood cell counts and health but does not directly involve methods for detecting viruses. While some viral infections can influence blood cell counts, hematology itself is not a method for identifying the presence of viruses.

Therefore, Option C is the correct answer.

44. (C): Respiratory Pathway Characteristics: The respiratory pathway, particularly the citric acid cycle (Krebs cycle), plays a dual role in cellular metabolism.

Amphibolic Nature:

- **Catabolic Functions:** It breaks down carbohydrates, fats, and proteins to produce energy in the form of ATP. This involves the oxidation of acetyl-CoA derived from these macromolecules.
- **Anabolic Functions:** It provides key intermediates that serve as building blocks for the synthesis of amino acids, nucleotides, and other essential biomolecules.

Integration of Metabolic Processes: The ability to participate in both the degradation and synthesis of molecules makes the respiratory pathway amphibolic, meaning it has both catabolic and anabolic roles.

Therefore, the respiratory pathway is best described as an amphibolic pathway due to its involvement in both the breakdown and synthesis of biomolecules.

45. (D): Detailed Associations:

(*a*) **Diabetes Insipidus – II. Water loss and dehydration:** Diabetes Insipidus is characterized by excessive urination (polyuria) and extreme thirst (polydipsia), leading to significant water loss and potential dehydration. It results from a deficiency of the hormone vasopressin or the kidneys' response to it.

(*b*) **Exophthalmic Goiter – III. Grave's Disease:** Exophthalmic Goiter refers to an enlarged thyroid gland (goiter) with protruding eyes (exophthalmos), commonly associated with Grave's Disease, an autoimmune disorder that leads to hyperthyroidism.

(*c*) **Acromegaly – IV. Disfigurement of face:** Acromegaly results from excess growth hormone, leading to the enlargement of bones and tissues, particularly noticeable in the face, hands, and feet, causing disfigurement.

(*d*) **Hyperglycemia – I. Dysregulation of glucagon:** Hyperglycemia refers to high blood glucose levels. While it is primarily associated with insulin dysfunction (as in *Diabetes mellitus*), glucagon dysregulation can also contribute to elevated blood glucose by increasing hepatic glucose production.

So, Option D is the correct answer.

46. (C): An equilateral triangle with side length 6 cm has its corners cut off to form a regular hexagon. To achieve this, each corner of the triangle is truncated such that the remaining figure is a regular hexagon with equal sides.

Determining the Side Length of the Hexagon: When the corners of an equilateral triangle are cut off to form a regular hexagon, each side of the hexagon corresponds to one-third of the original triangle's side.

Therefore, the side length of the hexagon, a is:

$$a = \frac{6}{3} = 2 \text{ cm}$$

Calculating the Area of the Regular Hexagon: The formula for the area of a regular hexagon with side length a is:

$$\text{Area} = \frac{3\sqrt{3}}{2}a^2$$

Substituting $a = 2$ cm:

$$\text{Area} = \frac{3\sqrt{3}}{2} \times (2)^2 = \frac{3\sqrt{3}}{2} \times 4$$

$$= 6\sqrt{3} \text{ cm}^2$$

Simplifying $6\sqrt{3}$ cm^2:

$$6\sqrt{3} \text{ cm}^2 = \frac{6}{\sqrt{3}} \times \frac{\sqrt{3}}{\sqrt{3}} = \frac{6\sqrt{3}}{3}$$

$$= 2\sqrt{3} \text{ cm}^2$$

Among the given options, Option C: $6\sqrt{3}$ cm^2 simplifies to $2\sqrt{3}$ cm^2, which matches our calculated area.

Therefore, the area of the regular hexagon is Option C.

47. **(A):** Let the length of the train be L meters and its speed be v meters per second.

Passing the Standing Man:

Time taken to pass the man = 6 seconds.

Distance covered while passing the man = Length of the train (L).

Therefore:

$$L = v \times 6 \quad \ldots(1)$$

Passing the Platform:

Length of the platform = 210 meters.

Time taken to pass the platform = 16 seconds.

Distance covered while passing the platform = Length of the train + Length of the platform (L + 210).

Therefore:

$$L + 210 = v \times 16 \quad \ldots(2)$$

Solving the Equations:

From equation (1):

$$L = 6v$$

Substitute L = $6v$ into equation (2):

$$6v + 210 = 16v$$

$$\Rightarrow \quad 210 = 10v$$

$$\Rightarrow \quad v = 21 \text{ m/s}$$

Now, find L: $L = 6v = 6 \times 21 = 126$ m

Therefore, the length and speed of the train are 126 meters and 21 m/s respectively, corresponding to Option A.

48. **(D):** Let the total number of votes be V.

Votes Received:

One candidate received 30% of the total votes = 0.3 V

The other candidate received the remaining 70% of the total votes = 0.7 V

Vote Difference:

The candidate who received 70% lost by 500 votes.

Therefore:

$$0.7V - 0.3V = 500$$

$$\Rightarrow \quad 0.4V = 500$$

$$\Rightarrow \quad V = \frac{500}{0.4} = 1250$$

Therefore, the total number of votes polled is Option D: 1250.

49. **(C):** Evaluate the expression step by step:

$$9.6 \times 3.6 \div 7.2 + 10.8 \times \frac{1}{18} - \frac{1}{10}$$

Multiplication and Division:

$$9.6 \times 3.6 = 34.56$$

$$34.56 \div 7.2 = 4.8$$

Multiplication:

$$10.8 \times \frac{1}{18} = 0.6$$

Subtraction:

$$4.8 + 0.6 - 0.1 = 5.3$$

Therefore, the value of the expression is Option C: 5.3.

50. **(A):** To determine when the traffic lights will change simultaneously again, find the Least Common Multiple (LCM) of their individual cycle times.

Cycle Times:

Traffic Light 1: 48 seconds

Traffic Light 2: 72 seconds

Traffic Light 3: 108 seconds

Prime Factorization:

$$48 = 2^4 \times 3^1$$
$$48 = 2^4 \times 3^1$$
$$108 = 2^2 \times 3^3$$

Calculating LCM:

Take the highest power of each prime factor:

$$\text{LCM} = 2^4 \times 3^3 = 16 \times 27$$
$$= 432 \text{ seconds}$$

Convert seconds to minutes and seconds:

$$432 \div 60 = 7 \text{ minutes and } 12 \text{ sec.}$$

Adding to Initial Time:

Initial time when all lights change simultaneously: 8:20:00 h

Adding 7 minutes and 12 seconds:

8:20:00 + 0:07:12 = 8:27:12 h

Therefore, the traffic lights will change simultaneously again at 8:27:12 h, corresponding to Option A.

51. **(D):** Let, x = smallest even number

$x + 2$, $x + 4$, $x + 6$ = the next three consecutive even numbers

y = smallest odd number

$y + 2$, $y + 4$ = the next two consecutive odd numbers

Sum of Even Numbers:

$$x + (x + 2) + (x + 4) + (x + 6)$$
$$= 4x + 12$$

Sum of Odd Numbers:

$$y + (y + 2) + (y + 4) = 3y + 6$$

Given Relationship:

$$4x + 12 = (3y + 6) + 107$$
$$\Rightarrow \quad 4x + 12 = 3y + 113$$
$$\Rightarrow \quad 4x = 3y + 101 \quad \ldots(1)$$

Given Sum of Smallest Numbers:

$$x + y = 55 \quad \ldots(2)$$

Solving the Equations:

From equation (2): $y = 55 - x$

Substitute into equation (1):

$$4x = 3(55 - x) + 101$$
$$\Rightarrow \quad 4x = 165 - 3x + 101$$
$$\Rightarrow \quad 7x = 266 \Rightarrow x = 38$$

Therefore, the smallest even number is 38, corresponding to Option D.

52. **(C):** The maximum distance between any two points inside or on a cube is the length of its space diagonal.

Formula for Space Diagonal: For a cube with side length a, the space diagonal d is:

$$d = a\sqrt{3}$$

Given: $a = 1$ cm

Calculating the Diagonal:

$$d = 1 \times \sqrt{3} = \sqrt{3} \text{ cm}$$

Therefore, the maximum distance is Option C.

53. **(B):** We need to form natural numbers using the digits 1, 2, 3, 4, 5, 6, 7 each exactly once, with digits 3, 4, and 5 always in the middle.

Understanding "Middle": Assuming "middle" refers to the central three positions in a 7-digit number.

Arrangements:

- **Digits 3, 4, 5:** These occupy the middle three positions. They can be arranged among themselves in 3! = 6 ways.

- **Remaining Digits (1, 2, 6, 7):** These occupy the first two and last two positions. They can be arranged in 4! = 24 ways.
- Total Number of Natural Numbers:

$$6 \times 24 = 144$$

Therefore, the number of such natural numbers is Option B: 144.

54. (C): To find the acute angle between the hour and minute hands at 02:15:

Minute Hand Position: At 15 minutes, the minute hand points directly at the 3, which is 90° from the 12 o'clock position.

Hour Hand Position:

- Each hour represents 30° (360°/12).
- At 02:15, the hour hand has moved $2 \times 30° = 60°$ plus an additional,

$$\frac{15}{60} \times 30° = 7.5°$$

- Total position of the hour hand:

$$60° + 7.5° = 67.5°$$

Calculating the Angle Between Them:

$$|90° - 67.5°| = 22.5°$$

Therefore, the acute angle between the hour and minute hands at 02:15 is Option C: 22.5°.

55. (C): To determine the total number of squares in a standard 8×8 chessboard:

Counting Squares of All Sizes:

1×1 squares: $8 \times 8 = 64$

2×2 squares: $7 \times 7 = 49$

3×3 squares: $6 \times 6 = 36$

4×4 squares: $5 \times 5 = 25$

5×5 squares: $4 \times 4 = 16$

6×6 squares: $3 \times 3 = 9$

7×7 squares: $2 \times 2 = 4$

8×8 squares: $1 \times 1 = 1$

Summing All Squares:

$64 + 49 + 36 + 25 + 16 + 9 + 4 + 1 = 204$

Therefore, the total number of squares on a chessboard is Option C: 204.

56. (D): For the quadratic equation $x^2 + ax + 1 = 0$ to have no real roots, the discriminant must be less than zero.

Discriminant Calculation: The discriminant D of a quadratic equation $ax^2 + bx + c = 0$ is given by:

$$D = b^2 - 4ac$$

For the given equation:

$$x^2 + ax + 1 = 0$$

$$D = a^2 - 4(1)(1)$$

$$= a^2 - 4$$

Condition for No Real Roots:

For no real roots: $D < 0$

$\Rightarrow \quad a^2 - 4 < 0$

$\Rightarrow \quad a^2 < 4$

$\Rightarrow \quad -2 < a < 2$

Option Analysis:

- Option A: $a \le 2$ — Includes $a = 2$, where D = 0 (one real root), which does not satisfy the "no real roots" condition.
- Option B: $a \ge 2$ — Does not satisfy $a^2 < 4$.
- Option C: $-2 \le a < 2$ — Includes $a = -2$, where D = 0 (one real root), which does not satisfy the "no real roots" condition.
- Option D: $-2 < a < 2$ — Correctly excludes $a = \pm 2$, ensuring $D < 0$.

Therefore, the correct range for a is $-2 < a < 2$, corresponding to Option D.

57. (B): To find the average age of the whole class, we need to consider the total ages of both boys and girls and divide by the total number of students.

Given Data:

Number of boys (B): 30

Average age of boys: 12 years

Number of girls (G): 60

Average age of girls: 10 years

Calculating Total Ages:

Total age of boys:

$$30 \times 12 = 360 \text{ years}$$

Total age of girls:

$60 \times 10 = 600$ years

Total age of the class:

$360 + 600 = 960$ years

Total Number of Students:

30 boys + 60 girls = 90 students

Average Age Calculation:

$$\text{Average age} = \frac{\text{Total age}}{\text{Total number of students}}$$

$$= \frac{960}{90} = 10.666 \ldots \text{ years} \approx 10.66 \text{ years}$$

Therefore, the average age of the whole class is 10.66 years, corresponding to Option B.

58. (A): To determine the side length of the rhombus, we can use the properties of rhombus diagonals.

Given Data: Diagonals of the rhombus: 16 cm and 12 cm

Properties of Rhombus Diagonals:

Diagonals of a rhombus bisect each other at right angles.

Each half-diagonal forms a right-angled triangle with the sides of the rhombus.

Calculating Half-Diagonals:

Half of first diagonal: $\frac{16}{2} = 8$ cm

Half of second diagonal: $\frac{12}{2} = 6$ cm

Using Pythagoras Theorem: In the right-angled triangle formed by half-diagonals and the side of the rhombus:

$$\text{Side} = \sqrt{8^2 + 6^2} = \sqrt{64 + 36} = \sqrt{100}$$

$$= 10 \text{ cm}$$

Therefore, the side of the rhombus is 10 cm, corresponding to Option A.

59. (A): To find the median of the given discrete values, follow these steps:

Given Values: The variable takes the following discrete values:

$x + 4, x - 3.5, x - 2.5, x - 3, x - 2, x + 0.5, x - 0.5, x + 5$

Arranging in Ascending Order:

$x - 3.5, x - 3, x - 2.5, x - 2, x - 0.5, x + 0.5, x + 4, x + 5$

Number of Values: There are 8 values (even number), so the median is the average of the 4th and 5th values.

Identifying the 4th and 5th Values:

4th value: $x - 2$

5th value: $x - 0.5$

Calculating the Median:

$$\text{Median} = \frac{(x-2)+(x-0.5)}{2}$$

$$= \frac{2x - 2.5}{2} = x - 1.25$$

Therefore, the median is $x - 1.25$, corresponding to Option A.

60. (A): When a salary is first increased and then decreased by the same percentage, the overall change is a net decrease.

Initial Salary: Let the initial salary be S.

After 5% Increase:

$$S_{\text{after increase}} = S + 0.05S = 1.05S$$

After 5% Decrease:

$$S_{\text{after decrease}} = 1.05S - 0.05 \times 1.05S$$
$$= 1.05S \times (1 - 0.05)$$
$$= 1.05S \times 0.95$$
$$= 0.9975S$$

Calculating the Change:

$$\text{Change} = 0.9975S - S$$
$$= -0.0025S$$
$$= -0.25\% \text{ of } S$$

The negative sign indicates a decrease in salary.

Therefore, the salary decreases by 0.25%, corresponding to Option A.

61. (D): Different types of antibodies (Immunoglobulins) play specific roles in the immune system.

IgA (Immunoglobulin A):

- **Location:** Predominantly found in mucosal areas such as the respiratory and gastrointestinal tracts.
- **Function:** Protects against inhaled and ingested pathogens by preventing their attachment and entry into mucosal surfaces. It is also present in secretions like saliva, tears, and breast milk.

Other Antibodies:

- **IgG:** Main antibody in blood and extracellular fluid, providing systemic immunity.
- **IgD:** Involved in B cell activation but has a less clearly defined role in immunity.
- **IgM:** First antibody produced in response to an infection, effective in agglutination and complement activation.

Therefore, the antibody type that protects against inhaled and ingested pathogens is IgA, corresponding to Option D.

62. (A): Hormones can be classified based on their mechanism of action, particularly in how they transmit signals within target cells.

Nuclear Receptors:

- **Definition:** Receptors located within the cell nucleus that directly influence gene expression by acting as transcription factors.
- **Hormones Utilizing Nuclear Receptors:**
  Steroid Hormones: Such as cortisol and estrogen.
  Thyroid Hormones: Such as thyroxine (T4) and triiodothyronine (T3).

Mechanism of Action for Thyroid Hormones:

- **Transport into the Cell:** Thyroid hormones are lipophilic and can easily diffuse through the cell membrane.
- **Binding to Receptors:** Once inside the cell, they bind to thyroid hormone receptors located in the nucleus.
- **Gene Regulation:** The hormone-receptor complex binds to specific DNA sequences, regulating the transcription of target genes involved in metabolism, growth, and development.

Other Options:

- **Follicle Stimulating Hormone (FSH), Luteinizing Hormone (LH), Insulin:** These hormones typically act via cell surface receptors and second messenger systems rather than nuclear receptors.

Therefore, the hormone that transmits its signal via nuclear receptors is the Thyroid hormone, corresponding to Option A.

63. (C): Each amino acid is encoded by one or more codons in the genetic code. The number of codons per amino acid varies:

Amino Acids and Their Codons:

- Valine: 4 codons
- Threonine: 4 codons
- Tryptophan: 1 codon
- Isoleucine: 3 codons

Unique Codon: Tryptophan is the only amino acid coded by a single codon, which is UGG.

Option Analysis:

- Option A (Valine): Multiple codons
- Option B (Threonine): Multiple codons
- Option C (Tryptophan): Single codon
- Option D (Isoleucine): Multiple codons

Therefore, the amino acid coded by a single codon is Tryptophan, corresponding to Option C.

64. (D): This question asks to identify the statement that is NOT correct regarding carbohydrates.

Analyzing Each Statement:

- **Option A:** Glucose is stored in animals as glycogen. Correct. Animals store glucose primarily as glycogen in the liver and muscles.

- **Option B:** Glucose is stored in plants as starch. Correct. Plants store glucose as starch in seeds and tubers.
- **Option C:** Cellulose is a polymer of only glucose. Correct. Cellulose is a polysaccharide composed exclusively of β-D-glucose units linked by β-1,4-glycosidic bonds.
- **Option D:** Hemicellulose is a polymer of only glucose. Incorrect. Hemicellulose is a heterogeneous group of polysaccharides that can include various sugars such as xylose, mannose, galactose, rhamnose, and others, not just glucose.

Conclusion: Therefore, the statement "Hemicellulose is a polymer of only glucose" is NOT correct, corresponding to Option D.

65. (D): This question requires matching organelles with their respective functions.

Correct Matching Based on Standard Functions:

(*a*) **Mitochondria** → III. Energy production

Function: Mitochondria are the powerhouses of the cell, generating ATP through cellular respiration.

(*b*) **Nucleolus** → IV. Ribosomal RNA synthesis

Function: The nucleolus is responsible for the synthesis and assembly of ribosomal RNA (rRNA) and the formation of ribosomes.

(*c*) **Golgi complex** → I. Protein processing and transport

Function: The Golgi apparatus modifies, sorts, and packages proteins for secretion or delivery to other organelles.

(*d*) **Endoplasmic reticulum** → II. Protein synthesis

Function: The endoplasmic reticulum (ER) comes in two forms:

Rough ER: Studded with ribosomes, involved in protein synthesis.

Smooth ER: Involved in lipid synthesis and detoxification (not directly relevant here).

66. (D): During glycoprotein synthesis, carbohydrates are added to proteins in the Golgi complex. The Golgi apparatus is responsible for modifying, sorting, and packaging proteins received from the endoplasmic reticulum. Specifically, it adds carbohydrate groups to proteins through a process called glycosylation, which is essential for proper protein folding, stability, and cell signaling functions.

- **Option A (Mitochondria):** Involved in energy production, not protein glycosylation.
- **Option B (Lysosome):** Involved in degradation of cellular waste, not glycoprotein synthesis.
- **Option C (Nucleus):** Site of DNA replication and transcription, not directly involved in adding carbohydrates to proteins.
- **Option D (Golgi complex):** Correct, as the Golgi is responsible for glycosylation of proteins, thereby forming glycoproteins.

Therefore, Option D: Golgi complex is the correct answer.

67. (D): When a single base in the middle of a protein-coding open reading frame is replaced with another base, it results in a point mutation. The possible outcomes include:

- **Option A:** It may not make any difference to the protein sequence. (Possible if it's a silent mutation.)
- **Option B:** It may cause a single amino acid mutation. (Possible if it leads to a missense mutation.)
- **Option C:** It may create a premature stop codon. (Possible if it leads to a nonsense mutation.)

- **Option D:** The mRNA will not be recruited for translation. (Unlikely unless the mutation triggers mRNA degradation mechanisms like nonsense-mediated decay, which is not a general outcome.)

Since Option D is not a likely general outcome of a single base substitution, it is the correct answer.

68. (B): Covaxin is an indigenous vaccine developed in India against SARS-CoV-2 (COVID-19). It is an inactivated whole virion vaccine. This means that the entire virus is killed (inactivated) so that it cannot cause disease but still elicits an immune response in the body.

- **Option A:** mRNA vaccines (e.g., Pfizer, Moderna).
- **Option B:** Correct. Covaxin uses inactivated whole virions.
- **Option C:** Purified viral proteins would be subunit vaccines.
- **Option D:** DNA-based vaccines are not used in Covaxin.

Therefore, Option B: inactivated whole virions is correct.

69. (A): Antibody diversity is achieved through gene rearrangement. The immune system generates a vast array of antibodies by rearranging gene segments (V, D, J genes) during B-cell development. This process allows for the creation of unique antigen-binding sites, enabling the immune system to recognize a wide variety of pathogens.

- **Option A:** Correct. Gene rearrangement leads to diverse antibodies.
- **Option B:** Domain swapping refers to structural rearrangements between proteins, not directly related to antibody diversity.
- **Option C:** Post-translational modification involves chemical changes to proteins after translation, not directly responsible for antibody diversity.
- **Option D:** Proteolytic processing refers to cleavage of proteins, not directly related to generating antibody diversity.

Therefore, Option A: Gene rearrangement is correct.

70. (D): The immunoglobulin that primarily crosses the placenta to provide passive immunity to the fetus is IgG. IgG is the only class of antibody that can cross the placenta in significant amounts, thereby protecting the fetus against infections during pregnancy.

- **Option A:** IgM cannot cross the placenta.
- **Option B:** IgM does not cross the placenta; only IgG does.
- **Option C:** IgA does not significantly cross the placenta; it is found in mucosal areas and secretions.
- **Option D:** Correct. IgG only crosses the placenta.

Therefore, Option D: IgG only is correct.

71. (A): Epimers are sugars that differ in configuration around only one specific carbon atom.

- **Option A:** Correct. D-Mannose and D-Glucose are epimers differing at C-2.
- **Option B:** D-Glucose and D-Glucose are identical, not epimers.
- **Option C:** D-Arabinose and L-Arabinose are enantiomers, not epimers.
- **Option D:** D-Glucose and D-Fructose are not epimers; they are structural isomers (aldohexose vs. ketohexose).

Therefore, Option A: D-Mannose and D-Glucose is correct.

72. (B): Matching the phases of mitosis with their functions:

(*a*) Prophase → III. Condensation of DNA into chromatids:

During prophase, chromosomes condense and become visible as chromatids.

(*b*) Metaphase → IV. Chromatids line up along an axis

In metaphase, chromatids align at the metaphase plate (center of the cell).

(*c*) Anaphase → II. Separation of the two daughter chromosomes

Anaphase involves the separation of sister chromatids to opposite poles.

(*d*) Telophase → I. Reformation of the nuclear envelope around daughter chromosomes

Telophase sees the reformation of nuclear envelopes around the separated chromatids.

Therefore, the correct matching is Option B: III, IV, II, I.

73. (A): The specificity of an antibody molecule is provided by the variable regions of its light and heavy chains. Specifically, the light chain variable region is crucial for binding to specific antigens. These regions form the antigen-binding sites, allowing antibodies to recognize and bind to unique molecular structures on pathogens.

- **Option A:** Correct. Light chain variable region.
- **Option B:** Light chain constant region is involved in effector functions, not specificity.
- **Option C:** Heavy chain constant region-I is involved in effector functions, not specificity.
- **Option D:** Hinge region provides flexibility but not specificity.

Therefore, Option A: Light chain variable region is correct.

74. (D): Calcium alginate-based synthetic seeds tend to lose water rapidly and become hard pellets. To prevent this, the capsules can be coated with Elvax 4260, a type of polymer that acts as a moisture barrier, thereby maintaining the integrity and viability of the synthetic seeds by reducing water loss.

- **Option A:** Polyethylene glycol is a plasticizer, not specifically used for preventing water loss in synthetic seeds.
- **Option B:** Airtight packaging is impractical for seed storage.
- **Option C:** Treating with sterile water before encapsulation does not directly address water loss.
- **Option D:** Correct. Coating with Elvax 4260 provides a protective barrier against moisture loss.

Therefore, Option D: Coating the capsules with Elvax 4260 is correct.

75. (A): Codon bias refers to the phenomenon where certain codons are preferred over others for encoding the same amino acid. This bias can influence the efficiency and accuracy of protein synthesis, as some codons correspond to more abundant tRNA molecules, facilitating faster translation.

- **Option A:** Correct. Some codons for a particular amino acid are used more frequently.
- **Option B:** Human bias in assigning codons is incorrect; the genetic code is universal.
- **Option C:** Codon bias exists in many organisms, including plants.
- **Option D:** While codon usage may vary across proteins, the primary concept is the preferential use of certain codons over others for the same amino acid.

Therefore, Option A: Some codons for a particular amino acid are used more frequently is correct.

76. (D): Quantum dots (QDs) are semiconductor nanocrystals that have unique optical and electronic properties due to their quantum mechanical characteristics. The statements regarding Quantum dots as fluorescent probes are as follows:

- **Option A:** They are highly resistant to photo-bleaching. *True*. Quantum dots exhibit excellent photostability and do not easily degrade or lose fluorescence upon prolonged exposure to light, making them superior to traditional fluorescent dyes.

- **Option B:** They can generate fluorescence of different emission wavelengths. *True*. The emission wavelength of Quantum dots can be precisely tuned by altering their size. Smaller QDs emit light at shorter wavelengths (blue), while larger QDs emit at longer wavelengths (red).
- **Option C:** They are nanocrystals of different sizes. *True*. Quantum dots are indeed nanocrystals, and their size can be controlled during synthesis to achieve desired optical properties.
- **Option D:** Their fluorescence properties do not depend on the size of the Quantum dots. *False*. The fluorescence properties of Quantum dots are highly dependent on their size. This size-dependence is a fundamental characteristic of Quantum dots, where quantum confinement effects dictate the energy levels and thus the emission wavelengths.

Therefore, Option D is NOT true for Quantum dots as fluorescent probes in fluorescence microscopy.

77. (C): The Central Dogma of molecular biology, proposed by Francis Crick in 1958, describes the flow of genetic information within a biological system. It outlines the directional processes through which information is transferred from DNA to RNA to protein. The main processes included in the Central Dogma are:

- **Replication:** The process by which DNA makes a copy of itself.
- **Transcription:** The synthesis of RNA from a DNA template.
- **Translation:** The synthesis of proteins based on the RNA template.

Reverse transcription is the process of synthesizing DNA from an RNA template, which was discovered later with the identification of retroviruses like HIV. This process is NOT covered in the original Central Dogma.

Therefore, Option C: Reverse transcription is the correct answer as it was NOT included in Francis Crick's Central Dogma.

78. (C): Beggiatoa is a genus of bacteria known for its unique metabolic capabilities. To classify its nutritional type, we analyze the given characteristics:

- **Organic carbon:** Indicates it requires organic compounds for carbon, making it a heterotroph.
- **Inorganic chemicals and inorganic electron donor:** Suggests it obtains energy through the oxidation of inorganic substances, classifying it as a chemolithotroph.

Combining these, Beggiatoa is classified as a chemolithoheterotroph, which means it uses inorganic compounds as electron donors and requires organic compounds for carbon.

Therefore, Option C: Chemolithoheterotroph is correct.

79. (C): Let's evaluate each option regarding microbial culture media:

- **Option A:** Mannitol salt agar is an enriched and differential media.

  Incorrect. Mannitol salt agar is both selective and differential, not enriched. It selects for *staphylococci* and differentiates based on mannitol fermentation.
- **Option B:** Selective components in MacConkey (MAC) agar are eosin Y and methylene blue which inhibits the growth of gram-positive bacteria.

  Incorrect. While MacConkey agar is selective for gram-negative bacteria, the selective agents are bile salts and crystal violet, not eosin Y and methylene blue.

- **Option C:** Blood agar is a differential media which is differentiated on the basis of bacterial ability to produce hemolysins.

  Correct. Blood agar differentiates bacteria based on hemolysis patterns (alpha, beta, gamma hemolysis).
- **Option D:** Bile salts and crystal violet present in the EMB agar media inhibits the gram positive bacteria growth and hence helps to differentiate between gram positive and gram negative.

  Incorrect. While bile salts and crystal violet are present in Eosin Methylene Blue (EMB) agar and selective for gram-negative bacteria, the primary differentiation in EMB is lactose fermentation, not merely gram status.

Therefore, Option C is the correct statement about microbial culture media.

80. **(B):** The described microscope characteristics match those of a Confocal Microscope:

- **Uses lasers to scan the specimen at a specific depth:** Confocal microscopes employ laser light to illuminate precise focal planes within the specimen.
- **Illuminates one area at a time:** They scan the specimen point-by-point or line-by-line, building up a detailed image.
- **Blocks stray light:** Uses a pinhole to eliminate out-of-focus light, enhancing image contrast and resolution.
- **Excellent contrast and resolution:** Confocal microscopy provides high-resolution, high-contrast images, especially useful in thick specimens.

Other Options:

- **Option A (DIC Microscope):** Uses optical tricks to enhance contrast but does not block stray light.
- **Option C (Scanning Electron Microscope):** Uses electrons, not lasers, and provides surface topography.
- **Option D (Phase Contrast Microscope):** Enhances contrast in transparent specimens without using lasers.

Therefore, Option B: Confocal Microscope is the correct answer.

81. **(A):** To calculate the Relative Centrifugal Force (RCF), use the formula:

$$\text{RCF} = 1.118 \times 10^{-5} \times r \times (\text{RPM})^2$$

Where:

r = radius of the rotor (in centimeters)

RPM = revolutions per minute

Given:

$$\text{RPM} = 20000$$

$$\text{Diameter of rotor} = 7 \text{ cm}$$

$$\Rightarrow \text{radius,} \quad r = 7/2 = 3.5 \text{ cm}$$

Plugging in the values:

$$\text{RCF} = 1.118 \times 10^{-5} \times 3.5 \times (20000)^2$$
$$= 1.118 \times 10^{-5} \times 3.5 \times 400000000$$
$$= 1.118 \times 10^{-5} \times 1.4 \times 10^{9}$$
$$= 1.118 \times 1.4 \times 10^{4}$$
$$= 1.566 \times 10^{4}$$
$$\approx 15660 \text{ g (or 15680 g)}$$

Therefore, the RCF is approximately 15680 g, corresponding to Option A.

82. **(B):** In centrifugation, the effective radius at which particles sediment (r_{min}, r_{max}, and r_{av}) should ideally be consistent for optimal separation. Vertical rotors hold the sample tubes in a strictly vertical orientation, thus minimizing the variation in radius between the inner and outer portions of the tube. This leads to a more uniform sedimentation path and minimal deviation among r_{min}, r_{max}, and r_{av}.

In contrast, fixed-angle and swing rotors present greater variations due to the angled or swinging positions of the tubes.

83. (A): Matching the molecules with their functions:

(*a*) mRNA → III. Provides template for translation

Function: mRNA carries the genetic information from DNA to the ribosome, serving as a template for protein synthesis during translation.

(*b*) tRNA → II. Carries amino acids for translation

Function: tRNA transports specific amino acids to the ribosome, matching them with the corresponding codons on mRNA during protein synthesis.

(*c*) snRNA → IV. Involved in RNA splicing

Function: Small nuclear RNA (snRNA) is involved in the processing and splicing of pre-mRNA in the nucleus.

(*d*) siRNA → I. Inhibits gene expression

Function: Small interfering RNA (siRNA) plays a role in RNA interference, leading to the degradation of specific mRNA molecules and thus inhibiting gene expression.

Therefore, the correct matching is: Option A: III, II, IV, I.

84. (A): The surface area to volume ratio (SA) is highest in cells that are small and lack complex internal structures. Among the given options:

- **Red Blood Cells (RBC):** Highly specialized with a biconcave shape, maximizing surface area relative to volume for efficient gas exchange. They lack a nucleus and most organelles, contributing to a high SA ratio.
- **Fibroblast:** Larger cells with extensive internal structures, resulting in a lower SA ratio.
- **Keratinocyte:** Relatively large and structured, with lower SA compared to RBCs.
- **Hepatocyte:** Large liver cells with extensive internal organelles, resulting in a low SA ratio.

Therefore, Option A: RBC has the highest surface area to volume ratio.

85. (B): To determine the grams of Albumin and Aspirin required for the reaction, we use the concept of moles and molecular weights.

Given:

- **Albumin:** Molecular weight = 67,000 Da (67,000 g/mol)
- **Aspirin:** Molecular weight = 180 Da (180 g/mol)
- **Reaction Stoichiometry:** 1 millimole (mmol) Albumin reacts with 0.5 mmol Aspirin

Calculations:

Albumin:

Moles required = 1 mmol = 0.001 mol

Mass = moles × molecular weight

Mass of Albumin

$= 0.001 \text{ mol} \times 67{,}000 \text{ g/mol} = 67 \text{ g}$

Aspirin:

Moles required = 0.5 mmol = 0.0005 mol

Mass = moles × molecular weight

Mass of Aspirin

$= 0.0005 \text{ mol} \times 180 \text{ g/mol} = 0.09 \text{ g}$

$= 90 \text{ mg}$

Therefore, 67 grams of Albumin and 90 milligrams of Aspirin are required, corresponding to Option B.

86. (B): Posttranslational modifications (PTMs) are chemical modifications that occur to proteins after their synthesis (translation). These modifications can alter protein function, localization, stability, and interactions. Among the amino acids listed, Serine (Ser), Threonine (Thr), and Tyrosine (Tyr) are notable for their ability to undergo PTMs, particularly phosphorylation.

Serine, Threonine, Tyrosine (Option B):

- **Phosphorylation:** These amino acids have hydroxyl (-OH) groups in their side chains, making them prime targets for phosphorylation by kinases. Phosphorylation plays a crucial role in regulating enzyme activity, signal transduction pathways, and cellular processes.
- **Other Modifications:**

  Glycosylation: While less common on Ser and Thr compared to Asn, they can still undergo glycosylation.

  O-GlcN-Acylation: A reversible modification involving the addition of N-acetylglucosamine.

Other Options:

- **Option A (Glycine, Leucine, Tryptophan):** Glycine and Leucine lack functional groups that are commonly modified posttranslationally. Tryptophan can undergo oxidation but is less involved in widespread PTMs.
- **Option C (Cysteine, Glutamine, Proline):** Cysteine can form disulphide bonds (a type of PTM), but Glutamine and Proline are less frequently modified.
- **Option D (Glutamic acid, Arginine, Methionine):** These amino acids can undergo modifications like methylation (Arginine) and oxidation (Methionine), but they are not as universally modified as Ser, Thr, and Tyr.

Therefore, Option B: Serine, Threonine, Tyrosine is the correct choice as these amino acids are prominently involved in various posttranslational modifications.

87. (B): The peptides listed are GKWLY, YLWKG, WGKLY, WLKGY. While their sequences differ, they are composed of the same set of amino acids: Glycine (G), Lysine (K), Tryptophan (W), Leucine (L), and Tyrosine (Y).

- **Option A (Same sequence):** Incorrect, as the sequences are different.
- **Option B (Same amino acid composition):** Correct. Each peptide contains the same amino acids in different orders.
- **Option C (Same conformation):** Not necessarily true; different sequences can adopt different conformations.
- **Option D (Same interactome):** This depends on the sequence and structure, not just composition.

Therefore, the common feature of the peptides is that they have the same amino acid composition, making Option B the correct answer.

88. (C): To determine the final concentration of NaCl upon mixing, use the concept of dilution:

$$C_1V_1 + C_2V_2 = C_f (V_1 + V_2)$$

Given:

$$C_1 = 10 \text{ mM}$$
$$V_1 = 10 \text{ ml}$$
$$C_2 = 10 \text{ mM}$$
$$V_2 = 990 \text{ ml}$$

Plugging in the values:

$$(10 \times 10) + (10 \times 990) = C_f(10 + 990)$$
$$100 + 9900 = C_f \times 1000$$
$$10000 = 1000C_f$$
$$C_f = \frac{10000}{1000}$$
$$= 10 \text{ mM} = 0.01 \text{ M}$$

Therefore, the final concentration of NaCl is 0.01 M, corresponding to Option C.

89. (C): Peptide Nucleic Acids (PNAs) are synthetic polymers structurally similar to DNA or RNA but where the sugar-phosphate backbone is replaced by a peptide-like backbone, typically involving N-(2-aminoethyl)glycine units. This change gives PNAs unique properties compared to traditional nucleic acids.

From the provided options, let's identify the Peptide Nucleic Acid (PNA):

- **Option A** represents a typical DNA molecule, with a sugar-phosphate backbone linking nucleotide bases through phosphodiester bonds.
- **Option B** shows a structure of a nucleotide, specifically illustrating a guanine base attached to a sugar (ribose) and phosphate group, which is a fundamental unit of RNA.
- **Option C** shows a structure with bases linked through a peptide-like backbone without a sugar or phosphate group. This resembles the structure of PNA, where the bases are attached to a backbone that mimics peptides rather than sugars.
- **Option D** is a simple polypeptide chain, a straightforward protein structure, not a nucleic acid analog.

Based on these descriptions, Option C is the correct choice for a Peptide Nucleic Acid (PNA), as it features the distinct peptide-like backbone characteristic of PNAs, with nucleobases attached in a sequence.

90. **(C):** Among the options provided:

- **Wheat:** Approximately 13-15% protein content.
- **Moong Dal (Green Gram):** Approximately 24% protein content.
- **Rice:** Approximately 7% protein content.
- **Bajra (Pearl Millet):** Approximately 11% protein content.

Moong Dal has the highest protein content among the listed options, making Option C the correct answer.

91. **(A):** Delivering genes to non-proliferating terminally differentiated cells (such as neurons or muscle cells) requires vectors that can efficiently enter these cells and sustain gene expression without integrating into the host genome, minimizing immune responses.

Adeno-associated viral particles (Option A):

- **Advantages:** They can infect non-dividing cells, have low immunogenicity, and can mediate long-term gene expression.
- **Use Case:** Ideal for gene therapy in non-proliferating cells.

Other Options:

- **Retroviral particles (Option B):** Generally require cell division for integration into the host genome, making them less effective for non-dividing cells.
- **Calcium chloride (Option C):** A chemical method suitable for prokaryotes, not efficient for eukaryotic gene delivery.
- **Lipofectamine (Option D):** A lipid-based transfection reagent effective for dividing cells but less so for non-dividing cells.

Therefore, Option A: Adeno-associated viral particle is the most effective strategy for delivering genes to non-proliferating terminally differentiated cells.

92. **(A):** Human adeno-associated virus (AAV) vectors can deliver single-stranded DNA into host cells, functioning as a vaccine strategy that typically reduces the need for multiple booster doses because of their stable integration and sustained antigen presentation. The reason (R) states that such vaccines are given along with an adenovirus or a herpesvirus to avoid multiple boosters, which aligns with the principle that co-infection or co-delivery with other viral vectors can enhance immune responses and durability. Thus, the reason correctly explains why AAV-delivered DNA vaccines are associated with reduced booster requirements.

93. (A): In prokaryotes, the start codon on the mRNA is recognized by the 16S ribosomal RNA (rRNA) component of the small ribosomal subunit. The Shine-Dalgarno sequence, a purine-rich sequence located upstream of the start codon, is complementary to a sequence on the 16S rRNA. This complementarity facilitates the alignment of the ribosome with the start codon, ensuring accurate translation initiation.

- **Option A (16S rRNA):** Correct. It interacts with the Shine-Dalgarno sequence to position the start codon for translation.
- **Option B (5S rRNA):** Part of the large ribosomal subunit; does not interact with the start codon.
- **Option C (28S rRNA):** Found in eukaryotes, not prokaryotes.
- **Option D (18S rRNA):** Also found in eukaryotes as part of the small ribosomal subunit.

Therefore, the 16S rRNA is the correct component involved, making Option A the right answer.

94. (A): Assertion (A): In the eukaryotic genes, TATA box aids in transcription.

Correct. The TATA box is a DNA sequence found in the promoter region of many eukaryotic genes. It serves as a binding site for transcription factors and the RNA polymerase II complex, facilitating the initiation of transcription.

Reason (R): The TATA box facilitates formation of pre-initiation complex for transcription initiation.

Correct. The TATA box binds the TATA-binding protein (TBP), a component of the transcription factor complex. This binding recruits other transcription factors and RNA polymerase II to form the pre-initiation complex, which is essential for the start of transcription.

Conclusion: Both the Assertion (A) and the Reason (R) are correct, and the Reason (R) correctly explains the Assertion (A).

Therefore, the correct answer is Option A: Both (A) and (R) are correct and (R) is the correct explanation of (A).

95. (A): Tattoo ink is retained in the skin primarily by macrophages, which are a type of white blood cell responsible for engulfing and digesting cellular debris and pathogens. When tattooing, the ink particles are injected into the dermis layer of the skin, where macrophages ingest and encapsulate these particles. However, some ink particles remain trapped within the macrophages, allowing the tattoo to be permanently visible.

- **Option A (Macrophages):** Correct. They retain and encapsulate tattoo ink.
- **Option B (Melanocytes):** Responsible for pigment production but do not retain tattoo ink.
- **Option C (Keratinocytes):** Skin cells that provide structure; do not retain ink.
- **Option D (Lymphocytes):** Involved in immune responses but not primarily responsible for retaining tattoo ink.

Therefore, Option A: Macrophages is the correct answer.

96. (B): Secondary structures in proteins are defined, recurring conformations stabilized by hydrogen bonds, primarily including alpha helices and beta sheets, along with turns and loops that connect these elements. While "alpha helix," "beta sheet," and "loop" are well-known secondary structure motifs, "twist" is not recognized as a canonical secondary structure. Beta sheets and alpha helices are classical secondary structures, while loops are commonly accepted as connecting regions. The term "twist" does not refer to a standard secondary structure.

97. (D): The question involves determining the number of cells after '*n*' cycles of cell division, starting from a single cell, with each cell dividing into two in each cycle.

- **Starting Point:** 1 cell.
- **Each Cycle:** Every existing cell divides into 2 cells.
- **After 1 cycle:** $2^1 = 2$ cells.
- **After 2 cycles:** $2^2 = 2$ cells.
- **After '*n*' cycles:** $2^n = 2$ cells.

Therefore, the number of cells after '*n*' cycles is 2^n, corresponding to Option D.

98. (A): During prophase, several key events occur to prepare the cell for division. Nuclear envelope breakdown is a critical step that allows spindle fibers to access chromosomes. The processes aiding in nuclear envelope breakdown include:

- **Phosphorylation of nuclear membrane proteins (Option B):** Phosphorylation leads to the disassembly of the nuclear lamina, facilitating nuclear envelope breakdown.
- **Cytoplasmic microtubule dynamics (Option C):** The formation and rearrangement of microtubules help in disassembling the nuclear envelope.
- **Nuclear lamina disassembly (Option D):** Directly contributes to nuclear envelope breakdown.
- **Option A (Extension of the filopodia):** Not involved. Filopodia are actin-rich protrusions involved in cell movement and sensing the environment, not in nuclear envelope breakdown.

Therefore, Option A: Extension of the filopodia is NOT involved in the breakdown of the nuclear envelope during prophase.

99. (C): During embryonic development, the three primary germ layers—ectoderm, mesoderm, and endoderm—give rise to different tissues and organs.

Derived from Ectoderm:

- **Nerve cells (Neurons):** Develop from the ectodermal layer.
- Skin (Epidermis), Hair, Nails
- Sensory organs

Other Options:

- **Option A (Muscle):** Derived from the mesoderm.
- **Option B (Bone):** Derived from the mesoderm.
- **Option D (Blood):** Derived from the mesoderm.

Therefore, the nerve is the correct cell type derived from the ectoderm, making Option C the right answer.

100. (A): Regulatory B cells (Bregs) are a subset of B cells that play a role in maintaining immune homeostasis and preventing autoimmune responses. They function primarily by secreting anti-inflammatory cytokines, with Interleukin-10 (IL-10) being the most significant.

IL-10:

- **Function:** IL-10 is an anti-inflammatory cytokine that inhibits the synthesis of pro-inflammatory cytokines, reduces antigen presentation, and modulates the activity of T cells and other immune cells.
- **Role in Bregs:** Secretion of IL-10 by Bregs helps in suppressing excessive immune responses and promoting tolerance.

Other Cytokines:

- **IL-2 (Option B):** Generally involved in T cell proliferation.
- **TNF-alpha (Option C):** A pro-inflammatory cytokine.
- **IFN-gamma (Option D):** Involved in activating macrophages and promoting Th1 responses.

Therefore, Option A: IL-10 is the correct cytokine secreted by regulatory B cells to mediate their immunosuppressive functions.

101. (D): The Polymerase Chain Reaction (PCR) amplifies DNA exponentially. The number of DNA copies after n cycles can be calculated using the formula:

Number of copies = Initial copies × 2^n

Given: After 30 cycles: 1.2 billion copies

To find the number of cycles required to obtain 300 million copies, we can set up the equation:

$$2^n = \frac{300 \text{ million}}{\text{Initial copies}}$$

Assuming the initial number of copies is 1 (for simplicity):

$$2^n = 300{,}000{,}000$$

Taking the logarithm base 2 of both sides:

$$n = \log_2(300{,}000{,}000)$$

$$\approx 28.2 \text{ cycles}$$

Since the number of cycles must be an integer, 28 cycles are needed to achieve approximately 300 million copies.

Therefore, the correct answer is Option D: 28 cycles.

102. (A): Neoschizomers are a type of restriction endonuclease that recognize the same DNA sequence (identical recognition sites) but cleave the DNA at different positions within that sequence (different cleavage sites). This allows them to produce different fragment patterns despite targeting the same recognition sequence.

- **Option A:** Identical recognition site but different cleavage sites. (Correct)
- **Option B:** Different recognition sites but same cleavage site.
- **Option C:** Different recognition site but producing same sticky ends.
- **Option D:** Identical recognition and cleavage sites.

Therefore, Option A accurately describes neoschizomers.

103. (C): Telomerase is an enzyme that extends the telomeres at the ends of chromosomes. It carries its own RNA template, which it uses to synthesize DNA repeats. This process requires the enzyme to synthesize DNA from an RNA template, classifying it as an RNA-dependent DNA polymerase (also known as reverse transcriptase).

- **Option A:** DNA-dependent DNA polymerase
- **Option B:** DNA-dependent RNA polymerase
- **Option C:** RNA-dependent DNA polymerase (Correct)
- **Option D:** RNA-dependent RNA polymerase

Thus, Option C is correct.

104. (D): During gastrulation, the three primary germ layers form:

1. **Endoderm:** Forms the lining of the digestive and respiratory tracts.
2. **Mesoderm:** Gives rise to muscles, bones, the circulatory system, and other internal structures.
3. **Ectoderm:** Develops into the nervous system and the skin.

- **Option A:** The lining of the digestive tract will be formed by the endoderm. (True)
- **Option B:** The bones will be formed by the mesoderm. (True)
- **Option C:** The nerves will be formed by the ectoderm. (True)
- **Option D:** The skin will be formed by the mesoderm. (False) — The skin is derived from the ectoderm.

Therefore, Option D is NOT true.

105. (C):

- **Option A:** Transcription takes place in the nucleus of eukaryotic cells. (True)
- **Option B:** In prokaryotes mRNA is not capped. (True)

- **Option C:** Translation in eukaryotes takes place in the nucleus. (False) — In eukaryotes, translation occurs in the cytoplasm on ribosomes, not in the nucleus.
- **Option D:** In prokaryotes, DNA is replicated in the cytoplasm. (True)

Thus, Option C is the statement that is NOT correct.

106. (D): Rancidity refers to the spoilage of fats and oils, leading to unpleasant odors and flavors. This process is primarily caused by the action of lipolytic microbes, which produce lipases that hydrolyze lipids into free fatty acids and glycerol, resulting in rancid products.

- **Option A:** Proteolytic enzymes — Involved in protein breakdown.
- **Option B:** Photosynthetic microbes — Do not directly cause rancidity.
- **Option C:** Saccharolytic microbes — Break down sugars.
- **Option D:** Lipolytic microbes — Correct

Therefore, Option D is correct.

107. (C): Circular dichroism (CD) spectroscopy is a technique used to analyze the secondary structure content of proteins, including the proportion of alpha-helices, beta-sheets, and random coils.

It measures the differential absorption of left-handed and right-handed circularly polarized light, providing insights into the helical content of proteins.

- **Option A:** Infrared spectrometer — Used for various molecular analyses but not specifically for secondary structure content.
- **Option B:** Fluorescence — Primarily used to study protein folding and dynamics.
- **Option C:** Circular dichroism — Correct
- **Option D:** UV-visible spectrophotometer — Measures overall absorbance but not detailed secondary structures.

Thus, Option C is correct.

108. (C): DNA can adopt several helical conformations:

- **A-DNA:** Right-handed helix, compact form, typically found in dehydrated samples.
- **B-DNA:** Right-handed helix, most common form in living cells.
- **Z-DNA:** Left-handed helix, formed under high salt conditions or certain sequences.
- **C-DNA:** Less common, another right-handed form.

Therefore, Option C: DNA Z is left-handed.

109. (C): Anther culture is a tissue culture technique used to produce double haploid plants. It involves culturing pollen grains (microspores) from the anthers, which can develop into haploid plants. These haploids can then undergo chromosome doubling to become double haploids, resulting in completely homozygous lines in a single generation.

- **Option A:** Protoplast fusion — Used for hybridization.
- **Option B:** Embryo rescue — Used to salvage embryos from interspecific crosses.
- **Option C:** Anther culture — Correct
- **Option D:** Meristem culture — Used for cloning and virus elimination.

Thus, Option C is correct.

110. (D): Auxins are a class of plant hormones that regulate various aspects of growth and development.

Common auxins include:

- Indole acetic acid (IAA) — Auxin
- Indole butyric acid (IBA) — Auxin

- 2,4-dichlorophenoxy acetic acid (2,4-D) — Synthetic auxin

6-Benzylaminopurine (BAP) is a cytokinin, another class of plant hormones involved in cell division and differentiation, not an auxin.

- **Option A:** IAA — Auxin
- **Option B:** IBA — Auxin
- **Option C:** 2,4-D — Synthetic Auxin
- **Option D:** 6-Benzylaminopurine (BAP) — Not an auxin

Therefore, Option D is the correct answer as it is NOT an auxin.

111. **(C):** BLAST (Basic Local Alignment Search Tool) is a heuristic algorithm used for comparing primary biological sequence information, such as amino-acid sequences of proteins or nucleotides of DNA sequences. Unlike dynamic programming algorithms like Needleman-Wunsch and Smith-Waterman, which guarantee finding the optimal alignment but are computationally intensive, BLAST uses heuristics to quickly identify regions of local similarity. This makes BLAST significantly faster while still providing biologically meaningful results, especially for large databases.

- **Option A (Needleman-Wunsch):** A dynamic programming algorithm used for global sequence alignment.
- **Option B (Smith-Waterman):** A dynamic programming algorithm used for local sequence alignment.
- **Option C (BLAST):** Correct. A heuristic algorithm that operates faster than dynamic programming approaches.
- **Option D (Gradient Descent):** An optimization algorithm not related to sequence alignment.

Therefore, Option C: BLAST is the correct answer.

112. **(B):** For an imaginary Martian species with three nucleotides (X, Y, and Z), the number of possible 3-letter codons can be calculated using the principle of permutations with repetition:

Number of codons $= 3^3 = 27$

Each position in the 3-letter codon can be occupied by any of the three nucleotides, leading to a total of 27 unique combinations.

- **Option A (64):** Incorrect. This would be the number of codons if there were four nucleotides.
- **Option B (27):** Correct. $3 \times 3 \times 3 = 27$
- **Option C (9):** Incorrect. This would be the number of 2-letter codons.
- **Option D (4):** Incorrect. This is irrelevant to the calculation.

Therefore, Option B: 27 is the correct answer.

113. **(B):** To estimate the number of amino acids in an 11 KDa (11,000 Da) protein, use the average molecular weight of an amino acid, which is approximately 110 Da.

Number of amino acids

$$= \frac{\text{Total molecular weight}}{\text{Average molecular weight per amino acid}}$$

$$= \frac{11{,}000 \text{ Da}}{110 \text{ Da/amino acid}} = 10$$

- **Option A (90):** Slightly underestimated.
- **Option B (100):** Correct. The closest approximation.
- **Option C (110):** Overestimated.
- **Option D (120):** Overestimated.

Therefore, Option B: 100 is the correct answer.

114. **(A):** In a Gaussian (normal) distribution, the spread of the distribution is quantified by the standard deviation (σ). Decreasing the spread means reducing σ, which consequently increases the height of the distribution curve to maintain the total area under the curve (which is always 1).

- **Option A (smaller value of standard deviation):** Correct. A smaller σ leads to a narrower and taller bell curve.
- **Option B (higher value of standard deviation):** Incorrect. This would increase the spread and decrease the height.
- **Option C (smaller value of mean):** Irrelevant to spread and height.
- **Option D (higher value of mean):** Irrelevant to spread and height.

Therefore, Option A: smaller value of standard deviation is the correct answer.

115. **(C):** Maximum Likelihood is a statistical method used in phylogenetics to infer evolutionary trees based on DNA or protein sequence data. It evaluates different possible trees and selects the one that has the highest probability of producing the observed data, making it a robust method for phylogenetic analysis.

- **Option A (BLAST):** Primarily used for sequence similarity searches, not phylogenetic tree construction.
- **Option B (OMNIBUS):** Not a recognized method in phylogenetics.
- **Option C (Maximum likelihood):** Correct. A widely used method for constructing phylogenetic trees.
- **Option D (DAVID):** A tool for gene functional annotation, not for phylogeny.

Therefore, Option C: Maximum likelihood is the correct answer.

116. **(B):** In non-competitive inhibition, the inhibitor binds to an enzyme at a site different from the active site. This binding changes the enzyme's conformation, reducing its activity regardless of the substrate concentration. As a result, the degree of inhibition is independent of the initial substrate concentration.

- **Option A (Un-competitive inhibition):** Inhibitor binds only to the enzyme-substrate complex, affecting both K_m and V_{max}.
- **Option B (Non-competitive inhibition):** Correct. Degree of inhibition remains the same regardless of substrate concentration.
- **Option C (Competitive inhibition):** Inhibition can be overcome by increasing substrate concentration.
- **Option D (Mixed inhibition):** Inhibitor can bind to both the enzyme and the enzyme-substrate complex, affecting both K_m and V_{max}.

Therefore, Option B: Non-competitive inhibition is the correct answer.

117. **(C):** According to the Michaelis-Menten kinetics, the V_{max} is directly proportional to the enzyme concentration, while K_m (Michaelis constant) is an intrinsic property of the enzyme's affinity for the substrate and remains unchanged with enzyme concentration.

Doubling enzyme concentration:

V_{max}: Will double because more enzyme molecules can catalyze the reaction.

K_m: Will remain the same as it reflects the affinity between enzyme and substrate.

Given that $[S_0] >> [E]$: Substrate concentration is much higher than enzyme concentration, ensuring that the reaction is near V_{max}.

- **Option A (Both K_m and V_{max} will remain same):** Incorrect. V_{max} depends on enzyme concentration.
- **Option B (Both K_m and V_{max} will increase):** Incorrect. K_m remains unchanged.
- **Option C (V_{max} will increase; K_m will remain same):** Correct.

- **Option D (K_m will increase; V_{max} will remain same):** Incorrect. K_m remains unchanged; V_{max} increases.

Therefore, Option C: V_{max} will increase; K_m will remain same is the correct answer.

118. **(A):** For a zero-order reaction where the rate is constant and independent of the substrate concentration:

$$[A] = [A_0] - kt$$

Given:

- $k = 10^{-3}$
- Reactor volume = 100 L
- Initial concentration, $[A_0] = 0.1$

Total moles of A in the reactor:

$$\text{Total moles} = [A_0] \times \text{Volume}$$
$$= 0.1 \text{ mol/L} \times 100 \text{ L}$$
$$= 10 \text{ mol}$$

Rate of reaction (moles/min):

$$\text{Rate} = k \times \text{Volume}$$
$$= 10^{-3} \text{ mol/min} \times 100 \text{ L}$$
$$= 0.1 \text{ mol/min}$$

Time to exhaust A:

$$t = \frac{\text{Total moles}}{\text{Rate}}$$
$$= \frac{10 \text{ mol}}{0.1 \text{ mol/min}} = 100 \text{ min}$$

- Option A (100 min): Correct.
- Option B (200 min): Incorrect.
- Option C (300 min): Incorrect.
- Option D (40 min): Incorrect.

Therefore, Option A: 100 min is the correct answer.

119. **(C):** To calculate the surface area of gas bubbles per liter of reactor, follow these steps:

Given:

Diameter of air bubbles = 2 mm → Radius (r) = 1 mm = 0.1 cm

Gas hold-up = 10% → Gas volume in 1 L = 0.1 L = 100 cm^3

Volume of a single bubble:

$$V_{bubble} = \frac{4}{3}\pi r^3 = \frac{4}{3}\pi(0.1 \text{ cm})^3$$
$$\approx 0.00419 \text{ cm}^3$$

Number of bubbles:

$$= \frac{\text{Total gas volume}}{\text{Volume per bubble}} = \frac{100 \text{ cm}^3}{0.00419 \text{ cm}^3}$$

≈ 23876 bubbles

Surface area of a single bubble:

$$A_{bubble} = 4\pi r^2 = 4\pi(0.1 \text{ cm})^2$$
$$\approx 0.1256 \text{ cm}^2$$

Total surface area:

Number of bubbles × A_{bubble}

≈ 23876 × 0.1256 cm^2 ≈ 3000 cm^2

- Option A (30 cm^2): Underestimated.
- Option B (300 cm^2): Underestimated.
- Option C (3000 cm^2): Correct.
- Option D (30000 cm^2): Overestimated.

Therefore, the surface area of gas bubbles per liter of reactor is Option C: 3000 cm^2.

120. **(C):** In ion exchange chromatography, resolution between two proteins depends on two main factors:

Difference in Binding Affinity: Large difference in binding affinities ensures that proteins elute at different times, enhancing separation.

Dispersion (Peak Broadening): Small dispersion minimizes overlap between the elution profiles of the two proteins, leading to clearer separation.

Therefore, for good resolution, proteins should have a large difference in binding affinity and experience small dispersion.

- **Option A (large difference in binding affinity and large dispersion):** Large dispersion would cause peak overlap, reducing resolution.

- **Option B (small difference in binding affinity and large dispersion):** Both factors would negatively impact resolution.
- **Option C (large difference in binding affinity and small dispersion):** Correct. Enhances separation and resolution.
- **Option D (small difference in binding affinity and small dispersion):** Limited separation due to similar binding affinities.

Therefore, Option C: large difference in binding affinity and small dispersion is the correct answer.

121. (A): Assertion (A): Bacterial lipopolysaccharide (LPS) on its own does not induce memory B-cell in humans.

Correct. In humans, LPS acts primarily as a B-cell mitogen by activating B-cells through Toll-like receptor 4 (TLR4). However, without T-cell help, LPS alone is insufficient to induce the formation of memory B-cells. T-cell help is essential for class-switch recombination and the generation of high-affinity, long-lived memory B-cells.

Reason (R): LPS does not activate T-cell.

Correct. LPS primarily targets B-cells and innate immune cells like macrophages and dendritic cells. It does not directly activate T-cells. The lack of direct T-cell activation by LPS means that the necessary T-cell help for memory B-cell formation is absent when LPS acts alone.

Therefore, the correct answer is Option A: Both (A) and (R) are correct and (R) is the correct explanation of (A).

122. (A): The allergic immune response is primarily mediated by the antibody class Immunoglobulin E (IgE).

Here's how it functions:

IgE:

- **Role in Allergies:** IgE binds to allergens and triggers mast cells and basophils to release histamine and other inflammatory mediators.
- **Mechanism:** Upon subsequent exposure to the same allergen, the allergen binds to IgE antibodies on mast cells, causing degranulation and the release of substances that lead to allergic symptoms such as itching, swelling, and bronchoconstriction.

Other Immunoglobulins:

- **IgA:** Primarily found in mucosal areas and secretions like saliva and tears.
- **IgG:** The most abundant antibody in blood and extracellular fluid, involved in systemic immune responses.
- **IgM:** The first antibody produced in response to an infection.

Conclusion: IgE is specifically associated with allergic reactions, making Option A: IgE the correct answer.

123. (D): Detecting antibodies in the serum of infected patients can be achieved through various immunological techniques. Among the options provided:

- **ELISPOT (Option A):** Primarily used to detect individual cells secreting a specific cytokine or antibody, not directly for detecting antibodies in serum.
- **PCR (Option B):** Polymerase Chain Reaction is used to amplify DNA sequences and is not applicable for detecting antibodies.
- **Northern blot (Option C):** Used for detecting specific RNA sequences in a sample, not antibodies.
- **Western blot (Option D):** Correct. Western blotting is a technique that separates proteins by gel electrophoresis and then uses antibodies to detect specific proteins within the sample. It is widely used to confirm the presence of specific antibodies in patient serum, especially in diagnostic tests for diseases like HIV.

Conclusion: The presence of antibodies in infected patients' serum is most effectively detected by Option D: Western blot.

124. (C): The thymus is a primary lymphoid organ crucial for the maturation of T-cells. Here's what happens when the thymus is removed:

Role of the Thymus:

- **T-cell Maturation:** The thymus provides an environment for T-cell precursors to mature and undergo selection processes to become functional T-cells capable of distinguishing self from non-self.

Impact of Thymus Removal:

- **T-cell Development:** Without the thymus, T-cell maturation is severely impaired. This leads to a deficiency in functional T-cells, compromising the adaptive immune response.
- **B-cell Maturation:** B-cells mature in the bone marrow and are largely unaffected by the removal of the thymus.

Conclusion: Removing the thymus from neonatal mice specifically impairs T-cell maturation, making Option C: T-cell maturation will be impaired the correct answer.

125. (C): Influenza viruses frequently undergo genetic changes that affect their surface antigens. These changes are categorized as:

Antigenic Drift (Option C):

Definition: A series of small, gradual mutations (point mutations) in the genes encoding viral surface proteins (hemagglutinin and neuraminidase).

Consequence: Leads to minor changes in the antigens, allowing the virus to evade the host immune system over time. This necessitates the annual reformulation of influenza vaccines.

Antigenic Shift (Option B):

Definition: A major change resulting from the reassortment of genetic material between different viral strains, potentially creating a novel influenza virus with significantly different antigens.

Consequence: Can lead to pandemics due to the lack of pre-existing immunity in the population.

Conclusion: The gradual accumulation of point mutations leading to changes in influenza virus surface antigens is known as Option C: Antigenic drift.

126. (A): *Myasthenia gravis* is an autoimmune disorder characterized by:

Autoantibodies Target:

- **Acetylcholine Receptors (AChR):** The immune system produces antibodies against AChRs located at the neuromuscular junction.

Effect:

- **Impaired Neuromuscular Transmission:** The binding of autoantibodies to AChRs blocks acetylcholine from binding, reduces receptor density, and leads to muscle weakness and fatigue.

Other Options:

- **NOD1, TLR11, RIG-I Proteins:** These are involved in innate immune responses and are not the primary targets in *myasthenia gravis*.

Conclusion: In *myasthenia gravis*, patients produce autoantibodies against their own acetylcholine receptor proteins, making Option A: Acetylcholine receptor protein the correct answer.

127. (D): Matching Toll-like receptors, T-helper cells, T-cytotoxic cells, and plasmacytoid dendritic cells with their respective functions:

(*a*) Toll-like receptor 9 (TLR9):

Function: Recognizes unmethylated CpG dinucleotides, which are more common in bacterial and viral DNA.

Match: I. Recognition of unmethylated CpG dinucleotide

(*b*) **T-helper cells (Th cells):**

Function: Recognize antigens presented with MHC II molecules and assist in activating other immune cells.

Match: II. Recognition of antigen with MHC II complex

(*c*) **T-cytotoxic cells (CTLs):**

Function: Recognize and kill infected or cancerous cells presenting antigens with MHC I molecules.

Match: III. Recognition of antigen with MHC I complex

(*d*) **Plasmacytoid dendritic cells (pDCs):**

Function: Produce large amounts of type I interferons (IFN) in response to viral infections.

Match: IV. Type I interferon (IFN) production.

128. (C): Major Histocompatibility Complex (MHC) molecules play a critical role in the immune system by presenting peptide fragments to T-cells. The length of peptides they bind varies between MHC classes:

MHC Class I:

- **Peptide Length:** Typically 8-11 amino acids
- **Function:** Presents endogenous antigens to CD8+ T-cells.

MHC Class II:

- **Peptide Length:** Typically 13-25 amino acids, commonly 15-20 amino acids
- **Function:** Presents exogenous antigens to CD4+ T-helper cells.

Given the options:

- **Option A (8-11 amino acids):** Pertains to MHC Class I.
- **Option B (21-27 amino acids):** Slightly longer than typical for MHC Class II.
- **Option C (15-20 amino acids):** Correct. This range is appropriate for MHC Class II peptide binding.
- **Option D (507 amino acids):** Unrelated to peptide binding in MHC.

Conclusion: Peptides binding to MHC Class II molecules are typically 15-20 amino acids in length, making Option C the correct answer.

129. (B): The tuberculosis (TB) vaccine, known as Bacillus Calmette-Guérin (BCG), is derived from a strain of *Mycobacterium bovis*. Here's the rationale:

- ***Mycobacterium tuberculosis* (Option A):** The causative agent of TB, but not used for the vaccine to avoid virulence.
- ***Mycobacterium bovis* (Option B):** Correct. The BCG vaccine is a live attenuated strain of *Mycobacterium bovis*, which is less virulent than *Mycobacterium tuberculosis* but still effective in inducing immunity against TB.
- **Tuberculin (Option C):** A purified protein derivative used in diagnostic skin tests (Mantoux test) for TB, not a vaccine.
- **Mycobacterial DNA (Option D):** Not used as the basis for the traditional BCG vaccine.

Conclusion: The TB vaccine is made using *Mycobacterium bovis*, making Option B the correct answer.

130. (A): CD4 is a glycoprotein found on the surface of helper T-cells, some dendritic cells, macrophages, and monocytes. It plays a crucial role in the immune response by assisting in the activation of other immune cells.

Cell Types:

- **B-cells (Option A):** Do not express CD4. They primarily express CD19, CD20, and other B-cell markers.
- **T-cells (Option B):** Specifically, Helper T-cells (CD4+) express CD4.

- **Macrophage cells (Option C):** Can express CD4.
- **Gamma-delta T cells (Option D):** Generally do not express CD4; they have a different T-cell receptor.

Conclusion: Among the options provided, B-cells are absent of the CD4 antigen, making Option A: B.cells the correct answer.

131. (D): To correctly match the viruses in List-I with their corresponding descriptions in List-II, let's analyze each pairing:

(*a*) HIV

Description: HIV is a retrovirus that possesses Reverse Transcriptase.

Match: IV. Reverse Transcriptase

(*b*) Influenza virus

Description: Influenza viruses have a Segmented RNA genome.

Match: III. Segmented RNA virus genome

(*c*) Hepatitis C virus

Description: Hepatitis C virus utilizes RNA-dependent RNA polymerase for replication.

Match: I. RNA-dependent RNA polymerase

(*d*) Pox virus

Description: Pox viruses are characterized as dsDNA viruses.

Match: II. dsDNA virus

Therefore, the correct matching sequence is IV, III, I, II, corresponding to Option D.

132. (D): Hepatitis B virus (HBV) has a unique genome structure:

- It possesses a partially double-stranded DNA (dsDNA) genome.
- During replication, HBV converts its partially dsDNA into covalently closed circular DNA (cccDNA) in the host cell nucleus using reverse transcriptase.

Therefore, the correct answer is Option D: partially dsDNA.

133. (A): Herpes Simplex Virus (HSV) establishes latency in:

- Neuronal cells, specifically in the sensory ganglia of neurons.
- During latency, the virus remains dormant within these neurons and can reactivate periodically, causing recurrent infections.

Therefore, the correct answer is Option A: Neuronal cells.

134. (A): Poxviruses, such as the Variola virus (which causes smallpox), have a distinctive replication strategy:

- Unlike many DNA viruses that replicate in the nucleus, poxviruses replicate entirely in the cytoplasm of the host cell.
- They carry all necessary enzymes for DNA replication and transcription within their viral particles.

Therefore, the correct answer is Option A: Cytoplasm.

135. (A): Among the listed viruses:

- Dengue virus is a plus-sense single-stranded RNA (ssRNA) virus belonging to the Flavivirus family.
- Rotavirus is a double-stranded RNA (dsRNA) virus.
- Adenovirus is a double-stranded DNA (dsDNA) virus.
- Influenza virus is a negative-sense single-stranded RNA (ssRNA) virus.

Therefore, the correct answer is Option A: Dengue virus.

136. (B): In N-linked glycosylation:

- Carbohydrate moieties are attached to the side chain of the amino acid Asparagine (Asn).
- The consensus sequence for N-linked glycosylation is Asn-X-Ser/Thr, where X can be any amino acid except proline.

Therefore, the correct answer is Option B: Asparagine.

137. (A): The catalytic triad of Chymotrypsin consists of:

- Aspartate (Asp)
- Serine (Ser)
- Histidine (His)

These three amino acids work together to facilitate the cleavage of peptide bonds in proteins.

Therefore, the correct answer is Option A: Asp, Ser, His.

138. (A): In Sephadex G-200 (Size-Exclusion Chromatography):

- Proteins elute in order of decreasing size; larger proteins elute first.
- Given the elution order A, B, C, D, Protein A is the largest, and Protein D is the smallest.

In SDS-PAGE:

- Electrophoretic mobility is inversely proportional to molecular weight.
- Protein A, being the largest, will have the minimum electrophoretic mobility.

Therefore, the correct answer is Option A.

139. (C): Matching List-I with List-II:

(*a*) β-Oxidation

Function: Fatty acid degradation.

Key Enzyme: Thiolase.

Match: IV. Thiolase

(*b*) Glycolysis

Function: Glucose breakdown for energy.

Key Enzyme: Phosphofructo kinase-I.

Match: II. Phosphofructo kinase-I

(*c*) Gluconeogenesis

Function: Glucose synthesis from non-carbohydrate sources.

Key Enzyme: Phosphoenolpyruvate carboxylase.

Match: III. Phosphoenolpyruvate carboxylase

(*d*) Calvin cycle

Function: Carbon fixation in photosynthesis.

Key Enzyme: Ribulose Bisphosphate Carboxylase (Rubisco).

Match: I. Ribulose Bisphosphate Carboxylase

Therefore, the correct matching sequence is IV, II, III, I, corresponding to Option C.

140. (C): Lipid rafts are specialized microdomains within the plasma membrane characterized by:

- High concentrations of cholesterol
- Sphingolipids (such as sphingomyelin)

These components facilitate the formation of ordered structures that organize cell signaling molecules and membrane proteins.

Therefore, the correct answer is Option C: sphingolipid and cholesterol.

141. (D):

(*a*) Carbon monoxide (CO) contributes indirectly to the greenhouse effect by influencing the concentrations of other greenhouse gases. Thus, CO is associated with I. Greenhouse effect.

(*b*) Hydrocarbons (unburnt or partially burnt fuel components) are key precursors to II. Photochemical smog, as they react with nitrogen oxides in the presence of sunlight.

(*c*) Oxides of nitrogen (NOx) react with water vapor and other chemicals in the atmosphere to form nitric acid, contributing to III. Acid rain.

(*d*) Ozone (O3) near the Earth's surface is a pollutant and can impair plant growth, so it is linked to IV. Impaired plant growth.

Hence, the correct matches are (*a*)-I, (*b*)-II, (*c*)-III, (*d*)-IV.

142. (B): Excess nutrients, particularly nitrogen and phosphorus, in aquatic systems lead to eutrophication, which promotes the rapid growth of algae, known as algal blooms. These blooms can deplete dissolved oxygen in the water, causing hypoxic conditions that are detrimental to aquatic life.

Option B: Algal bloom is correct.

143. (C): *Trichoderma viride* is a fungus renowned for its ability to produce cellulolytic enzymes, which efficiently break down cellulose into glucose. This makes it a common choice for converting cellulose-based raw materials into fermentable sugars.

Option C: *Trichoderma viride* is correct.

144. (B): DDT (Dichlorodiphenyltrichloroethane) is a persistent chlorinated hydrocarbon pesticide with a halflife ranging from 2 to 15 years in the environment, depending on conditions such as temperature, sunlight, and microbial activity.

Option B: 2-15 years is correct.

145. (A): The Upflow Anaerobic Sludge Blanket (UASB) reactor is a type of anaerobic digestion system that does not use any packing material. Instead, it relies on the internal recycling of biomass based on gravity, forming a dense sludge blanket that treats wastewater efficiently.

Option A: UASB is correct.

146. (A): A Biochemical Oxygen Demand (BOD) value of more than 50 ppm indicates a high level of organic pollution in the water. This high BOD leads to the consumption of dissolved oxygen (DO) by microorganisms decomposing the organic matter, resulting in DO levels dropping below 6 ppm, which is harmful to aquatic life.

Option A: The DO content would be less than 6 ppm is correct.

147. (A): Electrostatic precipitators are the most widely used industrial methods for removing particulate matter from gas streams. They work by charging particles in the gas and then collecting them on oppositely charged plates, effectively removing dust and pollutants.

Option A: Electrostatic precipitation is correct.

148. (A): Ocean acidification is primarily caused by the dissolution of carbon dioxide (CO_2) in seawater, which forms carbonic acid (H_2CO_3). This acidification adversely affects marine life, particularly organisms with calcium carbonate structures.

Option A: Carbonic acid is correct.

149. (B): Antarctica is the driest continent on Earth, classified as a cold desert. It receives extremely low precipitation, making it the driest among all continents.

Option B: Antarctica is correct.

150. (B): An enhanced concentration of CO_2 in the environment generally leads to increased photosynthesis, especially in C3 plants, due to more available carbon for fixation. However, higher CO_2 levels can also lead to increased water demand as plants may open their stomata wider to intake more CO_2, resulting in greater transpiration rates.

Option B: increased photosynthesis and increased water demand is correct.

151. (A): In the mycorrhizal association between plants and fungi, there is a mutualistic exchange of nutrients:

- **Plant to Fungi:** Carbon Supply: Plants provide carbohydrates (such as sugars) to the fungi. This carbon is essential for the fungi's energy and growth.
- **Fungi to Plant:** Mineral Uptake: Fungi enhance the plant's ability to absorb essential minerals from the soil, particularly phosphorus and nitrogen.

They achieve this by extending the root system through their hyphal networks, increasing the surface area for nutrient absorption.

Other Options:

- **Option B:** Incorrect. Fungi do not typically provide proteins to plants, nor do plants primarily receive water in this association.
- **Option C:** Incorrect. The primary exchange is carbon from plants for minerals from fungi, not the other way around.
- **Option D:** Incorrect. There is a significant nutrient exchange between plants and fungi in mycorrhizal associations.

Therefore, Option A is correct.

153. (C): Statement I: In general, a higher auxin : cytokinin ratio will induce root formation under in vitro culture conditions in plants. **Correct.** A higher ratio of auxin to cytokinin favors the development of roots in plant tissue cultures. Auxins promote root initiation, while cytokinins promote shoot formation. Adjusting their ratio allows for controlled differentiation.

Statement II: NAA is a cytokinin and BAP is an auxin.

Incorrect.

- **NAA (Naphthaleneacetic acid):** It is an auxin, not a cytokinin.
- **BAP (6-Benzylaminopurine):** It is a cytokinin, not an auxin.

Conclusion:

- Statement I is correct.
- Statement II is incorrect.

Therefore, the correct answer is Option C: Statement I is correct but Statement II is not correct.

154. (C): Agrobacterium-mediated Plant Transformation:

- **Process:** Utilizes the natural ability of *Agrobacterium tumefaciens* to transfer a portion of its DNA (T-DNA) into the plant genome.
- **Facilitation:** The use of selection marker genes (such as antibiotic or herbicide resistance genes) is crucial. These markers allow for the selection of successfully transformed cells by enabling only those cells that have integrated the foreign DNA to survive under selective conditions.

Other Options:

- **Option A:** Incorrect. The efficiency of Agrobacterium-mediated transformation can be influenced by the plant's genotype.
- **Option B:** Incorrect. Integration of multiple T-DNA copies can occur; it does not always lead to a single copy.
- **Option D:** Incorrect. While opine genes are part of the natural Agrobacterium system, they are not required for laboratory-mediated transgenic plant production.

Therefore, Option C is correct.

155. (B): Virus-Free Plant Production:

- **Technique:** Culture of Shoot Apical Meristem is highly effective in producing virus-free plants.
- **Reason:** The shoot apical meristem is often free from systemic viruses because viruses typically do not invade the very tips of growing tissues. By culturing and regenerating plants from these meristematic tissues, it is possible to obtain plants that are free from viral infections.

Other Options:

- **Option A (Protoplast culture):** Useful for genetic engineering but not specifically for virus elimination.
- **Option C (Somatic embryogenesis from calli of leaf explants):** Can help reduce viral load but is less reliable than meristem culture.

- **Option D (Production of cybrids):** Used for combining cytoplasmic genomes but not primarily for virus-free plant production.

Therefore, Option B is correct.

156. (B): Suicide Plasmid Vectors:

- **Definition:** Plasmid vectors designed to be lost from the host cell after a certain period or under specific conditions.
- **Lacking Feature:** They lack an origin of replication that is functional in the target host. Without a functional origin, the plasmid cannot replicate independently and is thus "suicidal," ensuring it does not persist in the host cell.

Other Components:

- **Option A (Antibiotic marker):** Often present to select for transformed cells.
- **Option C (Multiple cloning sites):** Present to facilitate the insertion of foreign DNA.
- **Option D (Site for integration):** May be present depending on the vector design.

Therefore, Option B is correct.

157. (C): Digestive Enzymes in Animals:

Enzymes Present:

- **Glycogen:** Digested by glycogen phosphorylase and other enzymes.
- **Starch:** Digested by amylase enzymes.
- **Proteins:** Digested by proteases like pepsin and trypsin.

Enzyme Absent:

- **Cellulose:** Animals lack cellulase, the enzyme required to break down cellulose into glucose units. Therefore, cellulose is indigestible to most animals and passes through the digestive system as fiber.

Therefore, Option C is correct.

158. (D): Foreign DNA can be introduced into a zygote by methods like microinjection (directly injecting DNA into the pronucleus), electroporation (using electric pulses to permeabilize the cell membrane), or transduction (using viral vectors to carry foreign DNA into cells). Conjugation, however, is a process largely restricted to bacteria, wherein genetic material is transferred via direct cell-to-cell contact, typically through a pilus. This mechanism is not applicable for introducing foreign DNA into a eukaryotic zygote, making conjugation the incorrect method in this context.

159. (D): Leptin Receptor:

- **Primary Location:** Neuronal tissues, especially in the hypothalamus, where leptin signaling regulates appetite and energy balance.

Other Tissues:

- **Hepatic (Liver):** Have leptin receptors but not the primary site.
- **Muscle:** Have leptin receptors but not the primary site.
- **Adipose:** Express leptin and its receptors, but the primary signaling occurs in the brain.

However, the most prominent location for leptin receptors is in the neuronal tissues of the hypothalamus.

Therefore, Option D is correct.

Previous Paper (Solved)

Graduate Aptitude Test–Biotechnology (GAT-B)

Entrance Exam–2023

(For M.Sc. Biotechnology and Allied Areas; M.Tech. Biotechnology and Allied Areas; M.Sc. Agricultural Biotechnology and M.V. Sc. Animal Biotechnology)

(Exam held on 13 May, 2023)

Section-A

1. The deficiency of which vitamin in the body causes Pernicious Anaemia?

A. Vitamin B_6 B. Vitamin B_1

C. Vitamin B_{12} D. Vitamin B_2

2. The S.I. unit of Luminous Intensity:

A. Joule B. Watt

C. Mole D. Candela

3. Match List-I with List-II.

List-I	List-II
(*a*) Restriction Enzymes	I. Recombinant protein
(*b*) Microinjection	II. Molecular scissors
(*c*) Biolistics	III. Animal cell
(*d*) Heterologous host	IV. Gene gun

Choose the most appropriate match from the options given below:

	(*a*)	(*b*)	(*c*)	(*d*)
A.	II	III	IV	I
B.	I	II	IV	III
C.	IV	III	I	II
D.	II	I	III	IV

4. The gravitational force exerted by earth on a particle as it travels from the earth's surface towards the core:

A. increases

B. is zero

C. decreases

D. remains same

5. The electric field due to a dipole along the line passing through its midpoint and perpendicular to its axis is proportional to:

A. $1/r^2$ B. $1/r^3$

C. $1/r$ D. r

6. Ramesh has four friends. In how many ways can he invite one or more of them for dinner?

A. 15 B. 11

C. 16 D. 20

7. Which of the following is not isoelectronic with Na^+?

A. O^{-2} B. Mg^{+2}

C. Al^{+3} D. F^-

8. The mean square speed ($<v^2>$) of the molecules of a gas at absolute temperature T is proportional to:

A. 1/T B. $\sqrt{T}$

C. T D. T^2

1. C	2. D	3. A	4. C	5. B	6. A	7. A	8. C

9. Match List-I with List-II.

List-I	List-II
(*a*) $Na_6P_6O_{18}$	I. Cellobiose
(*b*) $C_{17}H_{35}COONa$	II. Cryolite
(*c*) $C_{12}H_{22}O_{11}$	III. Calgon
(*d*) Na_3AlF_6	IV. Soap

Choose the most appropriate match from the options given below:

	(*a*)	(*b*)	(*c*)	(*d*)
A.	III	II	I	IV
B.	III	IV	I	II
C.	III	II	IV	I
D.	II	I	IV	III

10. Which two gases together contribute maximally to global warming?

A. $CO_2 + N_2O$ B. CH_4 + CFCs
C. CO_2 + CFCs D. $CO_2 + CH_4$

11. A pseudoplastic fluid is best characterized by:
A. Increase in apparent viscosity as the sheer rate increases
B. Decrease in apparent viscosity as the sheer rate increases
C. The structure breaks down and apparent viscosity decreases with continuous shear stress
D. The structure builds up and apparent viscosity decreases with continuous shear stress

12. Which is the site of attachment of spindle fibres to chromosomes during metaphase of mitosis?
A. Kinetoplast of chromosomes
B. Kinetoplast and chromatid arms
C. Sister chromatids of chromosomes
D. Kinetochores of chromosomes

13. The separation and purification of the biological product after completion of the biosynthetic stage in a bioprocess is known as:
A. Downstream processing
B. Upstream processing
C. Quality control testing
D. Sample processing

14. Pipe A can fill a tank three times faster than Pipe B. If together the two pipes can fill the tank in 36 minutes, then Pipe B alone will be able to fill the tank in:
A. 81 minutes B. 108 minutes
C. 144 minutes D. 192 minutes

15. Which one of the following is an autosomal dominant disorder?
A. Myotonic dystrophy
B. Phenylketonuria
C. Hemophilia
D. Thalassemia

16. Identify the correct statement with respect to linkage and recombination frequency.
A. If the genes are loosely linked they show low recombination frequency
B. Two genes with 1.3 percent recombination frequency are loosely linked
C. If the genes are tightly linked they show high recombination frequency
D. If the genes are tightly linked they show low recombination frequency

17. Consider the two compounds given below:

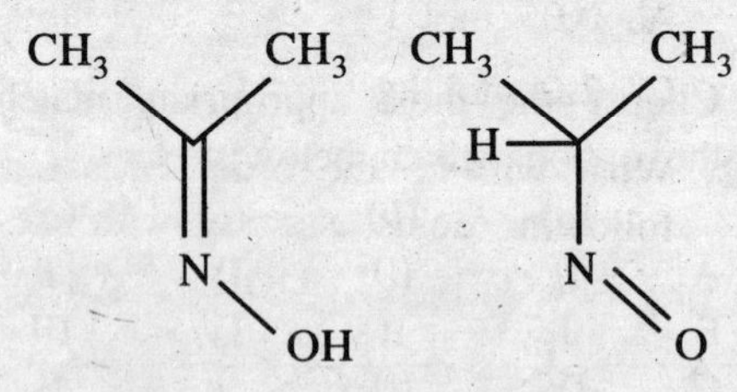

Which of the following best describes the relationship between them?
A. Resonance structures
B. Positional isomers
C. Functional isomers
D. Tautomers

9. B **10.** D **11.** B **12.** D **13.** A **14.** C **15.** A **16.** D **17.** D

18. Plants which are genetically identical to the original plant from which they are grown are called as:
A. Somatic hybrids B. Somaclones
C. Single cell explants D. Homozygous twins

19. Match List-I with List-II.

List-I	List-II
(*a*) Euchromatin	I. Densely packed DNA
(*b*) Heterochromatin	II. RNA degradation
(*c*) RNA	III. Transcriptionally active
(*d*) RNAse	IV. Poliovirus Genome

Choose the most appropriate match from the options given below:

	(*a*)	(*b*)	(*c*)	(*d*)
A.	III	I	IV	II
B.	I	II	IV	III
C.	III	I	II	IV
D.	IV	I	II	III

20. Two sources are called coherent if they produce waves:
A. of equal wavelength
B. of equal velocity
C. having a constant phase difference
D. having same shape of wavefront

21. Calculate the molality of 3.0 g acetic acid in 100 g of benzene.
A. 0.05 mol kg^{-1} B. 0.5 mol kg^{-1}
C. 2.0 mol kg^{-1} D. 0.2 mol kg^{-1}

22. What will be the order of acidity of the following acids, starting with the highest?

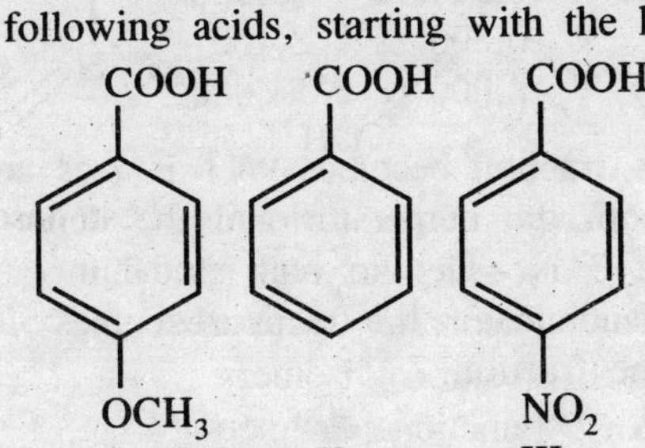

I II III

A. I > II > III
B. II > I > III
C. I = III > II
D. III > II > I

23. Identify A in the following reaction:

$$CH_3-\overset{O}{\overset{||}{C}}-CH_2-\underset{O}{\underset{||}{C}}-OCH_3 \xrightarrow{NaBH_4} A$$

A. $CH_3-\overset{OH}{\overset{|}{CH}}-CH_2-\underset{O}{\underset{||}{C}}-OCH_3$

B. $CH_3-\underset{OH}{\underset{|}{CH}}-CH_2-\underset{OH}{\underset{|}{CH}}-OCH_3$

C. $CH_3-\underset{OH}{\underset{|}{CH}}-CH_2-\underset{O}{\underset{||}{C}}-OCH_3$

D. $CH_3-\underset{OH}{\underset{|}{CH}}-CH_2-COOH$

24. A wedge of 60° representing one sixth portion of a circular disc of mass M and radius R is cut. It is rotated about a line perpendicular to its plane and passing through the centre of the original disc. Its momentum of inertia about the axis of rotation is:
A. 1/2 MR^2 B. 1/8 MR^2
C. $\sqrt{2}$ MR^2 D. 1/12 MR^2

25. Triangle ABC is a right angled triangle, right angled at B. If lengths of AB and BC are 60 cm and 80 cm respectively, and BD is an altitude of triangle ABC, then find the length of AD in cm.
A. 34 B. 36
C. 40 D. 42

18. B 19. A 20. C 21. B 22. D 23. C 24. D 25. B

26. Match List-I with List-II.

List-I	**List-II**
(*a*) Amylose	I. 2 units of α-D-glucose
(*b*) Glycogen	II. α-D-glucose + β-D-fructose
(*c*) Maltose	III. Animal starch
(*d*) Sucrose	IV. Plant starch

Choose the **correct** answer from the options given below:

	(*a*)	(*b*)	(*c*)	(*d*)
A.	IV	III	I	II
B.	IV	I	II	III
C.	III	II	I	IV
D.	III	I	II	IV

27. Half-life ($T_{1/2}$) of a radioisotope is related to decay constant (λ):

A. $T_{1/2} = 0.693/\lambda$ B. $T_{1/2} = e^{-\lambda}$

C. $T_{1/2} = -\lambda Ndt$ D. $T_{1/2} = -\lambda e^{-c}$

28. A basket contains 4 red, 5 blue and 3 green marbles. If 3 marbles are picked at random, what is the probability that either all are green or all are red?

A. 7/44 B. 7/12

C. 5/12 D. 1/44

29. Given below are two statements: one is labelled as Assertion (A) and the other is labelled as Reason (R).

Assertion (A): Morgan observed that the *Drosophila* genes for yellow body and white eye showed 1.3 percent recombination frequency.

Reason (R): Stritevant used the frequency of recombination between gene pairs on the same chromosome as a measure of the distance between genes.

In the light of the above statements, choose the **most appropriate** answer from the options given below:

A. Both (A) and (R) are correct and (R) is the correct explanation of (A)

B. Both (A) and (R) are correct, but (R) is not the correct explanation of (A)

C. (A) is correct, but (R) is not correct

D. (A) is not correct, but (R) is correct

30. A satellite is revolving around the Earth at some height (*h*). If R is the radius of orbit, then the time period of satellite is proportional to:

A. $(R + h)^3$ B. $(R + h)^{3/2}$

C. $(R + h)^{5/2}$ D. $(R + h)^6$

31. Which of the following is NOT a secondary lymphoid organ?

A. Lymph nodes B. Spleen

C. Peyer's patches D. Bone Marrow

32. Which one of the following is the working principle behind optical fibers?

A. Reflection

B. Refraction

C. Diffraction

D. Total internal reflection

33. The perimeter of the top of a rectangular table is 28 m, whereas its area is 48 m^2. What is the length of its diagonal?

A. 5 m B. 10 m

C. 12 m D. 12.5 m

34. The ratio of the incomes of two persons is 9 : 7 and the ratio of their expenditure is 4 : 3. If each of them manages to save ₹ 2000 per month, find their monthly income.

A. ₹ 9,000 & ₹ 7,000

B. ₹ 10,500 & ₹ 13,500

C. ₹ 10,800 & ₹ 8,400

D. ₹ 18,000 & ₹ 14,000

35. A fraction becomes 9/11 if 2 is added to both the numerator and the denominator. If 3 is added to both the numerator and denominator, the fraction becomes 5/6. Find the fraction.

A. 7/9 B. 2/3

C. 7/11 D. 8/9

26. A **27.** A **28.** D **29.** B **30.** B **31.** D **32.** D **33.** B **34.** D **35.** A

36. The potential energy U(r) of two point charges at a distance of r is proportional to:

A. $-1/r$ B. $1/r^2$
C. $1/r^3$ D. $1/r$

37. What type of modulation is used in television transmission?

A. amplitude modulation
B. frequency modulation
C. both amplitude and frequency modulation
D. no modulation is required

38. Which of the following alloys do NOT have copper as one of the constituent metal?

A. Brass B. German Silver
C. Bronze D. Solder

39. Which of the following is a component of smoke screens?

A. Phosphine
B. Calcium carbide
C. Copper phosphide
D. Calcium phosphate

40. A 10 m long ladder reaches a window 8 m above the ground. Find the distance of the ladder from the base of the wall.

A. 7 m B. 6 m
C. 10 m D. 5 m

41. Which of the following represents the equation of a straight line passing through origin?

A. $2x - 3y = 0$
B. $5x - 4 = 0$
C. $2y - 13 = 0$
D. $x + 2y + 9 = 0$

42. The moment of inertia of a uniform hollow sphere about it's diameter is:

A. MR^2 B. $MR^2/2$
C. 2/3 MR^2 D. 3/2 MR^2

43. Find the least number which when divided by 16, 18, 20 and 25 leaves 4 as remainder in each case but when divided by 7 leaves no remainder.

A. 8004 B. 13004
C. 18004 D. 18014

44. To obtain p-type Si semiconductor, we need to dope pure Si with:

A. Oxygen B. Aluminium
C. Phosphorous D. Germanium

45. Which two operation signs need to be interchanged to make the following equation correct?

$73 - 13 \times 42 \div 14 + 56 = 56$

A. + and × B. × and ÷
C. – and + D. – and ×

46. Blister copper is given its name because of the evolution of:

A. H_2 B. CO_2
C. SO_2 D. Cl_2

47. What is Totipotency?

A. The capacity to generate a whole seed from any cell or explant.
B. The capacity to generate plants with higher levels of vitamins and healthier fats.
C. The capacity to generate somatic hybrids from the naked protoplasts of two different cells.
D. The capacity to generate a whole plant from any cell or explant.

48. The conversion of FAD to $FADH_2$ is a:

A. One-step process with radical intermediates
B. Two-step process with ionic intermediates
C. Two-step process with radical intermediates
D. One-step process with ionic intermediates

36. D **37.** C **38.** D **39.** A **40.** B **41.** A **42.** C **43.** C **44.** B **45.** C
46. C **47.** D **48.** C

49. In areas with low air pollution, rain water generally has a pH of ~5-6 because of:

A. H_2SO_4 B. H_2CO_3
C. HNO_3 D. HNO_2

50. How are healthy plants recovered from diseased plants using plant tissue culture?

A. By growing floral buds of diseased plants *in vitro*
B. By growing leaf epidermis of diseased plants *in vitro*
C. By growing apical and axillary meristem of diseased plant *in vitro*
D. By growing phloem parenchyma cells of diseased plant *in vitro*

51. Robin could get equal number of ₹ 55, ₹ 85 and ₹ 105 category tickets for a movie. He spent ₹ 2940 for all the tickets. How many tickets of each category did he buy?

A. 12 B. 14
C. 16 D. 11

52. A man spends 75% of his income. If his income increases by 20% and his expenditure increases by 10%, then the percentage of increase in his saving would be:

A. 40% B. 30%
C. 50% D. 25%

53. The reaction between sodium hydroxide and ester of fatty acid is:

A. Esterification
B. Pyrophosphorolysis
C. Saponification
D. Neutralization reaction

54. Match List-I with List-II.

List-I	List-II
(*a*) Thalassemia	I. 44 + XXY chromosomes
(*b*) Turner's Syndrome	II. 45 + XX chromosomes
(*c*) Klinefelter's Syndrome	III. 44 + XO chromosomes
(*d*) Down's Syndrome	IV. Autosomal recessive

Choose the most appropriate match from the options given below:

	(*a*)	(*b*)	(*c*)	(*d*)
A.	IV	III	I	II
B.	I	III	II	IV
C.	II	IV	III	I
D.	IV	II	III	I

55. Rohit has some hens and some goats. If the total number of animals is 90 and the total number of animal feet is 248. What is the total number of goats Rohit has?

A. 32 B. 36
C. 34 D. 30

56. The reactive functional group in coenzyme Q is:

A. ester B. carboxylic acid
C. ketone D. thioester

57. Two fair cubical dice with faces numbered from 1 to 6 are rolled. What is the probability that the sum of the numbers on the two faces that appear on the top is 8, given that each of the two faces that appear on the top shows an odd number?

A. 1/18 B. 2/9
C. 5/36 D. 1/9

58. *Drosophila melanogaster* is suitable for genetic studies because:

(*a*) male and female flies are easily distinguishable
(*b*) the life cycle is completed in about three weeks
(*c*) a single mating produces large progeny
(*d*) phenotypic variations can be seen with low magnification microscope

Choose the **correct** answer from the options given below:

A. (*a*) and (*b*) only
B. (*c*) and (*d*) only
C. (*a*), (*c*) and (*d*) only
D. (*b*), (*c*) and (*d*) only

49. B **50.** C **51.** A **52.** C **53.** C **54.** A **55.** C **56.** D **57.** B **58.** C

59. You are supplied with a pain-relieving drug (MW 300) formulation at 5 mg/ml. A patient with a plasma volume of 2.5 L needs an initial dose of 20 μM for immediate relief. How much solution do you need to inject to the patient?

A. 1 ml B. 2 ml
C. 3 ml D. 4 ml

60. Which of the following are not electromagnetic waves?

A. Cosmic ray B. X-ray
C. β-ray D. α-ray

Section-B

61. Which of the following is NOT a preparative purification method for recovering protein product after fermentation?

A. Gel filtration chromatography
B. Ion exchange chromatography
C. Thin layer chromatography
D. Affinity chromatography

62. Match List-I with List-II.

List-I (Retrotransposons category)	List-II (Examples)
(*a*) LTR, autonomous	I. *Dasheng*
(*b*) Non-LTR, autonomous	II. *Alu*
(*c*) LTR, non-autonomous	III. copia
(*d*) Non-LTR, non-autonomous	IV. *I* factor

Choose the correct answer from the options given below:

	(*a*)	(*b*)	(*c*)	(*d*)
A.	II	III	IV	I
B.	III	IV	I	II
C.	IV	I	II	III
D.	III	II	I	IV

63. Which is NOT a part of mandatory design considerations of a Rotating Biological Contactor (RBC) system?

A. Staging of the RBC units and Secondary clarifier design
B. Loading criteria
C. Effluent characteristics
D. Rotation speed

64. Which one of the following conditions would, in general, be more conducive for the production of roots from callus tissue of plants under *in vitro* conditions?

A. Higher cytokinin/auxin ratio
B. Higher auxin/cytokinin ratio
C. Lower auxin/cytokinin ratio
D. Removal of both auxin and cytokinin from the medium

65. BCG vaccine is an example of:

A. Live attenuated vaccine
B. Heat killed vaccine
C. Recombinant vaccine
D. Subunit vaccine

66. Which of the following antibody is used by immune complexes for complement activation?

A. IgA B. IgM
C. IgE D. IgD

67. The first clinical use of gene therapy was done for treatment of:

A. Peptide deaminase deficiency
B. Adenosine deaminase deficiency
C. Cytosine decarboxylate deficiency
D. Adenosine demethylase deficiency

68. Which of the following is true for the free-energy change of a reaction (ΔG) catalyzed by an enzyme?

A. A reaction can take place spontaneously only if ΔG is negative.
B. In a system at equilibrium, there is a net change in the concentrations of the products and reactants and ΔG is positive.
C. The ΔG of a reaction is dependent of tire path of the transformation.
D. The ΔG provides information about the rate of a reaction.

59. C **60.** * **61.** C **62.** B **63.** D **64.** B **65.** A **66.** B **67.** B **68.** A

69. Which of the following bioprocess would yield maximum number of ATP per gram of the substrate consumed?
A. Anaerobic catabolism of amylose
B. Aerobic catabolism of glucose
C. Aerobic catabolism of methanol
D. Aerobic catabolism of acetate

70. The purity of a plasmid DNA preparation is determined by analyzing the ratio A_{260}/A_{280}. In a pure DNA preparation, the ratio will be:
A. close to 2
B. close to 1
C. less than 1
D. equal to 1

71. Which of the following is an RNA virus?
A. Dengue virus
B. Small pox virus
C. Human Papilloma virus
D. Hepatitis B virus

72. Which of the following is NOT a PCR based method?
A. Selective Amplification of Microsatellite Polymorphic Loci (SAMPL)
B. Sequence Tagged Site (STS) Amplification
C. Random Amplification of Polymorphic DNA (RAPD)
D. Restriction Fragment Length Polymorphism (RFLP)

73. Which of the following is normally used to determine the frequency of interferon gamma producing cells in a blood sample?
A. ELISPOT
B. ELISA
C. Immunoprecipitation
D. Confocal microscopy

74. Which of the following is the least abundant immunoglobulin class in a normal adult?
A. IgG B. IgE
C. IgM D. IgA

75. In which stage of HIV infection does one usually show symptoms of AIDS?
A. Within 15 days of sexual contact with an infected person
B. When the infected retrovirus enters the host cell
C. When HIV damages a large number of helper T-Lymphocytes
D. When the viral DNA is produced by reverse transcriptase

76. Information about the isoelectric point will be useful in the purification of a protein by which of the following chromatography techniques?
A. Gel Filtration chromatography
B. Ion exchange chromatography
C. Reverse phase chromatography
D. Hydrophobic interaction chromatography

77. Intine, the inner wall of pollen grain is made up of:
A. Cellulose and Pectin
B. Cellulose and Lignin
C. Pectin and Sporopollenin
D. Sporopollenin and Cellulose

78. Match List-I with List-II.

List-I	List-II
(*a*) IgM	I. Monomer; Mast Cell activation
(*b*) IgA	II. Naive B cell antigen receptor
(*c*) IgD	III. Monomer/Dimer/Trimer; Mucosal Immunity
(*d*) IgE	IV. Pentamer; Complement Activation

Choose the correct answer from the options given below:

	(*a*)	(*b*)	(*c*)	(*d*)
A.	III	I	II	IV
B.	IV	II	III	I
C.	II	III	IV	I
D.	IV	III	II	I

69. C **70.** A **71.** A **72.** D **73.** A **74.** B **75.** C **76.** B **77.** A **78.** D

79. Which of the following can be used for cell-line authentication?

A. Serotyping
B. Virus infection
C. PCR
D. Protein estimation

80. EST stands for:

A. Eukaryotic Sequencing Technology
B. Expressed Sequence Tag
C. Eukaryotic Sequence Test
D. Eukaryotic Sequence Tag

81. logP is a measure of drug-likeliness in Lipinsky's Rule of Five. P denotes

A. partition coefficient
B. polymer concentration
C. plasma coefficient
D. pyrogen concentration

82. The equilibrium sedimentation experiments performed by Meselson and Stahl in 1957 demonstrated that:

A. DNA replication takes place in a semiconservative fashion
B. DNA is double stranded
C. DNA chains are antiparallel
D. The nucleotides in DNA contain sugar

83. The theoretical RQ (respiratory quotient) for the complete oxidation of pyruvic acid will be:

A. 1.2
B. 0.6
C. 0.8
D. 1.0

84. Which of the following biochemical techniques does not use antibody?

A. ELISA
B. Western blotting
C. Isoelectric focusing
D. Immuno-affinity chromatography

85. Which of these viruses make use of RNA-dependent DNA polymerase during replication?

A. HIV
B. Dengue virus
C. Polio virus
D. Hepatitis E virus

86. Which of the following is a bacteria known for high yield of ethanol production?

A. *Saccharomyces cerevisiae*
B. *Pseudomonas aeruginosa*
C. *Zymomonas mobilis*
D. *Acetobacter aceti*

87. Which of the following is NOT an example of bioremediation?

A. Cleaning of oil spills by *Pseudomonas putida*
B. Decomposition of DDT by *Acetobacter aerogenus*
C. Removal of metal pollutants by *Pseudomonas* species
D. Depletion of ozone by *Aspergillus niger*

88. Which of the following statement is NOT true about the seed culture of Orchids?

A. They grow on salt poor medium as mineral requirement is low
B. Protocorm like structure is formed from the embryo
C. Seeds are very small and contain little or no food reserve
D. *In vitro* germination and development is very slow in Orchids

89. Which of the following is best suited to express large amount of glycosylated protein?

A. *E. coli*
B. *M. smegmatis*
C. Baculovirus
D. Lambda Phage

79. C **80.** B **81.** A **82.** A **83.** A **84.** C **85.** A **86.** C **87.** D **88.** D **89.** C

90. Which of the following is correct about the endoplasmic reticulum?

A. The smooth endoplasmic reticulum is involved in glycosylating the proteins

B. The rough endoplasmic reticulum is involved in synthesis of proteins that will either be inserted into cellular membranes or be secreted from the cell

C. The lumen of rough endoplasmic reticulum is highly reducing in nature

D. The smooth endoplasmic reticulum synthesizes globular proteins and rough endoplasmic reticulum synthesizes fibrillar proteins

91. Which one of the following amino acids contributes to the sulphur content in a protein?

A. Methionine B. Proline

C. Leucine D. Phenylalanine

92. In a reaction, if P (product) is at a lower internal energy level than S (substrate), then the reaction is:

A. Endothermic B. Exothermic

C. Homothermic D. Exomorphic

93. Cytokine released in response to virus infection is termed as:

A. Interleukin B. Interferon

C. Monokine D. Lymphokine

94. A humanized murine antibody is one in which the:

A. Heavy and light chain are from humans.

B. Heavy chain is from human and the light chain is from mouse

C. Light chain is from human and the heavy chain is from mouse

D. CDRs are from mouse and the other regions are from humans

95. Which of the following is a dominant genetic trait in humans?

A. Alkaptonuria

B. Sickle cell anemia

C. Huntington's disease

D. Phenylketonuria

96. Which is the most abundant protein in the entire biosphere?

A. Collagen

B. Haemoglobin

C. Proteases

D. Ribulose bisphosphate Carboxylase-Oxygenase

97. Which of the following is desirable in a primer for PCR?

A. Should always be more than 50 nucleotides in length

B. Secondary structures such as hairpins

C. Sequences with stretches of identical single nucleotides

D. Minimal complementarity between reverse and forward primers

98. Which of the following process involves the formation of nitrate from ammonia?

A. Ammonification B. Denitrification

C. Nitrification D. Nitrogen fixation

99. Neurotransmitters can be either excitatory or inhibitory, depending on where they are released, what receptors they bind to, and the ionic conditions that they encounter. Which of the following is a group of excitatory neurotransmitters?

A. Glycine and Glutamate

B. γ-aminobutyric acid (GABA) and Glycine

C. Glutamate, Acetylcholine and Serotonin

D. γ-aminobutyric acid (GABA) and Serotonin

100. A gene whose expression helps to identify transformed cell is known as:

A. Regulatory gene

B. Housekeeping gene

C. Structural gene

D. Selectable marker

90. B **91.** A **92.** B **93.** B **94.** D **95.** C **96.** D **97.** D **98.** C **99.** C **100.** D

101. Which of the following statements are NOT true for photochemical smog?
A. It is called oxidizing smog
B. Its main component is SO_2
C. It occurs in warm and dry climate
D. It is formed due to burning of fossil fuels

102. Trihydroxypropane is commonly known as:
A. Glycine
B. Glyceride
C. Glycerol
D. Glycolide

103. Anaerobic digestion proceeds in 4 steps. Which of the following represents the correct order of these four steps?
A. Hydrolysis → Acetogenesis → Acidogenesis → Methanogenesis
B. Hydrolysis → Acidogenesis → Acetogenesis → Methanogenesis
C. Methanogenesis → Acidogenesis → Acetogenesis → Hydrolysis
D. Acidogenesis → Hydrolysis → Acetogenesis → Methanogenesis

104. Given below are two statements:

Statement I: The T-DNA region of a naturally occurring (native) Ti plasmid of *Agrobacterium tumefaciens* contains genes for biosynthesis of plant hormones.

Statement II: The *virulence* genes of *Agrobacterium tumefaciens* are placed on a Helper Plasmid during development of a binary vector system for plant transformation.

In the light of the above statements, choose the **correct** answer from the options given below:
A. Both Statement I and Statement II are true
B. Both Statement I and Statement II are false
C. Statement I is correct, but Statement II is false
D. Statement I is incorrect, but Statement II is true

105. The distance between each turn in the helical strand of DNA is:
A. 20 Å B. 34 Å
C. 28 Å D. 42 Å

106. Which of the following is a hetero-polysaccharide?
A. Chitin B. Dextran
C. Agarose D. Glycogen

107. Which antibody is involved in Type I hypersensitivity reaction?
A. IgM B. IgG
C. IgE D. IgA

108. Which of the following is generally NOT a step in PCR?
A. Denaturation B. Elongation
C. Annealing D. Ligation

109. To combat a severe viral infection, what kind of immune response should be elicited?
A. Innate immune response only
B. Humoral response only
C. Cell mediated immune response only
D. Both humoral and cell mediated immune response

110. The UN Conference on Environment and Development (UNCED) is popularly known as:
A. Montreal protocol
B. Basel conference
C. Paris conference
D. Earth summit

111. Which of the following exhibits maximum mobility in an agarose gel?
A. Nicked Plasmid DNA
B. Supercoiled Plasmid DNA
C. Linear Plasmid DNA
D. Single stranded Plasmid DNA

101. B **102.** C **103.** B **104.** A **105.** B **106.** C **107.** C **108.** D **109.** D **110.** D **111.** B

112. Charles Richet was awarded Nobel Prize in Physiology or Medicine for his seminal discovery of:
A. immunoglobulins
B. hypersensitivity
C. mRNA vaccines
D. malaria vaccine

113. The structural gene *lac A* in the Lac operon is responsible for:
A. prevent entry of glucose into the bacterial cell
B. conversion of disaccharide sugar into their monosaccharides
C. conversion of lactose to allolactose
D. removal of toxic byproducts of lactose digestion

114. Match List-I with List-II.

List-I		List-II
(*a*) Nitrogen	I.	Carbon source
(*b*) Sucrose	II.	Macronutrient
(*c*) Manganese	III.	Organic supplement
(*d*) Myoinositol	IV.	Micronutrient

Choose the correct answer from the options given below that is the best possible match of various components of plant tissue culture media (List-I) and their role (List-II):

	(*a*)	(*b*)	(*c*)	(*d*)
A.	IV	III	II	I
B.	II	I	IV	III
C.	III	IV	I	II
D.	III	I	II	IV

115. Match List I with List II.

List-I (Antibiotic)	List-II (Mode of Action)
(*a*) Ampicillin	I. Inhibits translation by binding to 50S ribosomal subunit
(*b*) Kanamycin	II. Inhibits translation by binding to 30S ribosomal subunit
(*c*) Hygromycin-B	III. Inhibits cell wall synthesis by disrupting peptidoglycan cross-linking
(*d*) Chloramphenicol	IV. Inhibits translation by interfering with ribosomal translocation

Choose the correct answer from the options given below:

	(*a*)	(*b*)	(*c*)	(*d*)
A.	III	II	IV	I
B.	IV	III	II	I
C.	II	III	IV	I
D.	III	IV	I	II

116. Which one of the following would, in general, NOT be a preferred choice for transgenic plants?
A. Transgenic plants with higher levels of transgene expression
B. Transgenic plants with single-copy insertion of the transgene
C. Transgenic plants without any pleiotropic effects of the transgene
D. Transgenic plants with integration of multiple copies of the transgene

117. Most bacteria require vitamins as:
A. growth factors
B. energy source
C. carbon source
D. source of electron donors

118. Symptoms of an allergy attack can be prevented by:
A. Reducing the number of helper T cells in the body
B. Blocking the attachment of the IgE antibodies to the mast cells
C. Blocking the attachment of cytotoxic cells to dendritic cells.
D. Reducing the number of natural killer cells.

112. B **113.** C **114.** B **115.** A **116.** D **117.** A **118.** B

119. Which of the following is NOT a multiple sequence alignment program?

A. CLUSTALX

B. AMBER

C. CLUSTALW

D. PILEUP

120. The genome of Hepatitis B, a DNA virus, is unlikely to code for:

A. DNA dependent DNA polymerase

B. RNA dependent DNA polymerase

C. DNA dependent RNA polymerase

D. Envelope protein

121. Which of the following could be a recognition site of a Type IIp restriction endonuclease?

A. 5'-TTAACCGGAA-3'

B. 5'-GCTAATAGC-3'

C. 5,-TTTAATTAAA-3'

D. 5'-CCCTATAGG-3'

122. Which one of the following is **incorrect** about DNA microarray?

A. It is used for transcriptome analysis

B. It is used to identify expression of microRNAs only

C. All the expressed genes of a sample can be analyzed simultaneously

D. It can qualitatively as well as quantitatively analyze the expression of particular mRNA

123. Match List-I with List-II.

List-I	List-II
(*a*) *phytoene synthase*	I. DMH-11 mustard hybrid
(*b*) *barnase-barstar* system	II. Bt cotton
(*c*) *EPSPS*	III. Golden rice
(*d*) *Cry1A(c)*	IV. RoundUp Ready soybean

Choose the correct answer from the options given below:

	(*a*)	(*b*)	(*c*)	(*d*)
A.	IV	III	I	II
B.	IV	II	I	III
C.	III	I	IV	II
D.	III	IV	I	II

124. Given below are two statements: one is labelled as Assertion (A) and the other is labelled as Reason (R).

Assertion (A): Normally, the influenza virus changes its structure slightly for each passing year. Although, we are being exposed to these "modified" or slightly mutated and evolving influenza strains every year, we do not always come down with the flu-disease even if we have not received the yearly influenza vaccine. However, sometimes, we do get a really bad case of the flu, despite the fact that we have memory cells left from previous infections with the influenza virus.

Reason (R): This is a concept of original antigenic sin, which suggests that we only mount a primary response once we have exhausted the potential to use our memory cells to eradicate the infection. Since most of our first encounters with influenza will vary, the years in which "all" of the key influenza epitopes are significantly "new" to each of us will also vary. It is only in these years that we experience a new primary response to influenza virus, and therefore symptoms of the flu are most severe.

In the light of the above statements, choose the correct answer from the options given below:

A. Both (A) and (R) are true and (R) is the correct explanation of (A)

B. Both (A) and (R) are true, but (R) is NOT the correct explanation of (A)

C. (A) is true, but (R) is false

D. (A) is false, but (R) is true

119. B **120.** B **121.** C **122.** B **123.** C **124.** A

125. Xeroderma pigmentosum, an autosomal recessive disorder, is caused due to defects in:
A. Nucleotide Excision Repair Pathway
B. Base Excision Repair Pathway
C. Double Strand Break Repair Pathway
D. Homologous Recombination Repair Pathway

126. The cell mass of *E.coli* can be chemically represented as $CH_{1.77}O_{0.49}N_{0.24}$.

The degree of reduction for this biomass (based on available electrons) is:
A. 6.50 B. 2.50
C. 3.47 D. 4.07

127. Sterilization of fermentation media with steam is performed at:
A. 100°C for 30 minutes
B. 150°C for 5 minutes
C. 121°C for 10 minutes
D. 121°C for 20 minutes

128. Which of the following DNA, in its double-stranded form, will have the highest Tm?
A. ATGTTGGACCTTGAGTAATGCTA
B. AAAGGATTTCCCTTTGATCGTATG
C. ATTTACTAGATTACTAGTATTGATA
D. GCTTATGTATACCGGTTAGATCG

129. Which of the following is true for a water sample with BOD value of more than 50 ppm?
A. The DO content would be less than 6 ppm
B. The water is clean
C. The amount of organic matter is very less
D. The aquatic life will be thriving

130. Match List-I with List-II.

List-I (Enzymes)	List-II (Application)
(*a*) Alkalase	I. Animal feed
(*b*) Lipase	II. Maltose rich syrup
(*c*) β-amylase	III. Degreasing of leather hides
(*d*) Phytase	IV. Bioactive peptides

Choose the most appropriate match from the options given below:

	(*a*)	(*b*)	(*c*)	(*d*)
A.	I	II	III	IV
B.	II	III	I	IV
C.	IV	III	I	II
D.	IV	III	II	I

131. A monoclonal antibody was heterologously expressed in *E.coli* and Chinese hamster ovary (CHO) cell line, and the recombinant proteins obtained were purified to homogeneity and were found to be structurally identical. However, the protein expressed in *E.coli* was not functionally active. The reason could be:
A. The protein needs glycosylation for its function
B. *E.coli* introduced phosphorylation which was detrimental for function
C. CHO cells generally phosphorylate the heterologously expressed protein
D. There may be an inactivating ubiquitination modification in *E.coli*

132. Which of the following factors DO NOT affect the operation of a trickling filter?
A. organic loading
B. hydraulic flow rates
C. temperature of the water and air
D. linear flow velocity of the liquid

133. The pH indicator used for animal cell culture media is:
A. $CaCl_2$
B. Phenol red
C. Fetal Bovine Serum
D. Phenolphthalein

125. A	126. D	127. D	128. D	129. A	130. D	131. A	132. D	133. B

134. Match List-I with List II.

List-I	List-II
(*a*) Coenzyme A	I. Riboflavin
(*b*) FAD	II. Pantothenic acid
(*c*) Thiamine pyrophosphate	III. Niacin
(*d*) NAD^+	IV. Vitamin B_1

Choose the most appropriate match from the options given below:

	(*a*)	(*b*)	(*c*)	(*d*)
A.	I	II	III	IV
B.	II	I	IV	III
C.	III	IV	I	II
D.	IV	III	II	I

135. Fermentation of fruit juice to vinegar involves the use of:
A. Yeast only
B. Yeast with lactic acid bacteria
C. Yeast with butyric acid bacteria
D. Yeast with acetic acid bacteria

136. The electron transport system is present in the:
A. Inner mitochondrial membrane
B. Outer mitochondrial membrane
C. Outer chloroplast membrane
D. Mitochondrial matrix

137. Which of the following group of viruses generally infect plants?
A. Retroviruses
B. Riboviruses
C. Rheoviruses
D. Enteroviruses

138. The issue of pesticide contamination of carbonated drinks was brought into limelight by which of the following NGO?
A. Centre for Science and Environment
B. Bombay Natural History Society
C. Green Peace
D. Kalpavriksh

139. Find the odd one out with respect to pleotropic gene expression:
A. White eye gene in *Drosophila*
B. Flower colour gene in *Pisum sativum*
C. Marfan's Syndrome
D. Plumage colour in poultry

140. The addition of silver nitrate in plant tissue culture media is useful for managing and minimizing detrimental effects of which of the following plant hormones?
A. NAA
B. ethylene
C. GA-34
D. kinetin

141. In a typical bacterial growth curve, the phase in which cells adapt to growth condition and are metabolically active but not able to divide is:
A. Stationary phase
B. Lag phase
C. Exponential phase
D. Death phase

142. Given below are two statements:

Statement I: Protein-ligand docking involves binding of a small molecule to the protein receptor.

Statement II: Both AUTODOCK and LUDI are protein-ligand docking programs.

In the light of the above statements, choose the **correct** answer from the options given below:
A. Both Statement I and Statement II are true
B. Both Statement I and Statement II are false
C. Statement I is correct, but Statement II is false
D. Statement I is incorrect, but Statement II is true

134. B **135.** D **136.** A **137.** B **138.** A **139.** B **140.** B **141.** B **142.** A

143. Which of the following is NOT correct about mutagenesis using a chemical mutagen?
A. Low frequency of mutations in the gene of interest
B. Inability to confirm the location of the causal mutation even after getting the desired phenotype
C. Creation of point mutations at a specific position in the gene of interest
D. Analysis of a large number of offspring is required to isolate the desired mutant

144. Which of the following bacterium is generally used in plant genetic engineering?
A. *Thermus aquaticus*
B. *Pseudomonas syringae*
C. *Agrobacterium tumifaciens*
D. *Escherichia coli*

145. The "Microglial Cells" found in human brain, represent which of the following class of innate immune cells?
A. NK cells B. Macrophages
C. Eosinophils D. Neutrophils

146. Which of the following is an enveloped virus?
A. Dengue virus
B. Polio virus
C. Foot and mouth disease virus
D. Adenovirus

147. Vaccination in the last century had resulted in 100% reduction in active cases across the globe for:
A. small pox B. hepatitis B
C. tetanus D. tuberculosis

148. The study of various atmospheric parameters is called:
A. Geology
B. Hydrology
C. Meteorology
D. Environmental Biology

149. A culture of *E.coli* cells (1 ml) was diluted 10^6 fold and 100 μl of this was used for plating. After 24 hours incubation, the number of colony forming units (CFU) was found to be 180. The CFU count of the original culture is:
A. 1.8×10^9 CFU/ml
B. 1.8×10^6 CFU/ml
C. 1.8×10^7 CFU/ml
D. 1.8×10^8 CFU/ml

150. Given below are two statements:

Statement I: Orthologs are sequences from different organisms or genomes derived from speciation events while paralogs are sequences from the same organism or genome, which are derived from gene duplication events rather than speciation events.

Statement II: Both orthologs and paralogs are homologous sequences.

In the light of the above statements, choose the **correct** answer from the options given below:
A. Both Statement I and Statement II are true
B. Both Statement I and Statement II are false
C. Statement I is correct, but Statement II is false
D. Statement I is incorrect, but Statement II is true

151. Which of the following is a structurally independent three-dimensional unit associated with a particular functional role?
A. Domain B. Motif
C. Strand D. Helix

152. Which of the following is a multinucleated cell present in the bones of human body?
A. Cartilage B. Osteoclasts
C. Osteoblast D. Osteocyte

143. C **144.** C **145.** B **146.** A **147.** A **148.** C **149.** A **150.** A **151.** A **152.** B

153. Which of the following statement is NOT correct with respect to Cybrids?
A. They are the resultants of spontaneous fusion of protoplasts obtained from common callus culture
B. They allow the direct transfer of cytoplasmic male sterility(CMS) from donor to recipient
C. The process can bypass up to 12 backcrosses required in development of alloplasmic lines
D. It is a fusion of protoplast of one parent and cytoplast of the second.

154. The approximate length of the open reading frame coding for a 55 kDa bacterial protein will be:
A. 5000 nucleotides B. 5500 nucleotides
C. 1500 nucleotides D. 3000 nucleotides

155. Which of the following is generally NOT a primary metabolite?
A. Amino acids B. Polysaccharides
C. Antibiotics D. Vitamins

156. How much protein will you need to make 100 microlitres of 15 micromolar solution of a 95 kDa protein?
A. 0.1425 mg B. 1.425 mg
C. 0.01425 mg D. 14.25 mg

157. Harmful gases such as SO_2 can be removed from the industrial exhaust using
A. Catalytic convertor
B. Electrostatic precipitator
C. Earmuffs
D. Scrubber

158. Which of the following mineral elements play an important role in biological nitrogen fixation?
A. Copper
B. Magnesium
C. Molybdenum
D. Zinc

159. The degree of inhibition for an enzyme catalyzed reaction by an inhibitor is independent of substrate concentration. This is expected in case of:
A. Competitive inhibition
B. Un-competitive inhibition
C. Non-competitive inhibition
D. Allosteric inhibition

160. Which of the following viruses is likely to have the highest genomics mutation rate?
A. Small pox virus
B. Polio virus
C. Hepatitis B Virus
D. Adenovirus

Explanatory Answers

1. (C): Pernicious anemia is primarily caused by a deficiency in Vitamin B_{12}, also known as cobalamin. This essential vitamin is crucial for the production of red blood cells in the bone marrow. In pernicious anemia, the body is unable to properly absorb Vitamin B_{12} from the gastrointestinal tract due to a lack of intrinsic factor, a protein produced by the stomach lining that is necessary for B_{12} absorption. As a result, there is a decrease in the number of healthy red blood cells, leading to symptoms such as fatigue, weakness, shortness of breath, and neurological complications. Treatment often involves Vitamin B_{12} supplementation, typically administered through injections or oral supplements, to replenish the body's stores and alleviate symptoms.

153. A	154. C	155. C	156. A	157. D	158. C	159. C	160. B

2. **(D):** The SI unit of luminous intensity is the candela (cd). Luminous intensity refers to the amount of light emitted or reflected from a particular source in a specific direction. One candela is defined as the luminous intensity emitted in a specific direction by a source that emits monochromatic radiation of frequency 540×10^{12} hertz and that has a radiant intensity in that direction of 1/683 watt per steradian. Essentially, the candela measures the brightness of a light source as perceived by the human eye. This unit is crucial in various fields such as lighting design, optics, and photometry, where precise measurements of light intensity are required for practical applications.

3. **(A):**

(*a*) **Restriction Enzymes (Molecular scissors):** Restriction enzymes are often referred to as "molecular scissors" because they cut DNA at specific recognition sites, which is essential in genetic engineering for manipulating DNA sequences.

(*b*) **Microinjection (Animal cell):** Microinjection involves using a fine glass needle to inject substances, such as DNA, RNA, or proteins, directly into the cytoplasm or nucleus of a living cell. This technique is commonly used to introduce foreign genetic material into animal cells for various purposes, including transgenesis and gene editing.

(*c*) **Biolistics (Gene gun):** The gene gun, also known as biolistics or ballistic particle acceleration, is a method used to deliver DNA or RNA into the cells of a host organism. This technique involves coating tiny particles of gold or tungsten with the genetic material of interest and then propelling them into target cells using high-pressure helium or nitrogen gas. The gene gun is often used in plant transformation and genetic engineering of other organisms.

(*d*) **Heterologous host (Recombinant protein):** Heterologous host refers to an organism, typically a microorganism like bacteria or yeast that is used to produce recombinant proteins. Recombinant proteins are proteins that are engineered by inserting the gene encoding the desired protein into the host organism's genome. The host organism then synthesizes and expresses the recombinant protein, which can be harvested and purified for various applications in biotechnology, medicine, and research.

4. **(C):** As a particle travels from the Earth's surface towards the core the gravitational force exerted by the Earth decreases. This decrease occurs because as the particle moves closer to the center of the Earth, the distance between the particle and the majority of the Earth's mass decreases. According to Newton's law of universal gravitation, the force of gravity is inversely proportional to the square of the distance between two objects. Therefore, as the distance between the particle and the center of the Earth decreases, the gravitational force acting on the particle diminishes. Consequently, the gravitational force experienced by the particle progressively reduces as it moves towards the Earth's core.

5. **(B):** The electric field due to a dipole along the line passing through its midpoint and perpendicular to its axis is proportional to $1/r^3$. This relationship arises from the superposition principle and the geometry of the dipole. At points along the line passing through the midpoint and perpendicular to the dipole's axis, the contributions to the electric field from the positive and negative charges of the dipole have opposite directions and tend to cancel each other out more rapidly compared to points on the axis itself. This results in a faster decrease in the magnitude of the electric field with increasing distance (r) from the dipole leading to a proportionality of $1/r^3$.

6. **(A):** Required number of ways

$$= {}^4C_1 + {}^4C_2 + {}^4C_3 + {}^4C_4$$

$$= \frac{4!}{3!} + \frac{4!}{2! \times 2!} + \frac{4!}{3!} + \frac{4!}{4!}$$

$$= (4 + 6 + 4 + 1) = 15$$

7. **(A): O^{-2} (oxygen ion):** Oxygen normally has 8 electrons. With a 2- charge, it gains 2 electrons, totaling 10 electrons. This is not isoelectronic with Na^+.

Mg^{+2} (magnesium ion): Magnesium normally has 12 electrons. With a 2+ charge, it loses 2 electrons, totaling 10 electrons. This is isoelectronic with Na^+.

Al^{+3} (aluminum ion): Aluminum normally has 13 electrons. With a 3+ charge, it loses 3 electrons, totaling 10 electrons. This is isoelectronic with Na^+.

F^- (fluoride ion): Fluorine normally has 9 electrons. With a 1- charge, it gains 1 electron, totaling 10 electrons. This is isoelectronic with Na^+.

8. **(C):** The mean square speed ($<v^2>$) of gas molecules is directly proportional to the absolute temperature (T) of the gas. This relationship arises from the kinetic theory of gases, which states that the average kinetic energy of gas molecules is directly proportional to the temperature of the gas in Kelvin. Since kinetic energy is directly related to the square of the velocity (KE = 1/2 mv^2), the mean square speed, which represents the average of the squares of the velocities of all gas molecules, increases linearly with temperature. As the temperature of the gas increases, the average kinetic energy of the molecules increases, leading to higher velocities and, consequently, a greater mean square speed. Therefore, the mean square speed of gas molecules is proportional to the absolute temperature (T).

10. **(D):** Carbon dioxide (CO_2) and methane (CH_4) are two of the most significant greenhouse gases contributing to global warming. While CO_2 is the most abundant and persistent greenhouse gas emitted by human activities such as burning fossil fuels and deforestation, methane is much more potent at trapping heat in the atmosphere, albeit in lower concentrations. Methane is emitted from various sources, including livestock digestion, rice paddies, landfills, and natural gas production. Therefore, the combination of CO_2 and CH_4 emissions has a substantial impact on global warming.

11. **(B):** A pseudoplastic fluid, characterized by a decrease in apparent viscosity as the shear rate increases, exhibits shear-thinning behavior. This means that as the fluid experiences higher rates of shear stress, such as when it is stirred or flowed, its resistance to flow decreases, and it becomes less viscous. This behavior arises due to the breakdown or realignment of the internal structure of the fluid particles under shear stress, allowing them to flow more easily past each other. Common examples of pseudoplastic fluids include many types of polymer solutions, certain food products like ketchup and mayonnaise, as well as various cosmetics and paints. Understanding the rheological properties of pseudoplastic fluids is crucial in industries such as food processing, manufacturing, and cosmetics for controlling and optimizing processes involving these materials.

12. **(D):** During metaphase of mitosis, chromosomes align along the metaphase plate, a plane equidistant from the two spindle poles. At this stage, spindle fibers attach to the kinetochores, which are

protein structures located at the centromere region of chromosomes. The kinetochores serve as attachment sites for microtubules emanating from opposite spindle poles. The microtubules attached to the kinetochores exert tension on the chromosomes, aligning them at the metaphase plate. This alignment ensures that during the subsequent anaphase stage, sister chromatids are pulled apart and segregated into separate daughter cells. Therefore, the attachment of spindle fibers to the kinetochores is essential for the proper alignment and segregation of chromosomes during mitosis.

13. (A): Downstream processing in a bioprocess refers to the series of steps involved in separating, purifying, and recovering the desired biological product following its biosynthesis. After the biological product is synthesized by the microbial or cellular system during the upstream processing stage, downstream processing steps are employed to isolate the product from the fermentation broth or cell culture. These steps typically include cell removal or disruption, clarification, filtration, chromatography, and other purification techniques to separate the target product from impurities, such as cell debris, proteins, nucleic acids, and media components. The goal of downstream processing is to obtain a highly purified product with consistent quality and high yield, suitable for further formulation, characterization, and ultimately, commercial use in various industries, including pharmaceuticals, biotechnology, and food processing.

14. (C): Let x and $3x$ minutes taken by the two pipes. Then,

$$\frac{1}{x} + \frac{1}{3x} = \frac{1}{36}$$

$$\Rightarrow \quad \frac{4}{3x} = \frac{1}{36}$$

$$\therefore \quad x = \frac{36 \times 4}{3} = 48 \text{ minutes}$$

Hence, time taken by slower pipe

$= 3 \times 48$

$= 144$ minutes.

15. (A): Myotonic dystrophy is an autosomal dominant disorder characterized by progressive muscle wasting and weakness. It is caused by the expansion of a repetitive DNA sequence in the DMPK gene, located on chromosome 19. This expanded repeat leads to the formation of abnormal RNA molecules that accumulate in cells and disrupt normal cellular processes, particularly in muscle cells. The symptoms of myotonic dystrophy can vary widely and may include muscle stiffness, weakness, myotonia (prolonged muscle contractions), as well as involvement of other organs such as the heart, eyes, and brain. Because it is an autosomal dominant disorder, only one copy of the mutated gene is sufficient to cause the condition, and affected individuals have a 50% chance of passing the mutated gene to each of their offspring.

16. (D): The relationship between linkage and recombination frequency is crucial in understanding genetic inheritance patterns. When genes are tightly linked, meaning they are located close together on the same chromosome, they tend to be inherited together during meiosis with minimal or no crossover events occurring between them. Consequently, the recombination frequency between tightly linked genes is low because there is little opportunity for genetic material exchange through crossing over. On the other hand, when genes are loosely linked, they are farther apart on the chromosome, increasing the likelihood of crossover events during meiosis. This results in a higher recombination frequency, as there is a greater chance for genetic material exchange between the loosely linked genes. Therefore, the level of linkage between genes directly influences the recombination frequency, with tightly linked genes exhibiting low recombination frequency and loosely linked genes displaying higher recombination frequency.

18. (B): Somaclones are plants that are genetically identical to the original parent plant from which they are grown. They are produced through tissue culture techniques, such as somatic embryogenesis, where small pieces of plant tissue or cells are cultured in a nutrient-rich medium to regenerate into whole plants. Because these plants are derived from somatic cells of the parent plant rather than sexual reproduction, they maintain the exact genetic makeup of the parent plant. This allows for the propagation of desirable traits, such as disease resistance or high yield, without the variability associated with sexual reproduction. Somaclones have significant applications in agriculture, horticulture, and plant breeding, as they provide a reliable means of reproducing and preserving valuable plant traits.

19. (A):

(*a*) Euchromatin (Transcriptionally active): Euchromatin refers to the less condensed form of chromatin, characterized by loosely packed DNA and active gene expression. Within euchromatin, DNA is accessible to transcription factors and RNA polymerase enzymes, allowing for the transcription of genes into RNA molecules. Euchromatin regions often contain actively transcribed genes involved in various cellular processes.

(*b*) Heterochromatin (Densely packed DNA): Heterochromatin consists of highly condensed chromatin, with tightly packed DNA and transcriptionally inactive regions. It contains repetitive DNA sequences, silenced genes, and structural elements like centromeres and telomeres. Heterochromatin plays a role in maintaining chromosome structure and stability, and it can undergo dynamic changes in response to cellular signals and developmental cues.

(*c*) RNA (Poliovirus Genome): The genome of poliovirus is composed of RNA rather than DNA. RNA serves as the genetic material of the virus, encoding the viral proteins necessary for viral replication, assembly, and infection. Poliovirus is an RNA virus belonging to the Picornaviridae family and is the causative agent of poliomyelitis, a highly infectious viral disease.

(*d*) RNAse (RNA degradation): RNAse enzymes are responsible for catalyzing the degradation or breakdown of RNA molecules into smaller components. They play essential roles in regulating gene expression, RNA turnover, and RNA quality control within cells. RNAse enzymes can target and cleave RNA molecules at specific sites, leading to their degradation and eventual recycling of RNA components for cellular processes.

20. (C): Two sources are considered coherent if they produce waves with a constant phase difference. This means that the peaks and troughs of the waves maintain a consistent alignment or phase relationship over time. Coherence is essential for the phenomenon of interference, where waves overlap and interact with each other. When waves from coherent sources overlap, they can either reinforce each other (constructive interference) or cancel each other out (destructive interference), depending on their relative phase. Coherent sources are crucial in various areas of physics and engineering, including optics, acoustics, and quantum mechanics, where interference phenomena play a significant role in understanding and manipulating wave behavior.

25. (B):

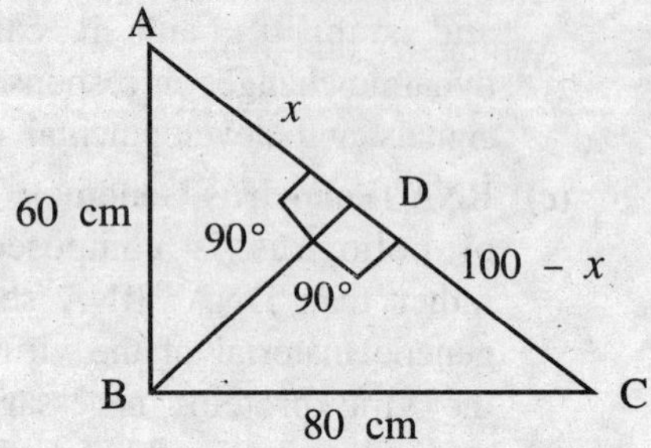

$\because \quad AC^2 = (60)^2 + (80)^2$

$= 3600 + 6400 = 10000$

$\therefore \quad AC = \sqrt{10000} = 100$ cm

$\because$ BD $\perp$ AC,

Let AD $= x$ and DC $= 100 - x$

In ΔABD,

$BD^2 = (60)^2 - x^2 \quad \ldots(i)$

In ΔBCD,

$BD^2 = (80)^2 - (100 - x)^2$

$\Rightarrow BD^2 = 6400 - (10000 + x^2 - 200x)$

$\Rightarrow BD^2 = 6400 - 10000 - x^2 + 200x \quad \ldots(ii)$

From (*i*) and (*ii*),

$3600 - x^2 = -3600 - x^2 + 200x$

$\Rightarrow \quad 200x = 7200 \Rightarrow x = 36$

Hence, length of AD = 36 cm.

26. (A):

(*a*) **Amylose (Plant starch):** Amylose is a component of plant starch and consists of linear chains of α-D-glucose units.

(*b*) **Glycogen (Animal starch):** Glycogen is the storage form of glucose in animals and is similar in structure to amylopectin, consisting of branched chains of α-D-glucose units.

(*c*) **Maltose (2 units of α-D-glucose):** Maltose is a disaccharide composed of two α-D-glucose molecules joined by a glycosidic bond.

(*d*) **Sucrose (α-D-glucose + β-D-fructose):** Sucrose is a disaccharide composed of one molecule of α-D-glucose and one molecule of β-D-fructose joined by a glycosidic bond.

29. (B): Assertion (A): This statement accurately describes Morgan's observation of a low recombination frequency for the genes affecting body and eye color in *Drosophila*. **Reason (R):** This statement correctly describes Sturtevant's use of recombination frequency to map genes, but it fails to directly explain Morgan's specific observation.

The connection between (A) and (R) is missing. While a low recombination frequency generally implies close proximity of genes, the reason (R) provides does not clarify why or how it specifically explains the 1.3% recombination observed by Morgan. The correct explanation for (A) likely involves the specific genetic linkage between the body and eye color genes, including details like their physical location on the chromosome and potential factors influencing their recombination behavior. (R), although a valid concept, simply highlights a broader principle without directly addressing the specifics of Morgan's observation.

30. (B): The time period (T) of a satellite orbiting around the Earth is determined by Kepler's third law, which states that the square of the orbital period of a planet (or satellite) is directly proportional to the cube of the semi-major axis of its orbit.

In this case, the semi-major axis of the satellite's orbit is the sum of the radius of the Earth (R) and the height of the satellite above the Earth's surface (h). Therefore, we have:

$T^2 \propto (R + h)^3$

Taking the square root of both sides, we get:

$T^2 \propto (R + h)^{3/2}$

Therefore, the time period of the satellite is proportional to: $(R + h)^{3/2}$.

31. (D): The bone marrow is not considered a secondary lymphoid organ. While it is crucial for the production and maturation of various immune cells, including B cells, which play a central role in adaptive immunity, it is primarily regarded as a primary lymphoid organ. Primary lymphoid organs are sites where immune cells are generated and mature, such as the bone marrow (for B cells) and the thymus (for T cells).

32. (D): Optical fibers work based on the principle of total internal reflection. When light enters an optical fiber, it encounters a boundary between the core (where light travels) and the cladding (a surrounding material with a lower refractive index than the core). Total internal reflection occurs when light traveling within the core strikes the boundary at an angle greater than the critical angle.

In this situation, instead of refracting out of the core, the light is completely reflected back into the core. This phenomenon allows light to propagate through the fiber by bouncing off the core-cladding boundary, effectively guiding the light along the length of the fiber with minimal loss. Total internal reflection is the key principle behind the transmission of light signals through optical fibers, making them essential components in telecommunications, fiber optic internet, and various optical technologies.

33. (B): $2(l + b) = 28$

$\therefore \quad l + b = 14 \qquad \ldots(i)$

$l \times b = 48$

$(l - b)^2 = (l + b)^2 - 4lb$

$= (14)^2 - 4 \times 48$

$= 196 - 192 = 4$

$l - b = 2 \qquad \ldots(ii)$

Adding (*i*) and (*ii*),

$l + b = 14$

$l - b = 2$

$2l = 16$

$\Rightarrow \quad l = 8$ m

$\therefore \quad b = 6$ m

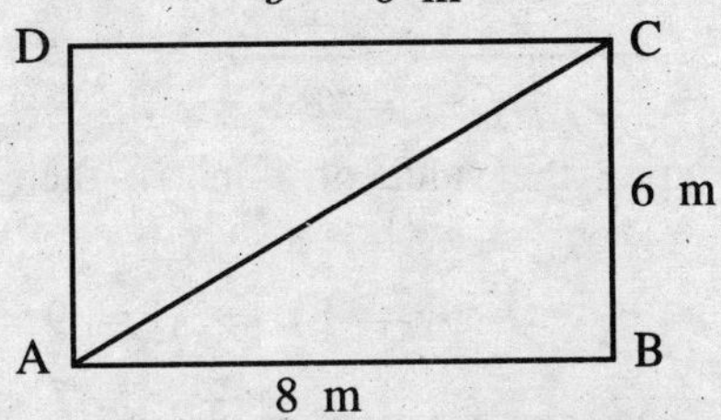

$\therefore$ Length of diagonal $= \sqrt{8^2 + 6^2}$

$= \sqrt{64 + 36} = \sqrt{100}$

$= 10$ m.

34. (D): Let the incomes of A and B be ₹ $9x$ and ₹ $7x$

and their expenditure be ₹ $4y$ and ₹ $3y$

Hence, $9x - 4y = 2000 \qquad \ldots(i)\ ] \times 3$

$7x - 3y = 2000 \qquad \ldots(ii)\ ] \times 4$

Subtracting (*ii*) from (*i*)

$27x - 12y = 6000$

$28x - 12y = 8000$

$- \quad + \quad -$

$x = 2000$

Hence, income of A = ₹ $9x$

$= 9 \times 2000 =$ ₹ 18000

and income of B = ₹ $7x$

$= 7 \times 2000 =$ ₹ 14000.

35. (A): Let the fraction $= \dfrac{x}{y}$

According to the question,

$$\frac{x+2}{y+2} = \frac{9}{11}$$

$\Rightarrow \quad 11x + 22 = 9y + 18$

$\Rightarrow \quad 11x - 9y = -4 \qquad \ldots(i)\ ] \times 5$

Now, $\dfrac{x+3}{y+3} = \dfrac{5}{6}$

$\Rightarrow \quad 6x + 18 = 5y + 15$

$\Rightarrow \quad 6x - 5y = -3 \qquad \ldots(ii)\ ] \times 9$

Subtracting (*ii*) from (*i*)

$$55x - 45y = -20$$
$$54x - 45y = -27$$
$$- \quad + \qquad +$$
$$x = 7$$

Putting the value of x in (*i*), then we get,

$77 - 9y = -4$

$\Rightarrow \quad 9y = 81 \Rightarrow y = 9$

Hence, the fraction $= \dfrac{x}{y} = \dfrac{7}{9}$.

37. (C): Television signals are complex waveforms that carry both video and audio information. In analog television transmission, the video signal is typically modulated using amplitude modulation (AM), where the amplitude of the carrier wave varies in accordance with the instantaneous amplitude of the video signal. Additionally, the audio signal is often modulated using frequency modulation (FM), where the frequency of the carrier wave varies in accordance with the instantaneous frequency of the audio signal. This combination of amplitude and frequency modulation allows for the transmission of both video and audio signals over the airwaves.

38. (D): Solder is an alloy commonly used for joining metals together, particularly in electronics and plumbing. It typically consists of a mixture of tin and lead, with varying proportions depending on the specific application. When heated, solder melts and forms a bond between the surfaces of the metals being joined. While solder may sometimes contain other elements for specific purposes, such as flux for improved wetting and cleaning properties, copper is not a typical constituent metal in solder. Instead, solder primarily relies on the combination of tin and lead to provide the desired melting temperature, mechanical properties, and ease of use for various soldering applications.

39. (A): Phosphine (PH_3) is a key component used in smoke screens for its ability to generate dense smoke when released. In smoke screen applications, phosphine is typically released from canisters or devices, where it reacts with oxygen in the air to form a cloud of fine smoke particles. This smoke effectively obscures visibility, making it difficult for adversaries or threats to observe or target specific areas. Phosphine-based smoke screens find applications in various fields, including military operations, firefighting, and emergency response scenarios, where temporary concealment of visibility is necessary for tactical advantage or protection.

40. (B):

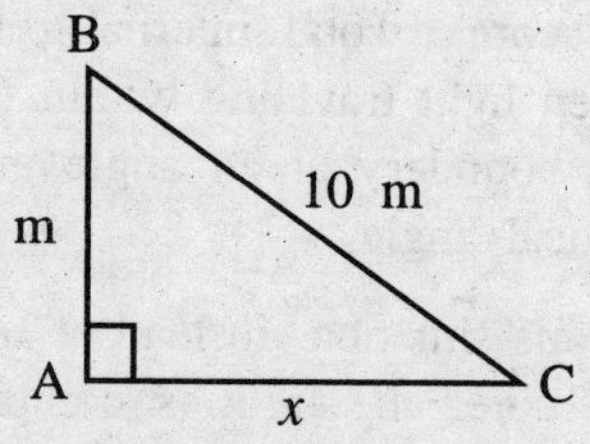

$x^2 = (10)^2 - (8)^2$

$= 100 - 64 = 36$

$\Rightarrow \quad x = 6$

Hence, distance of the ladder from the base of the wall = 6 m.

43. (C): L.C.M. of 16, 18, 20, 25 = 3600

Required number should be the form of 3600K + 4

Least value of K for which (3600 K + 4) is divisible by 7 i.e., K = 5

∴ Required number

$= 3600 \times 5 + 4 = 18004.$

45. (C): $73 + 13 \times 42 \div 14 - 56 = 56$

$\Rightarrow \quad 73 + 13 \times 3 - 56 = 56$

$\Rightarrow \quad 73 + 39 - 56 = 56$

$\Rightarrow \quad 112 - 56 = 56$

$\Rightarrow \quad 56 = 56$

Hence, – and + signs needs to be interchanged to make equation correct.

46. (C): Blister copper is produced through the smelting process of copper ores, which typically contain impurities such as sulphur in the form of sulphide minerals. When the copper ore is smelted at high temperatures, the sulphur in the ore combines with oxygen in the air to form sulphur dioxide (SO_2) gas. This gas evolves as bubbles or blisters on the surface of the molten copper, giving it a blistered appearance. The blistering occurs because the sulphur dioxide gas is released from the molten copper and rises to the surface, where it is released into the atmosphere. This process is characteristic of traditional copper smelting methods and is the reason why the resulting copper is known as blister copper.

47. (D): Totipotency is a remarkable property of plant cells, where each cell retains the potential to regenerate into a whole organism. This means that even a single cell from a plant, such as a cell from a leaf or a root, has the inherent capability to give rise to an entire plant under suitable conditions. This extraordinary feature underpins various techniques in plant biotechnology, such as tissue culture and cloning, where cells or tissues are cultured *in vitro* to produce genetically identical plants. Totipotency allows researchers to propagate plants efficiently, manipulate their genetic makeup, and produce new varieties with desired traits. Additionally, totipotent cells play a vital role in plant development and regeneration, enabling plants to repair injuries and regenerate damaged tissues throughout their lifecycle.

48. (C): The conversion of FAD to $FADH_2$ is a two-step process with radical intermediates. FAD, a coenzyme derived from riboflavin (vitamin B_2), undergoes reduction to form $FADH_2$ by accepting two electrons and two protons. This reduction occurs in two sequential steps, each involving the transfer of one electron and one proton. During these steps, radical intermediates are formed, where unpaired electrons are present, before ultimately yielding the fully reduced form of $FADH_2$. $FADH_2$ serves as an essential electron carrier in cellular respiration, shuttling electrons and protons to the electron transport chain, where they participate in ATP synthesis. This process is crucial for generating the energy currency ATP, which is utilized by cells for various biological processes.

49. (B): In areas with low air pollution, rainwater typically has a pH of approximately 5-6 due to the presence of carbonic acid (H_2CO_3). When rainwater falls through the atmosphere, it can absorb carbon dioxide (CO_2) from the air. This dissolved carbon dioxide reacts with water to form carbonic acid, a weak acid. Carbonic acid dissociates in water to release hydrogen ions (H^+) and bicarbonate ions (HCO^-_3), resulting in a slightly acidic solution. The acidity of rainwater, attributed to carbonic acid formation, is further enhanced by natural processes such as the dissolution of acidic gases like sulphur dioxide (SO_2) and nitrogen oxides (NOx) in the atmosphere, although these are typically present in lower concentrations in areas with minimal air pollution. Therefore, carbonic acid, originating from the dissolution of atmospheric CO_2, is the primary contributor to the acidity of rainwater in regions with low levels of air pollution.

50. (C): Plant tissue culture techniques allow for the propagation of plants from small tissue samples under controlled conditions in a laboratory setting. The apical and axillary meristems are regions of actively dividing cells located at the tips of stems and branches, respectively. These meristematic tissues contain undifferentiated cells with the potential to develop into various plant tissues and organs. By isolating and culturing the apical or axillary meristem tissues from diseased plants *in vitro*, it is possible to regenerate healthy plants. This process involves the growth of new shoots and roots from the meristematic cells, eventually leading to the development of whole plants free from disease.

51. (A): $55x + 85x + 105x = 2940$

$\Rightarrow \quad 245x = 2940$

$\Rightarrow \quad x = \frac{2940}{245} = \frac{588}{49} = 12$

Hence, required number of each category = 12.

52. (C): Let Man's income = ₹ 100

Expenditure = ₹ 75

Saving = 100 – 75 = ₹ 25

Now, Income = ₹ 120

Expenditure = $75 + \frac{10}{100} \times 75$

= 75 + 7.5 = ₹ 82.5

Saving = 120 – 82.5 = ₹ 37.5

Increase in saving = 37.5 – 25 = ₹ 12.5

Percentage increase in saving = $\frac{12.5}{25} \times 100$

= 12.5×4 = 50%.

53. (C): Saponification is a chemical reaction in which an ester, typically found in fats and oils, reacts with a strong base, such as sodium hydroxide (NaOH), to produce soap. In this reaction, the ester bond in the fat or oil molecule is hydrolyzed by the base, resulting in the formation of the sodium salt of the fatty acid (soap) and glycerol. The hydrolysis process breaks the ester linkage, releasing the fatty acid molecules, which then combine with the sodium ions from the base to form soap molecules. Glycerol, also known as glycerin, is a byproduct of the reaction and is typically recovered for use in other applications. Saponification is a key process in soap-making, where natural fats and oils are treated with a strong alkali to produce soap molecules, which have surfactant properties and can effectively clean dirt and grease from surfaces.

54. (A):

(*a*) **Thalassemia - Autosomal recessive:** Thalassemia is a genetic disorder caused by mutations in genes that are involved in the production of hemoglobin. It is inherited in an autosomal recessive manner, meaning that an individual needs to inherit two copies of the defective gene (one from each parent) to develop the disorder.

(*b*) **Tinner's Syndrome - 44 + XO chromosomes:** Turner syndrome, also known as Tinner's syndrome, is characterized by the presence of only one normal X chromosome in females (45, X), instead of the usual two (46, XX). This condition occurs due to abnormalities in the sex chromosomes.

(*c*) **Klinefelter's Syndrome - 44 + XXY chromosomes:** Klinefelter syndrome is characterized by the presence of an extra X chromosome in males, resulting in a karyotype of 47, XXY. This condition leads to various physical and developmental differences in affected individuals.

(*d*) **Down's Syndrome - 45 + XX chromosomes:** Down syndrome is caused by the presence of an extra copy of chromosome 21 (trisomy 21), resulting in a karyotype of 47, XX (or 47, XY in males). This extra genetic material leads to developmental delays and physical features characteristic of the syndrome.

55. (C): Let number of Hens $= x$
and number of Goats $= y$
According to the question,

$$x + y = 90 \quad \ldots(i)\ ] \times 2$$

and $$2x + 4y = 248 \quad \ldots(ii)\ ] \times 1$$

Subtractive (*ii*) from (*i*)

$$2x + 2y = 180$$
$$2x + 4y = 248$$
$$\underline{-\quad -\quad -}$$
$$-2y = -68$$
$$\Rightarrow \quad y = 34$$

Hence, total number of goats, $y = 34$.

56. (D): Coenzyme Q, or ubiquinone, is a vital molecule involved in cellular respiration and energy production within cells. Its reactive functional group is a thioester, characterized by a sulphur atom bonded to a carbonyl carbon. This thioester group is crucial for the role of coenzyme Q as an electron carrier in the mitochondrial electron transport chain. During cellular respiration, coenzyme Q shuttles electrons between various protein complexes, facilitating the transfer of electrons and protons across the inner mitochondrial membrane. This process ultimately generates a proton gradient that drives ATP synthesis. The thioester group undergoes reversible oxidation and reduction reactions, allowing coenzyme Q to accept and donate electrons efficiently, thereby facilitating ATP production. Additionally, the thioester group's reactivity enables coenzyme Q to participate in other cellular processes, including antioxidant defense mechanisms and lipid metabolism.

58. (C):

(*a*) *Drosophila melanogaster* male and female flies are easily distinguishable, primarily based on physical differences such as the size and shape of the abdomen and the presence of sex combs on the forelegs of males, making it easier to conduct genetic crosses and observe inheritance patterns.

(*c*) A single mating in *Drosophila melanogaster* produces large progeny, which is advantageous for genetic studies as it allows for the generation of a large number of offspring for analysis and experimentation.

(*d*) Phenotypic variations in *Drosophila melanogaster* can be observed with a low magnification microscope, enabling researchers to study and analyze various genetic traits and mutations efficiently. This feature facilitates the identification and analysis of genetic mutations and their effects on phenotype, contributing to the suitability of *Drosophila melanogaster* for genetic studies.

61. (C): Thin layer chromatography (TLC) is a chromatographic technique used primarily for analytical purposes rather than preparative purification. In TLC, a sample mixture is spotted onto a thin layer of adsorbent material, such as silica gel or alumina, which is coated onto a flat plate or sheet. The plate is then placed in a developing chamber containing a solvent system, and as the solvent moves up the plate through capillary action, it carries the sample components with it. Each component travels at a different rate based on its affinity for the stationary phase and the mobile phase, allowing for separation of the mixture into its individual constituents. TLC is commonly used for qualitative analysis, identification of compounds, and monitoring reactions, but it is not suitable for large-scale purification of proteins after fermentation, unlike preparative chromatography methods such as gel filtration, ion exchange, and affinity chromatography, which are specifically designed for this purpose.

62. (B):

(*a*) **LTR, autonomous - Copia:** Long Terminal Repeat (LTR) retrotransposons are autonomous elements that possess long terminal repeat sequences at their ends. Copia is an example of an LTR retrotransposon found in *Drosophila melanogaster*.

(*b*) **Non-LTR, autonomous - I factor:** Non-Long Terminal Repeat (Non-LTR) retrotransposons are also autonomous elements but lack long terminal repeat sequences. I factor is an example of a non-LTR retrotransposon found in yeast.

(*c*) **LTR, non-autonomous - Dasheng:** Dasheng is an example of a non-autonomous LTR retrotransposon. Non-autonomous retrotransposons lack the necessary genes for transposition and rely on other autonomous elements for their mobility.

(*d*) **Non-LTR, non-autonomous - Alu:** Alu elements are non-autonomous non-LTR retrotransposons found in the human genome. They do not possess long terminal repeats and are not capable of autonomous transposition.

63. (D): The rotation speed of the discs or media in an RBC system affects the treatment efficiency by influencing various factors such as mass transfer of oxygen, mixing of wastewater and biomass, and the contact time between the wastewater and the biofilm on the media surface. Optimal rotation speed is typically determined based on factors such as the specific design of the RBC units, the hydraulic loading rate, the characteristics of the wastewater being treated, and the desired treatment objectives. While rotation speed is crucial for the operational performance of the RBC system, it may not fall under the category of mandatory design considerations, as other factors such as the staging of RBC units, loading criteria, and effluent characteristics are often prioritized during the initial design phase to ensure the system's effectiveness and compliance with regulatory standards. However, rotation speed is still carefully monitored and adjusted during operation to optimize treatment efficiency and maintain proper functioning of the RBC system.

64. (B): In plant tissue culture, the hormonal balance between auxin and cytokinin is critical for regulating various developmental processes, including root formation from callus tissue. Auxin, such as indole-3-acetic acid (IAA), is known to promote root initiation and growth by stimulating cell division and differentiation in the root meristem. On the other hand, cytokinin, such as kinetin or benzyladenine (BA), has been shown to inhibit root development and promote shoot proliferation. Therefore, a higher auxin/cytokinin ratio in the growth medium creates an environment that favors root formation. This condition enhances the expression of genes associated with root development, leading to the induction of adventitious roots from callus tissue. By manipulating the auxin/cytokinin ratio, researchers can effectively control root morphogenesis and optimize the production of roots from callus tissue *in vitro* for various applications in plant biotechnology, such as micropropagation and genetic transformation.

65. (A): The BCG (Bacillus Calmette-Guérin) vaccine is a live attenuated vaccine widely used for the prevention of tuberculosis (TB). It is derived from a strain of *Mycobacterium bovis* that has been weakened through serial passages in culture. The process of attenuation involves modifying the pathogen in such a way that it retains its ability to stimulate an immune response but is significantly weakened in its ability to cause disease.

The BCG vaccine stimulates both cellular and humoral immune responses, leading to the production of antibodies and the activation of T cells that can recognize and eliminate Mycobacterium tuberculosis. Despite its efficacy in preventing severe forms of TB, the BCG vaccine may vary in its protective efficacy against pulmonary TB, which is the most common form of the disease. Nonetheless, it remains an essential tool in TB control programs, especially in regions with high TB burden, offering partial protection against severe forms of TB in vaccinated individuals.

66. (B): IgM antibodies are the first antibodies produced during the primary immune response to an antigen. They are pentameric in structure, with each unit containing an antigen-binding site, allowing IgM to efficiently bind to multiple antigens and form large immune complexes. These immune complexes, when recognized by the complement system, activate the classical pathway of complement activation. IgM-bound antigens serve as platforms for the binding and activation of complement proteins, leading to the formation of the membrane attack complex (MAC). The MAC creates pores in the membrane of target cells, ultimately leading to cell lysis and destruction. Therefore, IgM antibodies are crucial for the initiation of the complement cascade and subsequent elimination of pathogens or antigens.

67. (B): The first clinical use of gene therapy was pioneered for the treatment of Adenosine deaminase deficiency (ADA deficiency), a rare genetic disorder leading to severe combined immunodeficiency (SCID). Patients with ADA deficiency lack the enzyme adenosine deaminase, critical for the normal development and function of the immune system. The therapy involved the introduction of a functional ADA gene into the patient's hematopoietic stem cells using viral vectors. By restoring the deficient enzyme, gene therapy aimed to correct the underlying genetic defect and restore immune function. Although initial attempts faced challenges and limitations, this groundbreaking effort marked a significant milestone in the field of gene therapy, laying the foundation for future developments and applications in the treatment of genetic disorders.

68. (A): The statement "A reaction can take place spontaneously only if ΔG is negative" is consistent with the principles of thermodynamics. ΔG, the change in Gibbs free energy, determines the spontaneity of a reaction. If ΔG is negative, it signifies that the products have lower free energy than the reactants. In such cases, the reaction proceeds spontaneously, releasing energy in the process. This release of energy drives the reaction forward without the need for external energy input. Conversely, if ΔG is positive, the reaction is non-spontaneous, indicating that the products have higher free energy than the reactants. In such instances, the reaction will not occur spontaneously under normal conditions, as it would require an input of energy to proceed. Therefore, ΔG serves as a key determinant of whether a reaction will occur spontaneously or not.

69. (C): Methanol can be metabolized aerobically by certain microorganisms through a process known as methylotrophy. During aerobic catabolism of methanol, it is initially oxidized to formaldehyde by the enzyme methanol dehydrogenase. Formaldehyde is then further oxidized to formate by formaldehyde dehydrogenase. Formate is then converted to carbon dioxide and water, releasing electrons that enter the electron transport chain. This process allows for the production of ATP through oxidative phosphorylation.

70. (A): The A_{260}/A_{280} ratio is commonly used to assess the purity of nucleic acid preparations, including plasmid DNA. A ratio close to 2 indicates that the sample is relatively free from contaminants such as protein, RNA, or phenol. This ratio suggests that the DNA is of high purity because DNA absorbs light at 260 nm (A_{260}), while proteins and other contaminants absorb light at 280 nm (A_{280}). Therefore, a higher ratio indicates a higher purity of DNA in the sample, with less interference from contaminants.

71. (A): Dengue virus is an RNA virus. RNA viruses, unlike DNA viruses, contain RNA as their genetic material. Dengue virus, a member of the Flaviviridae family and the Flavivirus genus, is responsible for causing dengue fever, a mosquito-borne viral infection prevalent in tropical and subtropical regions. The virus is primarily transmitted to humans through the bite of infected Aedes mosquitoes. Dengue virus has a single-stranded RNA genome, which undergoes replication and transcription within host cells upon infection. Understanding the viral genome and its replication mechanisms is crucial for developing effective strategies for diagnosis, treatment, and prevention of dengue fever.

72. (D): Restriction Fragment Length Polymorphism (RFLP) is a molecular biology technique used to detect genetic variations in DNA sequences. It involves digesting genomic DNA with restriction enzymes that recognize specific DNA sequences and cleave the DNA at those sites. The resulting fragments are then separated by gel electrophoresis based on their size differences. Variations in DNA sequences, such as single nucleotide polymorphisms (SNPs) or insertions/deletions, can alter the restriction enzyme recognition sites, leading to differences in the size and number of restriction fragments produced. By comparing the patterns of DNA fragments obtained from different individuals or samples, RFLP analysis can reveal genetic differences and polymorphisms within the population. RFLP has been widely used in genetic mapping, linkage analysis, forensic DNA profiling, and disease association studies.

73. (A): ELISPOT (Enzyme-Linked Immunospot Assay) is a sensitive and widely used technique for detecting and quantifying the frequency of cells that secrete specific proteins, such as cytokines like interferon gamma (IFN-γ). In ELISPOT, cells are incubated on a membrane-coated plate, where they release IFN-γ upon stimulation. The secreted IFN-γ is captured by antibodies immobilized on the plate, forming spots that can be visualized and counted. ELISPOT allows for the enumeration of individual cytokine-secreting cells and provides valuable information about the immune response at the cellular level.

74. (B): Immunoglobulin E (IgE) is the least abundant immunoglobulin class in a normal adult due to its specialized role in the immune system. While IgG, IgA, and IgM are present in higher concentrations and have diverse functions in immune defense, IgE is primarily associated with allergic responses and defense against parasitic infections. IgE antibodies bind to specific receptors on the surface of mast cells and basophils, triggering the release of histamine and other inflammatory mediators upon exposure to allergens. This immune response leads to symptoms of allergic reactions such as itching, swelling, and inflammation. Although IgE plays a crucial role in certain immune responses, its low abundance reflects its specialized function and the tightly regulated nature of allergic reactions in the body.

75. (C): Symptoms of AIDS (Acquired Immunodeficiency Syndrome) typically appear during the advanced stage of HIV infection when the virus has caused significant damage to the immune system, particularly to CD4+ T-lymphocytes (helper T-cells). HIV targets and infects these cells, leading to their depletion and impairment of the immune response. As the CD4+ T-cell count declines, the body becomes increasingly susceptible to opportunistic infections and other complications characteristic of AIDS. Therefore, the onset of AIDS-related symptoms is closely associated with the progressive decline in CD4+ T-cell levels caused by HIV infection.

76. (B): Ion exchange chromatography separates proteins based on their net charge. In this technique, a stationary phase with charged groups attracts oppositely charged proteins, allowing for their selective retention. The pI of a protein, which is the pH at which it carries no net charge, is crucial in ion exchange chromatography because it helps determine the protein's behavior in different pH environments. At a pH above its pI, a protein carries a net negative charge and will bind to a positively charged ion exchange column (cation exchange chromatography). Conversely, at a pH below its pI, the protein carries a net positive charge and will bind to a negatively charged ion exchange column (anion exchange chromatography). Therefore, knowledge of the pI enables researchers to select the appropriate pH conditions for binding and elution of the target protein during ion exchange chromatography.

77. (A): The pollen grain's intine, the innermost wall crucial for germination and reproduction, exhibits a masterful interplay of structural robustness and dynamic flexibility. This delicate yet essential layer primarily comprises cellulose, providing the framework for strength and integrity, akin to the skeleton of a balloon. Pectin, acting as a molecular maestro, binds these fibers and imbues the intine with the elasticity needed for expansion and pollen tube formation. This synergistic partnership between cellulose and pectin orchestrates a finely tuned balance, enabling the intine to navigate the rigors of dispersal and germination, ultimately contributing to the perpetuation of life through successful plant reproduction.

78. (D):

(*a*) **IgM - Pentamer; Complement Activation:** IgM antibodies are typically found as pentamers, meaning they consist of five monomeric subunits joined together by a J chain. This structural arrangement allows IgM to have multiple antigen-binding sites, making it highly effective at agglutinating antigens and activating the complement system.

(*b*) **IgA - Monomer/Dimer/Trimer; Mucosal Immunity:** IgA antibodies can exist in different forms, including monomers, dimers, and trimers. In the mucosal immune system, IgA is primarily found as a dimer, with an additional protein called the secretory component. IgA plays a crucial role in mucosal immunity by neutralizing pathogens and preventing their adherence to mucosal surfaces.

(*c*) **IgD - Naive B cell antigen receptor:** IgD antibodies are expressed on the surface of naive B cells as antigen receptors. Along with IgM, IgD serves as a receptor for recognizing and binding antigens during the activation of B cells. This interaction is essential for initiating the adaptive immune response.

(*d*) **IgE - Monomer; Mast Cell Activation:** IgE antibodies are monomeric in structure and are known for their role in triggering allergic reactions. When IgE antibodies bind to allergens, they attach to the surface of mast cells and basophils, sensitizing them. Subsequent exposure to the allergen leads to the cross-linking of IgE molecules on these cells, triggering the release of inflammatory mediators like histamine, which causes allergic symptoms.

79. (C): PCR is a widely used method for cell-line authentication. It allows for the amplification of specific DNA sequences from the cells, which can then be compared to known genetic profiles or markers associated with the cell line. By targeting unique genetic markers, such as short tandem repeats (STRs) or single nucleotide polymorphisms (SNPs), PCR can confirm the identity of a cell line and detect any potential contamination or misidentification issues. Serotyping, virus infection, and protein estimation are not typically used for cell-line authentication purposes.

80. (B): An Expressed Sequence Tag (EST) is a short DNA sequence derived from a messenger RNA (mRNA) transcript. ESTs are generated by sequencing cDNA (complementary DNA) libraries constructed from mRNA samples of a particular tissue, cell type, or organism at a specific developmental stage or under certain conditions. These sequences represent fragments of expressed genes and can provide valuable information about gene expression patterns, alternative splicing events, and novel transcripts in a given biological sample. Therefore, ESTs are essential tools for gene discovery, transcriptome analysis, and functional genomics research.

81. (A): In Lipinski's Rule of Five, logP refers to the partition coefficient, which is a measure of the lipophilicity or hydrophobicity of a compound. Specifically, it represents the logarithm of the ratio of the concentrations of a compound in a nonpolar solvent (typically octanol) and an aqueous solution at equilibrium. The partition coefficient is a crucial parameter in drug design and pharmacology because it influences a compound's solubility, absorption, distribution, metabolism, and excretion (ADME) properties. Lipinski's Rule of Five uses logP as one of the criteria to evaluate the drug-likeness of potential drug candidates.

82. (A): Meselson and Stahl's famous experiment in 1957 provided strong evidence for the semiconservative replication of DNA. They labeled DNA with heavy isotopes of nitrogen (^{15}N) and then allowed the DNA to replicate in a medium containing only light isotopes (^{14}N). After one round of replication, the DNA molecules formed consisted of one old heavy strand and one newly synthesized light strand. This result supported the semiconservative model of DNA replication, where each daughter DNA molecule contains one parental strand and one newly synthesized strand. The experiment also ruled out the conservative and dispersive models of replication. Therefore, Meselson and Stahl's work played a crucial role in establishing the fundamental mechanism of DNA replication.

83. (A): In the theoretical calculation of the respiratory quotient (RQ) for the complete oxidation of pyruvic acid, it's essential to account for both the oxygen consumed during the reaction and the oxygen already present in pyruvic acid itself. When considering the oxygen consumed, 4 oxygen molecules are involved, but one of these molecules reacts with hydrogen to form water, leaving 3 oxygen molecules available for CO_2 production. Additionally, pyruvic acid contains 3 oxygen atoms, which contribute to the formation of 1.5 CO_2 molecules. Combining these sources of CO_2 production, the total CO_2 produced is 4.5 molecules. Dividing this by the 4 oxygen molecules consumed yields an adjusted RQ of approximately 1.2, indicating a balanced ratio of CO_2 production to oxygen consumption during the complete oxidation of pyruvic acid.

84. **(C):** Isoelectric focusing is the biochemical technique that does not typically involve the use of antibodies. Isoelectric focusing is a method used to separate and analyze proteins based on their isoelectric points (pI), which is the pH at which a protein carries no net electrical charge. During isoelectric focusing, proteins migrate towards their isoelectric point in a pH gradient until they reach a pH equal to their pI, where they become immobilized due to having no net charge. This technique relies on differences in the intrinsic properties of proteins rather than antibody-protein interactions, unlike the other options listed. ELISA (enzyme-linked immunosorbent assay), Western blotting, and immuno-affinity chromatography all involve the use of antibodies to detect, separate, or purify specific proteins.

85. **(A):** The virus that makes use of RNA-dependent DNA polymerase during replication is HIV (Human Immunodeficiency Virus), which belongs to the retrovirus family. Retroviruses like HIV use an enzyme called reverse transcriptase, which is an RNA-dependent DNA polymerase, to convert their RNA genome into DNA during replication. This process is called reverse transcription and is a crucial step in the replication cycle of retroviruses like HIV. Therefore, option A, HIV, is the correct answer. The other options, Dengue virus, Polio virus, and Hepatitis E virus, do not utilize RNA-dependent DNA polymerase during their replication cycles.

86. **(C):** *Zymomonas mobilis* is a bacterium known for its exceptional ability to ferment sugars into ethanol with high efficiency. This bacterium has several unique characteristics that make it well-suited for ethanol production. Firstly, *Zymomonas* mobilis possesses a high specific ethanol productivity, meaning it can produce ethanol rapidly per unit of biomass. Additionally, it has a high ethanol yield, meaning it can convert a significant portion of the sugar substrate into ethanol. *Zymomonas mobilis* also exhibits high tolerance to ethanol concentrations, allowing it to thrive and continue fermenting even in the presence of relatively high ethanol levels. Moreover, this bacterium has a simple metabolic pathway for ethanol production, which contributes to its efficiency in ethanol fermentation. Overall, *Zymomonas mobilis* is highly regarded in the biotechnology industry for its potential in biofuel production due to its remarkable ethanol-producing capabilities.

87. **(D):** Bioremediation involves the use of living organisms or their byproducts to degrade or remove pollutants from the environment. Options A, B, and C are examples of bioremediation processes: *Pseudomonas putida* breaking down oil spills, *Acetobacter aerogenes* decomposing DDT, and *Pseudomonas* species removing metal pollutants. However, option D, depletion of ozone by *Aspergillus niger*, does not align with bioremediation. Ozone depletion typically occurs due to chemical reactions involving synthetic compounds like chlorofluorocarbons (CFCs) and halons, not by biological organisms like *Aspergillus niger*. Thus, it's an incorrect choice in the context of bioremediation, which primarily involves biological processes for environmental cleanup.

88. **(D):** "*In vitro* germination and development is very slow in Orchids", is false. Orchids are known for their relatively fastidious growth requirements, and their *in vitro* germination and development can actually be relatively fast compared to some other plant species. Orchid seed germination and subsequent growth *in vitro* can be achieved under appropriate conditions, although it may require specific culture media and environmental conditions tailored to the needs of each Orchid species.

89. (C): Baculovirus, is best suited to express large amounts of glycosylated protein. Baculovirus expression systems, particularly using insect cell lines such as Sf9 or High Five cells, are widely used for the production of recombinant proteins, including glycosylated proteins. These systems offer advantages such as efficient protein folding and post-translational modifications, including glycosylation, which are important for the functionality and stability of many proteins. Additionally, Baculovirus expression systems can accommodate the expression of complex proteins that require extensive post-translational modifications, making them ideal for producing glycoproteins.

90. (B): The rough endoplasmic reticulum (RER) plays a crucial role in the synthesis of proteins destined for secretion or insertion into cellular membranes. Ribosomes attached to the RER synthesize these proteins, which are then translocated into the lumen of the RER. Inside the RER, the proteins undergo folding and post-translational modifications, such as glycosylation. These modifications are essential for the proper structure and function of the proteins. Once synthesized and modified, the proteins are packaged into transport vesicles and transported to various cellular destinations. In contrast, the smooth endoplasmic reticulum (SER) is involved in lipid metabolism, including synthesis of lipids and steroids, detoxification of drugs and toxins, and regulation of calcium levels in the cell. While both RER and SER are interconnected and share some functions, they have distinct roles in cellular physiology.

91. (A): Methionine is the amino acid that contributes sulphur to a protein. It is considered an essential amino acid as it cannot be synthesized by the human body and must be obtained from the diet. Methionine contains a sulphur atom in its side chain, which allows it to participate in the formation of disulphide bonds in proteins. Disulphide bonds play a crucial role in stabilizing the tertiary structure of proteins by forming covalent bonds between different regions of the protein chain. These bonds help maintain the overall structure and function of many proteins, contributing to their stability and integrity. Therefore, methionine is essential for the proper folding and stability of proteins in living organisms.

92. (B): In a reaction where the product (P) is at a lower internal energy level than the substrate (S), it signifies that energy has been liberated during the course of the reaction. This indicates an exothermic process, where heat is released to the surroundings. Exothermic reactions typically involve the formation of stronger bonds in the products compared to the bonds in the reactants. As a result, the products possess less internal energy than the reactants, leading to a decrease in energy overall. This release of energy can often be observed as heat, making the surroundings warmer. Thus, an exothermic reaction is characterized by the generation of energy, usually in the form of heat, rather than its absorption.

93. (B): Interferons are a diverse group of proteins that play a critical role in the body's defense against viral infections. When a virus infects a cell, it triggers a signaling cascade that leads to the production and release of interferons. These interferons then bind to receptors on neighboring cells, activating various signaling pathways that induce an antiviral state. This antiviral state involves the upregulation of genes that inhibit viral replication and spread, effectively limiting the spread of the infection to neighboring cells. Additionally, interferons also stimulate the immune system by enhancing the activity of immune cells, such as natural killer cells and macrophages, which helps in the clearance of infected cells. Overall, interferons serve as crucial mediators of the innate immune response to viral infections, providing an early line of defense against invading pathogens.

94. (D): A humanized murine antibody refers to an antibody that has been engineered to reduce its immunogenicity in humans while retaining the binding specificity and affinity of the original murine antibody. In this context, the term "humanized" typically refers to the framework regions of the antibody, which are the constant regions responsible for antibody structure and stability, being derived from human sequences. The complementarity-determining regions (CDRs), which are responsible for antigen binding, are derived from the murine antibody. By incorporating murine CDRs into a human framework, the resulting humanized antibody maintains its binding specificity to the target antigen while minimizing the risk of inducing an immune response when administered to humans.

95. (C): Huntington's disease (HD) is a neurodegenerative disorder characterized by progressive motor dysfunction, cognitive decline, and psychiatric symptoms. It follows an autosomal dominant inheritance pattern, meaning that an affected individual has a 50% chance of passing the mutated gene to each of their offspring. The mutation occurs in the HTT gene, leading to the production of an abnormal form of the huntington protein. This mutated protein accumulates in neurons, particularly in the brain's basal ganglia, leading to neuronal dysfunction and eventual cell death. Symptoms typically manifest in mid-adulthood, although the age of onset can vary widely. Currently, there is no cure for HD, and treatments focus on managing symptoms and improving quality of life for affected individuals and their families.

96. (D): Ribulose bisphosphate Carboxylase-Oxygenase (RuBisCO) is the most abundant protein in the biosphere. It is a key enzyme involved in photosynthesis, specifically in the fixation of carbon dioxide during the Calvin cycle. RuBisCO catalyzes the carboxylation of ribulose-1,5-bisphosphate (RuBP), a crucial step in carbon fixation, leading to the formation of two molecules of 3-phosphoglycerate (3-PGA). This process is fundamental for the synthesis of organic compounds and sustains life on Earth by converting atmospheric carbon dioxide into organic molecules, thus serving as the foundation of the food chain. Due to its pivotal role in carbon assimilation, RuBisCO is present in vast quantities across various ecosystems, making it the most abundant protein in nature.

97. (D): In PCR (Polymerase Chain Reaction), the design of primers is crucial for the success of the reaction. Primers are short, single-stranded DNA sequences that anneal to the target DNA template and provide a starting point for DNA polymerase to initiate DNA synthesis. It is desirable for primers to have minimal complementarity between reverse and forward primers to avoid primer-dimer formation, where the primers anneal to each other instead of the target DNA. Primer-dimers can interfere with amplification efficiency and specificity, leading to inaccurate results. Therefore, designing primers with minimal self-complementarity and minimal complementarity between reverse and forward primers is essential for successful PCR amplification.

98. (C): Nitrification is a crucial step in the nitrogen cycle, facilitating the conversion of ammonia to nitrate, which is a more stable form of nitrogen utilized by plants. This process occurs in two sequential oxidation reactions mediated by specific groups of bacteria. First, ammonia-oxidizing bacteria convert ammonia (NH_3) to nitrite (NO^-_2). Then, nitrite-oxidizing bacteria further oxidize nitrite to nitrate (NO^-_3). These bacteria play vital roles in soil ecosystems, where they contribute to the cycling of nitrogen, ultimately making it available to plants for growth. Nitrification is essential for maintaining soil fertility and supporting plant productivity in agricultural and natural environments.

99. (C): The group of excitatory neurotransmitters is primarily composed of molecules that facilitate the transmission of nerve impulses and promote neural activity. Among the options provided, the neurotransmitters Glutamate, Acetylcholine, and Serotonin are known to primarily exert excitatory effects. Glutamate is the most abundant excitatory neurotransmitter in the central nervous system, playing a crucial role in synaptic transmission and synaptic plasticity. Acetylcholine functions as an excitatory neurotransmitter at neuromuscular junctions and is also involved in cognitive functions such as learning and memory. Serotonin, although primarily associated with mood regulation and emotional processing, also modulates neuronal activity in various brain regions and can exert excitatory effects depending on the receptor subtype and neural circuitry involved.

100. (D): A selectable marker gene is utilized in genetic engineering to distinguish transformed cells from non-transformed ones. Typically, this gene confers resistance to a particular antibiotic or herbicide. After introducing foreign DNA into a host cell, the transformed cells are cultured in a medium containing the selective agent. Only the cells that have successfully integrated the selectable marker gene, along with the desired foreign DNA, will survive in the presence of the selective agent. This allows for the identification and isolation of cells that have undergone successful transformation, facilitating further experimentation or applications in biotechnology.

101. (B): The statement "Its main component is SO_2" is not true for photochemical smog. Photochemical smog, often referred to as oxidizing smog, is characterized by the presence of secondary pollutants such as ozone (O_3) and other oxidants. Unlike sulphurous smog, which is dominated by sulphur dioxide (SO_2) emissions, photochemical smog primarily arises from the interaction of nitrogen oxides (NO_x) and volatile organic compounds (VOCs) in the presence of sunlight. These precursor pollutants undergo complex chemical reactions, leading to the formation of a variety of secondary pollutants. Photochemical smog typically occurs in areas with warm and dry climates where sunlight is abundant, facilitating the photochemical reactions responsible for its formation. This type of smog is often associated with urban areas and industrial regions where fossil fuel combustion and vehicular emissions release NO_x and VOCs into the atmosphere, contributing to air pollution and the formation of photochemical smog.

102. (C): Trihydroxypropane, commonly known as glycerol, is a simple polyol compound that plays a crucial role in various biological processes and industrial applications. It consists of a propane backbone with three hydroxyl groups (-OH) attached to each carbon atom. Glycerol is a key component of lipids, where it acts as the backbone of triglycerides, the primary constituents of fats and oils. In addition to its role in lipid metabolism, glycerol serves as a humectant in skincare products, a sweetening agent in food and beverages, and a solvent in pharmaceuticals and cosmetics. Moreover, glycerol is utilized in the production of explosives, antifreeze, and various other industrial applications due to its non-toxic nature and versatility.

103. (B): Hydrolysis → Acidogenesis → Acetogenesis → Methanogenesis

Anaerobic digestion is a complex microbial process that breaks down organic matter in the absence of oxygen. The first step, hydrolysis, involves the breakdown of complex organic molecules into simpler compounds by enzymes secreted by hydrolytic bacteria. These simpler compounds, such as sugars and amino acids, are then fermented by acidogenic bacteria during acidogenesis, producing organic acids, alcohols, and other byproducts. Acetogenic bacteria further metabolize these compounds into acetic acid, hydrogen, and carbon dioxide during acetogenesis. Finally, methanogenic archaea convert the acetic acid, hydrogen, and carbon dioxide into methane and carbon dioxide through methanogenesis, generating biogas as a valuable energy source. Each step in this process is crucial for efficient organic matter degradation and biogas production in anaerobic digestion systems.

104. (A): Statement I is correct because the Ti plasmid (tumor-inducing plasmid) of *Agrobacterium tumefaciens* indeed contains genes for the biosynthesis of plant hormones, such as auxins and cytokinins. These hormones are produced by the bacterium to manipulate the growth of infected plant cells, leading to the formation of a tumor-like structure, which provides a nutrient-rich environment for the bacterium.

Statement II is also true. In the development of a binary vector system for plant transformation, the virulence (vir) genes of *Agrobacterium tumefaciens* are typically separated from the T-DNA region and placed on a separate helper plasmid. This binary system allows for more precise control over the transformation process, as the virulence genes are only activated when both the helper plasmid and the binary vector carrying the gene of interest are present in the bacterium. This separation ensures that the T-DNA region alone can be transferred to the plant cell without inducing tumor formation, facilitating the efficient genetic transformation of plants.

105. (B): The distance between each turn in the helical strand of DNA, also known as the pitch of the helix, is approximately 34 angstroms (Å), or 3.4 nanometers (nm). This measurement represents the distance the DNA molecule travels along its axis for each complete turn of the helix. It is an important characteristic of the DNA double helix structure and is determined by factors such as the number of base pairs per turn and the angle of rotation between adjacent base pairs.

106. (C): Heteropolysaccharides, such as agarose, are composed of more than one type of monosaccharide unit. In the case of agarose, it consists of repeating units of agarobiose, which is a disaccharide made up of D-galactose and 3,6-anhydro-L-galactopyranose. This structural complexity allows heteropolysaccharides to exhibit diverse properties and functions compared to homopolysaccharides. Agarose, for example, is widely used in gel electrophoresis due to its ability to form a stable gel matrix that separates biomolecules based on size. In contrast, homopolysaccharides like glycogen and dextran are simpler in structure, consisting of repeating units of the same monosaccharide, and typically serve as energy storage molecules or structural components in various organisms.

107. (C): The antibody involved in Type I hypersensitivity reactions is IgE (Immunoglobulin E). When an individual with allergies is exposed to an allergen, their immune system produces IgE antibodies specific to that allergen. These IgE antibodies bind to mast cells and basophils, sensitizing them. Upon subsequent exposure to the same allergen, the allergen cross-links with IgE antibodies bound to mast cells and basophils, triggering the release of inflammatory mediators like histamine. This cascade of events leads to the characteristic symptoms of Type I hypersensitivity reactions, such as sneezing, itching, and swelling.

108. (D): The step that is generally NOT a part of PCR (Polymerase Chain Reaction) is Ligation. PCR involves three main steps: Denaturation, Annealing, and Elongation (or Extension). During denaturation, the double-stranded DNA template is heated to separate it into two single strands. In the annealing step, the temperature is lowered to allow primers to bind specifically to their complementary sequences on the DNA template. Elongation involves the addition of nucleotides by a DNA polymerase enzyme, resulting in the synthesis of new DNA strands complementary to the template. Ligation, on the other hand, is a separate molecular biology technique used to join DNA fragments together using DNA ligase and is not typically a step in PCR.

109. (D): Both humoral and cell-mediated immune responses are required to combat a severe viral infection. The humoral immune response involves the production of antibodies by B cells, which can bind to viruses and neutralize them, preventing them from infecting host cells. Additionally, antibodies can also tag viruses for destruction by other immune cells. On the other hand, the cell-mediated immune response involves the activation of cytotoxic T cells, which can directly kill virus-infected cells by inducing apoptosis or secreting cytotoxic molecules. This response is crucial for eliminating cells that have been hijacked by the virus to produce more viruses. By coordinating both humoral and cell-mediated immune responses, the immune system can effectively control and eliminate viral infections.

110. (D): The UN Conference on Environment and Development (UNCED), widely recognized as the Earth Summit, convened in Rio de Janeiro, Brazil, in 1992. This landmark event brought together world leaders, policymakers, scientists, and activists to address pressing environmental and developmental challenges facing the planet. The Earth Summit resulted in significant agreements, including the Rio Declaration on Environment and Development, Agenda 21, and the establishment of the UN Framework Convention on Climate Change (UNFCCC) and the Convention on Biological Diversity (CBD). The conference emphasized the importance of sustainable development, biodiversity conservation, climate change mitigation, and equitable economic growth. The Earth Summit remains a pivotal moment in global environmental governance, setting the stage for ongoing international efforts to address environmental and developmental issues.

111. (B): In an agarose gel electrophoresis, the mobility of DNA molecules is influenced by their size, shape, and charge. Supercoiled plasmid DNA typically exhibits the maximum mobility in an agarose gel. This is because supercoiled DNA molecules are more compact and have a higher charge density compared to nicked or linear DNA molecules of the same size. As a result, they experience less frictional resistance as they migrate through the gel matrix in response to an electric field, allowing them to move faster and travel farther in the gel during electrophoresis.

112. (B): Charles Richet was awarded the Nobel Prize in Physiology or Medicine in 1913 for his seminal discovery of hypersensitivity, also known as anaphylaxis. His research elucidated the phenomenon of severe allergic reactions that occur upon re-exposure to certain antigens in sensitized individuals. Richet's work laid the foundation for understanding the complex mechanisms underlying immune hypersensitivity, paving the way for advancements in the diagnosis, treatment, and prevention of allergic conditions. His contributions have had a profound impact on the fields of immunology and medicine, highlighting the importance of recognizing and addressing immune-mediated hypersensitivity reactions.

113. (C): Allolactose is an isomer of lactose that serves as the inducer of the lac operon. The lac A gene encodes the enzyme β-galactosidase transacetylase, also known as galactoside O-acetyltransferase. This enzyme primarily catalyzes the conversion of lactose to allolactose by transferring an acetyl group from acetyl-CoA to the hydroxyl group at the C-3 position of the galactose moiety in lactose. Allolactose then binds to the lac repressor protein, causing a conformational change that leads to its release from the lac operator site. As a result, the lac operon becomes derepressed, allowing for the transcription of genes necessary for lactose metabolism, including lac Z (β-galactosidase) and lac Y (permease).

114. (B):

(*a*) **Nitrogen:** Nitrogen is an essential macronutrient required by plants in relatively large quantities. It is a vital component of proteins, nucleic acids, chlorophyll, and various other cellular compounds. In plant tissue culture media, nitrogen is often provided in the form of ammonium nitrate, ammonium sulphate, or other nitrogen-containing salts to support the growth and development of plant cells or tissues.

(*b*) **Sucrose:** Sucrose serves as a carbon source in plant tissue culture media. It provides the necessary energy for cellular metabolism and growth. Plant cells metabolize sucrose through glycolysis and other metabolic pathways to produce ATP, which is utilized for various cellular processes, including biosynthesis and cell division.

(*c*) **Manganese:** Manganese is a micronutrient essential for plant growth and development. It plays a crucial role in various enzyme activities, including photosynthesis, respiration, and nitrogen metabolism. In plant tissue culture media, manganese is often provided in the form of manganese sulphate or other manganese-containing salts to ensure optimal growth and development of plant cells or tissues.

(*d*) **Myoinositol:** Myoinositol is an organic supplement commonly added to plant tissue culture media. It serves multiple functions, including contributing to the osmotic balance of the medium and maintaining cell membrane integrity. Myoinositol also plays a role in signal transduction and lipid metabolism in plant cells, making it essential for their growth and development *in vitro*.

115. (A):

(*a*) **Ampicillin:** Ampicillin acts by inhibiting cell wall synthesis in bacteria. It specifically targets the enzyme transpeptidase, which is involved in cross-linking peptidoglycan chains in the bacterial cell wall. By interfering with this process, ampicillin weakens the bacterial cell wall, making it more susceptible to osmotic pressure and ultimately leading to cell lysis and death.

(*b*) **Kanamycin:** Kanamycin functions by inhibiting bacterial protein synthesis. It binds to the 30S ribosomal subunit of bacterial ribosomes, thereby preventing the initiation of translation and the assembly of the elongation complex. This disruption of protein synthesis interferes with bacterial growth and replication, ultimately leading to cell death.

(*c*) **Hygromycin-B:** Hygromycin-B interferes with ribosomal translocation during translation in bacteria. It binds to the A-site of the 30S ribosomal subunit and inhibits the movement of tRNAs from the A-site to the P-site, which is necessary for elongation of the polypeptide chain. This disruption of ribosomal function prevents the synthesis of essential proteins and impairs bacterial viability.

(*d*) **Chloramphenicol:** Chloramphenicol inhibits bacterial protein synthesis by binding to the 50S ribosomal subunit of bacterial ribosomes. It interferes with the peptidyl transferase activity of the ribosome, which is responsible for catalyzing the formation of peptide bonds between amino acids during translation. By blocking this enzymatic activity, chloramphenicol prevents the elongation of the polypeptide chain, leading to inhibition of protein synthesis and bacterial cell death.

116. (D): In general, transgenic plants with integration of multiple copies of the transgene would not be a preferred choice. While higher levels of transgene expression are desirable, and single-copy insertion reduces the risk of genetic instability and unpredictable expression patterns, multiple copies increase the likelihood of undesirable outcomes. Integration of multiple copies can lead to gene silencing, where excessive copies of the transgene interfere with its expression, resulting in reduced or no protein production. Additionally, it may cause positional effects, altering the regulation of nearby genes and potentially introducing unintended traits or phenotypic changes. Therefore, transgenic plants with a single-copy insertion of the transgene are typically preferred for more predictable and stable expression patterns.

117. (A): Vitamins serve as essential growth factors for most bacteria, playing crucial roles in various metabolic pathways necessary for their survival and proliferation. While bacteria possess the enzymatic machinery to synthesize many essential compounds, they often lack the biosynthetic pathways required for synthesizing certain vitamins. As a result, bacteria rely on external sources, such as their environment or host organisms, to obtain these vital nutrients. These vitamins act as coenzymes or cofactors in numerous enzymatic reactions involved in energy metabolism, nucleic acid synthesis, and cell wall biosynthesis, among other processes. Without these essential vitamins, bacteria may experience growth impairment or even cessation of growth, highlighting the significance of these molecules for bacterial physiology and survival.

118. (B): Allergy symptoms are primarily triggered by the release of histamine and other inflammatory mediators from mast cells, which are sensitized by IgE antibodies bound to their surface. Therefore, preventing the attachment of IgE antibodies to mast cells can effectively block the allergic response. This can be achieved through various strategies, such as using antihistamine medications, which interfere with histamine receptors, or through allergen-specific immunotherapy, which aims to desensitize the immune system to specific allergens and reduce the production of IgE antibodies. Blocking this attachment prevents the activation of mast cells and subsequent release of inflammatory molecules, thereby alleviating allergy symptoms.

119. (B): AMBER is not a multiple sequence alignment program; instead, it is a suite of programs primarily used in molecular dynamics simulations of biomolecules such as proteins and nucleic acids. CLUSTALX, CLUSTALW, and PILEUP, on the other hand, are all multiple sequence alignment programs commonly used to align sequences of nucleic acids or proteins. They are designed to identify regions of similarity between sequences and align them accordingly, facilitating the comparison of sequences and the identification of conserved motifs or domains.

120. (B): Hepatitis B virus (HBV) is a DNA virus that replicates its genome via a complex process involving reverse transcription. Unlike most DNA viruses, HBV lacks a DNA polymerase capable of directly synthesizing DNA from a DNA template. Instead, it relies on an RNA-dependent DNA polymerase, also known as reverse transcriptase, to convert its RNA genome into DNA. This enzyme catalyzes the synthesis of a DNA copy of the viral RNA genome, which is subsequently integrated into the host cell genome. This unique replication strategy enables HBV to persist in infected individuals and contributes to its pathogenicity.

122. (B): This statement is incorrect because DNA microarrays are not limited to the identification of microRNAs alone. DNA microarrays are powerful tools used for transcriptome analysis, allowing researchers to simultaneously analyze the expression levels of thousands of genes within a biological sample. This technology enables comprehensive profiling of gene expression patterns under different conditions or in different cell types. DNA microarrays are not limited to the identification of microRNAs; they can detect the expression of various types of nucleic acids, including mRNAs and microRNAs, providing insights into gene regulation and signaling pathways. Additionally, DNA microarrays offer both qualitative and quantitative analysis capabilities, allowing researchers to not only identify which genes are expressed but also measure the relative abundance of transcripts, facilitating the identification of differentially expressed genes and understanding of biological processes.

123. (C):

(*a*) **Phytoene synthase - Golden rice:** Phytoene synthase is an enzyme involved in the biosynthesis of carotenoids. Golden rice, a genetically modified rice variety, has been engineered to produce higher levels of beta-carotene (provitamin A) using phytoene synthase from other organisms, aiming to address vitamin A deficiency in regions where rice is a dietary staple.

(*b*) **Barnase-barstar system - DMH-11 mustard hybrid:** The barnase-barstar system is a gene-editing technique used to produce male-sterile lines in crops like mustard. DMH-11 mustard hybrid refers to a genetically modified mustard variety developed using this system to produce hybrid seeds with desirable traits such as high yield and resistance to pests.

(*c*) **EPSPS - RoundUp Ready soybean:** EPSPS (5-enolpyruvylshikimate-3-phosphate synthase) is an enzyme involved in the shikimate pathway, which is essential for the synthesis of aromatic amino acids in plants. RoundUp Ready soybean is a genetically modified soybean variety engineered to be resistant to the herbicide glyphosate, commonly known as RoundUp, through the introduction of a glyphosate-insensitive EPSPS gene.

(*d*) **Cry1Ac - Bt cotton:** Cry1Ac is a protein produced by the bacterium *Bacillus thuringiensis* (Bt) that is toxic to certain insect pests. Bt cotton is a genetically modified cotton variety that contains the gene encoding Cry1Ac, providing resistance to specific lepidopteran pests like bollworms.

124. (A): Assertion (A) correctly describes the phenomenon of original antigenic sin, which explains why individuals may experience severe symptoms of the flu despite having memory cells from previous exposures. Reason (R) provides a theoretical explanation for this phenomenon, stating that primary immune responses occur when individuals encounter significantly new influenza epitopes, leading to more severe symptoms. This aligns with the concept of original antigenic sin, where the immune system's response is influenced by previous exposures to related antigens, resulting in a less effective immune response against novel strains. Therefore, reason (R) appropriately explains the assertion (A).

125. (A): Xeroderma pigmentosum (XP) is a genetic disorder characterized by extreme sensitivity to sunlight, resulting in severe sunburns, skin lesions, and an increased risk of skin cancer. This condition is caused by mutations in genes involved in the Nucleotide Excision Repair (NER) pathway, which normally functions to repair DNA damage induced by ultraviolet (UV) light. Individuals with XP have impaired NER capacity, leading to the accumulation of DNA lesions and mutations in their skin cells. As a result, they are highly prone to developing skin cancers, including basal cell carcinoma, squamous cell carcinoma, and melanoma, often at a young age. Additionally, XP patients may exhibit neurological abnormalities and developmental delays due to DNA damage affecting other cell types. Management of XP involves strict sun protection measures and regular skin screenings to detect and treat skin cancers early.

127. (D): Sterilization of fermentation media with steam at 121°C for 20 minutes is a standard practice in biotechnology and microbiology laboratories. This method, known as autoclaving, utilizes high-pressure steam to achieve effective sterilization. At this temperature and duration, microbial cells, including spores, are destroyed, ensuring that the fermentation media is free from contamination. Autoclaving is preferred because it is relatively quick, reliable, and can be used for a wide range of materials, including liquids, solids, and heat-resistant equipment. This process is crucial for maintaining the purity of fermentation media and ensuring the success of microbial culture experiments and industrial fermentation processes.

128. (D): To determine which DNA sequence will have the highest melting temperature (Tm), we need to consider the stability of the DNA duplex formed by complementary base pairing. The Tm is influenced by the length of the DNA, its GC content, and the presence of specific sequences or motifs. Generally, sequences with higher GC content have a higher Tm because GC base pairs form three hydrogen bonds compared to the two hydrogen bonds formed by AT base pairs.

Let's analyze the sequences provided:

(A) ATGTTGGACCTTGAGTAATGCTA
GC content: 10/23 = 43.5%

(B) AAAGGATTTCCCTTTGATCGTATG
GC content: 7/23 = 30.4%

(C) ATTTACTAGATTACTAGTATTGATA
GC content: 7/23 = 30.4%

(D) GCTTATGTATACCGGTTAGATCG
GC content: 11/23 = 47.8%

Based on the GC content, sequence D has the highest GC content and is expected to have the highest Tm among the provided sequences.

129. (A): The Biological Oxygen Demand (BOD) value measures the amount of dissolved oxygen (DO) that aerobic microorganisms need to break down organic material present in a water sample over a specific period, usually five days. A BOD value of more than 50 ppm indicates a high level of organic pollution in the water sample. Therefore, this option is true because a high BOD value suggests that a significant amount of dissolved oxygen will be consumed during the decomposition of organic matter, resulting in low DO content. Option B is false because a high BOD value indicates pollution, not cleanliness. Option C is false because a high BOD value indicates a substantial amount of organic matter. Option D is false because high levels of organic pollution can lead to oxygen depletion, harming aquatic life rather than supporting its thriving.

130. (D):

(*a*) **Alkalase - Bioactive peptides:** Alkalase is an enzyme used in the production of bioactive peptides, which have various health benefits.

(*b*) **Lipase - Degreasing of leather hides:** Lipase is an enzyme that breaks down fats and oils, making it useful in degreasing processes such as degreasing leather hides.

(*c*) **β-amylase - Maltose rich syrup:** β-amylase is involved in the production of maltose-rich syrup from starch hydrolysis.

(*d*) **Phytase - Animal feed:** Phytase is an enzyme used in animal feed to improve phosphorus utilization from plant-based feed ingredients.

131. (A): Glycosylation is a post-translational modification common in eukaryotic cells like CHO cells but not in prokaryotic cells like *E. coli*. Many proteins, including antibodies, require proper glycosylation for their correct folding, stability, and biological activity. In this case, the monoclonal antibody expressed in *E. coli* lacked glycosylation, which likely resulted in incorrect folding or instability, leading to loss of function despite having the same structure as the protein expressed in CHO cells. The other options are less likely explanations as phosphorylation and ubiquitination are less common modifications for antibodies, and phosphorylation by *E. coli* would not necessarily be detrimental for function.

132. (D): Trickling filters are biological treatment systems used to remove organic matter and suspended solids from wastewater. While organic loading, hydraulic flow rates, and the temperature of water and air significantly influence the efficiency of trickling filters, the linear flow velocity of the liquid does not directly affect their operation. Instead, the key factors are the microbial activity within the filter media and the distribution of wastewater over the surface area. The velocity of the liquid within the filter media bed can vary but is generally not a critical parameter as long as adequate contact time and oxygen availability are maintained for microbial growth and organic matter degradation. Therefore, the linear flow velocity of the liquid is not a primary concern in the operation of trickling filters.

133. (B): Phenol red is an organic compound that serves as a pH indicator due to its ability to change colour in response to changes in pH. In cell culture media, it is often included to help monitor the pH of the environment in which cells are growing. The colour of phenol red shifts from red to yellow as the pH of the medium becomes more acidic. This colour change provides a visual indication to researchers, allowing them to quickly assess the pH of the cell culture medium. Maintaining the appropriate pH is essential for cell health and growth, as variations outside the optimal range can negatively impact cellular processes and viability. Therefore, the presence of phenol red helps ensure that the conditions for cell culture remain suitable for optimal cell growth and experimental outcomes.

135. (D): In the fermentation process of fruit juice to vinegar, yeast initially converts the sugars present in the fruit juice into alcohol through alcoholic fermentation. This step is carried out by yeast, which metabolizes sugars into ethanol and carbon dioxide. Subsequently, acetic acid bacteria, such as *Acetobacter* spp., oxidize the ethanol produced by yeast into acetic acid through acetic acid fermentation. This process gives vinegar its characteristic sour taste. Therefore, the fermentation of fruit juice to vinegar involves the cooperative action of both yeast and acetic acid bacteria.

136. (A): The inner mitochondrial membrane plays a crucial role in cellular respiration, housing the electron transport chain (ETC) along its folds called cristae. Within this membrane, the ETC transfers electrons derived from substrates like NADH and $FADH_2$ along a series of protein complexes, ultimately leading to the generation of a proton gradient across the membrane. This proton gradient drives ATP synthesis through ATP synthase, a process known as oxidative phosphorylation. Additionally, the inner mitochondrial membrane is impermeable to most ions and polar molecules, ensuring that the proton gradient is maintained and allowing for efficient ATP production. Thus, the inner mitochondrial membrane serves as a site for energy conversion, playing a central role in cellular metabolism.

137. (B): Riboviruses are the predominant type of viruses that infect plants, constituting the largest group of plant viruses. This category includes a diverse range of viruses, such as tobacco mosaic virus and tomato spotted wilt virus, which encompass single-stranded (positive or negative sense) and double-stranded RNA genomes. Their diversity allows them to adapt to different plant hosts and replication strategies. Riboviruses are often transmitted through vectors like insects, mechanical contact, or contaminated seeds, facilitating their spread and colonization in plant populations. Examples of riboviruses include positive-sense RNA viruses like tobacco mosaic virus and cucumber mosaic virus, negative-sense RNA viruses like tomato spotted wilt virus, and double-stranded RNA viruses like rice dwarf virus and wound tumor virus.

138. (A): The Centre for Science and Environment (CSE) brought the issue of pesticide contamination of carbonated drinks into the limelight. Through its research and advocacy efforts, CSE conducted studies revealing the presence of harmful pesticides in popular soft drink brands, raising significant concerns about public health and safety. The organization's findings sparked widespread public debate and prompted regulatory action to address the issue of pesticide residues in beverages. CSE's work underscored the importance of stringent quality standards and monitoring mechanisms to safeguard consumer well-being and ensure the safety of food and beverage products.

139. (B): The odd one out with respect to pleiotropic gene expression is the flower colour gene in *Pisum sativum* (pea plants). Unlike the other options, which involve genes known to have pleiotropic effects, such as the white eye gene in Drosophila, plumage colour in poultry, and Marfan's Syndrome, the flower colour gene in Pisum *sativum* (pea plants) typically exhibits Mendelian inheritance without significant pleiotropic effects. In *Pisum sativum,* the flower colour gene primarily determines the colour of the flowers (purple or white) without significantly influencing other traits or systems in the organism. Therefore, option B stands out as the odd one among the choices provided.

140. (B): In plant tissue culture, the addition of silver nitrate serves a crucial role in managing the detrimental effects of ethylene, a plant hormone involved in several physiological processes. Ethylene, while essential for various growth and developmental pathways, can also hinder tissue culture processes if present in excessive amounts. It can lead to issues such as the inhibition of shoot elongation and the induction of leaf abscission, ultimately compromising the success of tissue culture experiments. Silver nitrate acts as an ethylene antagonist by inhibiting ethylene responses in plant tissues, thereby mitigating its negative effects. By adding silver nitrate to the culture media, researchers can create a more conducive environment for tissue growth and development, ensuring better outcomes in plant tissue culture experiments. Therefore, the inclusion of silver nitrate is crucial for optimizing the conditions and enhancing the success rate of tissue culture techniques.

141. (B): In the bacterial growth curve, the lag phase represents the initial period when cells are adapting to the growth conditions but are not actively dividing. During this phase, bacteria are metabolically active, synthesizing enzymes, and preparing for growth. However, their population size remains relatively constant as they adjust to the new environment. This phase can vary in duration depending on factors such as the composition of the growth medium, the health of the inoculum, and the species of bacteria. Once bacteria have fully adapted to the conditions, they enter the exponential phase, characterized by rapid and continuous cell division.

142. (A): Statement I: Protein-ligand docking involves binding of a small molecule to the protein receptor. This statement accurately describes the fundamental principle of protein-ligand docking, which is the simulation of the interaction between a protein receptor and a small molecule ligand. In this process, the ligand is typically a potential drug candidate, and the protein receptor is often a target protein involved in a disease process. Protein-ligand docking aims to predict the most energetically favorable binding mode of the ligand within the protein's binding site, providing insights into the potential efficacy and mechanism of action of the ligand.

Statement II: Both AUTODOCK and LUDI are protein-ligand docking programs. This statement is also correct. AUTODOCK and LUDI are indeed software programs specifically designed for performing protein-ligand docking simulations. AUTODOCK employs a genetic algorithm to explore ligand binding conformations, while LUDI focuses on predicting ligand binding affinities based on their interactions with the protein receptor. These programs are widely used in computational drug discovery and have contributed significantly to the identification and optimization of potential drug candidates.

143. (C): Chemical mutagenesis involves treating organisms with chemicals that induce mutations in their DNA, potentially leading to changes in the organism's phenotype. While this method is effective in generating genetic diversity, it typically results in a low frequency of mutations in the gene of interest. Moreover, pinpointing the exact location of the causal mutation can be challenging, even after observing the desired phenotype. Unlike site-directed mutagenesis techniques, chemical mutagenesis does not create point mutations at specific positions in the gene; rather, it induces random mutations throughout the genome. Consequently, analyzing a large number of offspring is often necessary to isolate the desired mutant with the desired phenotype, making the process labor-intensive and time-consuming.

144. (C): *Agrobacterium tumifaciens* is commonly used in plant genetic engineering due to its natural ability to transfer DNA into plant cells. This bacterium contains a tumor-inducing (Ti) plasmid, which carries genes responsible for plant transformation. During infection, *A. tumifaciens* transfers a segment of its Ti plasmid, known as T-DNA, into the plant genome. Researchers have harnessed this natural mechanism to introduce desired genes into plants by replacing the tumor-inducing genes with genes of interest. Once inside the plant cell, the transferred DNA integrates into the plant genome, leading to the expression of the introduced gene. This approach has become a fundamental tool in plant biotechnology for introducing novel traits, enhancing crop yield, and conferring resistance to pests and diseases.

145. (B): Microglial cells found in the human brain belong to the class of innate immune cells known as macrophages. These specialized cells act as the first line of defense against pathogens and play crucial roles in immune surveillance, tissue repair, and maintenance of homeostasis within the central nervous system. Microglia are derived from myeloid progenitor cells and are capable of detecting and responding to various signals, including infection, injury, and inflammation, by migrating to the affected sites and phagocytosing pathogens or cellular debris. Their multifaceted functions highlight their importance in neuroinflammation and neurological disorders.

146. (A): An enveloped virus is characterized by having a lipid bilayer envelope surrounding its protein capsid. This envelope is typically derived from the host cell membrane during the process of viral budding. Dengue virus, the correct answer in this case, belongs to the family Flaviviridae and is indeed an enveloped virus. The lipid envelope plays a crucial role in the virus's ability to infect host cells, as it contains viral glycoproteins that facilitate attachment and entry into target cells. The envelope also helps the virus evade the host immune system by shielding viral proteins from detection.

147. (A): Vaccination against smallpox, a highly contagious and often fatal disease caused by the variola virus, has been one of the most successful public health interventions in history. Through widespread vaccination campaigns, smallpox was successfully eradicated worldwide by the end of the 20th century, making it the first and only human infectious disease to be eradicated through vaccination. The vaccine, made from a related virus called vaccinia, provided robust immunity against smallpox and played a crucial role in breaking the chain of transmission. The achievement of smallpox eradication stands as a testament to the effectiveness of vaccination programs in preventing and controlling infectious diseases on a global scale.

148. (C): Meteorology is the scientific study of the Earth's atmosphere, focusing on various atmospheric phenomena and processes such as weather patterns, climate, atmospheric composition, and interactions with the Earth's surface and oceans. Meteorologists utilize a combination of observational data, theoretical models, and computer simulations to understand and predict atmospheric conditions. They investigate factors like temperature, humidity, pressure, wind patterns, and precipitation to analyze and forecast weather events ranging from short-term phenomena like thunderstorms and hurricanes to long-term climate trends. Through meteorological research and analysis, scientists aim to enhance our understanding of the atmosphere and improve weather forecasting capabilities for societal benefit.

151. (A): A domain in protein structure refers to a distinct, independently folded three-dimensional unit that typically serves a specific functional role. Domains are often found within proteins and can fold autonomously, retaining their structure and function even when isolated from the rest of the protein. These structural units contribute to the overall function of the protein by performing specialized tasks such as catalysis, binding to other molecules, or providing structural stability. Domains can also exhibit modularity, where they are found in different proteins performing diverse functions, suggesting evolutionary conservation of their structural and functional properties. Understanding protein domains is crucial for deciphering the molecular mechanisms underlying various biological processes and for designing targeted therapeutics that interact with specific protein functions.

152. (B): Osteoclasts are multinucleated cells present in the bones of the human body. They are responsible for the resorption or breakdown of bone tissue, which is crucial for bone remodeling, repair, and maintenance of bone density. Osteoclasts are derived from monocyte/macrophage lineage cells and contain multiple nuclei, which aids them in their function of breaking down bone tissue through the secretion of enzymes and acids. This process of bone resorption by osteoclasts is tightly regulated and balanced with bone formation by osteoblasts to maintain skeletal integrity.

153. (A): Cybrids are hybrid cells resulting from controlled fusion techniques, not spontaneous events like those suggested in option A. These techniques involve deliberate transfer of cytoplasm from a donor cell to a recipient cell, typically through methods like cell fusion. Cybrids are particularly valuable in genetic research and breeding programs as they allow for the direct transfer of cytoplasmic traits, such as cytoplasmic male sterility in plants, from one organism to another. This controlled transfer facilitates the study and manipulation of cytoplasmic characteristics, providing insights into various biological processes and enabling the development of novel crop varieties with desired traits.

155. (C): Primary metabolites are molecules essential for basic cellular functions and are typically involved in processes such as energy production, growth, and development. Examples include amino acids, nucleotides, sugars, and lipids, which serve as building blocks for macromolecules and energy sources for cellular activities. Polysaccharides, like starch and glycogen, also fall under primary metabolites as they play crucial roles in energy storage and structural support. Vitamins, although required in small amounts, are considered primary metabolites as they participate in various metabolic reactions as coenzymes or precursors. However, antibiotics, such as penicillin or streptomycin, are generally not classified as primary metabolites. They are produced by microorganisms primarily as secondary metabolites, often serving as defense mechanisms or agents for interspecies competition rather than essential components of basic cellular metabolism.

157. (D): Scrubbers are devices used to remove harmful gases, such as sulphur dioxide (SO_2), from industrial exhaust. They work by passing the exhaust gas through a liquid (usually water or a chemical solution) that absorbs and neutralizes the pollutants present in the gas stream. This process helps to reduce air pollution and minimize the release of harmful gases into the atmosphere. Catalytic converters are primarily used in automobiles to convert harmful gases into less harmful emissions, while electrostatic precipitators are used to remove particulate matter from industrial exhaust. Earmuffs, on the other hand, are personal protective equipment used to protect hearing in noisy environments and are not related to gas pollution control.

158. (C): Molybdenum plays an essential role in biological nitrogen fixation as it is a component of nitrogenase enzymes. Nitrogenase enzymes are responsible for converting atmospheric nitrogen (N_2) into ammonia (NH_3), a process known as nitrogen fixation. Molybdenum is a cofactor in the active site of nitrogenase enzymes, facilitating the conversion of nitrogen gas into ammonia. Without sufficient molybdenum, biological nitrogen fixation would be impaired, leading to nitrogen deficiency in plants and ecosystems. Copper, magnesium, and zinc also play important roles in various biological processes, but they are not directly involved in nitrogen fixation.

159. (C): In non-competitive inhibition, the inhibitor binds to a site on the enzyme other than the active site, called the allosteric site, and this binding occurs regardless of whether the substrate is bound to the enzyme or not. As a result, the degree of inhibition is independent of substrate concentration. This is because the inhibitor and the substrate do not compete for the same binding site on the enzyme. In competitive inhibition, the inhibitor competes with the substrate for binding to the active site of the enzyme, so the degree of inhibition can be overcome by increasing the substrate concentration. Uncompetitive inhibition occurs when the inhibitor binds only to the enzyme-substrate complex, also leading to an increase in substrate concentration not overcoming inhibition. Allosteric inhibition involves the binding of an inhibitor to an allosteric site, but the degree of inhibition may or may not be independent of substrate concentration, depending on the specific characteristics of the allosteric enzyme.

160. (B): Among the options provided, polio virus (Poliovirus) is likely to have the highest genomics mutation rate. Poliovirus is an RNA virus, and RNA viruses generally exhibit higher mutation rates compared to DNA viruses due to the lack of proofreading mechanisms during RNA replication. This high mutation rate leads to genetic variability within the virus population, contributing to the rapid evolution of RNA viruses and the emergence of new variants. Smallpox virus, hepatitis B virus, and adenovirus are DNA viruses, which typically have lower mutation rates compared to RNA viruses like polio virus.

Previous Paper (Solved)

Graduate Aptitude Test–Biotechnology (GAT-B) Entrance Exam–2022*

(For M.Sc. Biotechnology and Allied Areas; M.Tech. Biotechnology and Allied Areas; M.Sc. Agricultural Biotechnology and M.V. Sc. Animal Biotechnology)

SECTION–A

1. The following structure represents:

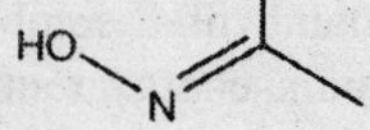

A. An oxime
B. A hydrazone
C. An imine
D. A Schiff's base

2. Given below are two statements:

Statement I: Activation energy is the difference between the energy level of substrate and its transition state during the reaction.

Statement II: Enzymes increase the activation energy enhancing the rate of reaction.

In the light of the above statements, choose the correct answer from the options given below.

A. Both Statement I and Statement II are true
B. Both Statement I and Statement II are false
C. Statement I is true but Statement II is false
D. Statement I is false but Statement II is true

3. What is the chance that a leap year, selected at random, will NOT have 53 Sundays as well as 53 Mondays?

A. $\frac{1}{7}$
B. $\frac{2}{7}$
C. $\frac{5}{7}$
D. 0

4. Which one of the following properties is primarily responsible for spreading of oil uniformly on water?

A. Viscosity
B. Partition coefficient
C. Dipole moment
D. Surface tension

5. What is the molarity of pure acetic acid? Assume density of acetic acid = 1 g/ml.

A. 12.7
B. 16.7
C. 20.7
D. 55.5

6. Histones are tightly bound to DNA because:

A. histones are positively charged proteins.
B. histones are negatively charged proteins.
C. histones get acetylated by the action of acyltransferases.
D. histones are highly hydrophobic in nature.

7. Given below are two statements:

Statement I: A single gene can affect multiple phenotypic expressions.

Statement II: In a polygenic trait, multiple genes contribute to a phenotypic expression.

In the light of the above statements, choose the *most appropriate* answer from the options given below.

A. Both I and II are correct
B. Both I and II are incorrect
C. I is correct but II is incorrect
D. I is incorrect but II is correct

8. The correct order of increasing acidic behaviour of the following compounds is:

$H_3C — CH_3$	$H_2C = CH_2$	$HC \equiv CH$
A	**B**	**C**

A. A > B > C
B. C > B > A
C. A > C > B
D. B > C > A

* *Exam held on 23/04/2022*

9. When 4 fair coins are tossed together what is the probability of getting at least 3 heads?

A. $\frac{1}{4}$ B. $\frac{3}{4}$
C. $\frac{5}{16}$ D. $\frac{3}{8}$

10. How many numbers between 100 and 999 can be formed with the digits 1, 4, 7, 8, and 9 if the digits are not repeated?

A. 20 B. 60
C. 12 D. 120

11. Which one of the following statements correctly describes type IIP restriction enzymes?

(*a*) They cut nucleotides only from the ends of the DNA.
(*b*) They act on only palindromic nucleotide sequences in the DNA.
(*c*) They recognize only GC rich sequences.
(*d*) They recognize only AT rich sequences.

Choose the **correct** answer from the options given below.

A. (*a*) and (*b*) only
B. (*b*) only
C. (*a*), (*b*) and (*c*) only
D. (*a*), (*b*) and (*d*) only

12. What is the sum of the first 10 terms of the sequence 1, 2, 4, 8, 16, ... ?

A. 1024 B. 1023
C. 1017 D. 1026

13. In a class of 35 students, 24 like to play cricket and 16 like to play football. Also, each student likes to play at least one of the two games. How many students like to play both cricket and football?

A. 11 B. 5
C. 8 D. 13

14. Interferons secreted by virus-infected cells induce:

A. antiviral response B. pro-viral response
C. apoptosis D. cell proliferation

15. If A is the set of even natural numbers less than 8 and B is the set of prime numbers less than 7, then the number of relations from A to B is:

A. 9 B. 9^2
C. 2^9 D. 2^3

16. Find the area of the triangle with vertices A(5, 4), B(–2, 4) and C(2, 14).

A. 34 sq. units B. 35 sq. units
C. 36 sq. units D. 32 sq. units

17. Energy of a photon of wavelength 410 nm is equal to the band gap between valence and conduction band of a semi-conductor. What is the minimum energy required to create an electron-hole pair? (Planck's constant (h) = 4.1×10^{-15} eV.s) (velocity of light = 3×10^8 m/s)

A. 0.03 eV B. 0.3 eV
C. 3.0 eV D. 30.0 eV

18. Which one of the following describes the relation between the acceleration of a body in space and the distance (r) of the body from the center of the earth?

A. $\frac{1}{r}$ B. $\frac{1}{r^3}$
C. $\frac{1}{r^2}$ D. $\sqrt{r}$

19. What is the net capacitance of the circuit given below?

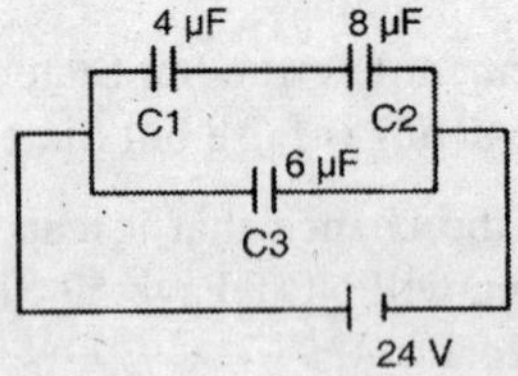

A. 8.6 μF B. 4.6 μF
C. 6.6 μF D. 3.0 μF

20. In a class of 60 students comprising girls and boys, 30% are boys. On a given day 10% of the class is absent and all the absentees are boys. What is the percentage of girls in this class on that given day?

A. 66.66 B. 77.77
C. 88.88 D. 100

21. Choose the correct order of increasing acidity for the following compounds:
(*a*) Benzoic acid
(*b*) 4-Methoxybenzoic acid
(*c*) 4-Nitrobenzoic acid
A. A > B > C B. B > C > A
C. C > B > A D. A > C > B

22. A box contains 5 pink, 3 green, and 2 yellow balls. Three balls are picked up randomly. What is the probability that none of the balls drawn is green?
A. $\frac{3}{16}$ B. $\frac{7}{24}$
C. $\frac{5}{13}$ D. $\frac{4}{23}$

23. Which one of the following is **NOT** due to total internal reflection?
A. Brilliance of a diamond.
B. Working of an optical fibre.
C. Mirage on a hot summer day.
D. Difference between apparent and real depth of a water body.

24. Three taps A, B, and C together can fill an empty cistern in 10 minutes, the tap A alone can fill it in 20 minutes and the tap B alone in 40 minutes. How long will the tap C alone take to fill it?
A. 40 minutes
B. 24 minutes
C. 32 minutes
D. 16 minutes

25. The ratio of the velocity of the hydrogen molecule to that of the oxygen molecule under identical conditions is:
A. 1 : 1 B. 1 : 2
C. 1 : 4 D. 1 : 16

26. Cystic fibrosis is caused due to:
A. mutation in a gene encoding G protein
B. deficiency of adenyl cyclase
C. mutation in a gene encoding an ABC transporter
D. over activation of the Ca++ ATPase pump

27. Which one of the following represents the correct electronic orbital configuration of Phosphorus (+3) ion (P^{3+})?
A. $1s^2\ 2s^2\ 2p^6\ 3s^2$ B. $1s^2\ 2s^2\ 2p^6\ 3s^1$
C. $1s^2\ 2s^2\ 2p^4\ 3s^2$ D. $1s^2\ 2s^2\ 2p^5\ 3s^1$

28. The limit of resolution of a light microscope is determined by:
A. wavelength of light
B. material of the lens
C. polarization of light
D. intensity of light vapor

29. What is the approximate vapor pressure of an isopropanol solution containing 9.2 g of a non volatile solute, glycerol ($C_3H_8O_3$) and 60 g of isopropanol (C_3H_8O) at 82°C? Given vapor pressure of isopropanol at 82°C is 100 kPa.
A. 81 kPa B. 91 kPa
C. 101 kPa D. 111 kPa

30. Which one of the following is NOT a property of a LASER?
A. It is monochromatic.
B. It is coherent.
C. It is directional.
D. It exhibits spontaneous emission.

31. Given below are two statements:

Statement I: The boiling point of *p*-nitrophenol is higher than the o-nitrophenol.

Statement II: Alcohols are less acidic than the Phenols.

In the light of the above statements, *choose* the *most appropriate* answer from the options given below.
A. Both Statement I and Statement II are correct.
B. Both Statement I and Statement II are incorrect.
C. Statement I is correct but Statement II is incorrect.
D. Statement I is incorrect but Statement II is correct.

32. Two trains running in opposite directions pass by a man standing on the platform in 20 seconds and 10 seconds respectively and they completely cross each other in 15 seconds. The ratio of their speeds is:

A. 1 : 3 B. 1 : 2
C. 1 : 1.5 D. 1 : 1

33. The alkaline nature of amines is due to the:

A. small size of nitrogen.
B. higher stability of nitrogen molecule at room temperature.
C. presence of unshared pair of electrons on nitrogen.
D. pyramidal geometry of nitrogen in amines.

34. The average of the first five multiples of 9 is:

A. 20 B. 27
C. 28 D. 30

35. Match items in List-I with items in List-II:

List-I	List-II
(*a*) Incomplete dominance	(*i*) ABO Blood types
(*b*) Recessive epistasis	(*ii*) Flower colour in 4 o'clock plant
(*c*) Dominant epistasis	(*iii*) Coat colour in mice
(*d*) Co-dominance	(*iv*) Fruit colour in summer squash

Choose the **correct** answer from the options given below:

	(*a*)	(*b*)	(*c*)	(*d*)
A.	(*ii*)	(*iv*)	(*iii*)	(*i*)
B.	(*i*)	(*iv*)	(*iii*)	(*ii*)
C.	(*i*)	(*iii*)	(*iv*)	(*ii*)
D.	(*ii*)	(*iii*)	(*iv*)	(*i*)

36. Find the modal age from the given data:

Age	10	11	12	14	15	16	18	20
No. of boys	12	9	8	6	6	5	4	3

A. 10 B. 12
C. 15 D. 20

37. What is the bond that can form between two cysteines in a protein?

A. Ionic bond B. Hydrogen bond
C. Disulfide bond D. Amide bond

38. If the price of the sugar is raised by 25%, find by how much per cent a homemaker should reduce the consumption of sugar so as not to increase his/her expenditure?

A. 10% B. 12%
C. 20% D. 15%

39. The boiling points of four liquids are shown below:

Compound	Boiling point
Acetone	56°C
Ethanol	78°C
Diethyl ether	35°C
Water	100°C

Which liquid has the highest vapor pressure at 25°C?

A. Acetone B. Ethanol
C. Diethyl ether D. Water

40. The rate of a reaction, $2A + B \rightarrow$ Products, is given by the rate equation

$$\text{Rate} = k[A]^2\,[B]$$

The value of the rate constant can be increased by:

A. increasing the concentration of A alone.
B. increasing the concentration of B alone.
C. increasing the concentrations of A and B simultaneously.
D. increasing the temperature.

41. Which one of the following force does not act on a cyclist when he/she bends on a horizontal turn?

A. Weight (Mg)
B. Centrifugal force (Mv^2/r)
C. Friction (f)
D. Torque ($\vec{\tau}$)

42. Lung cancer can be detected at an early stage by:

A. bone marrow test B. blood test
C. MRI scans D. gene sequencing

43. Point charges, +q each is placed at the corners of a triangle, square and regular pentagon. If the sides of the triangle, square and pentagon are a each, the electric field at the center will be:

A. zero in all three cases
B. maximum at the center of the pentagon
C. minimum at the center of the triangle
D. zero for square and positive for triangle and pentagon

44. Given below are two statements: one is labelled as Assertion (A) and the other is labelled as Reason (R).

Assertion (A): Inhibition of succinate dehydrogenase by malonate represents competitive inhibition.

Reason (R): Malonate and succinate bind at different sites of the enzyme.

In light of the above statements, choose the most appropriate answer from the options given below.
A. Both (A) and (R) are correct and (R) is the correct explanation of (A)
B. Both (A) and (R) are correct but (R) is not the correct explanation of (A)
C. (A) is correct but (R) is not correct
D. (A) is not correct but (R) is correct

45. Glycosidic linkages are the bonds between:
A. sugars
B. nucleotides
C. amino acids
D. lipids

46. Given below are two statements: one is labelled as Assertion (A) and the other is labelled as Reason (R).

Assertion (A): Bt toxin does not kill *Bacillus thuringiensis*.

Reason (R): The Bt toxin gets activated due to the alkaline pH of the insect gut.

In the light of the above statements, choose the *most appropriate* answer from the options given below.
A. Both (A) and (R) are correct and (R) is the correct explanation of (A)
B. Both (A) and (R) are correct but (R) is not the correct explanation of (A)
C. (A) is correct but (R) is not correct
D. (A) is not correct but (R) is correct

47. Starting with three different amino acids, how many different tripeptides can be made, assuming that each amino acid can be used more than once?
A. 3 B. 6
C. 9 D. 27

48. An organic compound contains 69.77% C, 11.63% H and 18.6% O. Its molecular formula is same as its empirical formula. The number of carbon atoms present per molecule would be:
A. 4 B. 5
C. 6 D. 7

49. Which one of the following is an autosomal dominant disorder?
A. Sickle cell anaemia
B. Cystic fibrosis
C. Tay-Sachs disease
D. Huntington's disease

50. What is the focal length (f) of a combination of lenses (convex and concave) as given below?

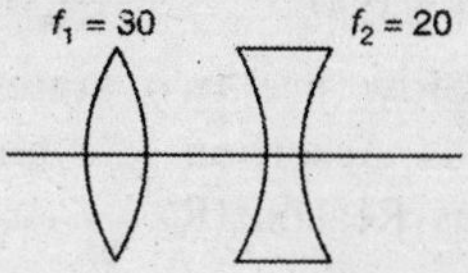

A. 30 cm B. 20 cm
C. 10 cm D. 60 cm

51. What will happen when the pressure is doubled in an isothermal process for a closed system containing a fixed amount of an ideal gas?
A. Volume will double
B. Volume will halve
C. Temperature will increase slightly
D. Temperature will decrease slightly

52. In uniform circular motion, which one of the following properties changes with time?
A. Momentum B. Speed
C. Mass D. Kinetic Energy

53. What is the half-life of a zero order reaction that takes 25 minutes for 80% completion?
A. 150.25 seconds B. 937.50 seconds
C. 950.50 seconds D. 156.25 seconds

54. The straight line $2x + 3y + 4 = 0$ touches the x-axis at:
A. –2.0 B. +2.0
C. –1.0 D. +1.0

55. The primary enzyme responsible for the rapid and accurate replication of the genome in *E. coli* is:
A. DNA Pol V B. DNA Pol II
C. DNA Pol III D. DNA Pol IV

56. The magnetic susceptibility is negative for:
A. paramagnetic materials.
B. ferromagnetic materials.
C. superconducting materials.
D. diamagnetic materials.

57. In the following reaction
$P_4(g) + 6\ Cl_2\ (g) = 4\ PCl_3\ (g)$
equilibrium is established by adding equal moles of $P_4(g)$ and $Cl_2(g)$; therefore, at equilibrium:
A. $[Cl_2] > [PCl_3]$ B. $[Cl_2] < [P_4]$
C. $[PCl_3] > [P_4]$ D. $[P_4] = [PCl_3]$

58. Given below are two statements: one is labelled as Assertion (A) and the other is labelled as Reason (R).

Assertion (A): Hematopoietic stem cells are considered multipotent precursors.

Reason (R): They have the ability to differentiate into different types of blood cells.

In light of the above statements, choose the *most appropriate* answer from the options given below.
A. Both (A) and (R) are correct and (R) is the correct explanation of (A)
B. Both (A) and (R) are correct but (R) is not the correct explanation of (A)
C. (A) is correct but (R) is not correct
D. (A) is not correct but (R) is correct

59. Which one of the following attributes of an electromagnetic wave remains constant when it enters from air to water?
A. Wavelength B. Velocity
C. Frequency D. Colour

60. Moving electron (velocity v) produces a magnetic field B such that:
A. B is perpendicular to v.
B. B is parallel to v.
C. Angle between B and v is 45 degrees.
D. B is opposite in direction to v.

SECTION–B

61. During protoplast fusion, the product has the nucleus of one of the parents in which the chromosome elimination took place in subsequent cell division and the cytoplasm was contributed by both the parents. Such a product will be called:
A. Symmetric hybrid
B. Asymmetric hybrid
C. Cybrid
D. Homokaryons

62. Which of the following can be a restriction site for type II restriction enzyme?
A. 5′ AAGG 3′
3′ TTCC 5′
B. 3′ AGTC 5′
5′ TCAG 3′
C. 5′ GGCC 3′
3′ CCGG 5′
D. 5′ ACCA 3′
3′ TGGT 5′

63. The selection of cells whose T-cell receptors respond to self-MHC is known as:
A. Negative selection
B. MHC restriction
C. Affinity maturation
D. Lineage commitment

64. In animal cell culture, a CO_2 enriched atmosphere in the incubation chamber is used:
A. for supplying CO_2 for the metabolic reaction.
B. to control pH by generating carbonic acid.
C. to control pH by generating bicarbonate.
D. to aid photosynthetic reaction.

65. How many clones must you get to make a representative genomic library of a plant with a genome of a billion base pairs (bp), provided your vector carrying capacity is 10 kbp. Remember to allow a redundancy of at least one order of magnitude.

A. Thousand B. Million
C. Lakh D. Hundred

66. Match the coefficients in List-I with their corresponding downstream processing steps given in List-II.

List-I	List-II
(*a*) Sedimentation coefficient	(*i*) aqueous two-phase extraction
(*b*) Partition coefficient	(*ii*) ultrafiltration
(*c*) Rejection coefficient	(*iii*) dialysis
(*d*) Activity coefficient	(*iv*) centrifugation

Choose the **correct** answer from the options given below.

	(*a*)	(*b*)	(*c*)	(*d*)
A.	(*iii*)	(*i*)	(*iv*)	(*ii*)
B.	(*ii*)	(*i*)	(*iv*)	(*iii*)
C.	(*iv*)	(*iii*)	(*i*)	(*ii*)
D.	(*iv*)	(*i*)	(*ii*)	(*iii*)

67. The following figure shows agarose gel electrophoresis of linear DNA, after digestion with *Bam*HI in lane C and *Eco*RI in lane D. The uncut DNA is shown in lane B.

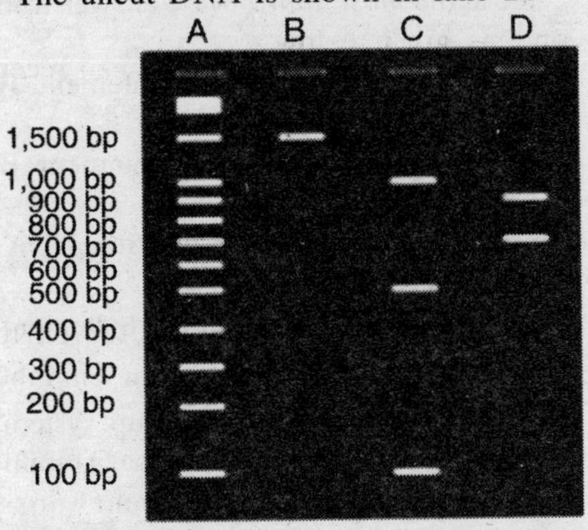

Statement I: This DNA has two restriction sites for *Bam*HI and one restriction site for *Eco*RI.

Statement II: Double digestion with *Bam*HI and *Eco*RI will yield 5 fragments.

In the light of the above statements, choose the *correct* answer from the options given below.

A. Both Statement I and Statement II are true
B. Both Statement I and Statement II are false
C. Statement I is true but Statement II is false
D. Statement I is false but Statement II is true

68. Which of the following hypersensitivity reactions involves the complement activation?

A. Both Type I and Type III hypersensitivity
B. Only Type I hypersensitivity
C. Both Type I and Type II hypersensitivity
D. Both Type II and Type III hypersensitivity

69. At the time of completion of protein synthesis, the sequence in which they leave the ribosome is:

A. Polypeptide chain, m-RNA, t-RNA
B. t-RNA, m-RNA, Polypeptide chain
C. Polypeptide chain, t-RNA, m-RNA
D. m-RNA, Polypeptide chain, t-RNA

70. Given below are two statements:

Statement I: In plants, the genes for smaller subunit of ribulose-1, 5-bisphosphate carboxylase/oxygenase (Rubisco) enzyme is present in the plastid genome.

Statement II: The genes for larger subunit of Rubisco enzyme is present in the nuclear genome.

Choose the *most appropriate* answer from the options given below:

A. Both Statement I and Statement II are correct.
B. Both Statement I and Statement II are incorrect.
C. Statement I is correct but Statement II is incorrect.
D. Statement I is incorrect but Statement II is correct.

71. Match items in List-I with items in List-II:

List-I	**List-I**
(*a*) Cytochrome oxidase	(*i*) Molybdenum ions
(*b*) DNA polymerase	(*ii*) Nickel ions
(*c*) Nitrate reductase	(*iii*) Magnesium ions
(*d*) Urease	(*iv*) Copper ions

Choose the **correct** answer from the options given below.

	(*a*)	(*b*)	(*c*)	(*d*)
A.	(*iv*)	(*iii*)	(*ii*)	(*i*)
B.	(*iv*)	(*iii*)	(*i*)	(*ii*)
C.	(*iii*)	(*i*)	(*iv*)	(*ii*)
D.	(*ii*)	(*iii*)	(*iv*)	(*i*)

72. Which is the class of Nitrogen containing plant secondary metabolites?

A. Terpenoids B. Flavonoids
C. Alkaloids D. Brassinosteroids

73. RNAi or post-transcriptional gene silencing is a conserved biological response to the presence of:

A. ssDNA B. dsDNA
C. ssRNA D. dsRNA

74. Given below are two statements:

Statement I: Sickle cell trait protects humans from malaria.

Statement II: Sickle cell trait can be transmitted even if both the parents are heterozygous for the β-globin gene.

In the light of the above statements, choose the *most appropriate* answer from the options given below.

A. Both Statement I and Statement II are correct
B. Both Statement I and Statement II are incorrect.
C. Statement I is correct but Statement II is incorrect.
D. Statement I is incorrect but Statement II is correct.

75. Lac repressor of *E. coli* contains ______ DNA binding motifs.

A. Helix turn helix B. Helix loop helix
C. Leucine zipper D. Zinc finger

76. The hormone that activates the enzyme adenylate cyclase is:

A. Epinephrine B. Insulin
C. Estrogen D. Progesterone

77. The 'Flavr Savr' tomato was the first genetically engineered crop product commercialized for human consumption. This transgenic tomato is known for:

A. increased bioactive compounds.
B. fortified Fe and Zn contents.
C. enhanced shelf-life.
D. bigger and pulpier fruits.

78. Which of the following phagocytic cells is NOT a tissue-specific macrophage cell?

A. Histiocyte B. Kupffer cell
C. Langerhans cell D. Dendritic cell

79. The following diagram represents a restriction enzyme site:

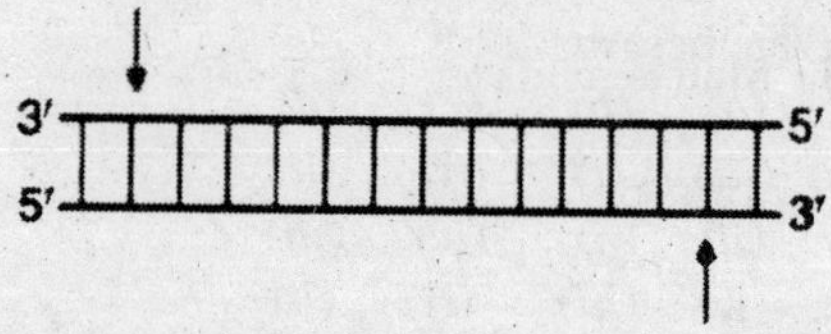

Statement I: After digestion cohesive ends will be formed.

Statement II: These cohesive ends will have phosphate groups at 3′ and 5′ ends.

Choose the *most appropriate* answer from the options given below.

A. Both Statement I and Statement II are correct
B. Both Statement I and Statement II are incorrect
C. Statement I is correct but Statement II is incorrect
D. Statement I is incorrect but Statement II is correct

80. In the human ABO blood group system, the alleles A and B are dominant to O, what will be the number of different possible genotypes?

A. 4 B. 8
C. 6 D. 12

81. Simple sequence repeats (SSRs) are best utilized for:

A. Phenotyping B. Genotyping
C. Karyotyping D. Gene Silencing

82. Increasing the agitation rate in a reactor breaks larger bubbles into smaller bubbles. If gas holdup does not change and the average bubble diameter reduces to half; then the gas-liquid interfacial area:

A. reduces by 50%
B. remains unchanged
C. increases by 100%
D. increases by 300%

83. A double-stranded genomic DNA has 40% (A+T) content. What is the average size (base pairs) of the fragments generated by digestion with the restriction enzyme Haelll (GGCC)?

A. 81 B. 123
C. 256 D. 4096

84. Match items in List-I with items in List-II:

List-I	List-II
(*a*) Silent mutation	(*i*) Amino acid codon is changed to a stop codon.
(*b*) Missense mutation	(*ii*) Amino acid sequence beyond the site of mutation is changed
(*c*) Nonsense mutation	(*iii*) Amino acid sequence is changed in one amino acid
(*d*) Frameshift mutation	(*iv*) Amino acid sequence is unchanged.

Choose the **correct** answer from the options given below:

	(*a*)	(*b*)	(*c*)	(*d*)
A.	(*i*)	(*ii*)	(*iii*)	(*iv*)
B.	(*iv*)	(*i*)	(*iii*)	(*ii*)
C.	(*iv*)	(*iii*)	(*i*)	(*ii*)
D.	(*iii*)	(*i*)	(*ii*)	(*iv*)

85. Which one of the following small RNAs in plants DO NOT regulate gene expression by cleavage of mRNAs?

A. MicroRNAs (miRNAs)
B. Short interfering RNA (siRNA)
C. Heterochromatin siRNAs (hcsiRNAs)
D. Trans-acting siRNAs (tasiRNAs)

86. Given below are two statements:

Statement I: Bisexual flowers promote self-pollination.

Statement II: Self-incompatibility and male-sterility prevent self-pollination.

Choose the *most appropriate* answer:

A. Both Statement I and Statement II are correct
B. Both Statement I and Statement II are incorrect
C. Statement I is correct but Statement II is incorrect
D. Statement I is incorrect but Statement II is correct

87. In power law model for fluid rheology, if the value of the flow behaviour index (n) is <1, then with increase in shear stress, the apparent viscosity:

A. remains unchanged.
B. increases.
C. decreases.
D. increases initially followed by a decrease.

88. Which one of the following techniques can be used to determine the copy number of a gene?

A. Chromatin Immuno-precipitation
B. Northern Blotting
C. ELISA
D. Real time PCR

89. The level of expression of recombinant protein can be gauged by:

A. number of transformants obtained.
B. nutrient utilization rate.
C. the level of resistance to antibiotic.
D. Western blotting.

90. Resolution power of transmission electron microscope is of the sub nanometer level. This is because:

A. the focal length of the electron microscope is significantly larger.

B. the contrast is enhanced by staining with atoms of heavy metal.
C. electron beams have much shorter wavelengths than visible light.
D. the electron microscope has a much greater ratio of image size to real size.

91. Given below are two statements: one is labelled as Assertion (A) and the other is labelled as Reason (R).

Assertion (A): Attenuation of the *trp* operon is possible in bacteria

Reason (R): Transcription and translation are coupled.

In the light of the above statements, choose the **correct** answer from the options given below.

A. Both (A) and (R) are true and (R) is the correct explanation of (A)
B. Both (A) and (R) are true but (R) is not the correct explanation of (A)
C. (A) is true but (R) is false
D. (A) is false but (R) is true

92. Corona virus is a:

A. positive sense single stranded RNA virus
B. negative sense single stranded RNA virus
C. double stranded RNA virus without segmented genome
D. double stranded RNA virus with segmented genome

93. Which of the following is NOT a muco-polysaccharide?

A. Chondroitin B. Inulin
C. Hyaluronic acid D. Heparin

94. Subtractive hybridization is useful to:

A. eliminates incomplete cDNA from a gene library.
B. create expression libraries based on genes that are currently being expressed.
C. identify and construct new probes for southern hybridization.
D. identify sets of genes that are only expressed under certain conditions.

95. Regulation of lac operon is negative as:

A. operon is expressed at low level to allow lactose entry.
B. lactose binds with the inducer to activate the operon.
C. expression of operon is blocked by an active repressor.
D. lactose needs a transporter to enter the cell.

96. Given below are two statements: one is labelled as Assertion (A) and the other is labelled as Reason (R).

Assertion (A): The major source of energy for the basic functioning of the cell is derived from oxidative metabolism.

Reason (R): Mitochondria oxidize substrates to CO_2, transferring the high energy electron from molecular oxygen to glucose.

In light of the above statements, choose the **correct** answer from the options given below.

A. Both (A) and (R) are true and (R) is the correct explanation of (A)
B. Both (A) and (R) are true but (R) is not the correct explanation of (A)
C. (A) is true but (R) is false
D. (A) is false but (R) is true

97. The rate of sedimentation of a particle is dependent on the:

A. density of the particle only.
B. size of the particle only.
C. viscosity of the medium only.
D. size and density of the particle along with the viscosity of the medium.

98. Given below are two statements: one is labelled as Assertion (A) and the other is labelled as Reason (R).

Assertion (A): A linear DNA molecule containing three recognition sites for a restriction enzyme will generate multiple DNA fragments on partial digestion.

Reason (R): Partial digestion by restriction enzymes leads to a heterogeneous population of DNA fragments.

Choose the *most appropriate* answer:
A. Both (A) and (R) are correct and (R) is the correct explanation of (A).
B. Both (A) and (R) are correct but (R) is not the correct explanation of (A).
C. (A) is correct but (R) is not correct.
D. (A) is not correct but (R) is correct.

99. Which one of the following are Professional Antigen Presenting Cells?
A. Fibroblasts B. Neutrophils
C. Dendritic cells D. Epithelial cells

100. Recombinants with insertional inactivation of β-galactosidase are identified by their inability to:
A. grow on the galactose-rich medium
B. grow on lactose rich medium
C. develop resistance to antibiotics
D. produce colour in the presence of a chromogenic substrate

101. The vector which is most commonly used in crop improvement is:
A. Bacterial Artificial Chromosome (BAC)
B. Agrobacterium
C. T4 phage
D. Cosmid

102. Given below are two statements:

Statement I: Virulence trait of *Agrobacterium tumefaciens* is borne on genomic DNA.

Statement II: The technique used to introduce genes in dicot is Ti-plasmid infection.

Choose the correct answer:
A. Both Statement I and Statement II are true.
B. Both Statement I and Statement II are false.
C. Statement I is true but Statement II is false.
D. Statement I is false but Statement II is true.

103. Once the cell passes the G1 checkpoint and enters S phase:
A. the cell cycle is completed bypassing the M phase.
B. the cell cycle is completed bypassing the G2 checkpoint.
C. completion of cell cycle will depend on G2 checkpoint.
D. the cell will next enter M phase bypassing the G2 checkpoint.

104. In Michaelis-Menten kinetics to increase the reaction velocity by 9 fold from 10% of V_{max} to 90% of V_{max} the substrate concentration has to increased by:
A. 3 fold B. 9 fold
C. 27 fold D. 81 fold

105. The stop codons lead to termination of translation because:
A. there are no tRNAs corresponding to stop codons.
B. the corresponding tRNA do not have an amino acid.
C. the corresponding tRNA cannot bind to ribosome binding site.
D. stop codon codes for pseudo amino acid.

106. The probability of mutation in a base-pair per round of replication is 5×10^{-10} and the genome length of bacteria is 5 Mb. The number of bacterial cells required so that there is one mutation on an average per replication cycle is:
A. 1000 B. 10000
C. 100000 D. 1000000

107. The melting point of fatty acids in plant derived oils is influenced by:
A. length and degree of unsaturation of hydrocarbon chain
B. degree of unsaturation only
C. length of hydrocarbon chain only
D. Number of side and branch chains present

108. Match List-I with List-II Match the industrial application of the following enzymes:

List-I	List-II
(*a*) Pencillinase	(*i*) Pharmaceutical
(*b*) Pectinase	(*ii*) Leather
(*c*) Trypsin	(*iii*) Wine
(*d*) Renin	(*iv*) Dairy

Choose the **correct** answer from the options given below.

	(a)	(b)	(c)	(d)
A.	(iv)	(iii)	(i)	(ii)
B.	(i)	(iii)	(ii)	(iv)
C.	(i)	(ii)	(iii)	(iv)
D.	(iv)	(ii)	(iii)	(i)

109. During replication, DNA polymerases distinguish between ribonucleoside and deoxyribonucleoside triphosphates (rNTPs and dNTPs) while synthesizing new DNA strand. Although rNTPs are present at approximately 10-fold higher concentration in the cell, they are incorporated at a rate that is more than 1000-fold lower than dNTPs. This is because:

A. DNA polymerases selectively cleave rNTPs.
B. rNTPs are cyclized by DNA polymerase.
C. Steric exclusion of rNTPs from the DNA polymerase active site.
D. dNTPs are more stable than rNTPs.

110. Self-splicing introns are an example of:

A. Spliceosomes B. Ribozymes
C. Mitrons D. Inteins

111. The primary function of the class I and class II MHC molecules is:

A. binding peptide antigens for recognition by antigen-specific receptors on T-cells.
B. mediation of T-independent B-cell responses.
C. helping in endocytosis of antigens by phagocytic cells.
D. aid opsonization of foreign particle.

112. NK cells are immunologically significant as they recognize and kill:

A. intracellular bacteria reservoirs
B. cells with upregulated MHC-I molecules.
C. cells with downregulated MHC-I molecules.
D. cells with self-marker on their surface.

113. Acid Fast bacteria:

A. have cell wall with mycolic acid
B. thrive well in acidic environment
C. are fastidious bacteria
D. cause acid depletion in soil

114. Given below are few statements related to plant tissue culture:

(*a*) Regeneration frequencies of different types of explants derived from the same plant would always be identical.
(*b*) A higher auxin: cytokinin ratio would induce production of roots under culture conditions.
(*c*) Embryos can be obtained from somatic tissues of plants under in vitro culture conditions.
(*d*) Since plants can photosynthesize, addition of a carbon source in tissue culture media is not advisable.

Choose the **correct** statements:

A. (*a*) and (*c*) only B. (*b*) and (*d*) only
C. (*b*) and (*c*) only D. (*a*) and (*d*) only

115. In which of the following genotypes of lac operon the enzyme synthesis would be constitutive?

(*a*) $I^+O^+Z^+Y^+/I^+O^+Z^+Y^+$
(*b*) $I^+O^cZ^+Y^+/I^+O^cZ^+Y^+$
(*c*) $I^-O^+Z^+Y^+/I^+O^+Z^+Y^+$
(*d*) $I^-O^+Z^+Y^+/I^-O^+Z^+Y^+$

Choose the **correct** answer from the options given below.

A. (*a*) and (*b*) only B. (*b*) and (*c*) only
C. (*b*) and (*d*) only D. (*c*) and (*d*) only

116. Which of the following statements is NOT TRUE about myeloma cells used in the hybridoma technique for the production of monoclonal antibodies?

A. These are cancerous B cells
B. They can divide indefinitely in a culture
C. Their HGPRT gene is non-functional
D. They can produce antibodies

117. Why is golden rice pale yellow in color?

A. It is rich in chlorophyll a.
B. It is rich in beta-carotene.
C. It is rich in chlorophyll b.
D. It is rich in phycobilins.

118. Given below are two statements:

Statement I: The recombination frequency of genes vary from 50% to 80%.

Statement II: Recombination can occur anywhere along the length of chromosome.

Choose the **correct** answer from the options given below:

A. Both statement I and statement II are correct

B. Both statement I and statement II are incorrect

C. Statement I is correct but statement II is not correct

D. Statement I is not correct but statement II is correct

119. In a family, if the father is affected by X-linked dominant trait and mother is normal, then:

A. some of the daughters can be normal.

B. all the daughters will be affected.

C. all the daughters and none of the sons will be affected.

D. all the sons and some of the daughters will be affected.

120. Which is a type of multigenic interaction?

A. Lethal allele

B. Co-dominance

C. Epistasis

D. ABO genotype

121. Arrange the following in correct sequence of steps in erythrocytic cycle of malarial parasite.

(*a*) Parasite feeds on the hemoglobin in the RBCs

(*b*) Parasite forms a schizont by multiple fissions

(*c*) The nucleus of the parasite moves on one side

(*d*) The food vacuole of the parasite increases in size

(*e*) The haemozoin granules are formed

Choose the **correct** answer from the options given below.

A. $(b) \to (c) \to (e) \to (d) \to (a)$

B. $(a) \to (d) \to (c) \to (e) \to (b)$

C. $(c) \to (a) \to (e) \to (b) \to (d)$

D. $(d) \to (e) \to (b) \to (c) \to (a)$

122. The fractional conversion from an immobilized enzyme continuous stirred tank reactor under steady state operation increases with an increase in agitation speed. This behaviour can be attributed to:

A. reduced internal pore diffusion.

B. reduced external film diffusion.

C. decrease in energy of activation for the enzyme.

D. enhanced enzyme activity due to mechanical shear.

123. RNA virus generally replicates in:

A. Cytoplasm B. Nucleus

C. Mitochondria D. Late endosomes

124. RNA Viruses are entirely dependent on the host for:

A. replication machinery

B. transcriptional machinery

C. RNA capping machinery

D. protein translation machinery

125. Given below are two statements: one is labelled as Assertion (A) and the other is labelled as Reason (R).

Assertion (A): Polyclonal antibodies bind to the same antigen but recognize different epitopes.

Reason (R): Polyclonal antibodies are produced by different B cell clones in the body.

In the light of the above statements, choose the *most appropriate* answer from the options given below.

A. Both (A) and (R) are correct and (R) is the correct explanation of (A)

B. Both (A) and (R) are correct but (R) is not the correct explanation of (A)

C. (A) is correct but (R) is not correct

D. (A) is not correct but (R) is correct

126. Given below are two statements:

Statement I: After immobilization, plant cells are subjected to stress to induce secondary metabolite production.

Statement II: Cells derived from different explants in culture always produce the same set of secondary metabolites.

Choose the **correct** answer:

A. Both Statement I and Statement II are true.

B. Both Statement I and Statement II are false.

C. Statement I is true but Statement II is false.

D. Statement I is false but Statement II is true.

127. In Michaelis-Menten kinetics what value of [S], as a fraction of K_m, is required to obtain a velocity equal to 80% V_{max}?

A. 2 K_m B. 4 K_m

C. 0.8 K_m D. K_m

128. Match items in List-I with items in List-II:

List-I	List-II
(*a*) *bar*	(*i*) virus-free plants
(*b*) meristem culture	(*ii*) herbicide resistance
(*c*) Chalcone Synthase	(*iii*) dwarf wheat
(*d*) *gar*	(*iv*) flower colour

Choose the **correct** answer from the options given below.

	(*a*)	(*b*)	(*c*)	(*d*)
A.	(*iii*)	(*i*)	(*iv*)	(*ii*)
B.	(*iv*)	(*iii*)	(*ii*)	(*i*)
C.	(*iii*)	(*iv*)	(*i*)	(*ii*)
D.	(*ii*)	(*i*)	(*iv*)	(*iii*)

129. What is the advantage of using the *neo* gene to disrupt the function of a gene in knockout mice?

A. The *neo* gene produces an antibiotic that kills unwanted cells.

B. The *neo* gene is the right size for disabling other genes.

C. The *neo* gene provides a selectable marker.

D. The *neo* gene produces a repressor that inhibits transcription of the target gene.

130. Which enzyme catalyzes strand separation at the replication fork?

A. Replication protein A

B. DNA helicase

C. DNA topoisomerase

D. DNA polymerase

131. Given below are two statements:

Statement I: BLAST compares protein as well as nucleotide sequences in order to identify areas of similarity.

Statement II: Genbank is a repository of data on the phenotypic results of gene knockouts in humans.

In light of the above statements, choose the *most appropriate* answer from the options given below:

A. Both Statement I and Statement II are correct

B. Both Statement I and Statement II are incorrect

C. Statement I is correct but Statement II is incorrect

D. Statement I is incorrect but Statement II is correct

132. Match items in List-I with items in List-II:

List-I	List-II
(*a*) *Rickettsia prowazekü*	(*i*) Relapsing fever
(*b*) *Variola major*	(*ii*) Epidemic typhus
(*c*) *Varicella zoster* Virus	(*iii*) Small pox
(*d*) *Borrelia recurrentis*	(*iv*) Shingles

Choose the **correct** answer from the options given below.

	(*a*)	(*b*)	(*c*)	(*d*)
A.	(*i*)	(*ii*)	(*iii*)	(*iv*)
B.	(*iv*)	(*i*)	(*ii*)	(*iii*)
C.	(*iii*)	(*iv*)	(*i*)	(*ii*)
D.	(*ii*)	(*iii*)	(*iv*)	(*i*)

133. Given below are two statements about hnRNA processing:

Statement I: Capping occurs by addition of methyl adenosine triphosphate to the 5′ end.

Statement II: During tailing, about 200-300 adenylate residues are added to its 3′ end.

Choose the **correct** answer:

A. Both Statement I and Statement II are true.

B. Both Statement I and Statement II are false.

C. Statement I is true but Statement II is false.

D. Statement I is false but Statement II is true.

134. Which of the following amino acids are both glucogenic and ketogenic?

(*a*) Isoleucine (*b*) Serine
(*c*) Tryptophan (*d*) Proline
(*e*) Phenylalanine

Choose the **correct** answer from the options given below.

A. (*a*), (*b*) and (*c*) only
B. (*a*), (*c*) and (*e*) only
C. (*c*) and (*e*) only
D. (*b*) and (*d*) only

135. Given below are two statements:

Statement I: Fatty acid synthase is an example of a tetramer of a multienzyme polypeptide.

Statement II: Each polypeptide harbors seven independent enzymatic functions.

In light of the above statements, choose the correct answer from the options given below.

A. Both Statement I and Statement II are true
B. Both Statement I and Statement II are false
C. Statement I is true but Statement II is false
D. Statement I is false but Statement II is true

136. Based on this phylogenetic tree, which statement is NOT correct?

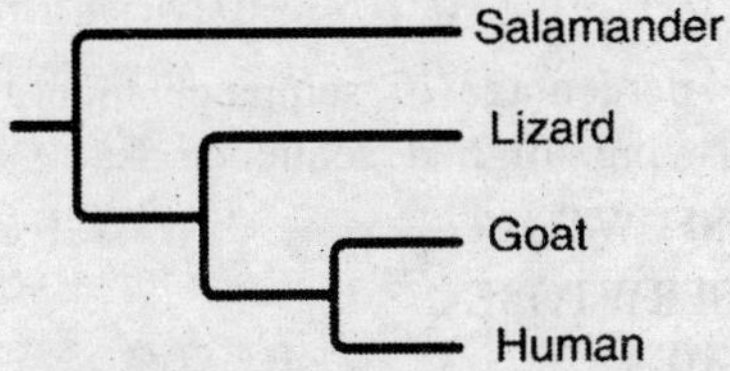

A. The salamander lineage is a basal taxon.
B. Salamanders are a sister group to the group containing lizards, goats, and humans.
C. Salamanders are as closely related to goats as to humans.
D. Lizards are more closely related to salamanders than to humans.

137. Immobilization of enzymes:

(*a*) increases the specificity of the enzyme in batch reaction
(*b*) leads to an increase in the apparent K_m.
(*c*) makes it unsuitable for its use in a continuous reactor system
(*d*) decrease the operational cost of the industrial process

Choose the **correct** answer from the options given below:

A. (*b*) and (*d*) only B. (*a*) and (*d*) only
C. (*a*) and (*c*) only D. (*c*) and (*d*) only

138. Which of the following represents the correct sequence of MAP kinase signal transduction pathway?

(*a*) Ras
(*b*) Raf
(*c*) ERK
(*d*) Mek

Choose the **correct** answer from the options given below.

A. (*a*)→(*b*)→(*c*)→(*d*)
B. (*a*)→(*b*)→(*d*)→(*c*)
C. (*a*)→(*c*)→(*b*)→(*d*)
D. (*b*)→(*a*)→(*c*)→(*d*)

139. Which is the correct order of progressively higher levels of chromatin organization?

(*a*) Nucleosome
(*b*) 30 nm chromatin fibre
(*c*) Looped domain

Choose the **correct** answer:

A. (*a*)→(*b*)→(*c*) B. (*c*)→(*b*)→(*a*)
C. (*c*)→(*a*)→(*b*) D. (*a*)→(*c*)→(*b*)

140. Which one the following is a protein prediction server based on artificial intelligence?

A. M-fold, B. Alpha-fold
C. PDB D. Swiss-Model

141. Which of the following is an inactivated vaccine?

A. Covishield
B. Sputnik V
C. Covaxin
D. Moderna mRNA-1273

142. The following statement is NOT true for the receptor binding site of a virus.

A. Binds specifically to a cellular receptor.
B. Cannot bind to neutralizing antibodies.
C. Interacts with the cellular receptor leading to activation of endocytosis.
D. Is composed of unique structural domains.

143. A peptide containing only aromatic amino acids (tryptophan, phenylalanine, and tyrosine) has a molecular weight of 11,000. The likely number of amino acid residues in the peptide is:

A. 60 B. 100
C. 125 D. 150

144. Knockout mice are usually created by:

A. chemically mutagenizing a mouse and selecting for mutant offspring.
B. creating a chimera by fusing cells from two different cell lines.
C. infecting the mouse with a retrovirus.
D. transfecting embryonic stem cells with an inactivated gene.

145. Which protein is required to seal a nick during Okazaki fragment maturation?

A. DNA Ligase
B. DNA helicase
C. DNA topoisomerase
D. DNA polymerase

146. The growth pattern that you will obtain for a media containing two different carbon sources one of which is preferentially utilized will be a:

A. diauxic growth.
B. exponential growth.
C. linear growth.
D. lag phase followed by log phase growth.

147. Match items in List-I with items in List-II:

List-I	List-II
(*a*) Inactivated Vaccines	(*i*) Based on the poisonous protein made by the pathogen.
(*b*) Live attenuated Vaccines	(*ii*) Based on the killed or altered disease-causing germ.
(*c*) Conjugate Vaccines	(*iii*) Based on the weakened form of the pathogen.
(*d*) Toxoid Vaccines	(*iv*) Based on the combination of the weak antigen coat of Pathogen and strong carrier proteins.

Choose the **correct** answer from the options given below:

	(*a*)	(*b*)	(*c*)	(*d*)
A.	(*ii*)	(*i*)	(*iii*)	(*iv*)
B.	(*ii*)	(*iii*)	(*i*)	(*iv*)
C.	(*iii*)	(*i*)	(*iv*)	(*ii*)
D.	(*ii*)	(*iii*)	(*iv*)	(*i*)

148. Match items List-I with items in List-II:

List-I	List-II
(*a*) Northern Hybridization	(*i*) Transcriptome analysis
(*b*) Next generation sequencing	(*ii*) Detection of proteins in specific cell types
(*c*) Sanger's DNA sequencing	(*iii*) Analysis of splice variants
(*d*) In-situ immunolocalization	(*iv*) Capillary sequencers

Choose the **correct** answer from the options given below:

	(*a*)	(*b*)	(*c*)	(*d*)
A.	(*ii*)	(*iii*)	(*iv*)	(*i*)
B.	(*iv*)	(*i*)	(*ii*)	(*iii*)
C.	(*iii*)	(*i*)	(*iv*)	(*ii*)
D.	(*iv*)	(*iii*)	(*i*)	(*ii*)

149. The percentage of sequence identity for the following aligned sequence is:

VKSFLWTQAL

VPSFRWTQSL

A. 10% B. 80%
C. 30% D. 70%

150. Which of the following mechanisms of antibiotic action is/are NOT correct?

(*a*) Tetracycline: Inhibits aminoacyl-tRNA binding to A-site.
(*b*) Rifampicin: Blocks correct positioning of A-site aminoacyl-tRNA for peptidyl transfer reaction

(*c*) Chloramphenicol: Inhibits EF-Tu function
(*d*) Ciproflaxicin: Inhibits Topoisomerases
(*e*) Puromycin: Causes premature chain termination during translation; mimics 3′ end of aminoacyl-tRNA in A-site and acts as acceptor.

A. (*b*) only
B. (*a*), (*c*) and (*d*) only
C. (*b*) and (*c*) only
D. (*d*) and (*e*) only

151. Given below are two statements:

Statement I: The fragments of DNA generated after cleavage with a restriction enzyme have a phosphate group on their 3′ ends and a –OH group on their 5′ ends.

Statement II: Cohesive ends generated by restriction digestion reanneal with complementary DNA fragment.

Choose the **correct** answer:

A. Both Statement I and Statement II are true.
B. Both Statement I and Statement II are false.
C. Statement I is true but Statement II is false.
D. Statement I is false but Statement II is true.

152. Assume that height in a plant is controlled by three genes (A, B and C) and each additive allele contributes equally. The base height of plant is 20 cm and the maximum possible height is 38 cm. The height of a plant with genotype AAbbCc will be:

A. 32 cm B. 29 cm
C. 19 cm D. 25.4 cm

153. Major limitation of using NMR to determine protein structure is:

A. resolution is poor.
B. resolved structure is not reliable.
C. difficult to interpret the data for larger proteins (>50 KDa).
D. difficult to interpret the data for smaller proteins (<15 KDa).

154. For a batch microbial process the specific rate of substrate utilization is 0.25 h^{-1} and specific product formation rate is 0.20 h^{-1}. The yield of product based on substrate consumed will be:

A. 0.80 B. 1.25
C. 0.05 D. 0.45

155. A plant species nearing its extinction due to virus infection has been given for tissue culture and micropropagation. What should scientists choose from the following four explants for culturing?

A. Shoot apical meristem
B. Internode
C. Leaf disc
D. Root tip

156. Identify the statement that is NOT applicable to enzyme catalyzed reaction.

A. The reaction proceeds with the conversion of the substrate to a higher energy transition state.
B. Enzymes alter the equilibrium constant of the reaction.
C. Enzymes decrease the energy of activation required for the reaction.
D. Some enzymes involve multiple steps of electron transfer.

157. Which of the following describes cyclin-dependent kinase (Cdk)?

(*a*) Cdk is inactive, or "turned off", in the presence of cyclin
(*b*) Cdk is present throughout the cell cycle
(*c*) Cdk is an enzyme that attaches phosphate groups to other proteins

Choose the **correct** answer from the options given below.

A. (*a*) and (*b*) only
B. (*b*) only
C. (*b*) and (*c*) only
D. (*c*) only

158. Penicillin consists of a thiazolidine ring fused to a β-lactam ring to which a variable R group is attached. This antibiotic inhibits cell wall synthesis by inhibiting:

A. β-lactamase
B. Phosphodiesterase
C. Transesterase
D. Glycopeptide transpeptidase

159. In a random mating population, the frequency of allele A1 and A2 in the gene pool is 0.8 and 0.2, respectively. In a population of 200 plants, how many will have the genotype A1A2?
A. 8
B. 128
C. 64
D. 32

160. A mutant bacterial cell has a defective aminoacyl synthetase that attaches a lysine to tRNAs with the anticodon AAA instead of the normal phenylalanine. During protein synthesis:
A. none of the proteins in the cell will contain phenylalanine.
B. lysine will partially replace phenylalanine at certain positions.
C. the cell will compensate for the defect by attaching phenylalanine to tRNAs with lysinespecifying anticodons.
D. lysine will replace phenylalanine at all amino acids positions.

ANSWERS

1. **(A)** **2.** **(C)** **3.** **(*)**
4. **(D)** **5.** **(B)** **6.** **(A)**
7. **(A)** **8.** **(B)**

9. **(C):** A sample space associated with the random experiment of tossing four coins is given by

S = {HHHH, HHHT, HHTH, THHH, HTHH, HHTT, TTHH, HTHT, HTTH, THTH, THHT, TTHT, HTTT, THTT, TTTH, TTTT}

∴ Required probability of getting at least 3 heads = $\frac{5}{16}$.

10. **(B):** Digits are 1, 4, 7, 8, 9 i.e., five in all. We have to form numbers between 100 and 999. Clearly they will be of three digits and their number will be

$$^5P_3 = \frac{5!}{3!} = 5 \times 4 \times 3 = 60$$

Hence, required number = 60

11. **(B)**

12. **(B):** 1, 2, 4, 8, 16, ... are in G.P.
Here, $a = 1$ and $r = 2$

$$S_n = \frac{a(r^n - 1)}{r - 1}$$

$$\therefore \quad S_{10} = \frac{1(2^{10} - 1)}{2 - 1} = \frac{2^{10} - 1}{1}$$

$$= \frac{1024 - 1}{1} = 1023.$$

13. **(B):** We have,

$$n(A \cup B) = n(A) + n(B) - n(A \cap B)$$
$$\Rightarrow \quad 35 = 24 + 16 - n(A \cap B)$$
$$\Rightarrow \quad 35 = 40 - n(A \cap B)$$
$$\Rightarrow \quad n(A \cap B) = 40 - 35 = 5$$

Hence, 5 students like to play both cricket and football.

14. **(A)**

15. **(C):** A = {2, 4, 6}
B = {2, 3, 5}
$n(A) = 3$ and $n(B) = 3$
$n(A \times B) = 3 \times 3 = 9 = 2$
Hence, there are 2 relations from A to B.

16. **(B):**

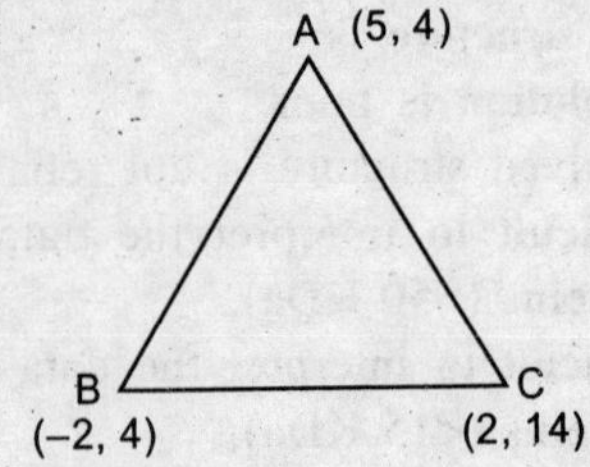

Area of triangle = $\frac{1}{2}[x_1(y_2 - y_3) + x_2(y_3 - y_1)$

$+ x_3(y_1 - y_2)]$

$= \frac{1}{2}[5(4 - 14) + (-2)(14 - 4)$

$+ 2\,(4 - 4)]$

$= \frac{1}{2}[-50 - 20 + 0]$

$= \frac{-70}{2}$

$= -35$

= 35 square units

17. (C) **18. (C)** **19. (A)**

20. (B): Number of boys = $\frac{30}{100} \times 60 = 18$

$\therefore$ Number of girls = 60 – 18 = 42

Number of absent = $\frac{10}{100} \times 60 = 6$

$\therefore$ Total number of students = 60 – 6 = 54

$\therefore$ % of girls = $\frac{42}{54} \times 100$

$= \frac{700}{9}$

= 77.77%.

21. (*) **22. (B)** **23. (D)**

24. (A): (A + B + C)'s 1 min work = $\frac{1}{10}$

(A + B)'s 1 min work = $\frac{1}{20} + \frac{1}{40}$

$= \frac{2+1}{40} = \frac{3}{40}$

C's 1 min work = $\frac{1}{10} - \frac{3}{40}$

$= \frac{4-3}{40} = \frac{1}{40}$

Hence, C alone can fill the tank = 40 minutes.

25. (*) **26. (C)** **27. (A)**

28. (A) **29. (B)** **30. (D)**

31. (A)

32. (D): Let x_1 and x_2 m/s be the speed of two trains and their lengths are l_1 and l_2 m, then

$l_1 = x_1 \times 20 = 20x_1$

$l_2 = x_2 \times 10 = 10x_2$

And also, $\frac{l_1 + l_2}{x_1 + x_2} = 15$

$\Rightarrow \quad \frac{20x_1 + 10x_2}{x_1 + x_2} = 15$

$\Rightarrow \quad 20x_1 + 10x_2 = 15x_1 + 15x_2$

$\Rightarrow \quad 5x_1 = 5x_2$

$\Rightarrow \quad \frac{x_1}{x_2} = \frac{5}{5} = \frac{1}{1}$

Hence, ratio of their speeds = 1 : 1.

33. (C)

34. (B): Sum of 9 + 18 + 27 + 36 + 45

= 9(1 + 2 + 3 + 4 + 5)

$= 9 \times \frac{5 \times 6}{2}$

= 135

Average = $\frac{135}{5} = 27.$

35. (D) **36. (A)** **37. (C)**

38. (C): Price of sugar increased = 25%

To retain the previous expenditure level, consumption of sugar should be reduced by

$= \left(\frac{100 \times 25}{100 + 25}\right)\%$

$= \frac{100 \times 25}{125}\% = 20\%$

39. (C) 40. (D) 41. (D)
42. (C) 43. (A) 44. (C)
45. (A) 46. (B) 47. (D)
48. (B) 49. (D) 50. (B)
51. (B) 52. (A) 53. (B)

54. (A): $2x + 3y + 4 = 0$

$\because$ Touches the x-axis

$\therefore \quad y = 0$

$2x + 0 + 4 = 0$

$\Rightarrow \quad 2x = -4$

$\Rightarrow \quad x = -2.$

55. (C) 56. (D) 57. (B)
58. (A) 59. (C) 60. (A)
61. (B) 62. (C) 63. (B)
64. (B) 65. (B) 66. (D)
67. (C) 68. (D) 69. (C)
70. (B) 71. (B) 72. (C)
73. (D) 74. (A) 75. (A)
76. (A) 77. (C) 78. (D)
79. (C) 80. (C) 81. (B)
82. (C) 83. (B) 84. (C)
85. (C) 86. (A) 87. (C)
88. (D) 89. (D) 90. (C)
91. (A) 92. (A) 93. (B)

94. (D) 95. (C) 96. (C)
97. (D) 98. (A) 99. (C)
100. (D) 101. (B) 102. (D)
103. (C) 104. (D) 105. (A)
106. (B) 107. (A) 108. (B)
109. (C) 110. (B) 111. (A)
112. (C) 113. (A) 114. (C)
115. (C) 116. (D) 117. (B)
118. (D) 119. (C) 120. (C)
121. (B) 122. (B) 123. (A)
124. (D) 125. (A) 126. (C)
127. (B) 128. (D) 129. (C)
130. (B) 131. (C) 132. (D)
133. (D) 134. (B) 135. (D)
136. (D) 137. (A) 138. (B)
139. (A) 140. (B) 141. (C)
142. (B) 143. (A) 144. (D)
145. (A) 146. (A) 147. (D)
148. (C) 149. (D) 150. (C)
151. (D) 152. (B) 153. (C)
154. (A) 155. (A) 156. (B)
157. (C) 158. (D) 159. (C)
160. (B)

Previous Paper (Solved)

M.Sc.—BIOTECHNOLOGY & BIOCHEMISTRY

Entrance Exam, 2021*

1. Select the correct option:

You may feel relieved when something bad ______.

A. happens B. does not happen
C. is remembered D. occurs

2. Select the most suitable synonym:

BARBARIAN

A. impolite B. unkind
C. uncivilized D. unlikeness

3. Which is the closest meaning of the phrase *to be hard of hearing*?

A. deaf B. deaf and dumb
C. careless D. arrogant

4. Identify the meaning of the idiom from the options given:

in black and white

A. useless B. in short
C. in writing D. in full swing

5. Select the most suitable antonym:

RAPTURE

A. despair B. elation
C. ecstasy D. hell

6. Complete the second sentence by selecting the appropriate option:

One way to improve your health is to eat more fruit and vegetables. Alternatively ______.

A. you could fall ill
B. you could go on diet
C. you could exercise more
D. you could increase risk of your heart disease

7. Fill in the blank with right word:

An employee might be given a privilege as a form of ______.

A. punishment B. reward
C. increment D. commendation

8. Select the correct option for use of 'the':

A. the Delhi
B. the city of Delhi
C. the southern India
D. the long island

9. Fill in the blank with correct preposition:

I am worried ______ exam.

A. on B. in
C. about D. of

10. Identify the meaning of the idiom from the options given:

Hit the road

A. end the journey
B. an accident
C. begin the journey
D. happy times

11. What will come in place of the question mark (?) in the expression?

$4003 \times 77 - 21015 = ? \times 116$

A. 2467 B. 2476
C. 2477 D. 2478

12. Given below are two statements, one is labelled as Assertion (A) and the other is labelled as Reason (R).

Assertion (A): For making statues, we are using bronze.

Reason (R) : Through the mixing of copper and tin, we produce bronze.

* Exam was conducted by NTA for CUET M.Sc Biotechnology/ Biochemisty. (Exam held on 24/09/2021)

In the light of the above statements, choose the ***most appropriate answer*** from the options given below.

A. Both (A) and (R) are correct and (R) is the correct explanation of (A) Reason (R)
B. Both (A) and (R) are correct, but (R) is NOT the correct explanation of (A)
C. (A) is correct, but (R) is not correct
D. (A) is not correct, but (R) is correct

13. The sum of five numbers is 260. The average of the first two numbers is 30 and the average of the last two numbers is 70. What is the third number?

A. 40 B. 50
C. 55 D. 60

14. Thirty-six workers can finish a piece of work in 14 days. If the work is to be completed in 8 days, how many extra workers are required?

A. 20 B. 23
C. 27 D. 29

15. What should come in the blank space in the following number series?

0, 2, ______, 18, 32, 50

A. 5 B. 6
C. 7 D. 8

16. Three of the following four are alike in certain way based on the English alphabetical series and so form a group. Which is the one that does not belong to that group?

A. WXST B. HCDE
C. QRMN D. IJEF

17. If every alternate Saturdays and all Sundays are holidays in a 30 days month beginning on Saturday, then how many working days are there in that month?

A. 21 B. 22
C. 23 D. 24

18. Pointing to a woman in the photograph, Neetu said, "Her son's father is the son-in-law of my mother". How is Neetu related to the woman?

A. Mother B. Sister
C. Aunt D. Mother-in-law

19. Suneeta travelled 5 km towards east direction. She turned right and went for 3 km. Then she turned west and travelled for 1 km. How far is she from the starting point?

A. 3 km B. 5 km
C. 9 km D. 10 km

20. In a certain code, TEMPORAL is written as OLDSMBSP. How is CONSIDER written in that code?

A. RMNBSFEJ B. BNMRSFEJ
C. RMNBJEFS D. TOPDQDCH

21. Whose creations are *Harshacharita* and *Kadambari*?

A. Kalhana B. Panini
C. Banabhatta D. Patanjali

22. Present-day Istanbul in Turkey was earlier known as:

A. Ankara B. Ottoman
C. Constantinople D. Byzantine

23. Space X is:

A. an Indian autogiant
B. a movie on space
C. name of a spy agency
D. an American private space company

24. UN day is celebrated on:

A. 26th November B. 24th November
C. 26th October D. 24th October

25. The famous novel, *Middlemarch* was written by:

A. George Eliot
B. William Wordsworth
C. William Shakespeare
D. Ruskin Bond

26. Which of the following Electronic transitions is not possible in UV-visible region?

A. n-π^* B. σ-σ^*
C. π-π^* D. None of these

27. Resolving power of a microscope mainly depends on:

A. Numerical aperture
B. Refractive index

C. Objective lens
D. Ocular lens

28. Which one of the following is not true according to Kasha's rule about fluorescence?
A. Bathochromic shift is observed in emission relative to absorption
B. Hypsochromic shift is observed in absorption relative to emission
C. Bathochromic shift is observed in absorption relative to emission
D. None of these

29. In reverse-phase liquid chromatography, stationary phase is:
A. Polar
B. Non-polar
C. Hydrophilic
D. Ionic

30. Diethylaminomethyl (DEAE) cellulose is an anion exchanger. It will prefer to bind with proteins having:
A. No charge
B. Positive charge
C. Negative charge
D. Binding does not related to charge

31. Increase in the half-cell potential (E°):
A. Decreases the oxidizing ability
B. Increases the reducing ability
C. Increases the oxidizing ability
D. Increases the reaction spontaneity

32. Lowest yield of ATP is achieved in:
A. Fermentation
B. Anaerobic respiration
C. Aerobic respiration
D. Both 1 and 2

33. Which of the following is a molecular motor?
A. Kinesin
B. Kinetin
C. Keratin
D. Dynactin

34. The key driving force for membrane transport is:
A. Membrane thickness
B. Distance
C. Molecular size
D. Concentration gradient

35. Sedimentation rate of a particle will be zero when:
A. Density of the particle is less than surrounding medium
B. Density of the particle is more than the surrounding medium
C. Density of the particle and the surrounding medium are equal
D. Density of the particle is twice as the surrounding medium

36. 70s ribosome is made up of:
A. 50s and 30s subunits
B. 50s and 40s subunits
C. 60s and 30s subunits
D. 60s and 40s subunits

37. For each ATP molecule breakdown, the Sodium-potassium pump causes movement of:
A. More Na^+ out and less K^+ in
B. K^+ out and Na^+ in on a one-for-one basis
C. Na^+ out and K^+ in on a one-for-one basis
D. K^+ and Na^+ in same direction

38. The number of a Nuclear pore in a cell depends on the:
A. Size of the cell
B. Transcriptional activity of the cell
C. DNA content of the cell
D. None of these

39. Which of the following is not a part of endomembrane system?
A. Golgi body
B. Lysosome
C. Ribosome
D. Vesicles

40. Michaelis-Menten constant (Km) is related to the:
A. Enzyme's affinity for the substrate
B. Enzyme's activity for the substrate
C. Enzyme's specificity for the substrate
D. Enzyme's selectivity for the substrate

41. The organelle absent in plant cells is:
A. Vacuoles
B. Lysosome
C. Chloroplast
D. Centriole

42. Which of the following is not associated with the extracellular matrix of the cells?
A. Proteoglycans B. Cellulose
C. Collagen D. Fibronectin

43. The enzyme responsible for removal of DNA supercoils is:
A. Helicase B. Topoisomerase
C. DNA ligase D. DNA polymerase

44. Which of the following is not a secondary messenger?
A. Calcium B. Diacylglycerol
C. cAMP D. AMP

45. Intramolecular rearrangements are catalyzed by:
A. Transferases B. Oxidoreductases
C. Isomerases D. Lyases

46. An enzyme without its required cofactor prosthetic group is referred as:
A. Coenzyme B. Holoenzyme
C. Apoenzyme D. Catalyst

47. Gram staining is based on the presence of ________ in the cell wall.
A. Phospholipid B. Peptidoglycan
C. Lipid A D. Glycolipid

48. Lipopolysaccharide present in Gram negative bacterial cell wall is considered as an endotoxin because of:
A. Lipid A B. Core polysaccharide
C. O side chain D. Integral proteins

49. Antibacterial action of penicillin could be attributed to:
A. Inhibition of the enzyme transpeptidase
B. Inhibition of the cross linking of peptidoglycan side chain
C. Hydrolysis of peptidoglycan
D. Both A and B

50. Severe acute respiratory syndrome Coronavirus 2 (SARS-CoV-2) is caused by a:
A. Single-stranded RNA virus
B. Double-stranded RNA virus
C. Single-stranded DNA virus
D. Double-stranded DNA virus

51. For exergonic reactions:
A. $\Delta G^{\circ\prime}$ (standard free energy change) is negative and K_{eq} (equilibrium constant) is more than one
B. $\Delta G^{\circ\prime}$ is positive and K_{eq} is more than one
C. $\Delta G^{\circ\prime}$ is negative and K_{eq} is less than one
D. $\Delta G^{\circ\prime}$ is positive and K_{eq} is less than one

52. 2-keto-3-deoxy-6-phosphogluconate is an intermediate of:
A. Pentose phosphate pathway
B. Entner-Doudoroff Pathway
C. Embden-Meyerhof Pathway
D. Tricarboxylic acid cycle

53. Product of dissimilatory nitrate reduction is:
A. ammonia B. nitrogen
C. nitrate D. nitric oxide

54. Which of the following is not a type of microtubule associated with mitotic spindle?
A. Polar B. Astral
C. Central D. Kinetochore

55. Which one of the following is the correct sequence of prophase-I stage?
A. Leptotene, Pachytene, Zygotene, Diplotene, Diakinesis
B. Leptotene, Zygotene, Pachytene, Diakinesis, Diplotene
C. Zygotene, Leptotene, Pachytene, Diplotene, Diakinesis
D. Leptotene, Zygotene, Pachytene, Diplotene, Diakinesis

56. Histones are rich in:
A. Lysine and Arginine
B. Lysine and Asparagine
C. Glycine and Lysine
D. Leucine and Arginine

57. Which one of the following is not true for euchromatin?
A. They are transcriptionally active
B. They are less condensed regions of chromosomes
C. Barr body is an example of euchromatin
D. All of these

58. Proteins responsible for structural maintenance of chromosomes are:
A. Condensin B. Cohesin
C. Collagen D. Both A and B

59. Activity of protein kinase A is regulated by:
A. Adenylyl cyclase
B. Phosphodiesterase
C. Phospholipase C
D. Phosphorylase kinase

60. Pectin is a polymer of:
A. Glucose
B. Glucuronic acid
C. Galacturonic acid
D. Iduronic acid

61. Triple-helical coiled coil structure is characteristic of:
A. Cellulose B. Collagen
C. Elastin D. Chymotrypsin

62. Which one of the following is not true for sliding filament model?
A. Shortening of I-band and H-zone with no effect on A-band
B. Shortening of I-band A-band and H-zone
C. No change in I-band and H-zone with shortening of A-band
D. Shortening of I-band and A-band with no effect on H-zone

63. Which of the following proteins is associated with amphibian development?
A. Collagen B. Catenin
C. Fibrin D. Keratin

64. Which of the following is not a pattern of cell movement during gastrulation?
A. Ingression B. Epiboly
C. Invasion D. Invagination

65. Amphibian eggs undergo:
A. Equal radial holoblastic cleavage
B. Unequal radial holoblastic cleavage
C. Equal radial meroblastic cleavage
D. Unequal radial meroblastic cleavage

66. Which of the following is not true for master regulatory transcription factors?
A. Must be expressed when the specification of a cell type begins
B. Must regulate the expression of genes specific to that cell type
C. Must be able to redirect a cell's fate to this cell type
D. Must be expressed when the specification of a cell type ends

67. Difference between chromatin regulation by high and low CpG promoters is:
A. High CpG content promoters repressed gene by DNA methylation
B. Low CpG content promoters repressed gene by histone methylation
C. High CpG content promoters repressed gene by histone methylation
D. Low CpG content promoters repressed gene by histone acetylation

68. NRSE stands for:
A. Notochord restrictive silencer element
B. Neural ectodermal specification enhancer
C. Neural restrictive specification enhancer
D. Neuron-restrictive silencer elements

69. Which substance determines the type of cleavage during development?
A. Nucleus B. Cytoplasm
C. Yolk D. Both A and B

70. Which of the following are not true for Karl Ernst von Baer's principles?
(*a*) General features of a large group of animals appear earlier in development as compared to the specialized features of a smaller group
(*b*) General features of a large group of animals appear later in development as compared to the specialized features of a smaller group
(*c*) Early embryo of a higher animal is look like a lower animal
(*d*) Early embryo of a higher animal is look like embryo of a lower animal
A. (*a*) and (*c*) B. (*b*) and (*c*)
C. (*b*) and (*d*) D. (*a*) and (*d*)

71. Enzymes involved in RNA interference are:
A. Dicer B. Cohesin
C. Bicoid D. Maskin

72. CRISPR is:
A. Clustered Regularly Interspaced Short Palindromic Repeats
B. Coordinated Regularly Interspaced Short Palindromic Repeats
C. Clustered Restrictive Interspaced Short Palindromic Regions
D. Coordinated Restrictive Interspaced Short Palindromic Repeats

73. Ramachandran plot is:
A. Used for determination of stable protein structures
B. Also called as $\phi-\psi$ plot
C. Provide information about sterically favorable conformations
D. All of these

74. Dextrans are:
A. Polymeric glucose
B. Polymeric galactose
C. Polymeric fructose
D. Polymeric lactose

75. According to Bohr effect the oxygen transport ability of hemoglobin is more when:
A. pH decreases
B. pH increases
C. pH remains stable
D. It is not affected by pH

76. Cells are small in size because:
A. It provides more surface area to volume ratio
B. It provides less surface area to volume ratio
C. Provides better regulation of cytoplasm by nucleus
D. Both A and B

77. Protein purification in 2-dimentional gel electrophoresis is based on:
A. Molecular weight
B. Charge
C. Number of subunits
D. Both A and B

78. _______ enzyme catalyzes conversion of fructose to fructose-6-phosphate requires _______ as cofactor.
A. Phosphofructokinase-I, Mg^{+2}
B. Phosphofructokinase-II, Mg^{+2}
C. Phosphofructokinase-I, Mn^{+2}
D. Phosphofructokinase-II, Mn^{+2}

79. Which of the following is not a possible catabolic fate of pyruvate?
A. Lactic acid fermentation
B. Citric acid cycle
C. Ethanol fermentation
D. Glycolysis

80. Allosterically regulated enzymes are
A. Oligomeric proteins
B. Possess regulatory and catalytic domains
C. Shows sigmoidal behaviour
D. Both B and C

81. Which of the following regulation occurs in slowest time frame?
A. Feedback regulation
B. Allosteric regulation
C. Regulation through covalent modification
D. Enzyme synthesis through gene induction

82. The unit of maximum specific growth rate (μ_{max}) is:
A. $Hours^{-1}$
B. Hours
C. Hours/litre
D. Hours/cell

83. Mammals do not have ability to convert fatty acids to glucose because:
A. They cannot undergo glyoxylate cycle
B. There is no need for such conversion
C. They can obtain glucose from other pathways
D. Both A and C

84. Formation of glycogen cannot be initiated in the absence of enzyme:
A. Glucokinase
B. Glycogenin
C. Glycogen phosphorylase
D. Glycogen kinase

85. Glucose sensors utilizes ______ as bioreceptor.
A. Glucose oxidase
B. Glucose kinase
C. Glucose dehydrogenase
D. Both A and C

86. Optical density is based on:
A. Light refraction
B. Light adsorption
C. Light diffraction
D. Light scattering

87. Purity of DNA samples can be checked by measuring the ratio of absorbances at 260 and 280 nm. For a pure DNA sample the ratio should be:
A. 1.8 B. < 1.8
C. < 1.0 D. > 2

88. ATP molecules produced by oxidation of one molecule of palmitoyl-CoA to CO_2 and H_2O are:
A. 38 B. 108
C. 148 D. 40

89. Carbamoyl phosphate synthetase I (first enzyme in urea cycle) is allosterically regulated by:
A. N-acetylglutamate
B. N-acetylglucosamine
C. N-acetylgalactosamine
D. N-acetylglucomate

90. Phenylketonuria is caused by dysfunction of:
A. Phenylalanine hydroxylase
B. Phenylalanine monooxygenase
C. Phenylalanine mutase
D. Phenylalanine dioxygenase

91. Key protein involved in complement activation is:
A. C5 B. C8
C. C9 D. C3

92. Separation of components in gas chromatography is based on:
A. Molecular weight
B. Partition coefficient
C. Molarity
D. Conductivity

93. Pervaporation is:
A. Ion removal
B. Separation of liquids
C. Separation of solids
D. Water purification

94. Obligate anaerobes are not able to survive in the presence of oxygen because of:
A. Absence of both catalase and superoxide dismutase
B. Presence of both catalase and superoxide dismutase
C. Presence of catalase but absence of superoxide dismutase
D. Presence of low levels of both catalase and superoxide dismutase

95. Which of the following sentences is not true for pyrosequencing?
A. It is based on detection of pulses of chemiluminescence
B. Requires the activity of enzyme sulfurylase
C. Requires addition of dideoxynucleotides
D. None of these

96. Match the List-I with List-II:

List-I	List-II
(*a*) Translation	I. Nucleus
(*b*) Replication	II. Mitochondria
(*c*) ATP synthesis	III. Stroma
(*d*) Dark reaction of photosynthesis	IV. Cytoplasm

	(*a*)	(*b*)	(*c*)	(*d*)
A.	III	I	II	IV
B.	II	I	IV	III
C.	IV	II	III	I
D.	IV	I	II	III

97. Impermeability of mitochondrial inner membrane to most ion, electrons and protons is due to the presence of:
A. Cardiolipin
B. Collagen
C. Chitin
D. Cellulose

98. Lateral root primordia are developed from:
A. Pericycle B. Endodermis
C. Ectodermis D. Xylem

99. *Agrobacterium tumefaciens* is used for transformation in plant tissue culture because:
A. Both *Agrobacterium* and plants have similar genetic make-up
B. *Agrobacterium* genomic DNA can transfer its genetic material to plants
C. *vir* genes of *Agrobacterium* aid in T-DNA transfer to plants
D. All of these

100. Mutation in the class A genes in ABCDE model of flower development leads to which pattern of floral organization?
A. Sepal-sepal-carpel-petal
B. Petal-carpel-sepal-sepal
C. Carpel-stamen-stamen-carpel
D. Stamen-petal-petal-sepal

ANSWERS

1	2	3	4	5	6	7	8	9	10
B	C	A	C	A	C	B	B	C	C
11	**12**	**13**	**14**	**15**	**16**	**17**	**18**	**19**	**20**
B	B	D	C	D	B	B & C	B	B	A
21	**22**	**23**	**24**	**25**	**26**	**27**	**28**	**29**	**30**
C	C	D	D	A	B	A	C	B	C
31	**32**	**33**	**34**	**35**	**36**	**37**	**38**	**39**	**40**
C	A	A	D	C	A	A	B	C	A
41	**42**	**43**	**44**	**45**	**46**	**47**	**48**	**49**	**50**
D	B	B	D	C	C	B	A	D	A
51	**52**	**53**	**54**	**55**	**56**	**57**	**58**	**59**	**60**
A	B	A	C	D	A	C	D	A	C
61	**62**	**63**	**64**	**65**	**66**	**67**	**68**	**69**	**70**
B	A	B	C	B	D	C	D	C	B
71	**72**	**73**	**74**	**75**	**76**	**77**	**78**	**79**	**80**
A	A	D	A	*	A	D	A	D	D
81	**82**	**83**	**84**	**85**	**86**	**87**	**88**	**89**	**90**
C	A	A	B	D	D	A	B	A	A
91	**92**	**93**	**94**	**95**	**96**	**97**	**98**	**99**	**100**
D	B	B	A	C	D	A	A	C	C

Previous Paper (Solved)

M.Sc.—Biochemistry Entrance Exam, 2020*

1. What would be the effect on the net reaction catalyzed by glyceraldehyde 3-phosphate dehydrogenase if phosphate was replaced by arsenate?
A. Rate of reaction will increase
B. Rate will be decreased
C. No effect on reaction rate
D. Uncoupling of phosphorylation

2. Cellular membranes are self sealing in nature—if they are punctured or disrupted mechanically they quickly and automatically reseal. What properties of such sealing are responsible for this feature?
A. hydrophobic effect of membrane lipids
B. hydrophilic effect of membrane lipids
C. charge-charge interaction among lipids
D. protein-lipid interactions

3. What type of chemical reaction is involved in conversion of isocitrate to α-ketoglutarate?
A. Carboxylation
B. Oxidative decarboxylation
C. Reducing decarboxylation
D. Oxido-reduction

4. Individuals can have relatively high levels of pyruvate in their blood due to:
A. Vitamin B deficiency
B. Vitamin D deficiency
C. Thiamine deficiency
D. Alcohol intake

5. Mammalian liver can carry out gluconeogenisis using starting material known as:
A. Oxaloacetate B. Acetyl-CoA
C. Citric acid D. Aspartate

6. Maple syrup urine disease is due to a metabolic defect in the pathway of degradation of:
A. Branched chain fatty acids
B. Cholesterol
C. Nucleotide
D. Branched chain amino acids

7. The specificity or stringency of a PCR reaction can be controlled by altering the reaction:
A. volume
B. target sequence
C. temperature and salt concentration
D. template copies

8. Principle regulation point in the biosynthesis of fatty acids is:
A. Acetyl-CoA carboxylase
B. β-Ketoacyl-ACP synthase
C. Citrate dehydrogenase
D. β-Lactamase

9. Denaturation of a protein or nucleic acid can be studied by:
A. SDS-PAGE
B. Isoelectric focusing
C. Spectrophotometry
D. Gel filtration

10. Folding of a protein is primarily governed by:
A. Ionic strength of solution
B. Presence of branched chain amino acids
C. Primary structure of a protein
D. Presence of hydrophobic amino acids

11. The biochemical products obtained after hydrolysis of glycolipids are:
A. Sugar, fatty acids, phosphoric acid
B. Sugar, fatty acids, nitrogen base
C. Sugar, fatty acid, glycerol
D. Sugar, fatty acid, sphingosine

12. D-glucose and D-mannose are:
A. Anomers B. Epimers
C. Optical isomers D. Diastereomers

* Exam was conducted by NTA for Admission in Delhi University M.Sc Biochemisty. (Exam held on 08/09/2020)

13. Collagen is rich in:
A. Glutamic acid and glycine
B. Glycine and glutamine
C. Glycine and proline
D. Glycine and alanine

14. The following amino acid is least likely to be found in a α-helix structure:
A. Alanine B. Cystine
C. Histidine D. Proline

15. Determine the pH of a 0.01N NaOH solution:
A. 3.0 B. 11.0
C. 10.0 D. 12.0

16. A HCl solution of 1 mM was diluted to 10^6. What would be the pH of the resulting solution?
A. 5.0 B. 7.0
C. 6.0 D. 2.0

17. The ligand used for affinity chromatography of RNA containing poly(A) sequence is:
A. Avidin B. 5' AMP
C. Oligo dT D. Lysine

18. Signal hypothesis for protein trafficking was proposed by:
A. Tom Rapoport B. Paul Nurse
C. Timothy Hunt D. Gunter Blobel

19. In SDS-PAGE, the migration of protein is effected by ....
A. Charge of the protein
B. Size of the protein
C. Both charge and size of the protein
D. Number of cysteine residues in protein

20. Glycerol is added to protein samples before loading them on the PAGE. What is the role of glycerol:
A. Provide stability to protein
B. Helps to bind SDS to the protein
C. Provide density to the protein sample
D. Helps in denaturing the disulphide bonds

21. What is the effect of urea and formamide on DNA:
A. Decrease the T_m of the DNA
B. Increase the T_m of the DNA
C. Helps in reannealing of the DNA
D. Stabilized DNA structure

22. A low auxin : cytokinin ratio leads to:
A. Shoot formation
B. Root formation
C. Fruit formation
D. Increased cell division

23. Megaloblastic anemia is caused due to deficiency of ....
A. Cobalamin B. Pyridoxine
C. Folic acid D. Niacin

24. Lineweaver-Burk plot is also known as ....
A. Hanes-Woolf plot
B. Double reciprocal plot
C. Eadie-Hofstee plot
D. Steady-state equation

25. Which of the following can be used to construct a linkage map of the Hfr chromosome?
A. frequency of recombination
B. time of entry
C. locus of mutation
D. transfer of F factor

26. Which of the following conditions decreases the level of denitrification?
A. Abundance of organic matter
B. Elevated temperatures
C. Availability of oxygen
D. Acidic pH

27. Name the class of immunoglobulin which takes part in hypersensitivity reaction?
A. IgG B. IgE
C. IgA D. IgM

28. In Phase contrast microscopy, the rate at which light passes through objects is ....
A. Inversely proportional to their refractive indices
B. Constant
C. Directly proportional to their refractive indices
D. Exponentially related to their refractive indices

29. The biosynthesis of proline takes place from ....
A. aspartic acid
B. glutamic acid
C. pyruvic acid
D. alpha-keto glutaric acid

30. The density of a solution prepared by dissolving 120 g of urea (mol.mass = 60) in 1000 g of water is 1.15 g/mL. The molarity of this solution is:
A. 1.02 M B. 2.05 M
C. 0.50 M D. 1.78 M

31. The third strand of triple helix is paired in which scheme?
A. Intermolecular base pair scheme
B. Hoogsteen base pair scheme
C. Intramolecular base pair scheme
D. G-quartet scheme

32. The first three bases of the 6-base recognition cleavage site of HindIII are AAG. What is the complete sequence of this 6 bp site?
A. AAGAAG B. AAGCTT
C. AAGGAA D. AAGCUU

33. What leads to the activation of protein kinase C?
A. Release of intracellular Ca^{+2} + diacylglycerol
B. Release of intracellular Mg^{+2} + diacylglycerol
C. Release of intracellular Ca^{+2} + glycerol
D. Release of intracellular Ca^{+2} + triacylglycerol

34. Trans bilayer diffusion is also called:
A. Facilitated diffusion
B. Lateral diffusion
C. Flip flop
D. Simple diffusion

35. The first step in the payoff phase of glycolysis is:
A. Reduction of 1, 3-bisphosphoglycerate to glyceraldehyde 3-phosphate
B. Oxidation of glyceraldehyde 3-phosphate to 1, 3-bisphosphoglycerate
C. Reversible conversion of dihydroxy-acetone phosphate to glyceraldehyde 3-phosphate
D. Irreversible conversion of dihydroxy-acetone phosphate to glyceraldehyde 3-phosphate

36. What region of antibody binds to protein A during affinity purification?
A. Heavy chain within the Fc region
B. Heavy chain within the variable region
C. Light chain within the Fab region
D. Light chain within the Fc region

37. If the oxidative phosphorylation was uncoupled in the mitochondria then there is a/an:
A. Decreased concentration of ADP in the mitochondria
B. Decreased oxidative rate
C. Increased inorganic phosphate in the mitochondria
D. Decreased production of heat

38. The enzyme responsible for the removal of supercoiling in replicating DNA ahead of the replication fork is:
A. Topoisomerase B. Primase
C. DNA polymerase D. Helicase

39. Which of the following are not DNA viruses?
A. Hepatitis B virus
B. Influenza A virus
C. CMV virus
D. Parvovirus

40. Formation of one molecule of glucose from pyruvate requires:
A. 4 ATP, 2 GTP and 2 NADH
B. 3 ATP, 2 GTP and 2 NADH
C. 4 ATP, 1 GTP and 2 NADH
D. 2 ATP, 2 GTP and 2 NADH

41. Cyanogen bromide is used for cleavage of proteins. The target site for cleavage is:
A. C-terminal end of Asparagine residue
B. C-terminal end of Methionine residue
C. C-terminal end of Glycine residue
D. C-terminal end of Proline residue

42. Who won the Noble prize in medicine in 2018 for their discovery of cancer therapy by inhibition of negative immune regulation?
A. James P. Allison, Tasuku Honjo
B. Michael W. Young, Michael Rosbash, Jeffrey C. Hall
C. William G. Kaelin, Gregg L. Semenza, Peter J. Ratcliffe
D. Shinya Yamanaka, John Gurdon

43. The biological role of restriction enzymes in bacteria is to:
A. repair DNA
B. induce DNA crossover
C. cleave foreign DNA
D. recombine DNA

44. Which of the following DNA sequences contains a 4-8 base palindromic site? (**Note:** Only one strand is shown.)
A. CAGTCC B. GCATATGC
C. CGATTAGC D. GAGAGAGA

45. Which of the following components is not a constituent of a typical A-tailing reaction?
A. Klenow exo-
B. ATP
C. Taq DNA polymerase
D. Blunt end DNA

46. Presence of salt during gel filtration helps to:
A. Allow separation of proteins of same molecular weight
B. Allow separation of proteins on basis of pI along with molecular weight
C. Reduce non-specific interaction of proteins with gel matrix
D. Reduce the proteolytic degradation of proteins during purification

47. HAT medium used for hybridoma production contains:
A. Thymidylate synthase
B. Thymidine kinase
C. Thymidine
D. Thiamine

48. Which of the following class of antibodies are expected to be immuno-precipitated predominantly using anti-J chain antibodies?
A. IgG B. IgM
C. IgD D. IgE

49. Which of the following antibodies is most efficient in causing agglutination?
A. IgM B. IgD
C. IgG D. IgE

50. An unknown bacteriophage has a base composition of 23% A, 36% T, 21% G, and 20% C. Its genome is likely to be:
A. Single stranded RNA
B. Single stranded DNA
C. Double stranded RNA
D. Double stranded DNA

51. During the growth of mammalian cells, the growth media was supplemented with radioactive amino acids. Which of the following molecules will be labelled?
A. Proteins B. DNA
C. RNA D. Glycolipids

52. An Indian student applied for post doctorate fellowship in Singapore and was asked to undergo test for Tuberculosis. He went to AIIMS, New Delhi for testing. The Tuberculin skin test (1st test) turned out to be positive, however, culture-based confirmation test (2nd test) revealed that he was negative for tuberculosis. What is the most likely reason for this observation?
A. The student had autoimmune antibodies
B. The student was vaccinated with BCG
C. The 1st test was not performed correctly
D. The 2nd test was not performed correctly

53. Why are Met and Trp often used to design DNA probes from amino acid sequences?
A. They do not have degenerate codons
B. Met is the first amino acid in the protein chain
C. Both are used often in proteins
D. They are hydrophobic

54. Malaria is caused by:
A. *Staphylococcus aureus*
B. *H. Influenza*
C. *Plasmodium*
D. *HIV*

55. The kind of covalent modification that occurs on both histones and DNA is:
A. Phosphorylation B. Methylation
C. Acetylation D. Sumoylation

56. A combination vaccine against three infectious diseases is:
A. Rifampicin B. BCG
C. DPT D. Dukoral

57. Which of the following is a non-sulfated glycosaminoglycan?
A. Hyaluronan B. Vimentin
C. Collagen D. Chondroitin S

58. Which of the following is the correct combination of marker enzymes used to identify different organelles during subcellular fractionation of eukaryotic tissue?
A. Cytosol-Lactate Dehydrogenase; Mitochondria-Succinate Dehydrogenase; Lysosome-Acid phosphatase; Peroxisome-Catalase
B. Cytosol-Succinate Dehydrogenase; Mitochondria-Lactate Dehydrogenase; Lysosome-Acid phosphatase; Peroxisome-Catalase
C. Cytosol-Acid phosphatase; Mitochondria-Succinate Dehydrogenase; Lysosome-Lactate Dehydrogenase; Peroxisome-Catalase
D. Cytosol-Catalase; Mitochondria-Succinate Dehydrogenase; Lysosome-Acid phosphatase; Peroxisome-Lactate Dehydrogenase

59. A patient diagnosed with Urticaria will have elevated levels of:
A. IgA B. IgG
C. IgE D. IgM

60. Dolly sheep was created by:
A. Artificial insemination
B. Somatic cell nuclear transfer
C. Embryonic stem cell mediated gene transfer
D. Pronuclear microinjection

61. Which of the following organisms is exploited for transfer of genes in plants?
A. *Agrobacterium tumefaciens*
B. *Staphylococcus aureus*
C. *Escherichia coli*
D. *Clostridium perfringens*

62. Which of the following is an example of attenuated vaccine?
A. Yellow fever B. Tetanus
C. Hepatitis B D. Meningococcal

63. Which of the following methods is not employed for affinity maturation of antibodies?
A. Hotspot mutagenesis
B. Error-prone PCR
C. High fidelity PCR
D. Chain shuffling

64. Intrinsic fluorescence of GFP is contributed by:
A. Cyclization and oxidation of residues: Ser-Tyr-Gly
B. Cyclization and oxidation of residues: Ser-Pro-Gly
C. Cyclization and oxidation of residues: Tyr-Gly-Pro
D. Cyclization and oxidation of residues: Ser-Tyr-Pro

65. Which of the following sequences are not palindromic?
A. AGCGAATTCGCT
B. TTAAGGATCCTTAA
C. GGCCAATTGGCCAA
D. ATGCATATGCAT

66. In eukaryotic cells, a protein containing oligosaccharide linked to manose-6-phosphate is destined to which of the following organelle?
A. Lysosomes B. Nucleus
C. Mitochondria D. Peroxisomes

67. Which of the following describe the phenomenon of antigenic drift in case of influenza virus?
A. A series of spontaneous point mutations that occur gradually, resulting in minor changes in HA and NA

B. Sudden emergence of a new subtype of influenza whose HA and possibly also NA are considerably different from that of the virus present in a preceding epidemic
C. A series of mutations that result in loss of antigenic HA and NA
D. A series of mutations that result in emergence of new antigenic components other than HA and NA

68. Passive administration of antibodies is employed as a mechanism for providing immediate protection against several toxins and pathogens. Which of the following is treated by passive immunization?
A. Tuberculosis B. Tetanus
C. Typhoid D. Leprosy

69. TA cloning is one of the most commonly employed technique for cloning inserts in desired vectors. Which of the following enzymes can be employed for preparing inserts for TA cloning?
A. Pfu DNA polymerase
B. Vent DNA polymerase
C. Adenylate kinase
D. Klenow exo-

70. The three-dimensional structure of tRNA is:
A. L-shape
B. Cloverleaf
C. Twisted triple helix
D. Double helix

71. What is the direction of translation of m-RNA?
A. Bidirectional B. 5' to 3'
C. 3' to 5' D. C to N terminus

72. The 'committed step' in the biosynthesis of cholesterol from acetyl CoA is:
A. Formation of acetoacetyl CoA from acetyl CoA
B. Formation of mevalonate from HMG CoA
C. Formation of HMG CoA from acetyl CoA and acetoacetyl CoA
D. Formation of squalene by squalene synthetase

73. Riboflavin is a coenzyme in the reaction catalyzed by the enzyme:
A. Acyl CoA synthetase
B. Acyl CoA dehydrogenase
C. Beta-Hydroxy acyl CoA
D. Enoyl CoA dehydrogenase

74. Which of the following pair of amino acids has more than one chiral center?
A. Lysine, Arginine
B. Aspartate, Glutamate
C. Serine, Tyrosine
D. Isoleucine, Threonine

75. Glucose enters muscle cells mostly by which of the following mechanism?
A. Simple diffusion
B. Facilitated diffusion using a specific glucose transporter
C. Co-transport with sodium
D. Co-transport with amino acids

76. Isoenzymes are:
A. Chemically, immunologically and electrophoretically different forms of an enzyme
B. Different forms of an enzyme similar in all properties
C. Able to catalyse different reactions
D. Biomolecules with different quaternary structures

77. Genes cannot be inserted into eukaryotic cells by:
A. Viruses
B. Chemical treatment
C. Microinjection
D. Splicing

78. Which of the following promotes glucose and amino acid uptake by muscle?
A. Adrenaline B. Insulin
C. Glucagon D. Cortisol

79. Angiotensin converting enzyme inhibitor are used to treat:
A. Diabetes B. Hypertension
C. Hyperthyroidism D. Obesity

80. The rate limiting step of urea cycle is mediated by:
A. Ornithine transcarbamoylase
B. Carbamoyl phosphate synthetase I

C. Arginosuccinate synthetase
D. Arginase

81. The active site of chymotrypsin consisting of a catalytic triad is composed of which of the following amino acid residues?
A. Serine, histidine and aspartate
B. Serine, histidine and glutamate
C. Threonine, histidine and aspartate
D. Methionine, histidine and aspartate

82. Which of the following is a transition mutation?
A. A-T → G-C
B. A-T → C-G
C. A-T → T-A
D. G-C → C-G

83. Outer and inner membrane of the bacteria can be separated by:
A. Electrophoresis
B. Sucrose density gradient centrifugation
C. Sonication
D. Gel filtration chromatography

84. Which of the following sequences is inversely palindromic?
A. 5′ GCATGC 3′
B. 5′ GCAACG 3′
C. 5′ GCAT 3′
D. 5′ GCAACGC 3′

85. Which of the following is not a dietary antioxidant?
A. Vitamin E B. Lipoic acid
C. Vitamin K D. Beta-carotene

86. The trigger to initiate the contractile process in skeletal muscle is:
A. Potassium binding to myosin
B. Calcium binding to tropomyosin
C. ATP binding to the myosin cross bridges
D. Calcium binding to troponin

87. NADPH:
A. Accepts 2 electrons and 2 hydrogen ions
B. Accepts 2 electrons and 1 hydrogen ions
C. Accepts 1 electron and 1 hydrogen ion
D. Transfers electrons in reductive biosynthesis

88. Photolysase functions to:
A. Repair pyrimidine dimers
B. Remove damaged bases
C. Ligate single-strand breaks
D. Ligate double stranded breaks

89. Which of the following is a vasodilator?
A. Norepinephrine B. Angiotensin II
C. Vasopressin D. Bradykinin

90. Tachycardia is a condition in which:
A. Heart beats slower than normal
B. Heart beats faster than normal
C. Heart stops beating
D. Heart collapses

91. When the resting membrane potential becomes less negative, the phenomenon is known as:
A. Hyperpolarization of the membrane
B. Depolarization of the membrane
C. Semi-polarization of the membrane
D. Repolarization of the membrane

92. Gastric inhibitory peptide (GIP) is secreted by:
A. Pancreas B. Gall bladder
C. Small intestine D. Rectum

93. A peptide which acts as potent smooth muscle hypotensive agent is:
A. Glutathione B. Bradykinin
C. Tryocidine D. Gramicidin-s

94. RNA polymerase I transcribes the genes for
A. mRNA precursors
B. 18S, 5.8S, and 28S rRNA
C. most tRNA
D. repair enzymes

95. Which of the following is a non reducing sugar?
A. Maltose B. Lactose
C. Trehalose D. Cellobiose

96. Caffeine:
A. Decreases cAMP levels
B. Increases cAMP levels
C. Increase potassium ions
D. Decreases potassium ions

97. What is Phenylketonuria (PKU)?
A. A rare metabolic disease that prevents the breakdown of phenylalanine
B. A rare metabolic disease that prevents the breakdown of all amino acids
C. A disorder of the skin that causes rashes and blistering
D. A disease that causes the body to make too much phenylalanine

98. Which of the following is an example of C3 plants?
A. Sugarcane B. Cactus
C. Wheat D. Orchids

99. How many number of ATP molecules are produced by one glucose molecule in aerobic respiration?
A. 39 B. 45
C. 34 D. 36

100. Down syndrome is a genetic disorder caused by the presence of all or part of a third copy of:
A. Chromosome 21
B. Chromosome 20
C. Chromosome 18
D. Chromosome 14

ANSWERS

1	**2**	**3**	**4**	**5**	**6**	**7**	**8**	**9**	**10**
D	A	B	C	A	D	C	A	C	C
11	**12**	**13**	**14**	**15**	**16**	**17**	**18**	**19**	**20**
C	B	C	D	D	B	C	D	B	C
21	**22**	**23**	**24**	**25**	**26**	**27**	**28**	**29**	**30**
A	A	C	B	B	D	B	A	B	B
31	**32**	**33**	**34**	**35**	**36**	**37**	**38**	**39**	**40**
B	B	A	C	B	A	C	A	B	A
41	**42**	**43**	**44**	**45**	**46**	**47**	**48**	**49**	**50**
B	A	C	B	B	C	C	B	A	B
51	**52**	**53**	**54**	**55**	**56**	**57**	**58**	**59**	**60**
A	B	A	C	B	C	A	A	C	B
61	**62**	**63**	**64**	**65**	**66**	**67**	**68**	**69**	**70**
A	A	C	A	C	A	A	B	D	A
71	**72**	**73**	**74**	**75**	**76**	**77**	**78**	**79**	**80**
B	B	B	D	B	A	D	B	B	B
81	**82**	**83**	**84**	**85**	**86**	**87**	**88**	**89**	**90**
A	A	B	A	C	D	D	A	D	B
91	**92**	**93**	**94**	**95**	**96**	**97**	**98**	**99**	**100**
B	C	B	B	C	B	A	C	D	A

Previous Paper (Solved)

M.Sc.—Biochemistry Entrance Exam, 2019*

1. Which of the following motif is present in calcium binding proteins?

A. Ig superfold B. EF hand

C. Solenoid D. Zinc fingers

2. Which of the following acts as an ATP buffer in human body?

A. AMP

B. Adenosine

C. Phosphocreatine

D. Phosphatase

3. Which of the following is a polar amino acid?

A. Threonine B. Alanine

C. Proline D. Methionine

4. Which of the following is a positively charged matrix for ion exchange chromatography?

A. CM cellulose

B. Phosphocellulose

C. Nitrilotriacetic acid

D. DEAE cellulose

5. Which of the following is not an ionophore?

A. Monensin

B. Valinomycin

C. Kanamycin

D. Gramicidin A

6. Which of the following is not true about insulin?

A. Insulin has 3 chains linked with disulphide bonds

B. Insulin increases the activity of glycogen synthase

C. Insulin lowers blood glucose

D. Insulin is degraded by Insulinase

7. Which of the following is not a peptide hormone?

A. Glucagon

B. Oxytocin

C. Estrogen

D. Vasopressin

8. Which of the following is the main immunoglobulin in the gut and secretions (saliva, milk, tears) and is important in mucosal immunity?

A. IgA B. IgM

C. IgD D. IgG

9. Which of the following is anorexigenic?

A. Endocannabinoids

B. Cortisol

C. Leptin

D. Ghrelin

10. Which of the amino acid is coded by a single codon?

A. Phenylalanine

B. Arginine

C. Tryptophan

D. Lysine

11. Which sugar will give a positive result with Seliwanoff test?

A. Mannose B. Glucose

C. Galactose D. Sucrose

12. Which immunoglobulin(s) is/are predominant in the primary (early) immune response?

A. IgG & IgM B. IgM

C. IgE D. IgG

13. Which group of a fully protonated glycine (NH_3+ -CH_2- COOH) will release the proton first when titrated against $-OH^-$ ions?

A. Amino group

B. Cannot be predicted

C. Both at the same time

D. Carboxyl group

* Exam was conducted by NTA for Admission in Delhi University M.Sc Biochemisty. (Exam held on 08/07/2019)

14. Which one of the following is not a protein estimation method?
A. Bicinchonic acid method
B. Biuret method
C. Barfoed method
D. Lowry method

15. Which is **not** an autoimmune disorder?
A. Grave's disease
B. Alzheimer's disease
C. Multiple sclerosis
D. Celiac disease

16. Which amino acid of a protein can make Schiff's base with reducing sugars?
A. Glycine
B. Lysine
C. Tryptophan
D. Phenylalanine

17. Which hormone is also known as hyperglycemic hormone?
A. Glucagon
B. Cholecystokinin
C. Insulin
D. Glycolase

18. Prof. Howard Temin is famous for:
A. Restriction Endonuclease
B. DNA Polymerase
C. DNA Ligase
D. Reverse Transcriptase

19. 4-hydroxy proline (a derivative of proline) is abundantly present in:
A. Keratin
B. Collagen
C. Myoglobin
D. Haemoglobin

20. Prof. Har Gobind Khorana is known to first synthesize:
A. Polynucleotide
B. Amino Acid
C. Protein
D. SiRNA

21. The classic complement pathway is initiated by interaction of C1 with
A. antigen-IgG complexes
B. antigen
C. factor B
D. bacterial lipopolysaccharides

22. Which one is a true membrane protein?
A. Glycophorin
B. Isocitrate lyase
C. Pyruvate kinase
D. Hemoglobin

23. Which of the following is a non-reducing sugar?
A. Trehalose
B. Cellobiose
C. Maltose
D. Lactose

24. Scientist(s) who proved that DNA and not protein as the genetic material?
A. Pasteur
B. Jenner
C. Hershey and Chase
D. Beadle and Tatum

25. Which of the following reagents are not a part of a typical polymerase chain reaction?
A. Forward and reverse primers
B. DNA polymerase
C. Magnesium chloride
D. Nucleotide triphosphates

26. Role of Glucagon is to:
A. Stimulate the formation of cyclic ATP in the liver cells
B. Stimulate the formation of cyclic AMP in the liver cells
C. Inhibit the formation of cyclic ATP in the liver cells
D. Inhibit the formation of cyclic AMP in the liver cells

27. In a competitive enzymatic reaction
A. K_m & V_{max} decreases
B. K_m & V_{max} increases
C. K_m increases - V_{max} shows no change
D. K_m shows no change & V_{max} decreases

28. Which vitamin helps in clotting of the blood?
A. Vitamin C
B. Vitamin K

C. Vitamin A
D. Vitamin E

29. Which enzyme carries the incision of the 3′ position of the lesion in nucleotide excision repair?
A. Urv B B. Urv C
C. Urv D D. Urv A

30. Which protein strictly interacts with phosphatidylcholine?
A. Phosphorylase
B. Phosphatase
C. Phospholipase
D. Alkaline phosphatase

31. Which of the following is an essential amino acid for humans?
A. Serine
B. Alanine
C. Isoleucine
D. Aspartic acid

32. Which of the following is a keto triose?
A. Glyceraldehyde
B. Erythrose
C. Dihydroxyacetone
D. Ribulose

33. Surface marker for helper T cells:
A. CD8 B. CD4
C. CD48 D. CD38

34. Which of the following would be the enzyme of choice to carry out nick translation reaction?
A. *E.coli* DNA polymerase I
B. T4 DNA polymerase
C. T7 DNA polymerase
D. Klenow fragment

35. Which is the main nitrogenous waste in humans?
A. Urea
B. Uric acid
C. Ammonia
D. Ammonium nitrate

36. An integral membrane protein spanning lipid bilayer of the membrane must have only:
A. No hydrophobic domains
B. Two hydrophobic domains
C. Three hydrophobic domains
D. One hydrophobic domains

37. Dihydrofolate reductase catalyses the conversion of:
A. Dihydrofolic acid to tetrahydrofolic acid, using ATP as electron donor
B. Dihydrofolic acid to tetrahydrofolic acid, using NADPH as electron donor
C. Tetrahydrofolic acid to Dihydrofolic acid, using NADPH as electron donor
D. Tetrahydrofolic acid to Dihydrofolic acid, using ATP as electron donor

38. Kupffer cells are a specialized type of:
A. Lymphocytes
B. Macrophages
C. Dendritic cells
D. Monocytes

39. Process of production of glucose from amino acids and the glycerol portion of fats is known as:
A. Gluconeogenesis
B. Glycogenolysis
C. Glycerogenesis
D. Glucosis

40. The vitreous humor of the eye is composed of:
A. Chondroitin sulfate
B. Keratan sulfate
C. Hyaluronic acid
D. Heparin

41. Lipid bilayer membrane model was proposed by
A. Danielle and Davson
B. Singer
C. Handerson and Nigel Unwin
D. Robertson

42. Very last step of protein secretion from live cells is:
A. Sumoylation
B. Endocytosis

C. Membrane Fusion
D. Protein Sorting

43. The protein responsible for the formation of fibrin monomers is:
A. Thrombin
B. Prothrombin
C. Heparin
D. Plasminogen

44. The precursor to IAA (indole 3-acetic acid) biosynthesis in plant is
A. Tyrosine
B. Phenylalanine
C. Tryptophan
D. Methionine

45. Human intestinal flora cannot digest:
A. Cellulose B. Lignin
C. Starch D. Pectin

46. Bacterial conjugation was discovered by
A. Lederberg and Tatum
B. Beadle and Tatum
C. Griffith
D. Robert Koch

47. Urea and Formamide can
A. Decrease Tm of the DNA
B. Does not effect denaturation of DNA
C. Increase Tm of the DNA
D. Helps faster re-annealing of DNA

48. An example for non-ionic detergent is:
A. Triton X-100
B. Benzalkonium chloride
C. Sodium dodecyl sulphate
D. Cetyl trimethyl ammonium bromide

49. Lectins are known to bind to only:
A. Ceramide
B. Cholesterol
C. Phospholipids
D. Glycoconjugates

50. Excessive intake of which of the following element causes mottled teeth and enlarged bones?
A. Magnesium B. Fluorine
C. Phosphate D. Calcium

51. Absence of functioning peroxisomes results in:
A. Wolman disease
B. Zellweger syndrome
C. Hunter syndrome
D. Gaucher disease

52. A single calmodulin protein can bind to how many calcium molecules?
A. 4 B. 1
C. 3 D. 2

53. Propagation of the action potential through the heart is fastest in the:
A. AV node
B. Purkinje fibers
C. SA node
D. Arterial muscle

54. During aerobic exercise, blood flow remains relatively constant within:
A. The brain
B. The heart
C. The skin
D. The skeletal muscles

55. During electron transport, protons are pumped out of the mitochondrion at each of the major sites except for
A. complex II
B. complex IV
C. complex III
D. complex I

56. Receptor molecule on cell surface for cholera toxin is:
A. Glycophorin
B. EGFR
C. Cholesterol
D. Ganglioside

57. Isoelectric focusing is a technique routinely used to resolve proteins based on their:
A. Hydrophobicity
B. Mass
C. pI
D. Length

58. Abzymes are:
A. Also referred to as zymogens
B. Enzymes that are highly specific like antibodies
C. Enzymes that hydrolyze antibodies
D. Antibodies that have catalytic activities

59. Acid phosphatase can be used as a marker enzyme for:
A. Lysosomes
B. Peroxisomes
C. Nucleus
D. Mitochondria

60. Pyridoxal phosphate is the functional form of:
A. Vitamin B_6 B. Vitamin B_3
C. Vitamin B_{12} D. Vitamin B_2

61. Tay-Sachs disease is caused by the deficiency of:
A. Glucocerebrosidase
B. Galactosidase A
C. Hexosaminidase A
D. Sphingomyelinase

62. During DNA replication, which of the following enzymes is involved in removing supercoiling ahead of the replication fork?
A. Primase
B. Polymerase
C. Helicase
D. Topoisomerase

63. Cell-cell fusion is exhibited by formation of:
A. Multinucleated cells
B. Microvilli
C. Elevated numbers of endosomes
D. Numerous vacuoles

64. Vitamin B_{12} is absorbed primarily in the:
A. Jejunum B. Duodenum
C. Stomach D. Ileum

65. Hybridoma technology uses splenocytes for production of
A. T-cell receptors
B. MHC complex
C. Monoclonal antibodies
D. Viral antigens

66. RNA lacks which of the following?
A. O-Glycosidic bond
B. N-Glycosidic bond
C. Purines
D. Ester bond

67. Grafts between genetically identical individuals (i.e., identical twins)
A. are not rejected if a kidney is grafted, but skin grafts are rejected
B. are not rejected, even without immunosuppression.
C. are subject to hyperacute rejection.
D. are rejected slowly as a result of minor histocompatibility antigens.

68. Multidrug resistance of cancer cells are because of appearance of:
A. High rate of protein secretion
B. Increase of endocytotic capacity
C. Elevated number of GM1 on cell surface
D. Special transporter protein on plasma membrane

69. Enzyme with highest turnover number:
A. Lysozyme
B. G3PDH
C. Carbonic anhydrase
D. LDH

70. In Counter immunoelectrophoresis
A. the antibody will migrate towards anode
B. the antibody will migrate towards cathode
C. electrophoresis will drive the antibody and antigen toward each other
D. electrophoresis will drive the antibody and antigen parallel to each other

71. Antigen-presenting cells that activate helper T cells must express which one of the following on their surfaces?
A. IgE
B. Class I MHC antigens
C. Class II MHC antigens
D. Gamma interferon

72. Sphingomyelinase catalyses the conversion of sphingomyelin into:
A. Phosphatidylinositol and ceramide

B. Phosphatidylserine and ceramide
C. Phosphocholine and ceramide
D. Acetylcholine and ceramide

73. Different form of a plasmid, on an agarose gel, after electrophoresis may be in which of the following sequence in decreasing molecular weight?
A. Linear plasmid, open circular, super coiled plasmid
B. Super coiled, open circular, linear plasmid DNA
C. Open circular, super coiled, linear plasmid DNA
D. Open circular, linear plasmid, super coiled plasmid

74. Mode of action of Penicillin on killing bacterial cells is through its action on:
A. Ribosome
B. Plasma membrane
C. Cell wall
D. DNA

75. AP endonuclease is responsible for
A. Double strand break repair
B. Base excision repair
C. Nucleotide excision repair
D. Mismatch repair

76. At room temperature (25°C) an unsaturated fatty acid will have
A. Solid consistency
B. Waxy consistency
C. Liquid consistency
D. Cannot be predicted

77. All of the following statements are true about lysozyme except which one?
A. It hydrolyzes peptidoglycan.
B. It destroys gram-positive cell walls.
C. It is an organelle in white blood cells.
D. It is an enzyme.

78. pH of an aqueous solution of 1M NaCl:
A. Acidic
B. Neutral
C. Alkaline
D. Cannot be determined

79. Formation of a disulphide bond between two cysteine residues requires which of the following?
A. Oxidising agent
B. Adenosine triphosphate
C. Reducing agent
D. Ammonium sulphate

80. What is the average resting potential of a mammalian cell?
A. +50 mV B. +60 mV
C. –70 mV D. –50 mV

81. What is approximatively the half life of 32-Phosphorus?
A. 30 days B. 42 days
C. 14 days D. 28 days

82. Sucrose is composed of
A. Mannose and glucose
B. Galactose and mannose
C. Fructose and galactose
D. Fructose and glucose

83. Muscular dystrophy is a result of mutation in the
A. Mutation in glycophorin A
B. Mutation in Ankyrin
C. Mutation in band 3 protein
D. Mutation in dystrophin

84. The immunoglobulin fold primarily comprises of:
A. Gamma turn
B. Alpha helices
C. Beta sheets
D. Unstructured loop

85. The disease commonly called 'LOCK JAW" is caused by
A. *Clostridium tetani*
B. *Mycobacterium leprea*
C. *Treponema pallidum*
D. *Bacillus pertussis*

86. The extinction coefficient of a protein at 280 nm is 20 $mM^{-1}cm^{-1}$. The absorbance of the protein solution at 280 nm, appropriately diluted, measured in a cuvette with a path length of 1 cm is 100. What is the concentration of the protein solution?

A. 5 mM B. 0.2 M
C. 0.2 mM D. 5 M

87. The value of ionic product of water at 25°C is:

A. 1×10^{-7}
B. 1×10^{14}
C. 1×10^{-14}
D. 1×10^{7}

88. The effect IP_3 activation is:

A. Increase in intracellular calcium levels
B. Increase in extracellular calcium levels
C. Closure of calcium channels in ER
D. Activation of calmodulin

89. The mammalian cells are routinely cultured under *in vitro* conditions at:

A. 37°C, 5% O_2
B. 37°C, 5% CO_2
C. 25°C, 5% O_2
D. 25°C, 5% CO_2

90. The role of cholecystokinin is to:

A. Stimulate gastric motility
B. Stimulate enzyme secretion from pancreas
C. Inhibit enzyme secretion from pancreas
D. Inhibit gastric motility

91. The salt which is commonly used for salting out effect during extraction of proteins is:

A. NH_4NO_3
B. $(NH_4)_2SO_4$
C. K_2SO_4
D. NaCl

92. The most abundant phospholipid in bacterial cell membrane is

A. Phosphatidylcholine
B. Phosphatidylserine
C. Phosphatidylethanolamine
D. Cholesterol

93. The only enzyme belonging to Krebs cycle that is attached to inner mitochondrial membrane

A. Isocitrate dehydrogenase
B. Malate dehydrogenase
C. NADH dehydrogenase
D. Succinate dehydrogenase

94. The cooperativity phenomenon of protein-ligand interaction was first determined in:

A. Hemoglobin-O_2
B. Peroxidase-H_2O_2
C. ATCase-Aspartate
D. Myoglobin-O_2

95. The bond angle between H-O-H in water molecule is:

A. 55° B. 105.4°
C. 104.5° D. 110°

96. The class of immunoglobulin present in highest concentration in the blood of a human newborn is

A. IgA B. IgM
C. IgE D. IgG

97. Chemical messengers that are secreted by cells into the extracellular fluid and affect neighbouring target cells of a different type are typically known as

A. Autocrines
B. Neurotransmitters
C. Paracrines
D. Endocrines

98. A regulatory control site of a gene

A. Must be very close to the gene it affects.
B. Must lie upstream of the gene it affects.
C. Is *trans*-acting
D. Is *cis*-acting

99. IgM antibody can bind to:

A. Two epitopes
B. Ten epitopes
C. Four epitopes
D. One epitope

100. Calcitonin is secreted by:

A. Parathyroid hormone
B. Pituitary gland
C. Adrenal gland
D. Thyroid hormone

ANSWERS

1	2	3	4	5	6	7	8	9	10
B	C	A	D	C	A	C	A	C	C
11	**12**	**13**	**14**	**15**	**16**	**17**	**18**	**19**	**20**
D	B	D	C	B	B	A	D	B	A
21	**22**	**23**	**24**	**25**	**26**	**27**	**28**	**29**	**30**
A	A	A	C	D	B	C	B	A	C
31	**32**	**33**	**34**	**35**	**36**	**37**	**38**	**39**	**40**
C	C	B	A	A	D	B	B	A	C
41	**42**	**43**	**44**	**45**	**46**	**47**	**48**	**49**	**50**
A	C	A	C	A	A	A	A	D	B
51	**52**	**53**	**54**	**55**	**56**	**57**	**58**	**59**	**60**
B	A	B	A	A	D	C	D	A	A
61	**62**	**63**	**64**	**65**	**66**	**67**	**68**	**69**	**70**
C	D	A	D	C	A	B	D	C	C
71	**72**	**73**	**74**	**75**	**76**	**77**	**78**	**79**	**80**
C	C	A	C	B	C	C	B	A	C
81	**82**	**83**	**84**	**85**	**86**	**87**	**88**	**89**	**90**
C	D	D	C	A	A	C	A	B	B
91	**92**	**93**	**94**	**95**	**96**	**97**	**98**	**99**	**100**
B	C	D	A	C	D	C	D	B	D

Previous Paper (Solved)

M.Sc.—Biotechnology Entrance Exam, CET (PG), 2018

1. Which organism is known as work horse of Biotechnology?

A. Drosophila B. Bacillus
C. Mouse D. *E. coli*

2. Zebra fish is the best model organism to study:

A. Development studies
B. Protein-protein interaction studies
C. DNA-protein interaction studies
D. Protein production

3. Ames test is used to study:

A. Cell division B. DNA replication
C. Mutagenicity D. Pollution

4. The nick translation is carried out by:

A. DNA polymerase I
B. DNA polymerase II
C. DNA polymerase III
D. DNA topoisomerase

5. Hoogsteen base pairing was recognized in:

A. B DNA B. Z DNA
C. Triple helix D. RNA

6. DNA as a transforming agent was confirmed by:

A. Grifith
B. Avery *et al.*
C. Hershey and Chase
D. Watson and Crick

7. White eye of drosophila is X linked recessive character. If a red eyed male is mated with white eyed female, what percentage of the males will be white eyed?

A. 0% B. 25%
C. 50% D. 100%

8. Hardy and Weinberg equation is used to study:

A. Epistasis
B. Abberations
C. Bacterial genetics
D. Population genetics

9. FISH technique is used to:

A. DNA sequencing
B. Northern blotting
C. Localize gene on chromosome in *in situ* condition
D. Identify a chromosome

10. In native PAGE the migration of protein depends on:

A. Size of protein
B. Size and shape of protein
C. Charge of protein
D. Size, shape and charge of protein

11. Biomarker is:

A. Colour coding of chromosomes
B. A molecule specifically produced under diseased condition
C. A marker that is genetically inherited
D. A molecule in the genome

12. The absorbance of a protein sample at 280 nm can be measured by:

A. Colorimeter
B. Spectrophotometer
C. pH meter
D. Electrophoresis

13. After centrifugation which type of rotor will collect pellet in the centre of the tube?

A. Vertical rotor
B. Spinwin
C. Swinging bucket
D. Angle head rotor

14. The ampicillin resistance gene on a vector allows bacteria containing vector to grow in the presence of ampicillin as it:

A. does not allow ampicillin to enter in the bacteria
B. degrades ampicillin
C. proliferates cell growth
D. changes the target of ampicillin

15. The DNA synthesis takes place during:
A. G1 phase B. S phase
C. G2 phase D. M phase

16. The distance between -10 and -35 region of prokaryotic promoter is important for it's proper binding with:
A. DNA polymerase
B. RNA polymerase
C. Sigma factor
D. Transcription factor

17. In the Meselson-Stahl DNA replication experiment, what percent of the DNA was composed of one light strand and one heavy strand after two generation of growth in N-14 containing growth media?
A. 100 B. 75
C. 50 D. 25

18. For the DNA strand 5'-TTCGCAATGC-CATTA-3' the correct complementary strand will be:
A. 5'-TAATGGCATTGCGAA-3'
B. 3'-TAATGGCATTGCGAA-5'
C. 5'-UAAUGGCAUUGCGAA-3'
D. 3'-UAAUGGCAUUGCGAA-5'

19. The overall source of energy for photosynthesis is:
A. Energy of electron transport in the thylakoid membrane
B. Energy released when water is oxidized and oxygen is produced
C. Energy from the hydrolysis of ATP
D. Light energy from the sun

20. Holliday junctions are formed during:
A. Replication B. Transcription
C. Recombination D. Conjugation

21. Isoscizomers are the restriction enzymes which:
A. Show 50% homology in the recognition site
B. Restriction enzymes specific to the same recognition sequence
C. Show no homology in recognition sequence
D. Are GC rich

22. First human health care product from recombinant DNA technology to reach the market was developed and licensed by:
A. Glaxosmithkline and Genentech
B. Genentech
C. Glaxosmithkline
D. Eli Lilly and Genentech

23. The process of bacteria taking up naked DNA from environment is known as:
A. Transformation B. Transduction
C. Replication D. Transcription

24. The absorbance of an undiluted DNA sample at A_{260} is 2.0. What is the concentration of DNA of the sample?
A. 20 µg/ml B. 50 µg/ml
C. 100 µg/ml D. 200 µg/ml

25. EF-Tu is a factor used in:
A. Replication B. Transcription
C. Translation D. Termination

26. Km of an enzymatic reaction remains same as in control in the presence of:
A. Non competitive inhibitor
B. Competitive inhibitor
C. Uncompetitive inhibitor
D. Irreversible inhibitor

27. In lysogenic cycle:
A. A bacteriophage genome causes break down of host cell
B. A bacteriophage genome gets integrated into host genome
C. A bacteriophage secretes DNA degrading enzyme
D. A bacteriophage gets replicated in high number in the host

28. An overall result of photosynthesis in plants is the use of electrons from water to reduce:
A. Glucose B. Carbon dioxide
C. Oxygen D. Chlorophyll

29. The purpose of heat inactivation of serum in animal cell culture is:
A. To activate growth hormones
B. To inactivate any bacteria present
C. To inactivate complement system
D. To remove unwanted proteins

30. The BOD of waste water is related to the presence of:
A. Organic material B. Inorganic material
C. Bacteria D. Fungus

31. Quoram sensing system:
A. Senses signals from environment
B. Senses compounds secreted by bacteria itself
C. Senses photosynthesis
D. Senses respiration

32. The cell surface marker on cytotoxic T cell is:
A. CD4 B. CD7
C. CD8 D. CD9

33. Mycoplasma is a:
A. Bacteria B. Fungus
C. Algae D. Protozoa

34. Transposase facilitate:
A. Replication
B. Transcription
C. General recombination
D. Site specific recombination

35. Self reactive antibodies are not found in serum due to the:
A. Self reactive B cells are killed by CD8 T cells
B. Self reactive B cells are changed into IgE and secreted out
C. Self reactive B cells are not generated
D. B cells stimulated via surface bound antibody in the absence of T cell commit suicide

36. Anaphylaxis can be triggered by cross-linking of IgE receptors on:
A. Monocytes B. Mast cells
C. B-cells D. Eosinophil

37. Which of the following is caused by DNA virus?
A. Typhoid B. Measles
C. Small pox D. Yellow fever

38. Whole genome shotgun sequencing depends on:
A. Rapidly sequencing of thousands of small randomly cloned fragments
B. Sequence only large fragments
C. Sequence only fragments cloned in BAC
D. Sequence only fragments cloned in YAC

39. Hybrid dysgenesis in drosophila refers to the high rate of mutation in germ line cells resulting from a cross of:
A. P strain females (females with autonomous P elements) with P strain males
B. males with autonomous P elements (P Strain/P cytotype) and M strain females
C. P strain females (females with autonomous P elements) with M strain males (lack P element)
D. M strain male and M strain female

40. The acquisition energy by glucose fermentation requires:
A. Electron transport of electrons from NADH
B. Substrate level phosphorylation
C. Long chain fatty acid oxidation
D. NADPH

41. RNA interference (RNAi) or Post-Transcriptional Gene Silencing (PTGS) is a conserved biological response to:
A. DNA-RNA hybrid
B. Methylated double stranded DNA
C. Single stranded DNA
D. Double-stranded RNA

42. A culture of *E. coli* was grown in the presence of ^{32}P isotope. After 24 h the radioactivity will be present in which fraction of the cell?
A. DNA B. Protein
C. Lipid D. Enzyme

43. Phage display is a technique in which:
A. Gene of interest is cloned in phage in front of strong promoter
B. Gene encoding a protein of interest is inserted into a phage coat protein gene
C. Gene is cloned in a phagemid
D. Gene is cloned in a cosmid

44. Choose the correct statement for continuous reactor at steady state:
A. Substrate, product and biomass change with time
B. Substrate, product and biomass do not change with time

C. Product and biomass change with time
D. Only product change with time

45. Secondary structure of protein can be determined by:
A. CD spectroscopy
B. Fluorescence spectroscopy
C. Fluorescence microscopy
D. 2D gel electrophoresis

46. A plasmid can be eliminated from a cell by:
A. Transformation B. Curing
C. Electroporation D. Transduction

47. Homologous DNA undergo…. recombination:
A. Site specific B. Replicative
C. General D. Mutagenic

48. The environmental pollutants might result in death of river bed due to the:
A. Excess of sediments at bottom
B. Abundance of toxic material
C. Excess algae
D. Depletion of oxygen

49. Milk is pasteurized by keeping it at:
A. 53°C for 30 min B. 63°C for 30 min
C. 73°C for 30 min D. 83°C for 30 min

50. DNA repair machinery distinguishes the newly synthesized DNA from parental DNA by:
A. New strands are methylated while old strands are not
B. New strands are acetylated while old strands are not
C. Old strand is methylated while new is not
D. Old strand is acetylated while new is not

51. Bacterial rRNA sequencing approach gives information about:
A. Morphology of bacteria
B. Drug resistance
C. The species to which this rRNA containing bacteria is closely related to
D. Types of ribosomes produced by bacteria

52. RNA can be used as a template in PCR reaction after reverse transcription into:
A. cDNA B. tRNA
C. rRNA D. Ribozyme

53. tRNA donates aminoacid to the growing chain at:
A. A site B. P site
C. M site D. B site

54. The function of molecular chaperones is:
A. Translocation of protein
B. Termination of protein synthesis
C. Post translational modifications
D. Protein folding

55. Tetracycline inhibits the growth of bacteria by binding to:
A. 40S ribosomal subunit
B. 60S ribosomal subunit
C. 30S ribosomal subunit
D. 50S ribosomal subunit

56. Inactive protein form of an enzyme is:
A. Apoenzyme B. Co enzyme
C. Holoenzyme D. Cofactor

57. RNA guided genome editing is possible through:
A. RNAi B. CRISPR-Cas
C. Activation tagging D. Knock outs

58. International transport of GMOs is regulated by:
A. Cartegena protocol to the convention on biological diversity
B. Universal declaration on bioethics and human rights
C. Geneva declaration
D. Declaration of human genome and human rights

59. In plant tissue culture, embryoid formation from pollens is due to:
A. Cellular totipotency
B. Organogenesis
C. Double fertilization
D. Test tube culture

60. A trangenic crop which may help is solving night blindness in developing countries is:
A. Bt soyabean
B. Golden rice
C. Flavr savr tomatoes
D. Starlink maize

61. A short fragment of DNA has 50 A and 50 C bases. Total number of nucleotide in the DNA fragment is:
A. 50 B. 100
C. 150 D. 200

62. A nucleoside is:
A. A nucleotide without sugar
B. A nucleotide without phosphate
C. A nucleotide without nitrogenous base
D. Nitrogenous base with phosphate

63. All publicly available sequence of protein is available in:
A. GenBank B. EML
C. DDBJ D. TrEMBL

64. The use of information that is of commercial value are protected by:
A. Institutional law
B. Public forum
C. Intellectual property rights
D. Institutional property rights

65. DNA protein binding can be studied with help of:
A. Southern hybridization
B. FISH
C. Pulse field gel electrophoresis
D. Gel shift assay

66. Reversed phase HPLC utilizes:
A. A hydrophobic stationary phase and a polar mobile phase
B. A hydrophobic stationary phase and a non-polar mobile phase
C. A hydrophilic stationary phase and a non-polar mobile phase
D. A hydrophilic stationary phase and a polar mobile phase

67. Conformational variation between B and Z forms of DNA is partially due to:
A. Rotation of glycosidic bond
B. Loss of hydrogen bond
C. Lack of hydrophobic interaction
D. Increase in humidity

68. Which of the following growth media would you expect to result in synthesis of high levels of mRNA for the enzymes of the *E. coli* lac operon?
A. High glucose, high lactose
B. Low glucose, low lactose
C. High glucose, low lactose
D. No glucose, high lactose

69. TATA box and Pribnow box are components of:
A. Operators B. Promoters
C. Enhancers D. Activators

70. Microarray is used to analyze:
A. The genome sequence
B. Differential expression of genes
C. Proteome
D. Metabolism

71. When a chromosomal DNA is digested with a restriction enzyme, the digested fragments will appear as?
A. Discrete single band
B. Nearly 3-5 sharp bands
C. Smear
D. Ladder

72. Which of these cannot be used for structural proteomics?
A. X-ray crystallography
B. NMR spectroscopy
C. 2D gel electrophoresis
D. Mass spectrometry

73. Glycosylation of newly synthesized protein takes place in:
A. Mitochondria
B. Nucleus
C. Vacuole
D. Endoplasmic reticulum

74. The electron transport chain is located predominantly in the:
A. Outer membrane of the mitochondria
B. Inter membrane space of the mitochondria
C. Inner membrane of the mitochondria
D. Matrix of the mitochondria

75. DNA-RNA hybridization takes place in:
A. Southern hybridization
B. Northern hybridization
C. Western hybridization
D. Eastern hybridization

ANSWERS

1	2	3	4	5	6	7	8	9	10
D	A	C	A	C	B	D	D	C	D
11	**12**	**13**	**14**	**15**	**16**	**17**	**18**	**19**	**20**
B	B	C	B	B	C	C	A	D	C
21	**22**	**23**	**24**	**25**	**26**	**27**	**28**	**29**	**30**
B	D	A	C	C	A	B	B	C	A
31	**32**	**33**	**34**	**35**	**36**	**37**	**38**	**39**	**40**
B	C	A	D	D	B	C	A	B	B
41	**42**	**43**	**44**	**45**	**46**	**47**	**48**	**49**	**50**
D	A	B	B	A	B	C	D	B	C
51	**52**	**53**	**54**	**55**	**56**	**57**	**58**	**59**	**60**
C	A	B	D	C	A	B	A	A	B
61	**62**	**63**	**64**	**65**	**66**	**67**	**68**	**69**	**70**
D	B	A	C	D	A	A	D	B	B
71	**72**	**73**	**74**	**75**					
C	C	D	C	A					

EXPLANATORY ANSWERS

3. The Ames test is a widely employed method that uses bacteria to test whether a given chemical can cause mutations in the DNA of the test organism. More formally, it is a biological assay to assess the mutagenic potential of chemical compounds. A positive test indicates that the chemical is mutagenic and therefore may act as a carcinogen, because cancer is often linked to mutation. The test serves as a quick and convenient assay to estimate the carcinogenic potential of a compound because standard carcinogen assays on mice and rats are time-consuming (taking two to three years to complete) and expensive. However, false-positives and false-negatives are known.

4. If a single-stranded DNA molecule is placed in contact with a complementary single-stranded DNA sequence, the two molecules will associate with one another by hydrogen bonding between the bases on their respective strands. This association, or hybridisation, forms the basis of very powerful techniques for detecting and quantifying specific nucleic acid sequences. Whether the hybridisation is done in solution or to nucleic acid immobilised on filters, a radioactively labelled probe will be required. The enzyme DNA polymerase I from *E. coli* catalyses a reaction which can be used to replace existing unlabelled nucleotides in DNA with radioactive ones. The reaction has been called nick translation, and is very widely used in molecular biology. It has nothing whatever to do with the translation of messenger RNA into a polypeptide chain on the ribosomes.

5. **Triple-stranded DNA** (also known as **H-DNA** or **Triplex-DNA**) is a DNA structure in which three oligonucleotides wind around each other and form a triple helix. In triple-stranded DNA, the third strand binds to a B-form DNA (via Watson–Crick base-pairing) double helix by forming Hoogsteen base pairs or reversed Hoogsteen hydrogen bonds.

8. The Hardy-Weinberg equation is a mathematical equation that can be used to calculate the genetic variation of a population at equilibrium. In 1908, G. H. Hardy and

Wilhelm Weinberg independently described a basic principle of population genetics, which is now named the Hardy-Weinberg equation. The equation is an expression of the principle known as Hardy-Weinberg equilibrium, which states that the amount of genetic variation in a population will remain constant from one generation to the next in the absence of disturbing factors.

9. Cytogenetics entered the molecular era with the introduction of *in situ* hybridization, a procedure that allows researchers to locate the positions of specific DNA sequences on chromosomes. Since the first *in situ* hybridization experiments in 1969 (Gall & Pardue, 1969), many variations of the procedure have been developed, and its sensitivity has increased enormously. Today, most *in situ* hybridization procedures use fluorescent probes to detect DNA sequences, and the process is commonly referred to as FISH (fluorescence *in situ* hybridization). A variety of FISH procedures are available to cytogeneticists, who use them to diagnose many types of chromosomal abnormalities in patients. The success of FISH, and all other methods of *in situ* hybridization, depends on the remarkable stability of the DNA double helix.

11. Biomarker: A biological molecule found in blood, other body fluids, or tissues that is a sign of a normal or abnormal process, or of a condition or disease. A biomarker may be used to see how well the body responds to a treatment for a disease or condition. Also called molecular marker and signature molecule.

12. Spectrophotometry can be used to estimate DNA or RNA concentration and to analyze the purity of the preparation. Typical wavelengths for measurement are 260 nm and 280 nm. In addition measurements at 230 nm and 320 nm can provide further information. Purines and pyrimidines in nucleic acids naturally absorb light at 260 nm. For pure samples it is well documented that for a pathlength of 10 mm, an absorption of 1A unit is equal to a concentration of 50 μg/ml DNA and 40 μg/ml for RNA. For oligonucleotides the concentration is around 33 μg/ml but this may vary with length and base sequence.

15. S phase (Synthesis Phase) is the phase of the cell cycle in which DNA is replicated, occurring between G_1 phase and G_2 phase. Since accurate duplication of the genome is critical to successful cell division, the processes that occur during S-phase are tightly regulated and widely conserved.

20. Holliday junction, cross-shaped structure that forms during the process of genetic recombination, when two double-stranded DNA molecules become separated into four strands in order to exchange segments of genetic information. This structure is named after British geneticist Robin Holliday, who proposed the original model for homologous (or general) recombination in 1964.

Homologous recombination occurs during meiosis and is characterized by the exchange of genes between a maternal chromatid and a paternal chromatid of a homologous chromosome pair. The two parent DNA molecules, which have long stretches of similar base sequences, are separated into single strands, resulting in base pairing that leads to a four-stranded DNA structure. The Holliday junction travels along the DNA duplex by "unzipping" one strand and reforming the hydrogen bonds on the second strand.

25. EF-Tu (elongation factor thermo unstable) is a prokaryotic elongation factor responsible for catalyzing the binding of an aminoacyl-tRNA (aa-tRNA) to the ribosome. It is a G-protein, and facilitates the selection and binding of an aa-tRNA to the A-site of the ribosome. As a reflection of its crucial role in translation, EF-Tu is one of the most abundant and highly conserved proteins in prokaryotes. It is found in eukaryotic mitochondria as TUFM.

30. **Biochemical Oxygen Demand** (**BOD**, also called **Biological Oxygen Demand**) is the amount of dissolved oxygen needed (*i.e.* demanded) by aerobic biological organisms to break down organic material present in a given water sample at certain temperature over a specific time period. The BOD value is most commonly expressed in milligrams of oxygen consumed per litre of sample during 5 days of incubation at 20 °C and is often used as a surrogate of the degree of organic pollution of water. BOD can be used as a gauge of the effectiveness of wastewater treatment plants. BOD is similar in function to chemical oxygen demand (COD), in that both measure the amount of organic compounds in water. However, COD is less specific, since it measures everything that can be chemically oxidized, rather than just levels of biodegradable organic matter.

33. Mycoplasma is a mollicute genus of bacteria that lack a cell wall around their cell membranes. This characteristic makes them naturally resistant to many common antibiotics such as penicillin or other beta-lactam antibiotics that target cell wall synthesis. They can be parasitic or saprotrophic. Several species are pathogenic in humans, including *M. pneumoniae*, which is an important cause of atypical pneumonia and other respiratory disorders, and *M. genitalium*, which is believed to be involved in pelvic inflammatory diseases. *Mycoplasma* species are the smallest bacterial cells yet discovered, can survive without oxygen, and come in various shapes.

49. The process of pasteurization was named after **Louis Pasteur** who discovered that spoilage organisms could be inactivated in wine by applying heat at temperatures below its boiling point. The process was later applied to milk and remains the most important operation in the processing of milk.

Ontario Pasteurization Regulations

Milk: 63° C for not less than 30 min., 72° C for not less than 16 sec., or equivalent destruction of pathogens and the enzyme phosphatase as permitted by Ontario Provincial Government authorities. Milk is deemed pasteurized if it tests negative for alkaline phosphatase.

63. The GenBank sequence database is an open access, annotated collection of all publicly available nucleotide sequences and their protein translations. This database is produced and maintained by the National Center for Biotechnology Information (NCBI) as part of the International Nucleotide Sequence Database Collaboration (INSDC). The National Center for Biotechnology Information is a part of the National Institutes of Health in the United States. GenBank and its collaborators receive sequences produced in laboratories throughout the world from more than 100,000 distinct organisms. The database started in 1982 by Walter Goad and Los Alamos National Laboratory. GenBank has become an important database for research in biological fields and has grown in recent years at an exponential rate by doubling roughly every 18 months.

69. In molecular biology, the **TATA box** (also called the **Goldberg-Hogness box**) is a sequence of DNA found in the core promoter region of genes in archaea and eukaryotes. The prokaryotic homolog of the TATA box is called the Pribnow box which has a shorter consensus sequence. The TATA box was first identified in 1978 as a component of eukaryotic promoters. Transcription is initiated at the TATA box in TATA-containing genes. The TATA box is the binding site of the TATA-binding protein (TBP) and other transcription factors in some eukaryotic genes. Gene transcription by RNA polymerase II depends on the regulation of the core promoter by long-range regulatory elements such as enhancers and silencers. Without proper regulation of transcription, eukaryotic organisms would not be able to properly respond to their environment.

Previous Paper (Solved)

M.Sc.—Biotechnology Entrance Exam, CET (PG), 2017*

1. The hydrolytic enzymes found in lysosomes work most effectively under conditions of:
A. Low pH B. High pH
C. High saline D. Neutral pH

2. Autoclave is primarily used for:
A. Lyophilization B. Crystallization
C. Polarization D. Sterilization

3. Endotoxins produced by gram negative bacteria is present in:
A. Inner cell membrane
B. Lipopolysacharide
C. Ribosomes
D. Peptidoglycan

4. In an exponentially growing batch culture of *Saccharomyces cerevisiae,* the cell density is 30 gl^{-1} (DCW), the specific growth rate (μ) is 0.5 and substrate uptake rate (v) is 20 $gl^{-1}h^{-1}$. The cell yield coefficient, $Y_{x/s}$, will be:
A. 0.05 B. 0.25
C. 0.50 D. 0.75

5. Which one of the following is a sulphur containing amino acid?
A. Tryptophan B. Valine
C. Cysteine D. Arginine

6. A hapten is:
A. A small chemical group which reacts with preformed antibodies
B. A strong immunogen
C. An antibody
D. A region of antibody

7. Adenine and Thymine in complementary DNA strands are joined by .... hydrogen bonds.
A. One B. Two
C. Three D. Four

8. The bond in subsequent amino acids responsible for primary structure of proteins is:
A. Ionic Bond
B. Phosphodiester Bond
C. Glycosidic Bond
D. Peptide Bond

9. Which of the following is NOT present in RNA?
A. Uracil B. Adenine
C. Thymine D. Cytosine

10. Vitamins are:
A. Organic substances that cannot be synthesized by animals
B. Inorganic substances that cannot be synthesized by animals
C. Inorganic substances that can be synthesized by animals
D. Organic substances that can be synthesized by animals

11. Which of the following is the nucleotide database?
A. BLASTn B. PROSITE
C. CATH D. GenBank

12. Which of the following CANNOT be grown on artificial culture medium?
A. *Escherichia coli*
B. Tobacco Mosaic Virus
C. *Saccharomyces cerevisiae*
D. *Aspergillus niger*

13. *Penicillium chrysogenum* is industrially used to produce the antibiotic penicillin. In a batch culture of *Penicillium chrysogenum,* the maximum penicillin synthesis occurs during the:
A. Lag phase B. Exponential phase
C. Stationary phase D. Death phase

14. During bacterial conjugation the genetic material is transferred from the donor bacteria to the recipient through:
A. Cell wall B. Pili
C. Medium D. Capsule

* Conducted by Punjab University.

15. According to Oparin's Theory, atmosphere of earth just before the origin of life consisted of:
A. Water vapours, reducing environment (CH_4, NH_3, H_2) without oxygen
B. Water vapours, reducing environment (CH_4, NH_3, H_2) with oxygen
C. Reducing environment (CH_4, NH_3, H_2) without oxygen and water vapours
D. Reducing environment (CH_4, NH_3, H_2) with oxygen and without water vapours

16. The bacterial cells having F plasmid integrated in their chromosome are termed as:
A. P^+ B. F^-
C. Hfr D. Hbr

17. Which of the following organelle is called "suicide bags" of cells?
A. Nucleolus B. Ribosome
C. Chloroplast D. Lysosome

18. α–D-Glucose and β–D–glucose are .... of each other:
A. Epimers B. Anomers
C. Multirotation pair D. Ketoenol pair

19. The unit of chromosome map is:
A. Millimendel B. Decibarr
C. Centimorgan D. Femtostern

20. The process for the synthesis of RNA from a DNA template is known as:
A. Replication B. Translation
C. Transcription D. Mutation

21. An organism which has two identical alleles of a gene is called:
A. Dominant B. Hybrid
C. Heterozygous D. Homozygous

22. Which of the following is the SI unit of Absorbance (A) of light by a solution?
A. Absorbance (A) is a unitless quantity.
B. Candela
C. Candela/m^2 (Candela per square meter)
D. Candela/s (Candela per second)

23. Which of the following molecular genetic techniques is used to identify protein-protein interactions?
A. Southern hybridization analysis
B. Polymerase chain reaction
C. Fluorescence in situ hybridization
D. Yeast two-hybrid system

24. At a pH below its isoelectric point, an amino acid exists as:
A. Anion
B. Cation
C. Undissociated molecule
D. Zwitterion

25. Which of the following is the immediate source of energy for active transport?
A. Carbohydrates B. Lipids
C. ATP D. NADP

26. Which of the following is NOT a method to introduce genetic variation into bacteria?
A. DNA amplification
B. Transduction
C. Transformation
D. Directed Mutagenesis

27. Match the vitamins in Group-I with the processes/reactions in Group-II.

Group-I	*Group-II*
P. Pantothenic acid	1. Electron transport
Q. Riboflavin (Vitamin 2)	2. Transfer of 1-C units
R. Vitamin B6	3. Decarboxylation
S. Folic acid	4. Fatty acid metabolism
	5. Hydrolysis

A. P-5, Q-2, R-4, S-1
B. P-4, Q-1, R-3, S-2
C. P-4, Q-2, R-1, S-5
D. P-2, Q-1, R-3, S-5

28. Which of the following represents a dihybrid?
A. WsSs B. WsSS
C. WWSs D. wwSs

29. Which technique is most frequently employed for repeated use of enzymes in bioprocesses?
A. Polymerization B. Isomerization
C. Ligation D. Immobilization

30. Match the entries in Group-I with the most suitable methods of sterilization in Group-II.

Group-I	*Group-II*
P. Serum	1. Autoclave
Q. Luria broth	2. Membrane filtration

R. Biological safety cabinets	3. UV irradiation
	4. Gamma irradiation

A. P-4, Q-2, R-3　　B. P-1, Q-4, R-2
C. P-2, Q-1, R-3　　D. P-1, Q-1, R-4

31. Which of the following processes is involved during the identification of blood groups?
A. Precipitation　　B. Immunization
C. Opsonization　　D. Agglutination

32. Independent assortment of genes occurs due to the orientation of chromosomes at:
A. Metaphase I of meiosis
B. Metaphase II of meiosis
C. Metaphase of mitosis
D. Telophase of mitosis

33. Which one of the following is a known protoplast fusion inducing agent?
A. Nucleotides
B. ATP
C. Polyethylene glycol
D. Colchicine

34. Given is the sequence of one strand of DNA-5' TCGATC 3'. Which of the following is the sequence of the complementary strand?
A. AGCTAG　　B. TCGATC
C. CTAGCT　　D. ACTAGC

35. A solute binds with an adsorbent with a dissociation constant K_D. The retention time of the solute in a chromatography column containing this adsorbent:
A. Influences K_D
B. Increases with increasing K_D
C. Decreases with increasing K_D
D. Is independent of K_D

36. Heat inactivation of serum is carried out to inactivate:
A. Prions
B. Pathogenic bacteria
C. Mycoplasma
D. Complement

37. Nude mice are a type of laboratory animal with a genetic mutation that results in an inhibited immune system due to a greatly reduced number of T cells. Nude mice refers to:
A. Mice without thymus
B. Mice without skin
C. Knockout mice
D. Transient mice

38. Which of the following protects cells by preventing crystallization of water during cryopreservation of mammalian cells?
A. Dimethyl Sulphoxide
B. Polyvinyl Alcohol
C. Ethanol
D. Ethylene oxide

39. Which of the following organelle is found only in plants but not in animals?
A. Nucleus　　B. Plastids
C. Mitochondria　　D. Ribosomes

40. The catalytic efficiency for an enzyme is defined as:
A. K_{cat}　　B. V_{max}/K_{cat}
C. K_{cat}/K_m　　D. K_{cat}/V_{max}

41. Endogenous antigens are presented on to the cell surface along with:
A. MHC-I　　B. MHC-II
C. Fc receptor　　D. Folic acid receptor

42. In steady state fermentation, the washout stage occurs if:
A. Dilution rate is less than the maximum specific growth rate
B. Dilution rate is higher than the maximum specific growth rate
C. Cell concentration becomes maximum
D. Substrate concentration becomes minimum

43. In a thin layer chromatography experiment using a silica gel plate, the solvent front and a compound showed migration of 20 cm and 15 cm, respectively. What would be the R_f value for the compound?
A. 0.25　　B. 0.33
C. 0.75　　D. 1.33

44. Prior exposure to pathogens is known to increase resistance to future pathogen attacks. This phenomenon is known as:
A. Systemic acquired resistance
B. Hypersensitive response
C. Innate immunity
D. Antibody mediated response

45. Match the entries in the Group-I with the elution conditions in Group-II.

Group-I	*Group-II*
P. Ion-exchange chromatography	1. Isocratic solvent
Q. Hydrophobic column chromatography	2. Ampholytes
R. Gel filtration chromatography	3. Increasing gradient of salt
S. Chromatofocusing	4. Decreasing gradient of polarity

A. P-4, Q-1, R-2, S-3 B. P-4, Q-3, R-1, S-2
C. P-3, Q-4, R-2, S-1 D. P-3, Q-4, R-1, S-2

46. During sterilization of a fermentation medium in a given bioreactor, $\nabla_{heating} = 12.56$, $\nabla_{cooling} = 7.48$ and the total value of ∇ required for whole sterilization process is 52, where $\nabla\nabla$ is the design criteria. What is the value of $\nabla_{holding}$?

A. 31.96 B. 72.04
C. 46.92 D. 57.08

47. Cell lysis needs to be carried out as a down-stream process if the product of interest is:

A. Extracellular B. Intracellular
C. Toxic D. Heat labile

48. Which of the following features is NOT required in a prokaryotic expression vector?

A. Origin of replication
B. Selection Marker
C. Ribosome binding site
D. CMV promoter

49. Which of the following is a sequence alignment tool?

A. BLAST B. PRINT
C. NCBI D. EMBL

50. For an enzymatic reaction, K_m is the defined as the substrate concentration at which:

A. The reaction rate is double of the maximum
B. The reaction rate is one half of the maximum
C. The enzyme is completely saturated with substrate
D. The enzymatic reaction stops

51. The helix content of a protein can be determined using:

A. A high resolution electron microscope
B. A fluorescence spectrometer
C. A circular dichroism spectrometer
D. A UV-Visible spectrophotometer

52. Which of the following is a protein sequence database?

A. GenBank B. PIR
C. DDBJ D. FASTA

53. When a cell is about to divide, before cell division, the amount of DNA in cell becomes:

A. One fourth B. Half
C. Less than Half D. Double

54. The use of living organisms to degrade the pollutants present in environment is called:

A. Clarification B. Centrifugation
C. Bioremediation D. Amplification

55. The process of introducing DNA into cells by exposing them to high voltage electric pulse is known as:

A. Electrofission B. Electrofusion
C. Electrosonication D. Electroporation

56. The effect of a reversible competitive inhibitor on an enzyme can be nullified by:

A. Increasing the concentration of reactant
B. Increasing the concentration of product
C. Increasing the temperature of reaction
D. Increasing the salt concentration

57. The specific activity of an enzyme in a crude extract of *S. Cerevisiae* is 10 units/mg of protein. The extract is passed through an ion-exchange column. After this purification step, the specific activity of the enzyme changes to 68 units/mg of protein. The fold purification for this step is:

A. 0.14 B. 6.8
C. 680 D. 58

58. An apoenzyme is:

A. Vitamin B. Amino Acid
C. Carbohydrate D. Inactive Protein

59. Protoplasts are the cells which are lacking in:

A. Cell wall B. Cell Membrane
C. Cytoplasm D. Periplasm

60. Primary mediator in anaphylaxis reactions is:
A. Histamine B. Seratonin
C. Heparin D. Aspargine

61. Proteomics is the study of:
A. Regulatory proteins of an organism
B. Structural proteins of an organism
C. Catalytic proteins of an organism
D. All proteins of organism

62. Which one of the following is an RNA dependent DNA synthetase (an enzyme that synthesizes DNA using RNA as a template)?
A. DNA Polymerase I
B. DNA Polymerase II
C. Reverse Transcriptase
D. Deoxy ribonuclease

63. Using monod's model, determine the dilution rate at the cell wash out condition for a chemostat with the given parameters $\mu_{max} = 1\ h^{-1}$; $Y_{x/s} = 0.55\ gg^{-1}$; $K_s = 0.2\ gL^{-1}$; $S_0 = 10\ gL^{-1}$.
A. 0.98 h^{-1} B. 1.98 h^{-1}
C. 0.49 h^{-1} D. 1.49 h^{-1}

64. Association of nitrogen fixing bacteria with the leguminous roots is an example of:
A. Parasitism B. Neutralism
C. Mutualism D. Communalism

65. IPR stands for:
A. Indian Patent Rights
B. International Patent Rights
C. International Property Rights
D. Intellectual Property Rights

66. Which statement is *true* of both prokaryotic and eukaryotic cells?
A. Prokaryotic cells are generally much larger than eukaryotic cells
B. Eukaryotic cells have ribosomes but prokaryotic cells do not
C. Both have DNA as their primary genetic material
D. Eukaryotic cells have plasma membrane and prokaryotic cells do not

67. Which of the following biomolecules is likely to have buffering action?
A. Carbohydrates B. Proteins
C. DNA D. Lipids

68. Resolving power of a microscope is determined by:
A. Wavelength of light source
B. Intensity of light source
C. Power of light source
D. Does not depend on any feature of light source

69. The soil bacteria most commonly used for transformation in plants is:
A. *Rhizobacterium*
B. *Escherichia coli*
C. *Agrobacterium tumifaciens*
D. *Mycorrhizae*

70. During fermentation, the top portion inside the reactor is left free without broth. This portion is called:
A. Shaft B. Head space
C. Impeller D. Sparger

71. The sugar moiety present in DNA is:
A. Deoxyribose B. Ribose
C. Fructose D. Ribulose

72. The micro-organism most commonly used for industrial production of amylase enzyme is:
A. *Saccharomyces cerevisiae*
B. *Aspergillus niger*
C. *Bacillus subtilis*
D. *Penicillium chrysogenum*

73. Analysis of protein antigen is carried out by which of the following techniques:
A. Southern blot B. Northern blot
C. Eastern blot D. Western blot

74. Batch fermentation is an example of:
A. Closed system
B. Open system
C. Fed-batch system
D. Equilibrium system

75. The enzyme usually used in Polymerase Chain Reaction is:
A. RNA polymerase
B. Taq polymerase
C. Ribonuclease
D. Endonuclease

ANSWERS

1	2	3	4	5	6	7	8	9	10
A	D	B	D	C	A	B	D	C	A
11	**12**	**13**	**14**	**15**	**16**	**17**	**18**	**19**	**20**
D	B	C	B	A	C	D	B	C	C
21	**22**	**23**	**24**	**25**	**26**	**27**	**28**	**29**	**30**
D	A	D	B	C	A	B	A	D	C
31	**32**	**33**	**34**	**35**	**36**	**37**	**38**	**39**	**40**
D	A	C	A	C	D	A	A	B	C
41	**42**	**43**	**44**	**45**	**46**	**47**	**48**	**49**	**50**
A	B	C	A	D	A	B	D	A	B
51	**52**	**53**	**54**	**55**	**56**	**57**	**58**	**59**	**60**
C	B	D	C	D	A	B	D	A	A
61	**62**	**63**	**64**	**65**	**66**	**67**	**68**	**69**	**70**
D	C	A	C	D	C	B	A	C	B
71	**72**	**73**	**74**	**75**					
A	C	D	A	B					

EXPLANATORY ANSWERS

2. An **autoclave** is a pressure chamber used to carry out industrial processes requiring elevated temperature and pressure different from ambient air pressure. Autoclaves are used in medical applications to perform sterilization and in the chemical industry to cure coatings and vulcanize rubber and for hydrothermal synthesis. They are also used in industrial applications, especially regarding composites, see autoclave (industrial).

3. **Lipopolysaccharides** (**LPS**), also known as lipoglycans and **endotoxins**, are large molecules consisting of a lipid and a polysaccharide composed of O-antigen, outer core and inner core joined by a covalent bond; they are found in the outer membrane of Gram-negative bacteria, and elicit strong immune responses in animals. The term *lipooligosaccharide* ("LOS") is used to refer to a low-molecular-weight form of bacterial lipopolysaccharides.

5. Methionine, cysteine, homocysteine, and taurine are the 4 common sulphur-containing amino acids, but only the first 2 are incorporated into proteins. Sulphur belongs to the same group in the periodic table as oxygen but is much less electronegative. This difference accounts for some of the distinctive properties of the sulphur-containing amino acids. Methionine is the initiating amino acid in the synthesis of virtually all eukaryotic proteins; N-formylmethionine serves the same function in prokaryotes. Within proteins, many of the methionine residues are buried in the hydrophobic core, but some, which are exposed, are susceptible to oxidative damage. Cysteine, by virtue of its ability to form disulphide bonds, plays a crucial role in protein structure and in protein-folding pathways. Methionine metabolism begins with its activation to S-adenosylmethionine. This is a cofactor of extraordinary versatility, playing roles in methyl group transfer, 5′-deoxyadenosyl group transfer, polyamine synthesis, ethylene synthesis in plants, and many others.

7. The chemical DNA was first discovered in 1869, but its role in genetic inheritance was not demonstrated until 1943. In 1953, James Watson and Francis Crick determined that the structure of DNA is a double-helix polymer, a spiral consisting of two DNA strands wound around each other. Each strand is composed of a long chain of monomer nucleotides. The nucleotide of DNA consists of a deoxyribose sugar molecule to which is attached a phosphate group and one of four nitrogenous bases: two purines (adenine and guanine) and two pyrimidines (cytosine and thymine). The nucleotides are joined together by covalent bonds between the phosphate of one nucleotide and the sugar of the next, forming a phosphate-sugar backbone from which the nitrogenous bases protrude. One strand is held to another by hydrogen bonds between the bases; the sequencing of this bonding is specific—*i.e.,* adenine bonds only with thymine, and cytosine only with guanine.

9. **RNA**, abbreviation of **ribonucleic acid**, complex compound of high molecular weight that functions in cellular protein synthesis and replaces DNA (deoxyribonucleic acid) as a carrier of genetic codes in some viruses. RNA consists of ribose nucleotides in strands of varying lengths. The structure varies from helical to uncoiled strands. One type, transfer RNA (tRNA), sometimes called soluble, or activator, RNA, contains fewer than 100 nucleotide units; other types contain thousands of units. The nitrogenous bases in RNA are adenine, guanine, cytosine, and uracil.

11. The **International Nucleotide Sequence Database Collaboration** (**INSDC**) consists of a joint effort to collect and disseminate databases containing DNA and RNA sequences. It involves the following computerized databases: DNA Data Bank of Japan (Japan), GenBank (USA) and the European Nucleotide Archive (UK). New and updated data on nucleotide sequences contributed by research teams to each of the three databases are synchronized on a daily basis through continuous interaction between the staff at each the collaborating organizations.

19. In genetics, a **centimorgan** (abbreviated **cM**) or **map unit** (m.u.) is a unit for measuring genetic linkage. It is defined as the distance between chromosome positions (also termed loci or markers) for which the expected average number of intervening chromosomal cross overs in a single generation is 0.01. It is often used to infer distance along a chromosome. However, it is not a true physical distance.

20. **Transcription** is the first step of gene expression, in which a particular segment of DNA is copied into RNA (especially mRNA) by the enzyme RNA polymerase. Both DNA and RNA are nucleic acids, which use base pairs of nucleotides as a complementary language. During transcription, a DNA sequence is read by an RNA polymerase, which produces a complementary, antiparallel RNA strand called a primary transcript.

25. Active transport is the movement of molecules across a membrane from a region of their lower concentration to a region of their higher concentration—in the direction against the concentration gradient. Active transport requires cellular energy to achieve this movement. There are two types of active transport – primary active transport that uses ATP, and secondary active transport that uses an electrochemical gradient. An example of active transport in human physiology is the uptake of glucose in the intestines.

31. **Agglutination** is the process that occurs if an antigen is mixed with its corresponding antibody called isoagglutinin. This term is commonly used in blood grouping. This occurs in biology in three main examples: The clumping of cells such as bacteria or red blood cells in the presence of an antibody or complement.

36. The practice of heat inactivating serum was originally developed when only serum from adult animals was available for cell culture. Adult serum contains various immune factors, particularly serum complement, which may inhibit or destroy cells under certain conditions. Heating serum is intended to inactivate serum complement. Today, serum is often heat-inactivated without any evidence of beneficial effect, simply because an earlier protocol calls for such action. However, certain applications demand the inactivation of complement; e.g., it is often required to preserve the integrity of immunoassays. Heat inactivation will increase precipitates and consequent turbidity in the serum.

39. Most organelles are common to both animal and plant cells. However, plant cells also have features that animal cells do not have: a **cell wall**, a large **central vacuole**, and plastids such as chloroplasts. Plants have very different lifestyles from animals, and these differences are apparent when you examine the structure of the plant cell. Plants make their own food in a process called **photosynthesis**. They take in carbon dioxide (CO_2) and water (H_2O) and convert them into sugars.

44. Systemic acquired resistance (SAR) is an induced immune mechanism in plants. Unlike vertebrate adaptive immunity, SAR is broad spectrum, with no specificity to the initial infection. An avirulent pathogen causing local programmed cell death can induce SAR through generation of mobile signals, accumulation of the defense hormone salicylic acid, and secretion of the antimicrobial PR (pathogenesis-related) proteins. Consequently, the rest of the plant is protected from secondary infection for a period of weeks to months. SAR can even be passed on to progeny through epigenetic regulation.

49. In bioinformatics, **BLAST** for **B**asic **L**ocal **A**lignment **S**earch **T**ool is an algorithm for comparing primary biological sequence information, such as the amino acid sequences of proteins or the nucleotides of DNA sequences. A BLAST search enables a researcher to compare a query sequence with a library or database of sequences, and identify library sequences that resemble the query sequence above a certain threshold.

58. **An apoenzyme** is an inactive enzyme, activation of the enzyme occurs upon binding of an organic or inorganic cofactor. Holoenzyme– **An apoenzyme** together with its cofactor. A holoenzyme is complete and catalytically active. Most cofactors are not covalently bound but instead are tightly bound.

69. ***Agrobacterium*** is a genus of Gram-negative bacteria established by H.J. Conn that uses horizontal gene transfer to cause tumors in plants. *Agrobacterium tumefaciens* is the most commonly studied species in this genus. *Agrobacterium* is well known for its ability to transfer DNA between itself and plants, and for this reason it has become an important tool for genetic engineering.

The *Agrobacterium* genus is quite heterogeneous. Recent taxonomic studies have reclassified all of the *Agrobacterium* species into new genera, such as *Ahrensia*, *Pseudorhodobacter*, *Ruegeria*, and *Stappia*, but most species have been controversially reclassified as *Rhizobium* species.

71. The backbone of the DNA strand is made from alternating phosphate and sugar residues. The sugar in DNA is 2-deoxyribose, which is a pentose (five-carbon) sugar. The sugars are joined together by phosphate groups that form **phosphodiester** bonds between the third and fifth carbon atoms of adjacent sugar rings.

75. Taq **polymerase**. Like DNA replication in an organism, **PCR** requires a DNA **polymerase enzyme** that makes new strands of DNA, using existing strands as templates. The DNA **polymerase typically used in PCR** is called Taq **polymerase**, after the heat-tolerant bacterium from which it was isolated (Thermus aquaticus).

Previous Paper (Solved)

M.Sc.—Biotechnology Entrance Exam, CET (PG), 2016

1. What would be the isoelectric point of phenylalanine if pKa for COOH group is 1.54 and the NH_3^+ group is 8.54?
A. 1.54
B. 5.04
C. 8.54
D. 10.08

2. Which is a correct statement?
A. A protein will have +ve charge at pH below its isoelectric point
B. A protein will have +ve charge at pH above its isoelectric point
C. pH does not affect charge on a protein
D. A protein will have –ve charge at pH below its isoelectric point

3. Which of the following statements is correct?
A. An aqueous solution of a strong base will dissociate to form conjugate acid
B. The pH of a buffer does not change with dilution
C. The pH of a buffer does not change with temperature
D. A strong electrolyte solution always contain high concentration of acid or base

4. Peptide bond in a protein absorbs UV at:
A. 280 nm B. 260 nm
C. 230 nm D. 200 nm

5. Tertiary structure of globular proteins is stabilized primarily by:
A. Ionic interactions
B. vander Waal's forces
C. Hydrophobic interactions
D. Disulphide bonds

6. Which of the following is possible with SDS PAGE?
A. Determination of isoelectric point of a protein
B. Determine biological activity of a protein
C. Separate proteins on the basis of charge
D. Separate proteins on the basis of molecular mass

7. Which disease is not related to protein misfolding?
A. Alzheimer
B. Kuru
C. Ataxia telangiectasia
D. Creutzfeld-Jakob disease

8. In a helix turn helix motif of a repressor which amino acid likely to interact with DNA?
A. Glycine B. Lysine
C. Glutamate D. Aspartate

9. Why is collagen assembly sensitive to substitution of glycine with any other amino acids in its primary sequence?
A. Glycine is cross linked to lysine in the triple helix
B. Collagen is glycosylated at the glycine residues
C. Side chains of glycine fit well into the space available in the triplex helix of collagen
D. Interchain hydrogen bonding is done by glycine

10. In prokaryotes translation is guided by:
A. TATA box
B. Pribnow-Schaller box
C. Shine Dalgarno sequence
D. Kozak sequence

11. RNA has maximum stability at which pH?
A. Acidic pH
B. Alkaline pH
C. Neutral pH
D. Not stable at both acid and alkaline pH

12. What is the function of peroxisome?
A. Oxidation of fatty acids with production of energy
B. Oxidation of fatty acids with production of hydrogen peroxide
C. Oxidation of sugars without ATP production
D. Production of acid hydrolases

13. Which of the following is likely to happen during purification of a protein by column chromatography?
A. Protein with less charge will elute at high salt concentration
B. Protein with high charge will elute at low salt concentration
C. Large protein will elute slowly from gel filtration column
D. Small protein will elute slowly from gel filtration column

14. A functional unit of a protein that can fold independently:
A. Motif
B. Fold
C. Domain
D. Module

15. Antibiotic tetracycline blocks protein synthesis by inhibiting:
A. Binding of amino acyl *t*RNA to ribosome
B. Peptidyl transferase
C. Translocase
D. Binding of large subunit of ribosome *m*RNA

16. Which initiation facto prevents charged *t*RNA other than initiator *t*RNA to be associated with small subunit of ribosome?
A. IF 1 B. IF 2
C. IF 3 D. IF 4

17. Which of the following characteristics is not shared by haemoglobin and myoglobin?
A. Presence of heme as the prosthetic group
B. Multi subunit protein
C. Reversible binding of oxygen
D. Involvement of histidine residue in oxygen binding

18. Which of the following is observed after series of successive purification steps for a protein?
A. Specific activity and percentage yield of the protein will decrease
B. Specific activity will remain same but percentage yield of the protein will increase
C. Specific activity will decrease but percentage yield of the protein will increase
D. Specific activity will increase but percentage yield of the protein will decrease

19. Which glycolytic enzyme is not allosterically controlled?
A. Hexokinase
B. Phosphofructo kinase
C. Phosphoglycerate kinase
D. Pyruvate kinase

20. Which enzyme combination will give maximum alcohol production from starch?
A. α-amylase and glucose isomerase
B. α-amylase and β-amylase
C. β-amylase and glucoamylase
D. α-amylase and glucoamylase

21. When do you expect highest level of transcription from *E. coli* lac promoter?
A. Low glucose high lactose concentration in medium
B. High glucose high lactose concentration in medium
C. High glucose low lactose concentration in medium
D. Low glucose low lactose concentration in medium

22. Which of the following industrial enzymes is not involved in the activity listed?
A. α–amylase in beer making
B. renin in cheese making
C. papain as meat tenderizer
D. protease in detergent

23. Based on the given data on enzyme kinetics which is the best substrate for the enzyme:

	Km (μM)	**Kcat (Sec^{-1})**
A.	0.01	2×10^{-2}

B. 0.1	4×10^{-4}
C. 1.0	2×10^{2}
D. 10	7×10^{1}

24. Which statement is true for *E. coli* tac promoter?

A. It is a strong constitutive promoter used for high level of expression of heterologous proteins in *E. coli*
B. It is a weak promoter used for expression of toxic heterologous proteins in *E. coli*
C. It contains –10 region of trp promoter and –35 of lac promoter
D. It is repressed by lac repressor and derepressed by IPTG

25. Tryptophan acts as ———— in regulating tryptophan operon.

A. Inducer B. Repressor
C. Co-repressor D. Enhancer

26. Which enzyme is not associated with gene silencing?

A. Histone methyl transferase
B. Histone deacetylase
C. Histone acetyl transferase
D. DNA methyl transferase

27. Which of the following factor will not activate lytic phase of lamda phage?

A. Abundance of CI protein
B. High level of cro protein
C. Activation of rec A
D. Environmental stress

28. Which virus is associated with micro-cephalopathy?

A. Chikungunya
B. Zika
C. Ebola
D. Hunta

29. Which is wrongly paired?

A. Infection through aerosols — Polio
B. Nosocomial infection — *Pseudomonas aeruginosa*
C. Food intoxication — *Staphylococcus aureus*
D. Zoonosis — Ebola

30. Which statements are correct with respect to hepatitis?

1. Type A and E hepatitis viruses spread through oral fecal route.
2. High prevalence of Hepatitis C virus is seen in patients with hemophilia.
3. Vaccination cannot prevent infection by Hepatitis A virus.
4. All hepatitis viruses have DNA as the genome.

A. 1 and 2 B. 1 and 3
C. 2 and 3 D. 2 and 4

31. Mark the wrong pair:

A. Ethanol — *Zymomonas mobilis*
B. Insulin — *Saccharomyces cerevisiae*
C. Citric acid — *Aspergillus flavus*
D. Lysine — *Corynebacterium glutamicum*

32. Which characteristic is not shown by Archeae bacteria?

A. Presence of S-layer on surface
B. Presence of N-acetyltalosaminuronic acid in cell wall
C. Pseudomurein cross linked by D amino acids
D. Membrane lipid with ether linkage

33. A light microscope has resolution upto:

A. 2 nm B. 0.2 μm
C. 2.0 μm D. 20 μm

34. Identify the correct pairing:

(*a*) Phagocytes	1. Gut epithelium
(*b*) M Cell	2. Allergy
(*c*) Basophils	3. Malignant cells
(*d*) Natural killer cells	4. Toll like receptors

A. (*a*)–2, (*b*)–3, (*c*)–1, (*d*)–4
B. (*a*)–4, (*b*)–1, (*c*)–2, (*d*)–3
C. (*a*)–3, (*b*)–4, (*c*)–2, (*d*)–1
D. (*a*)–1, (*b*)–2, (*c*)–4, (*d*)–3

35. Mark the correct statement:

A. Eosin Methylene Blue medium can differentiate between *E. coli* and *Enterobacter*
B. MacConkey Agar is selective for Gram positive lactose fermenters

C. Yeast Extract Mannitol Agar is used to isolate *Azotobacter* from soil
D. No growth on blood agar is indicative of *Staphylococcus* aureus

36. Which of the following is most heat resistant?
A. *Coxiella burnetii*
B. *Mycobacterium tuberculosis*
C. *Mycobacterium bovis*
D. *Staphylococcus aureus*

37. Which scientist gave the method for phylogenetic analysis of bacteria?
A. Carl Woese B. Carig Venter
C. Kary Mulis D. James Watson

38. Which of the following features are exclusive to Gram positive bacteria?
(*i*) Presence of peptidoglycan in cell wall
(*ii*) Reproduction by binary fission
(*iii*) Teichoic acid in cell wall
(*iv*) Formation of endospores
A. (*i*) and (*iii*) B. (*ii*) and (*iv*)
C. (*ii*) and (*iii*) D. (*iii*) and (*iv*)

39. Which of the following carcinogens act directly with DNA?
A. Ethyl methyl sulphonate
B. Benzo(a) pyrene
C. Dimethylnitrosoamine
D. Aflatoxin B

40. Thymine-Thymine dimers formed in DNA by UV exposure are repaired by bacteria through:
A. Mismatch repair
B. Excision repair
C. End Joining repair
D. SOS repair

41. How many chromosomes are present in *Agrobacterium tumefaciens?*
A. 1 B. 2
C. 3 D. 4

42. Rolling circle replication is not shown by:
A. M13 phage B. Lamda phage
C. Adenovirus D. Herpesvirus

43. Which of the following can degrade cell wall of non dividing bacteria?
A. Penicillin B. Ampicillin
C. Lysozyme D. Chitinase

44. Mark the correct order on the bases of genome size :
A. *E. coli* > *S. cerevisiae* > *H. influenza* > Φ × 174
B. *H. influenza* > *E. coli* > *S. cerevisiae* > Φ × 174
C. *S. cerevisiae* > *E. coli* > *H. influenza* > Φ × 174
D. Φ × 174 > *S. cerevisiae* > *E. coli* > *H. influenza*

45. Which BLAST program would be used to search translated nucleotide data base using a protein query?
A. Blast X B. tblast X
C. tblast n D. Blast p

46. Which of the ozone depleting substance has not been phased out from India?
A. CFC B. HCFC
C. Halons D. CTC

47. In 2015, the meeting on climate change was held in:
A. Paris B. Brazil
C. Copenhagen D. Brussels

48. Which is not true about nitrobacter?
A. It is a nitrifying bacteria
B. Nutritionally it is a chemolithoautotroph
C. It can oxidize ammonia
D. It uses oxygen as the final acceptor of electrons

49. Which enzyme plays an important role in degradation of xenobiotics?
A. Superoxide dismutase
B. Methyltransferase
C. Oxidoreductase
D. Hydrolase

50. *In situ* cleaning up of contaminated soil cannot be done by:
A. Bioaugmentation B. Biostimulation
C. Co-metabolism D. Dredging

51. Major source of sulphur dioxide is:
A. Power plants
B. Industries
C. Automobiles
D. Volcanic emissions

52. Mark the incorrect statement:
A. Recombinant cosmids are single copy plasmids that can be packaged into lamda phage heads
B. A recombinant lamda phage cannot package DNA insert less than 38 kb.
C. BACs are high capacity but single copy vectors
D. Transformation efficiency of recombinant plasmid decreases when insert is more than 15 kb.

53. Which of the following enzyme does not show proof reading activity?
A. Klenow fragment
B. *E. coli* DNA Polymerase 1
C. Taq DNA polymerase
D. T7 DNA polymerase

54. Mark the wrong statement:
A. YEp based on 2 μ plasmid have been extensively used for production of heterologous proteins
B. *Pichia pastoris,* a methylotrophic yeast has been used for production of biopharmaceuticals
C. Yeast cells can carry out post translational modifications of eukaryotic proteins
D. Yeast cells can be used for efficient expression of heterologous genes under inducible alcohol dehydrogenase promoter

55. Which is the simplest technique to check the differential expression of genes?
A. Genomes sequencing
B. Real time PCR
C. Microarray
D. Transcriptome analysis

56. Major drawback of current gene therapy vector system is:
A. Immunogenicity
B. Lack of cell specificity
C. Delivery into cell nucleus
D. Degradation of therapeutic gene

57. Which statement is incorrect?
A. Bt brinjal has improved crop yield in India
B. Bt brinjal is resistant to lepidopterans
C. Bt cotton seeds are supplied by Mahyco Biotech
D. Lepidopterans have shown resistance to Bt cotton

58. For nick translation of DNA identify the requirements:
A. DNA polymerase, DNase, primers, dNTPs, ligase
B. DNA polymerase, S1 nuclease, labelled dNTPs, ligase
C. DNA dependent RNA polymerase, DNase, labelled dNTPs, ligase
D. DNA polymerase 1, DNase, labelled dNTPs, ligase

59. Horizontal gene transfer can occur through all except:
A. conjugation
B. binary fission
C. transposons
D. phages

60. *Pfu* polymerase is used in PCR because of its:
A. High thermal stability
B. High fidelity
C. High processivity
D. Low RNase activity

61. The landmark discovery that dsRNA can repress gene expression was done in :
A. *Drosophila*
B. *Zebra fish*
C. *Caenorhabditis elegans*
D. *Arabidopsis thaliana*

62. Under which of the following growth conditions, recombinant culture would give least biomass?
A. A continuous reactor with low concentration of glucose
B. A fed batch reactor with low glucose concentration
C. A batch fermentation with low initial glucose concentration
D. A batch fermentor with high initial glucose concentration

63. Which statement about cDNA library is incorrect?
- A. RNA is used as the template
- B. It enriches coding sequences
- C. It is bigger than genomic library
- D. It is useful for cloning of eukaryotic genes

64. Which enzyme is used for labelling of 5' end of DNA?
- A. Ligase
- B. Terminal transferase
- C. Alkaline phosphatase
- D. Polynucleotide kinase

65. Flavr Savr tomatoes were created by RNAi technology using:
- A. Electroporation
- B. Gene gun
- C. Protoplast fusion
- D. *Agrobacterium tumefaciens*

66. Yeast cells can be transformed by:
- A. Lithium chloride
- B. Calcium chloride
- C. Electroporation in the presence of sodium chloride
- D. Calcium chloride in the presence of PEG

67. Who is the Father of Green Revolution?
- A. Norman Borloug
- B. David Petrick
- C. Paul Berg
- D. Nathan Arber

68. Which method is most suitable to produce alkaloids in plant tissue culture?
- A. Pollen culture
- B. Callus culture
- C. Embryo culture
- D. Suspension culture

69. Binary vector used for transformation of plants does not contains:
- A. Vir genes
- B. Transgene
- C. Left and right border sequences
- D. *E. coli* origin of replication

70. In plant tissue culture high Auxin to cytokine ratio will:
- A. Support growth of callus cells
- B. Cause death of callus cells
- C. Support shoots formation
- D. Support root formation

71. Which of the following is not an X linked disease?
- A. Colour blindness
- B. Haemophilia
- C. Albinism
- D. Duchenne muscular dystrophy

72. Which of the following statements are correct?
- (*i*) Class I MHC proteins are found on all nucleated cells
- (*ii*) Class I MHC proteins process antigens of intracellular pathogens
- (*iii*) Bacterial antigens are processed by class I MHC molecules and presented by macrophages
- (*iv*) Class I MHC molecules present antigens to helper T cells

- A. (*i*) and (*iv*)
- B. (*i*) and (*ii*)
- C. (*ii*) and (*iv*)
- D. (*iii*) and (*iv*)

73. Which of the following statements is incorrect?
- A. Release of cytochrome C from mitochondria activates caspases
- B. Caspases are serine proteases that act on cytoskeleton proteins
- C. They selectively cleave proteins at C terminal to aspartate
- D. Binding of neural growth factor to receptor on neurons block caspase activation

74. A cell expressing telomerase enables it to synthesize DNA:
- A. Independent of DNA or RNA template
- B. Using DNA template that is part of the enzyme complex
- C. Using RNA template that is part of the enzyme complex
- D. Has no role in DNA replication

75. What is the mechanism of inactivation of cyclins at the end of each phase of cell cycle?
- A. Dephosphorylation
- B. Ubiquitinylation
- C. Farnesylation
- D. Glycosylation

ANSWERS

1	2	3	4	5	6	7	8	9	10
B	A	B	D	C	D	C	B	C	C
11	**12**	**13**	**14**	**15**	**16**	**17**	**18**	**19**	**20**
A	B	D	C	A	B	B	D	C	D
21	**22**	**23**	**24**	**25**	**26**	**27**	**28**	**29**	**30**
A	B	C	D	C	C	A	B	A	A
31	**32**	**33**	**34**	**35**	**36**	**37**	**38**	**39**	**40**
C	C	B	B	A	A	A	D	A	B
41	**42**	**43**	**44**	**45**	**46**	**47**	**48**	**49**	**50**
B	C	C	C	C	B	A	C	C	D
51	**52**	**53**	**54**	**55**	**56**	**57**	**58**	**59**	**60**
A	*	C	D	B	A	A	D	B	B
61	**62**	**63**	**64**	**65**	**66**	**67**	**68**	**69**	**70**
C	D	C	D	D	A	A	D	A	D
71	**72**	**73**	**74**	**75**					
C	B	B	C	B					

Note : *An '*' in the key indicates that either the question is ambiguous or it has printing mistake. All candidates will be given credit for this question.*

EXPLANATORY ANSWERS

2. Isoelectric focusing is the separation of proteins with different isoelectric points using electrophoresis. A pH gradient is established within the gel slab so that one end is basic and the other end is acidic. At any pH value below its isoelectric point, a protein will have a net positive charge and move toward the cathode. At pH values above its isoelectric point, a protein carries a net negative charge and will move toward the anode. At its isoelectric point, the protein is electrically neutral and does not move.

3. The Henderson-Hasselbalch equation predicts that the pH of a buffer should not change with dilution, but this is only true over a small range due to non-ideal behaviour of ions in solution.

4. In proteins, absorption of the peptide bond occurs at around 190 nm. Because changes in absorption in this region are ascribed to conformational changes of the backbone of the molecule, they are potentially of great importance.

5. A protein's tertiary structure is its three-dimensional shape. Tertiary structures are mostly determined by hydrophilic and hydrophobic interactions between R groups of amino acids. Hydrophobic residues prefer to be on the interior of proteins, which reduces their proximity to water. Hydrophilic N–H and C=O bonds found in the polypeptide chain get pulled in by these hydrophobic residues.

10. In prokaryotes the so-called Shine-Dalgarno sequence which is recognised by the 16S ribosomal RNA is located within 10 bases upstream of the AUG sequence at which translation is initiated.

12. β-oxidation of fatty acids in peroxisomes is the major pathway for the production of hydrogen peroxide.

28. Virus

Chikungunya	Fever, muscle aches, rash
Ebola virus	Ebola hemorrhagic fever
H1N1	H1N1 (swine) flu
Hantavirus	Hemorrhagic fever associated with severe pulmonary syndrome
Hepatitis C virus	Parenterally transmitted hepatitis
Hepatitis E virus	Enterically transmitted hepatitis
Human immunodeficiency virus (HIV)	HIV infection and AIDS
Human herpesvirus 6 (HHV-6)	Roseola subitum
Human herpesvirus 8 (HHV-8)	Associated with Kaposi sarcoma in immunosuppressed patients, including people with HIV infection
West Nile virus	West Nile fever
Zika virus	Fever, rash, muscle aches, possible microcephaly in children born to infected mothers

33. The theoretically possible resolution in light microscopy is approximately 0.2 μm. The resolving power of an objective lens is the ability to show two object details separately from each other in the microscope image. The higher the numerical aperture and the shorter the wavelength, the better the resolving power.

35. Levine's eosin-methylene blue agar (LEMB) is used for the differentiation of *E. coli* and *Enterobacter aerogenes.* Colonies of *E. coli* show a typical greenish metallic sheen and dark purple centers. Some biotypes of *E. coli,* however, do not produce typical colonies with a green sheen, and slow or nonlactose fermenters produce colourless colonies.

47. The UNFCCC Conference of the Parties (COP) 21, to be held in Paris at the end of 2015, is potentially a very important milestone in international climate change policy. It is at this conference that parties will come together to try to reach a new international climate agreement (the Agreement) to replace Kyoto when it ends in 2020.

51. The main source of sulphur dioxide into the atmosphere is from the burning of fossil fuels in power stations, oil refineries and industrial plants. Vehicles and domestic boilers also release sulphur dioxide. It is also produced naturally as a result of forest fires and from volcanoes.

60. The four DNA fragments (A, B, C, D) encompassing the full-length HIV-1 genome were first amplified individually in a high fidelity PCR by using *Pfu* DNA polymerase. Each of the fragments shared a segment of an overlapping sequence with one or two other fragments.

Previous Paper (Solved)

M.Sc.—Biotechnology Entrance Exam, CET (PG), 2015*

1. The first-ever amino acid decoded by Nirenberg and Matthaei using a RNA composed of poly-Uracil was:

A. Phenylalanine B. Methionine
C. Tyrosine D. Glycine

2. Which is the common metal ion present in Cytochrome C, Cytochrome P450 and ribonucleotide reductase?

A. Fe B. Mo
C. Ni D. Cu

3. The Hardy-Weinberg distribution is a relationship between:

A. Gene frequencies and Genotype frequencies
B. Gene frequencies and Phenotype frequencies
C. Gene pool and Genotype frequencies
D. Gene pool and Phenotype frequencies

4. An increase in total absorbance at 260 nm for a DNA in solution is the result of:

A. Denaturation B. Renaturation
C. Destruction D. Precipitation

5. Meselson-Stahl experiment helped to decode information about:

A. Distinguishing between Conservative and Dispersive replication only
B. Distinguishing between Closed circular DNA and Linear DNA replication
C. Distinguishing between Conservative and Semi-conservative replication
D. Distinguishing between Semi-conservative and Dispersive replication only

6. All of the following statements are true for eukaryotic RNA polymerase, except:

A. Catalyzes the synthesis of RNA
B. One RNA polymerase for all three types of RNA
C. Three types of RNA polymerase for all three types of RNA
D. In all the reactions DNA is the template

7. Identify the correct subunit composition of *E. coli* RNA polymerase core enzyme.

A. $Alpha_2$, Beta, Beta′, Omega
B. $Alpha_2$, Beta, Delta, Omega
C. $Alpha_2$, Beta′, Sigma, Omega
D. Alpha, Beta, Sigma, Omega

8. Diakinesis in meiosis is marked by all of the following, except:

A. Centromere splitting
B. Chiasmata visible
C. Chromatids visible
D. Chromosomes detach from envelope

9. During base excision repair which enzymes combination work to repair damaged base:

A. Glycosylases and polymerases
B. Polymerases and lyases
C. Polymerases and isomerases
D. Glycosylases and lyases

10. A polyribosome is a:

A. Complex of ribosomes
B. Complex of ribosomes and *m*RNA
C. Complex of ribosomes, *t*RNA and *m*RNA
D. Complex of ribosomes and *t*RNA

11. The following neuro-transmitters are derivatives of tyrosine except:

A. Dopamine B. Epinephrine
C. Norepinephrine D. Histamine

12. Which of the following are produced during mixed acid fermentation?

A. An acid and Alcohol
B. An acid and Ester
C. An acid and Ketone
D. An acid and Aldehyde

* *Held on 14 June, 2015*

13. Actinomycin is an inhibitor of the process of transcription by:

A. Intercalates between adjacent G-C base pairs through phenoxazone ring

B. Intercalates between adjacent A-T base pairs through phenoxazone ring

C. Intercalates between adjacent A-C base pairs through phenoxazone ring

D. Inrercalates between adjacent G-T base pairs through phenoxazone ring

14. The linking number in the super helix topology of DNA can be defined as:

A. Number of times that one strand winds around the other

B. Number of supercoil of DNA

C. Number of times that one strand links the other in 5′→3′ direction

D. Number of times that one strand links the other in 3′→5′ direction

15. The only prokaryotic enzyme that can introduce negative supercoils:

A. DNA gyrase B. DNA helicase

C. DNA lyase D. DNA polymerase

16. The Klenow fragment has the following activities:

A. DNA polymerase only

B. DNA polymerase and 3′→5′ exonuclease

C. DNA polymerase and 5′→3′ exonuclease

D. Both DNA 3′→5′ and 5′→3′ exonuclease

17. The enzyme Telomerase is a:

A. Protein B. Modified protein

C. Ribonucleic acid D. Ribonucleoprotein

18. The international sequences repository GenBank is:

A. Genetic sequence database

B. Genetic sequence analysing program

C. Genetic sequence creating program

D. Genetic sequence analysis funding agency

19. Blood group (A, B and O types) of a person is determined by sugars attached to:

A. Proteins and Bilirubin

B. Proteins and lipids

C. Lipids and Myoglobin

D. Lipids and Haemoglobin

20. The enzymes of lysosomes share which of the following important property?

A. All have an optimal activity at a neutral pH

B. All have an optimal activity at an acid pH

C. All have an optimal activity at an alkaline pH

D. All have optimal activity at any pH

21. All of the following are true for Vitamin-E except:

A. Most important location where vitamin E exerts its antioxidant effect is lungs

B. Red and white blood cells are also benefitted by its protective effect

C. Four major forms of vitamin E exist

D. Some forms of vitamin E are water soluble

22. Which of the following features help to identify the cells in metaphase during mitotic cell division?

A. Chromosomes fully extended and located at equatorial planes

B. Chromosomes fully condensed and located at equatorial planes

C. Chromosomes fully extended and located at polar regions

D. Chromosomes fully condensed and located at polar regions

23. The most abundant protein in the world is:

A. Ribulose bisphosphate carboxylase

B. Haemoglobin

C. Keratin

D. Collagen

24. The carbon dioxide acceptor in photosynthetic reaction is:

A. Pyruvate

B. Glycerol 1, 3 bisphosphate

C. Ribulose 1, 5 bisphosphate

D. Glyceraldehyde

25. During Edman degradation phenyl isothiocynate reacts with which of the following group in a peptide?

A. N-terminal amino group

B. C-terminal carboxyl group

C. Side chain thiol group

D. Side chain hydroxyl group

26. Type I hypersensitivity is mediated by:
A. Mast cells and IgE
B. Mast cells and IgG
C. B cells and IgA
D. B cells and IgG

27. All of the following are inhibitors of electron transport chain except:
A. Rotenone B. Cyanide
C. Antimycin A D. Nystatin

28. Myosin is a motor protein because it can:
A. Convert chemical energy of ATP hydrolysis into mechanical energy
B. Convert chemical energy of GTP hydrolysis into mechanical energy
C. Convert mechanical energy into chemical energy of ATP synthesis
D. Convert mechanical energy into chemical energy of GTP synthesis

29. In alcoholic fermentation the steps of Pyruvate degradation are:
A. Pyruvate → Acetaldehyde + CO_2 → Ethanol
B. Pyruvate → Lactate → Ethanol + CO_2
C. Pyruvate → Phosphoenolpyruvate → Ethanol + CO_2
D. Pyruvate → Acetic acid + CO_2 → Ethanol

30. The oxygen binding affinity of Haemoglobin is reduced in the presence of:
A. 1, 2-Bisphosphate glycerate
B. 1, 3-Bisphosphate glycerate
C. 2, 3-Bisphosphate glycerate
D. 1, 2, 3-Trisphosphate glycerate

31. Chitin, the major structural polysaccharide in the exoskeleton of crabs is a polymer of:
A. N-Acetyl fructosamine
B. N-Acetyl Glucosamine
C. N-Acetyl Galactosamine
D. N-Acetyl Mannosamine

32. Identify the cellular location of enzymes of citric acid cycle:
A. Nucleus
B. Cytoplasm
C. Mitochondria
D. Endoplasmic reticulum

33. The nick translation is brought about by:
A. Peptidyl transferase
B. RNA polymerase
C. DNA polymerase I
D. Riboenzyme

34. Which word best describes the reaction catalysed by a Restriction endonuclease?
A. Hydrolysis
B. Redox reaction
C. Phosphoryl group transfer
D. Isomerization

35. During therapeutic cloning which of the following step is performed?
A. Nucleus from egg is replaced by nucleus of sperm and then embryo is generated
B. Nucleus from sperm is replaced by nucleus of somatic cell and then embryo is generated
C. Nucleus from egg is replaced by nucleus of somatic cell and then embryo is generated
D. Somatic cell nucleus from testes and ovary are used to generate embryo

36. Which chemical reactivity best describes the action of DNA ligases?
A. Synthesis of DNA polymers
B. Synthesis of RNA polymers
C. Formation of a phosphodiester bond
D. Formation of a peptide bond

37. The first genome sequence project was accomplished in which of the following species?
A. *H. influenzae*
B. *D. melanogaster*
C. *S. cerevisae*
D. *A. thaliana*

38. Which of the following is key component of hot start PCR?
A. Use of radioactive primers
B. Starting the PCR in hot laboratory
C. Keeping Taq polymerase inactive till the primers are annealed properly
D. Keeping the PCR tubes at 90°C for two hrs before starting PCR

39. The superfamily of heme containing proteins functioning to detoxify xenobiotics are:
A. Cytochrome P460
B. Cytochrome P450
C. Cytochrome P440
D. Cytochrome P420

40. All of the following enzymes equip the cells with antioxidant mechanism except:
A. Superoxide dismutase
B. Catalase
C. Glutathione peroxidase
D. Malate dehydrogenase

41. The purine rich sequences-10 nucleotides upstream from start codon of prokaryotic *m*RNAs, helping to position the ribosome to initiate translation:
A. CCAAT box
B. TATA box
C. Shine-Dalgamo sequence
D. Kozak sequence

42. The "orphan" Genes are the genes with:
A. No known function
B. Arbitrary function
C. Opposite function in prokaryotes and eukaryotes
D. Variant function in variant conditions

43. The following list represents the moderately repetitive sequences in the Human Genome except:
A. Long inter-spersed nuclear elements
B. Short inter-spersed nuclear elements
C. Long terminal repeats
D. Gene cluster elements

44. Which of the following statements are not true for the antibodies?
A. Each antibody has light as well as heavy chain
B. Each light chain has variable and constant regions
C. Variable regions of light chains only form the antigen binding sites
D. Existence of J region increases the amount of diversity

45. The cells that have undergone genetic modifications to grow indefinite are called:
A. Primary cell culture
B. Secondary cell culture
C. Cell strain
D. Cell line

46. During the technique of Two-Dimensional gel electrophoresis:
A. Isoelectric focusing is followed by SDS-PAGE
B. Isoelectric focusing is followed by Native PAGE
C. SDS-PAGE is followed by Isoelectric focusing
D. Native-PAGE is followed by Isoelectric focusing

47. To perform enzymatic amplification of genomic DNA by Polymerase chain reaction all of the following are required except:
A. Taq polymerase
B. Reverse transcriptase
C. Primers
D. Mixture of de-oxyribonucleotides

48. An individual who has both the alleles of his phenotypically different parents for a trait is:
A. Homozygous
B. Heterozygous
C. Dominant
D. Recessive

49. The cells which cannot undergo meiotic division are:
A. Diploid cells
B. Sperm forming cells
C. Ovum generating cells
D. Haploid cells

50. RNA editing is best represented as:
A. Discrete change to specific nucleotide sequences within a RNA molecule after it has been generated by RNA polymerase by removal of introns
B. Discrete changes to specific nucleotide sequences within a RNA molecule after it has been generated by RNA polymerase by addition of a modified G nucleotide at 5' end

C. Discrete changes to specific nucleotide sequences within a RNA molecule after it has been generated by RNA polymerase by addition of poly A tail at 3' end
D. Discrete changes to specific nucleotide sequences within a RNA molecule after it has been generated by RNA polymerase by mutating a base within RNA

51. The *t*RNA has the anticodon UAC the *m*RNA codon usage would be:
A. GUA B. AUG
C. UUA D. GUG

52. Which of the following sets of recombinant DNA technology tool incorrectly paired with its applications?
A. DNA polymerase—Copies DNA sequences in PCR
B. Restriction endonuclease—production of DNA fragments for cloning purpose
C. Reverse transcriptase—production of *c*DNA
D. DNA topoisomerase—Creating sticky ends by specific cuts

53. The commercial availability of Single Cell Proteins (SCP), in spite of being attractive source of nutrients for human, is ailing due to:
A. Increased formation of phosphoric acid from high Nucleic acid contamination of SCP
B. Increased formation of uric acid from high Nucleic acid contamination of SCP
C. Increased formation of hydrogen peroxide from high Nucleic acid contamination of SCP
D. Increased formation of Glycolic acid from high Nucleic acid contamination of SCP

54. Which of the following gene is not a part of *lac operon* of *E. coli?*
A. Genes for inducible enzymes of lactose metabolism
B. Genes for the repressor a regulatory protein
C. Gene for RNA polymerase
D. The promoter site where RNA polymerase binds

55. What is the use of aminopterine in HAT selection of monoclonal antibodies?
A. To block synthesis of purines by *De novo* pathway in hybridoma clones
B. To block synthesis of pyrimidines by salvage pathway in hybridoma clones
C. To block synthesis of purines by *De novo* pathway in spleen cells
D. To block synthesis of purines from salvage pathway in mycloma cells

56. The most commonly used dyes to detect signals in DNA microarray analysis are:
A. Ethidium bromide dye
B. Propidium iodide dye
C. Cyanine dyes
D. Coomassie brilliant blue dye

57. The long DNA molecules or ~10 Mb can be resolved using which of the following forms of electrophoresis?
A. Standard Agarose gel electrophoresis
B. SDS-polyacrylamide gel electrophoresis
C. Native polyacrylamide gel electrophoresis
D. Pulsed field gel electrophoresis

58. Isolation of certain form of bacteria can be performed by following physical methods, except:
A. Heat treatment
B. Incubation temperature
C. pH of medium
D. Use of inhibitory chemicals

59. The Gram-positive bacteria are different from Gram-negative bacteria in following aspects except:
A. Gram-positive have a much greater amount of peptidoglycan in their cell walls
B. The walls of some of the Gram-positive bacteria contain teichoic acid
C. Gram-positive can easily be destroyed by lysozyme
D. Cytoplasmic membrane is absent in Gram-negative but present in Gram-positive

60. The B form of DNA exhibits the following characteristics except:
A. It is a left-handed helix

B. The strands of the helix are wrapped around an axis with a diameter of 20Å
C. Building blocks are purines and pyrimidine bases
D. Every complete turn of the helix covers a distance of 34Å

61. The measurement of bacterial growth can be done by following methods except:
A. Turbidimetric
B. Nitrogen determination
C. Microscopic count
D. Transformation

62. During anaerobic state, the main purpose of fermentation involves which of the following?
A. Regeneration of NAD′ from NADH permitting continuous glycolysis
B. Glucose oxidation to generate reducing equivalents
C. Regeneration of NADP′ from NADPH permitting continuous glycolysis
D. Regeneration of FAD from FADH2 permitting continuous glycolysis

63. Karyotype not compatible with survival of human embryo represent:
A. Total number of genes
B. Ploidy level of species
C. Frequency of induced nutations
D. Single nucleotide polymorphism

64. Cancer cells show all the following except:
A. Unlimited number of cell divisions
B. Polyploidy
C. Growth without external signals
D. Severe contact inhibition

65. *Bacillus thuringiensis* is used for production of:
A. Fungicidal proteins
B. Bactericidal proteins
C. Insecticidal proteins
D. Interferon gamma

66. Iodine used during gram staining acts as:
A. A stain
B. A mordant
C. A fixative
D. A dehydrant

67. Fredric Sanger won the nobel prize for which of the discoveries?
A. For sequencing of ΦX174 virus and nucleotide sequencing of insulin
B. For sequencing of ΦX174 virus and amino acid sequencing of insulin
C. For amino acid sequencing of somatostatin and amino acid sequencing of insulin
D. For nucleotide sequencing of somatostatin and nucleotide sequencing of insulin

68. All of the following species are petroleum degrading species except:
A. *Penicillium*
B. *Nocardia*
C. *Candida*
D. *Micrococcus*

69. All the following techniques are used in determining DNA protein interactions, except:
A. Electrophoretic mobility shift assay
B. Chromatin yeast one hybrid system
C. DNA foot printing
D. DNA finger printing

70. The genetic change that occurs in more than 1 per cent of the population is represented as:
A. Gene Polymorphism
B. Gene Duplication
C. Gene Redundancy
D. Genetic Mutation

71. Which of the following best represent Proto-oncogene in normal cells?
A. Code for proteins those lead to halting of cellular proliferation
B. Code for proteins those initiate cellular apoptosis
C. Code for proteins those are involved in stimulating cell division
D. Code for proteins those suppress cell cycle at G1/S transition

72. More than 1000 repeats of CTG triplet would be seen in severely affected individuals with:
A. Fragile X-Syndrome
B. Retinoblastoma
C. Myotonic Dystrophy
D. Xeroderma pigmentosum

73. Biotechnology product Streptokinase produced by several species of *S treptococci* is available commercially for:
A. Causing transfer of phosphate group on to human proteins
B. Causing thrombolysis in humans
C. Causing spermicidal action in humans
D. Causing mucolytic action in humans

74. Which of the following represents batch fermentation?
A. An open system
B. Submerged system
C. Closed system
D. Feeder batch system

75. The medium used to differentiate lactose and non-lactose fermenters is:
A. Sugar medium
C. MacConkeys medium
B. Citrate medium
D. Stuart's medium

ANSWERS

1	2	3	4	5	6	7	8	9	10
A	A	A	A	C	B	A	A	A	B
11	**12**	**13**	**14**	**15**	**16**	**17**	**18**	**19**	**20**
D	A	A	A	A	B	D	A	B	B
21	**22**	**23**	**24**	**25**	**26**	**27**	**28**	**29**	**30**
D	B	A	C	A	A	D	A	A	C
31	**32**	**33**	**34**	**35**	**36**	**37**	**38**	**39**	**40**
B	C	C	A	C	C	A	C	B	D
41	**42**	**43**	**44**	**45**	**46**	**47**	**48**	**49**	**50**
C	A	D	C	D	A	B	B	D	D
51	**52**	**53**	**54**	**55**	**56**	**57**	**58**	**59**	**60**
A	D	B	C	D	C	D	D	D	A
61	**62**	**63**	**64**	**65**	**66**	**67**	**68**	**69**	**70**
D	A	B	D	C	B	B	C	D	A
71	**72**	**73**	**74**	**75**					
C	C	B	C	B					

Previous Paper (Solved)

JNU-M.Sc. Biotechnology
Combined Entrance Exam, 2014

PART-A

Answer All Questions.

1. Microbial capsules are stained by:
A. acid-fast staining B. negative staining
C. simple staining D. gram staining

2. Proteins produced by bacteria to inhibit growth of other strains of the same organism are called:
A. antibiotics B. B factors
C. R factors D. bacteriocins

3. Which one of the following marine bacteria is capable of bioluminescence?
A. *Azotobacter* B. *Salmonella typhi*
C. *Vibrio fisheri* D. *Thiothrix nivea*

4. Gene for which of the following processes is ***not*** encoded in any virus for its propagation?
A. Replication of the genome
B. Modulation of host defenses
C. Membrane biosynthesis
D. Assembly and packaging of the genome

5. Nerve impulses are normally carried forward to another neuronal cell body by:
A. synaptic cleft B. dendrites
C. axon D. myelin sheaths

6. The target tissue for luteinizing hormone is:
A. liver B. kidney
C. pancreas D. gonad

7. The excretory organs of insects are called:
A. flame cells
B. nephridia
C. nephrons
D. malpighian tubules

8. Hemophilia is a sex-linked recessive trait in humans. If a father and a son are both hemophiliacs in a family but the mother is normal, her genotype must be:
A. X^hX^h B. X^HX^h
C. X^HX^H D. X^HO

9. In diploid organisms, the specialized cell that undergoes meiosis is:
A. meiocyte B. oocyte
C. antherozoid D. ovum

10. The exine of a pollen grain is made of:
A. lignocellulose B. pectin
C. cellulose D. sporopollenin

11. Pollination which occurs in closed flower is known as:
A. allogamy B. cleistogamy
C. protogyny D. protandry

12. The genotypic ratio of F_2 in a monohybrid cross is:
A. 3 : 1 B. 15 : 1
C. 1 : 2 : 1 D. 9 : 6

13. The complete set of chromosomal and extra-chromosomal genes of an organism is called:
A. genome B. gene pool
C. gene bank D. gene library

14. A nucleoside is formed of:
A. pentose sugar, phosphate and nitrogen base
B. phosphate and nitrogen base
C. pentose sugar and phosphate
D. pentose sugar and nitrogen base

15. In forensic science, which of the following techniques is most frequently used?
A. Bacterial cloning
B. DNA footprinting
C. DNA fingerprinting
D. DNA cloning

16. Anticodon occurs in:
A. *t*RNA B. *m*RNA
C. *mt*RNA D. *r*RNA

17. Isoelectric point of a protein is the pH at which net charge on the molecule is:
A. one B. zero
C. positive D. negative

18. In prokaryotes, the genomic DNA is:
A. located in nucleus
B. wrapped in histones
C. located as nucleoid body
D. located in endoplasmic reticulum

19. Which of the following is an unbranched polysaccharide?
A. Cellulose B. Glycogen
C. Starch D. Sucrose

20. Which amino acid is a major neurotransmitter in the brain?
A. Tyrosine B. Glutamate
C. Tryptophan D. Serine

21. Calculate the energy in kcal per mole associated with a radiation of a wavelength 300 nm ($h = 1.58 \times 10^{-37}$ kcal-s, $c = 3.0 \times 10^{8}$ m s^{-1}).
A. 90.49 B. 94.09
C. 99.04 D. 109.49

22. The magnetic quantum number is related to:
A. size B. shape
C. orientation D. spin

23. $KMnO_4$ acts as an oxidant in neutral medium and gives MnO_2. The equivalent weight of $KMnO_4$ in neutral medium will be:
A. molecular weight / 7
B. molecular weight / 4
C. molecular weight / 3
D. molecular weight / 2

24. In the reaction $Cl_2 \rightarrow Cl^- + ClO^{3-}$, the chlorine is:
A. oxidized
B. reduced
C. both oxidized and reduced
D. neither oxidized nor reduced

25. The movement of the solute particles under the influence of an applied electric potential is called:
A. electrolysis B. electroosmosis
C. electrophoresis D. electroplating

26. Hydrogenation of vegetable oil to yield Vanaspati Ghee (general name) is an example of:
A. autocatalysis
B. negative catalysis
C. homogeneous catalysis
D. heterogeneous catalysis

27. The unit of rate constant for zero-order reaction is:
A. lit sec^{-1} B. lit $mole^{-1}$ sec^{-1}
C. mole lit^{-1} sec^{-1} D. mole sec^{-1}

28. Of the four halogens, the one with the highest electron affinity is:
A. fluorine B. chlorine
C. bromine D. iodine

29. The cation smallest in size is:
A. Na^+ B. Mg^{2+}
C. Al^{3+} D. Si^{4+}

30. Which one of the following does ***not*** hydrolyze?
A. CCl_4 B. $SiCl_4$
C. $GeCl_4$ D. $SnCl_4$

31. The gas that cannot be collected over water is:
A. N_2 B. O_2
C. SO_2 D. PH_3

32. The process of converting hydrated alumina into anhydrous alumina is called:
A. roasting B. smelting
C. dressing D. calcination

33. In C_2H_6, carbon atom is hybridized as:

A. sp^3 B. sp^2
C. sp D. sp^3d^2

34. Formaldehyde reacts with NH_3 to give:

A. formalin
B. formamide
C. hexamethylenetetramine
D. methylamine

35. Hydroboration is the reduction of alkynes to:

A. alkenes giving *cis*-addition products
B. alkenes giving *trans*-addition products
C. alkenes giving mixture of *cis*- and *trans*-addition products
D. alkanes

36. The number of solutions of $x^2 - y - 1 = 0$ and $x^2 + y - 1 = 0$ is:

A. 0
B. unique
C. finite but not unique
D. infinite

37. The value of $\lim_{y\to\infty} y \sin\left(\frac{1}{y}\right)$ is:

A. 0 B. 1
C. infinity D. y

38. The vectors $\hat{i} + \hat{j}$ and $-\hat{i} - \hat{j}$ are:

A. parallel
B. perpendicular
C. scalar multiples of each other
D. unit vectors

39. In how many points does the pair of lines $x = 0$ and $y = 0$ intersect?

A. They do not intersect
B. In only one point
C. In infinite points
D. In two points

40. If $A = \begin{pmatrix} 1 & i \\ -i & 1 \end{pmatrix}$, then $A^2 - 2A$ is:

A. $\begin{pmatrix} 0 & 0 \\ 0 & 0 \end{pmatrix}$ B. $\begin{pmatrix} 1 & 1 \\ 1 & 1 \end{pmatrix}$

C. $2\begin{pmatrix} 1 & 1 \\ 1 & 1 \end{pmatrix}$ D. A

41. The function $f(x) = x^3 - 1$ has the real root(s):

A. –1, 0 and 1 B. –1 and 1
C. 0 and 1 D. 1

42. The function $\sin|x|$ is:

A. continuous and differentiable at $x = \pi$
B. continuous but not differentiable at $x = \pi$
C. differentiable but not continuous at $x = \pi$
D. neither continuous nor differentiable at $x = \pi$

43. Consider the solutions $x(t)$ to the differential equation $\frac{dx}{dt} = -x$. Then

A. $x(t)$ is always positive
B. $x(t)$ is always negative
C. $x(t)$ can never be equal to 0
D. $x(t)$ can be 0 only if x (t = 0) = 0

44. Consider the salaries (in hundreds of rupees) of five persons:
5, 5, 5, 5, 1000
Then

A. the mean salary is much larger than the median salary
B. the median salary is much larger than the mean salary
C. the sample size is too small to allow the computation of a mean or median
D. every person earns 204 (the average salary)

45. The function $\sin x + \sin\frac{x}{2} + \sin(2x)$ has period:

A. 2π B. $\frac{\pi}{2}$
C. π D. 4π

46. A rock thrown horizontally from a 180 m high cliff strikes the ground 108 m from the base of the cliff. Calculate the speed at which it was thrown. Assume g = 10 m/s^2.

A. 180 m/s B. 9 m/s
C. 18 m/s D. 10·8 m/s

47. The average speed of blood flow in the aorta (r = 1.0 cm) is about 40 cm/s. The viscosity of blood is 4 × 10^{-3} Pa s. Calculate the Reynolds number assuming density of blood to be equal to that of water.

A. 1000 B. 4100
C. 400 D. 2000

48. Sound intensity of a jet plane at a distance of 1 km is 80 W/m^2. What will be the intensity at 2 km?

A. 20 W/m^2 B. 40 W/m^2
C. 10 W/m^2 D. 1 W/m^2

49. What is the total number of degrees of freedom of 1 mole of diatomic gas? Assume that the distance between the atoms is absolutely fixed.

A. 6 × 10^{23} B. 36 × 10^{23}
C. 5 D. 30 × 10^{23}

50. Which of the following plots of a function represents most accurately the decay of a radioactive material?

A.

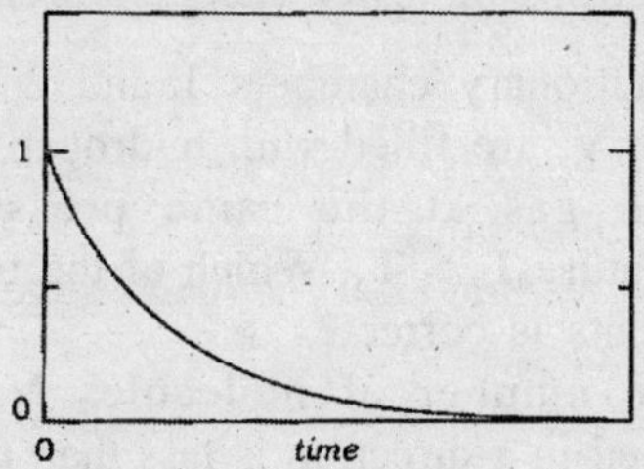

B.

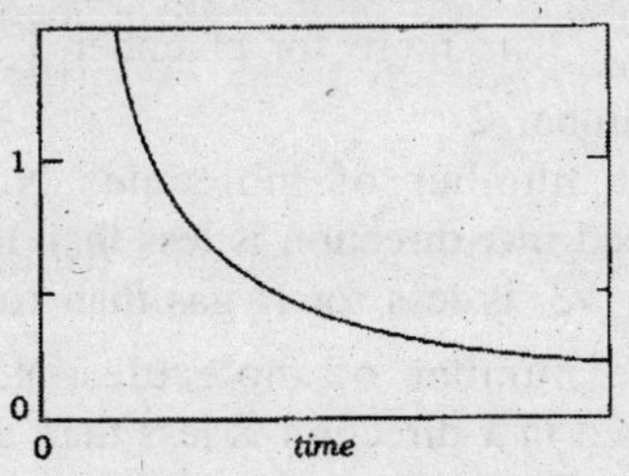

C.

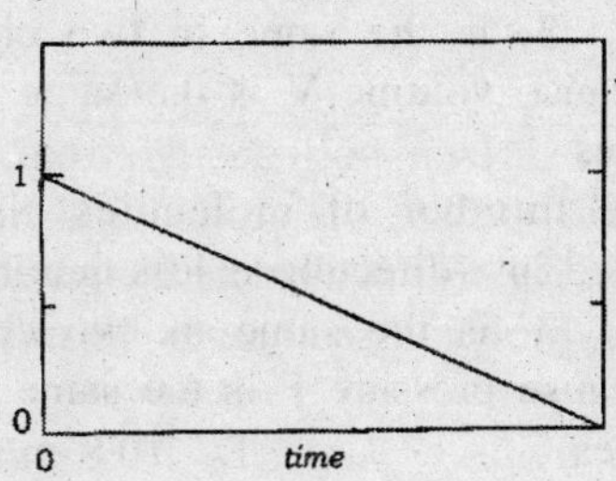

D.

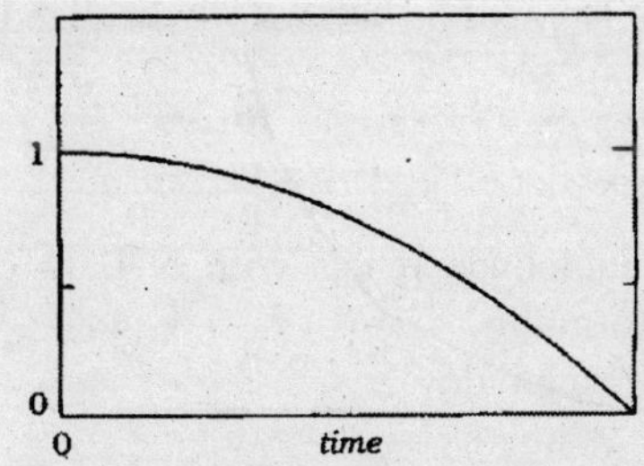

51. A ball of mass M at rest and positioned at A at the top of a hill starts rolling down from A towards B and then reaches its final position C at the bottom of the hill. See the figure below:

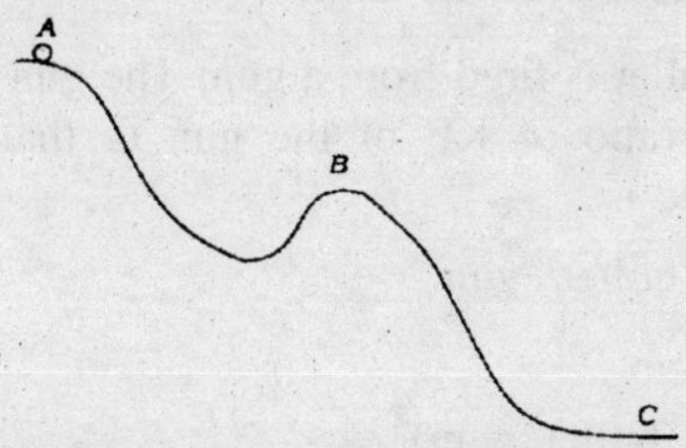

Which figure shown below best captures the potential energy as the ball goes from A to B to C?

A.

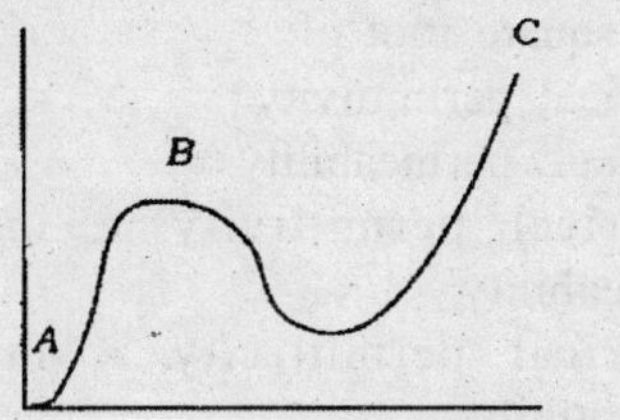

B.

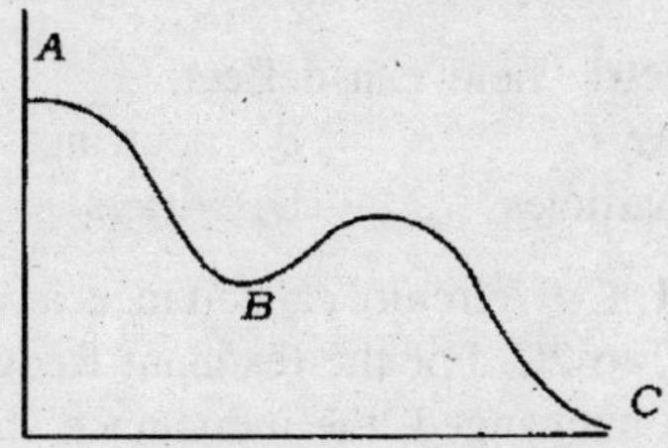

C.

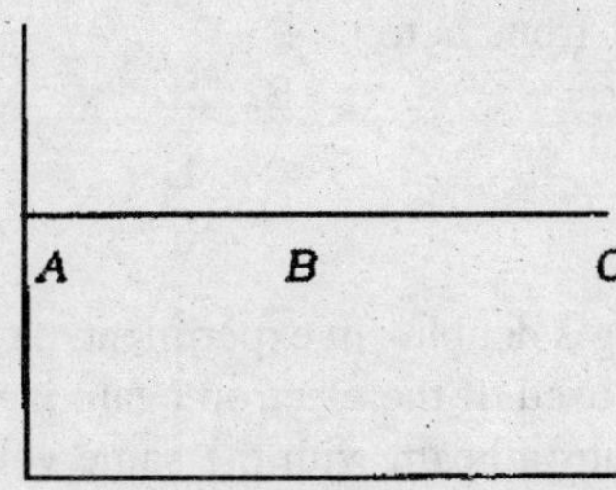

D. 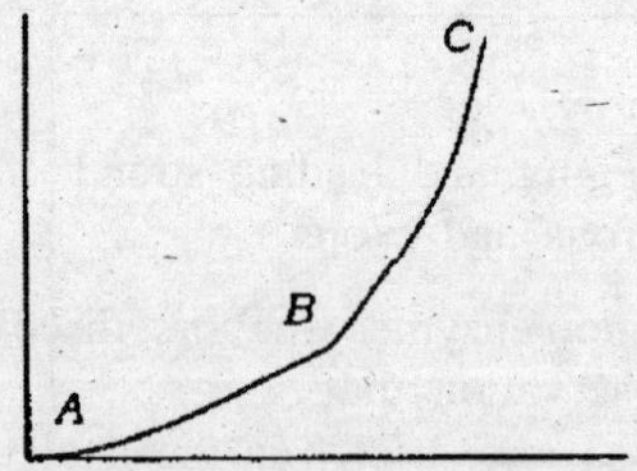

52. As a planet orbits the sun, which of the following must remain constant?
A. Velocity
B. Angular momentum
C. Gravitational force
D. Radius of the orbit

53. A bullet is fired from a gun. The gun recoils. The ratio of KE of the gun to that of the bullet is:

A. m_{bullet}/m_{gun}

B. 1

C. $(m_{bullet}/m_{gun})^2$

D. $1/(m_{bullet}/m_{gun})$

54. Refractive index of materials is approximately equal to square root of:
A. electrical permittivity
B. magnetic permeability
C. electrical permittivity × magnetic permeability
D. electrical permittivity × magnetic susceptibility

55. An electric field can deflect:
A. X-rays B. neutrons
C. β-particles D. γ-rays

56. In an L-C-R circuit, capacitance is changed from C to 2C. For the resonant frequency to remain unchanged, the inductance should be changed from L to:
A. 4L B. 2L
C. $\frac{L}{2}$ D. $\frac{L}{4}$

57. In Young's double-slit experiment, an electron beam is used. If the electron beam is replaced by a neutron beam with the same velocity as that of electrons, we can maintain the same fringe width by:
A. there will be no interference pattern in the first place
B. increasing the distance between the slits
C. increasing the distance between slits and the detector
D. the fringe width does not change on the nature of particles

58. The colour of the star indicates its:
A. weight
B. size
C. distance
D. temperature

59. When yellow light is incident on a surface, no electrons are emitted while green light can emit? If blue light is incident on the surface, then:
A. no electrons are emitted
B. electrons of the same energy as when green light was incident are emitted
C. electrons of higher energy are emitted
D. electrons of lower energy are emitted

60. Two stationary chambers 1 and 2, each of volume V, are filled with hydrogen gas and oxygen gas at the same pressure but temperature $T_1 > T_2$. Which of the following statements is correct?
A. The number of molecules N, whose speed in x-direction is less than the value $c_{r.m.s.}/3$, is more for chamber 1 than for chamber 2
B. The number of molecules N, whose speed in x-direction is less than the value $c_{r.m.s.}/3$, is less for H gas than for O gas
C. The number of molecules N, whose speed in x-direction is less than the value $c_{r.m.s.}/3$, is the same in two chambers, because volume V is the same for both cases
D. The number of molecules N, whose speed in x-direction is less than the value $c_{r.m.s.}/3$, is the same in two chambers, because pressure P is the same for both cases

PART-B

Answer *any* sixty questions

61. Streptomycin and neomycin inhibit bacterial protein synthesis by:
A. inhibiting binding of aminoacyl *t*RNA to ribosomes
B. preventing binding of $tRNA^{fmet}$ to the P site
C. inhibiting peptidyl transferase activity
D. preventing formation of 70S ribosomes

62. Some prokaryotes produce discrete layers external to cell wall. Which one of such layers will be used by bacteria as means to evade host defence system?
A. Capsule
B. Sheath
C. Slime layer
D. Protein jacket or S layer

63. The severity of disease caused by virus can vary greatly between infected individuals. What is the most important factor that determines the severity of viral disease?
A. Nature of the host cell receptor used by virus for entry
B. Nature of viral genome to be replicated inside the host cell
C. Permissiveness of the cell to support viral infection
D. Relationship that virus establishes with the host immune system

64. Which of the following is ***not*** the characteristic of bacterial growth curve?
A. Shows development of microbial population under relatively stable environmental conditions
B. Plotted with logarithmic numbers
C. Graphs number of microbes versus time
D. Each growth curve consists of four distinct phases

65. Which of the following is ***not*** associated with prokaryotic cells?
A. Semiconservative replication
B. Inducible operons
C. Lagging and leading strand
D. Introns and exons

66. Restriction enzymes were first discovered with the observation that:
A. DNA is restricted in the nucleus
B. phage DNA is destroyed in host cells
C. foreign DNA is kept out of cells
D. foreign DNA is restricted to the cytoplasm

67. Which one of the following is a function of pyridoxine in nutrition of bacteria and archaea?
A. It is a precursor of folic acid, a coenzyme involved in one carbon unit transfer
B. It is a precursor of NAD and NADP, which are coenzymes involved in hydrogen transfer
C. It is a component of the coenzyme for transaminase and amino acid decarboxylase
D. It is a coenzyme involved in molecular rearrangements

68. In viral RNA, when a single-stranded loop region base pairs with a complementary sequence outside the loop creating a secondary structure, it is called:
A. multibranched loop
B. pseudoknot
C. interior loop
D. bulge loop

69. Respiration in endoparasite cestodes is:
A. aerobic
B. anaerobic
C. both aerobic and anaerobic
D. partially aerobic

70. Monophyodont and absence of corpus callosum are characteristic features of:
A. rabbit B. human
C. kangaroo D. rat

71. The term 'humor' used in animal biology refers to:
A. tissue protein
B. tissue fat and protein

C. blood and lymph
D. plasma and lymph

72. Parkinson's disease is associated with which of the following neurotransmitters?
A. Acetylcholine B. Dopamine
C. GABA D. Endorphin

73. How many teeth grow twice in the life of man?
A. 4 B. 12
C. 20 D. 28

74. Acetylcholinesterase is:
A. a neurotransmitter
B. an enzyme that breaks down a neurotransmitter
C. a triggering agent that initiates action potential
D. an enzyme which stops propagation of neuronal impulse

75. In a polarized neuron at rest
A. the inside of the neuron is more negatively charged than the outside
B. outside of the neuron is more negatively charged than the inside
C. either of the above can be true
D. there is no charge at all

76. An extra finger in humans is rare but is due to a dominant gene. When one parent is normal and the other parent has an extra finger but is heterozygous for the trait, what is the probability that the first child will be normal?
A. 0% B. 25%
C. 50% D. 75%

77. LOD scores are used to predict:
A. crossover frequency
B. gene sequence
C. gene linkage
D. number of genes involved in the expression of a given trait

78. Hairy root cultures for secondary metabolite production are induced by transforming plant cells with:
A. *Bacillus subtilis*
B. *Agrobacterium tumefaciens*
C. *Bacillus thuringiensis*
D. *Agrobacterium rhizogenes*

79. The symbiotic relationship between fungi and roots of higher plants is called:
A. lichen B. helotism
C. mutualism D. mycorrhiza

80. Heterocysts are found in:
A. Anabaena B. Synechocystis
C. Spirulina D. Oscillatoria

81. Which of the following hormones stimulates ethylene release?
A. Auxin B. Gibberellin
C. Cytokinin D. Abscisic acid

82. Aerenchyma formation in wetland species especially in rice is induced by:
A. cytokinin B. auxin
C. ABA D. ethylene

83. Culturing of cells in liquid agitated medium is called:
A. micropropagation B. liquid culture
C. agar culture D. suspension culture

84. Which of the following is the best method for the production of virus-free plants?
A. Embryo culture B. Meristem culture
C. Ovule culture D. Anther culture

85. Treatment of DNA with base like 0.1 N NaOH:
A. has no effect
B. disrupts hydrogen bonds, resulting in single-stranded polynucleotides
C. hydrolyzes phosphodiester bonds, resulting in mononucleotides
D. the effect will depend on the ratio of the DNA and NaOH

86. Unstable base that can be readily deaminated and leads to incorporation of wrong base during transcription is:
A. adenine B. guanine
C. cytosine D. thymine

87. Protein coding gene in higher eukaryotes lacking introns is that encoding:
A. α-globin
B. α-collagen
C. immunoglobulin light chain
D. histones

88. In agarose gel, bromophenol blue migrates at the same rate:
A. equivalent to 300 bp linear double-stranded DNA
B. equivalent to 600 bp linear double-stranded DNA
C. equivalent to 1 kb linear double-stranded DNA
D. equivalent to 100 bp linear double-stranded DNA

89. Gene knockout involves:
A. heterologous repair
B. homologous repair
C. heterologous recombination
D. homologous recombination

90. Wild type gene : ATG ACC AGG TG
⇓
Mutant : ATG AC A CAG GTG
The above mutation is an example of:
A. transition B. transversion
C. frameshift D. inversion

91. Restriction enzymes do not destroy the DNA of the same organism that makes them, because:
A. their genome does not contain the restriction site for the particular enzyme
B. restriction enzymes are secreted out and they are not localized in the cell
C. their genome is protected by *de novo* methylation
D. restriction enzyme will remain inactive in the same organism

92. Denaturing gradient gel electrophoresis separates DNA fragments on the basis of:
A. length
B. sequence difference
C. both length and sequence difference
D. GC content

93. The final number of base pairs of a DNA segment that wrapped around the octameric disc of histones in nucleosomal core particle is:
A. 142 B. 144
C. 146 D. 148

94. While comparing the expression levels of a gene 'X' in two experimental samples by PCR, the expression levels of housekeeping gene in both would be:
A. controlled
B. standard
C. exogenous normalizing variable
D. endogenous normalizing variable

95. The mechanism by which some cells make discrete changes to specific nucleotides within *m*RNA molecule after it has been generated by RNA polymerase is:
A. RNA editing B. splicing
C. capping D. polyadenylation

96. A ribonuclease that degrades the RNA portion of DNA-RNA hybrid molecule is:
A. RNase A B. RNase T_1
C. RNase H D. RNase III

97. What extra property does a shuttle vector have as compared to a standard cloning vector?
A. Longer multiple cloning sites with more restriction sites
B. Two genes for antibiotic resistance instead of one
C. Two origins of replication instead of one
D. A promoter that can be easily controlled experimentally

98. Which of the following statements is ***not*** true of telomerase?
A. It has ribozyme activity
B. It adds repeat sequences to the 5′ end of DNA
C. It is present in eukaryotes
D. Its activity decreases with the cell age

99. The role of methylation of DNA is now viewed as:
A. interfering with DNA transcription by blocking base pairing between cytosine and guanine
B. complexing with enhancers to prevent transcription
C. preventing mutations
D. insuring that genes that are turned off/stay off

100. RNA virus that changes proto-oncogene into oncogene upon integration of its genome in host chromosomal DNA is:
A. Retro virus B. Corona virus
C. Measles virus D. Rubella virus

101. In a globular protein, hydrophobic amino acids are generally found:
A. on the surface of the molecule
B. in the interior of the molecule
C. both on the surface and in the interior of the molecule
D. in the hinge region of the molecule

102. Proper folding of proteins following synthesis is assisted by:
A. polysomes
B. chaperones
C. acyl carrier proteins
D. histidyl-*cis-trans*-isomerase

103. Which of the following is a glycosaminoglycan?
A. Glycogen B. Collagen
C. Hyaluronic acid D. Chitin

104. α-Subunits of G-protein contain the binding site for:
A. cAMP B. cGMP
C. GTP D. ATP

105. Transcription in prokaryotes is inhibited by:
A. nalidixic acid B. cycloheximide
C. tetracycline D. rifampicin

106. Which of the following is a signalling receptor?
A. Mannose receptor B. Toll-like receptor
C. Vitamin D receptor D. Thyroid receptor

107. For performing SDS-gel electrophoresis, an excess of β-mercaptoethanol is added to protein samples for the purpose of:
A. disruption of disulphide bonds of proteins
B. maintaining native structure of proteins
C. maintaining buffering capacity of the solution
D. hydrolysis of proteins

108. Which chemical force is disrupted by urea when it is used for solubilizing the aggregated proteins?
A. van der Waals B. Ionic
C. Hydrophobic D. Covalent

109. Blocking of enzyme action by blocking its active sites occurs during:
A. allosteric inhibition
B. competitive inhibition
C. feedback inhibition
D. non-competitive inhibition

110. Which one of the following statements is true with reference to enzymes?
A. Apoenzyme = Holoenzyme + Co-enzyme
B. Holoenzyme = Apoenzyme + Co-enzyme
C. Co-enzyme = Apoenzyme + Holoenzyme
D. Holoenzyme = Co-enzyme – Apoenzyme

111. Nucleic acids in a solution can be estimated by measuring their absorption at 260 nm. Their absorption in this range is due to:
A. phosphate groups
B. purine and pyrimidine bases
C. ribose and deoxyribose sugars
D. purines only

112. A 20-year-old anaemic man is found to have an abnormal form of β-globin that is 172 amino acids long, rather than the 141 found in the normal protein. Which of the following point mutations is likely to be associated with this abnormality?
A. UAA → CAA B. UAA → UAG
C. CGA → UGA D. GAU → GAC

113. Protein concentration estimated by Lowry's method is dependent on the contents of which of the following amino acids in the protein?
A. Leucine and isoleucine
B. Lysine and histidine
C. Tryptophan and tyrosine
D. Aspartic acid and glutamic acid

114. Vasopressin is a:
A. dipeptide B. nonapeptide
C. tripeptide D. hexapeptide

115. A Biochemistry student isolates all the enzymes of the TCA cycle and adds OAA, acetyl CoA, appropriate energy precursors, cofactors and water. Which of the following will not be a direct product of his/her experiment?

A. ATP B. GTP
C. NADH D. $FADH_2$

116. Uncouplers are those which dissociate:
A. electron transport from oxidative phosphorylation
B. electron transport from TCA cycle
C. F0 subunit of ATP synthase from Fl subunit
D. glycolysis from TCA cycle

117. Cellular organelles containing hydrolytic enzymes are called:
A. microsomes B. lysosomes
C. chloroplasts D. ribosomes

118. Cytokine released from cell A stimulates an adjoining cell B. This is an example of:
A. autocrine signal B. paracrine signal
C. endocrine signal D. cytocrine signal

119. Which of the following statements is ***not*** true in the context of mitochondrial electron transport and oxidative phosphorylation?
A. Complex V is ATP synthase as well as ATPase
B. ATP synthase is a proton pump
C. Cytochrome C is a part of complex IV
D. $FADH_2$ is fed to the electron transport chain through complex II

120. Which of the following statements is correct concerning prion disease?
A. It is a disease process in which proteins are pathophysiologic entity
B. It is a disease process in which a messenger RNA secondary structure appears to be the pathophysiologic entity
C. It is found only in humans
D. The disease process is readily reversible

121. The armature of a 60 Hz a.c. generator rotates in a magnetic field of 0.15 T. If the area of the coil is 2.0×10^{-2} m^2, how many loops must the coil contain if the peak output is to be E_0 = 170 V?
A. 50 turns B. 100 turns
C. 150 turns D. 300 turns

122. The isotope C^{14} has a half-life of 5730 years. If sometime a sample contains 1×10^{22} C^{14} nuclei, what is the number of decays per second?
A. 4×10^{12} B. 1.7×10^{18}
C. 1.7×10^{-34} D. 3.8×10^{10}

123. The output of the logic gate

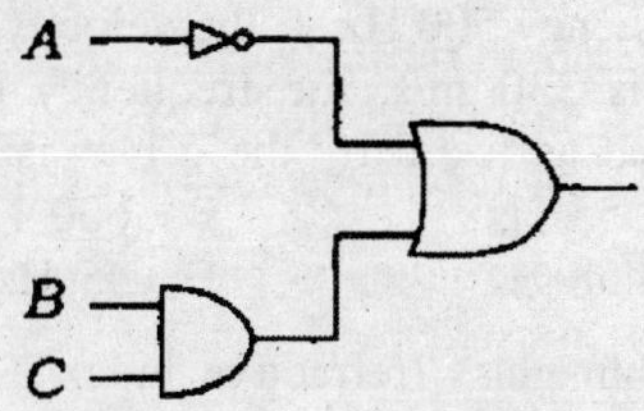

will be 1, when:
A. A = 1, B = 0, C = 0
B. A = 1, B = 1, C = 0
C. A = 1, B = 0, C = 1
D. A = 0, B = 1, C = 1

124. A parallel-plate capacitor is made by stacking n equally spaced plates connected alternately to the positive and negative terminals of a battery. If the capacitance between any two adjacent plates is C, then the resultant capacitance is:
A. C/(n – 1) B. (n – 1)C
C. nC D. C/n

125. A block of mass m and having charge $+q$ is placed on a frictionless horizontal surface and is connected with a spring of spring constant k as shown in the figure. A horizontal electric field of magnitude E is applied parallel to the spring. The maximum amplitude of the SHM of the block is:

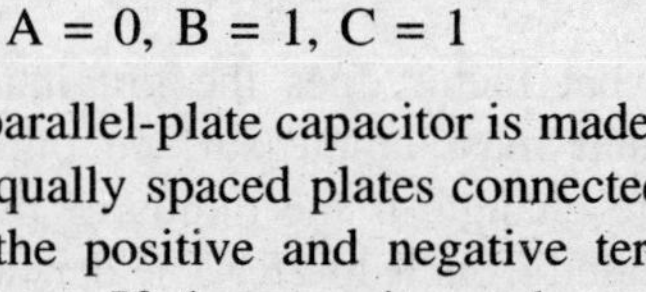

A. $\frac{m}{k}$ B. $\frac{qm}{k}$
C. $\frac{qE}{k}$ D. $\frac{mE}{k}$

126. A mercury thermometer reads +0.5°C when placed in a melting ice and 100.8°C when placed in a steam. What is the true reading if the reading of the thermometer is 20°C?

A. 19.45°C B. 20.56°C
C. 21.56°C D. 18.98°C

127. The driver of a car travelling with a speed 30 m/s towards a hill sounds a horn of frequency 500 Hz. If the velocity of sound in air is 330 m/s, the frequency of reflected sound as heard by the driver is:

A. 555 Hz B. 600 Hz
C. 458 Hz D. 45 Hz

128. A thin glass (refractive index 1.5) lens has optical power of –5 D in air. Its optical power in a liquid medium with refractive index 1.4 will be:

A. 1 D B. 25 D
C. –25 D D. –1 D

129. By what factor does the gravitational force per unit mass at the surface of a planetary sphere change, if the radius of the sphere is increased from R to 2R? The density of matter of the planet remains constant.

A. $\frac{1}{2}$ B. 4
C. 2 D. $\frac{1}{4}$

130. Three capacitors, each of capacitance C and of breakdown voltage V, are joined in series. The capacitance and breakdown voltage of the combination will be:

A. 3C, V/3 B. C/3, 3V
C. 3C, 3V D. C/3, V/3

131. The magnetic force acting on a charged particle of charge –2 μC in a magnetic field of 2 T acting in *y*-direction, when the particle velocity is $(2\hat{i} + 3\hat{j}) \times 10^6 \text{ m s}^{-1}$, is:

A. 4 N in *z*-direction B. 8 N in *y*-direction
C. 8 N in *z*-direction D. 8 N in –*z*-direction

132. A solid metallic sphere of radius R is melted and remade into a solid cylinder of radius R. I_S and I_C are the moment of inertia of the sphere, and the cylinder around the axis passing through its centre and parallel to its length. The ratio of moment of inertia of the sphere and that of cylinder, *i.e.*, I_S/I_C, is:

A. 4/5 B. 2/3
C. 1/5 D. 7/5

133. Zeeman splitting is the splitting of energy levels:

A. on application of magnetic field
B. by thermal energy
C. on application of electric field
D. by mechanical energy

134. The fingerprint region of an infrared spectrum is usually represented by:

A. 4000 cm^{-1} to 3000 cm^{-1}
B. 3000 cm^{-1} to 2500 cm^{-1}
C. 1500 cm^{-1} to 500 cm^{-1}
D. 4000 cm^{-1} to 1500 cm^{-1}

135. Keesom forces are due to interaction between:

A. permanent dipole and induced dipole
B. induced dipole and induced dipole
C. permanent dipole and permanent dipole
D. charge and charge

136. What is the correct order of increasing energy for the following spin-spin pairs?

A. $\alpha\alpha < \alpha\beta = \beta\alpha < \beta\beta$
B. $\alpha\alpha = \beta\beta < \alpha\beta = \beta\alpha$
C. $\beta\beta < \alpha\beta = \beta\alpha < \alpha\alpha$
D. $\alpha\beta = \beta\alpha < \alpha\alpha = \beta\beta$

137. Beckmann rearrangement is an example of:

A. 1,2-shift where migration origin is carbon and the migration terminus is nitrogen
B. 1,3-shift where migration origin is carbon and the migration terminus is nitrogen
C. 1,2-shift where migration origin is nitrogen and the migration terminus is carbon
D. 1,3-shift where migration origin is nitrogen and the migration terminus is carbon

138. In a process, nitrogen is heated to 576 K in a vessel of constant volume. If it enters the vessel at 180 K and 55 atm, what pressure would it exert at the working temperature if it behaves as a perfect gas?

A. 138 atm B. 176 atm
C. 137.5 atm D. 167 atm

139. Assuming the energy difference between 4*s* and 3*d* orbitals to be small, following 8 electronic arrangements have been listed for the outer electronic configuration of chromium:

	4*s*	3*d*				
(*i*)	↑↓	↑	↑	↑	↑	
(*ii*)	↑↓	↑↓	↑	↑		
(*iii*)	↑↓	↑↓	↑↓			
(*iv*)	↑	↑	↑	↑	↑	↑
(*v*)	↑	↑↓	↑	↑	↑	
(*vi*)		↑↓	↑	↑	↑	↑
(*vii*)		↑↓	↑↓	↑	↑	
(*viii*)		↑↓	↑↓	↑↓		
	1	2	3	4	5	6

Which of these arrangements are ***not*** possible according to Aufbau principle?

A. (*ii*), (*v*), (*vii*) B. (*v*), (*vii*), (*viii*)
C. (*iii*), (*v*), (*viii*) D. (*iii*), (*v*), (*vii*)

140. The correct arrangement of the following molecules in the increasing bond order is:

A. $He_2 < O_2 < N_2 < C_2$
B. $He_2 < O_2 \approx N_2 < C_2$
C. $He_2 < O_2 \approx C_2 < N_2$
D. $He_2 < N_2 < O_2 < C_2$

141. In wurtzite structure of ZnS, the sulphide (S^{2-}) ions are arranged in a hexagonal closed packed (hcp) arrangement and zinc ion occupies half of the tetrahedral sites. Therefore, the coordination numbers of Zn^{2+} and S^{2-} ions in wurtzite structure respectively are:

A. (6 : 4) B. (4 : 4)
C. (4 : 6) D. (6 : 6)

142. Sulphamic acid, $H_2N(SO_2)OH$, behaves as a dibasic acid in liquid ammonia, whereas it acts as a monobasic acid in aqueous solution. This is because:

A. NH_2 group cannot donate a proton to water
B. it is difficult to undergo protolysis in ammonia
C. it requires a solvent that can give up a proton
D. NH_2 can readily donate a proton to water

143. Alkali metals are typically soft, show low densities, and have low melting and boiling points. This is because they have:

A. only one valence electron
B. large atomic size
C. negative values of standard reduction potentials
D. negligible electron gain enthalpy

144. The ionization enthalpy of hydrogen atom is 1.312×10^6 J mol^{-1}. The energy required to excite the electron in the atom from $n = 1$ to $n = 2$ is:

A. 8.51×10^5 J mol^{-1} B. 6.56×10^5 J mol^{-1}
C. 7.56×10^5 J mol^{-1} D. 9.84×10^5 J mol^{-1}

145. The expected spin-only magnetic moments of complexes of a transition metal like nickel or cobalt depend on:

A. oxidation state of the metal
B. nature of the ligand
C. geometry of the complex
D. number of unpaired electrons in free transition metal ion

146. $[Ti(H_2O)_6]^{3+}$ complex gives purple colour when dissolved in water. This is attributed to:

A. promotion of t_{2g} electron to e_g level
B. ligand to metal charge transfer
C. metal to ligand charge transfer
D. $\pi \rightarrow \pi^*$ transition

147. A negative E° means that:

A. the redox couple is a stronger reducing agent than H^+/H_2 couple
B. the redox couple is a weaker reducing agent than H^+/H_2 couple
C. the reaction is not a spontaneous oxidation reduction reaction
D. negative E° value is not a determinant of oxidation reduction reaction

148. If E°_{cell} for the reaction

$Mg(s) + 2Ag^+ (0.0001\ M) \rightleftharpoons Mg^{2+} (0.130\ M) + 2Ag(s)$

is 3.17 V, calculate its E_{cell}.

A. 2.96 V B. 3.17 V
C. 3.38 V D. 0.00 V

149. A solution of $CuSO_4$ was electrolyzed for 10 minutes with a current of 1.5 amperes. What is the mass of Cu deposited at the cathode? Faraday's constant is 96487 C/mol. 1 mol of Cu weighs 63 g.

A. 0.2938 g B. 0.2938 mg
C. 0.1469 g D. 0.1469 mg

150. For a reaction, A + B + C → P, rate law is given by rate = k [A] [B]. What is the order of the reaction?

A. 0 B. 1
C. 2 D. 3

151. Which one of the following has bacteriostatic effect on microbes?

A. Penicillin B. Aminoglycoside
C. Ofloxacin D. Tetracycline

152. The shape of XeO_2F_2 is:

A. square planar
B. distorted trigonal bipyramidal
C. square pyramidal
D. distorted tetrahedral

153. If tan A and tan B are the roots of the equation $x^2 - px + q = 0$, then the value of tan (A + B) equals:

A. $\frac{p}{q}$ B. $\frac{p}{1-q}$
C $\frac{q}{1-p}$ D. $\frac{p}{1+q}$

154. A set contains $2n + 1$ elements. The number of subsets of the set having at most n elements is:

A. 2^n B. 2^{n+1}
C. 2^{2n} D. 2^{2n-1}

155. The equation $x^2 + 4xy + 4y^2 + 5x + 10y + 4 = 0$ represents:

A. a parabola
B. a pair of intersecting lines
C. a pair of parallel lines
D. an ellipse

156. If $A = \begin{bmatrix} a & b \\ b & a \end{bmatrix}$ and $A^2 = \begin{bmatrix} p & q \\ q & p \end{bmatrix}$, then:

A. $p = a^2 + b^2$, $q = 2ab$
B. $p = a^2 + b^2$, $q = a^2 - b^2$
C. $p = 2ab$, $q = a^2 + b^2$
D. $p = a^2 + b^2$, $q = ab$

157. The differential equation $\frac{dy}{dx} + \frac{y}{x} = 0$ represents

A. a pair of lines passing through the origin
B. a family of concentric circles
C. a family of ellipses
D. a family of hyperbolas

158. Let,

$$f(x) = \begin{cases} x^2 & \text{if } 0 \le x < 1 \\ 2x + a & \text{if } 1 \le x \le 2 \\ b - x & \text{if } 2 < x \le 3 \end{cases}$$

be a continuous function on [0, 3], then

A. $a = -1, b = 5$ B. $a = 1, b = 5$
C. $a = -1, b = -5$ D. $a = 1, b = -5$

159. Let, $f(x) = |x - 2| + |5 - x|$. Then the minimum value of the function is:

A. 0 B. 1
C. 2 D. 3

160. $\lim_{x \to 0} \frac{1 - \cos^3 x}{x \sin x \cos x}$ is equal to:

A. $\frac{2}{5}$ B. $\frac{3}{5}$
C. $\frac{3}{2}$ D. $\frac{3}{4}$

ANSWERS

1	2	3	4	5	6	7	8	9	10
D	D	C	C	C	D	D	B	B	D
11	**12**	**13**	**14**	**15**	**16**	**17**	**18**	**19**	**20**
A	C	A	A	C	A	B	B	A	B

21	22	23	24	25	26	27	28	29	30
B	C	C	C	C	D	C	B	D	A
31	**32**	**33**	**34**	**35**	**36**	**37**	**38**	**39**	**40**
C	D	A	C	A	B	B	A	B	A
41	**42**	**43**	**44**	**45**	**46**	**47**	**48**	**49**	**50**
D	A	C	D	D	C	D	A	D	A
51	**52**	**53**	**54**	**55**	**56**	**57**	**58**	**59**	**60**
B	B	A	C	C	C	D	D	C	B
61	**62**	**63**	**64**	**65**	**66**	**67**	**68**	**69**	**70**
A	A	*	C	D	B	A	B	B	C
71	**72**	**73**	**74**	**75**	**76**	**77**	**78**	**79**	**80**
C	B	C	B	A	C	C	D	D	A
81	**82**	**83**	**84**	**85**	**86**	**87**	**88**	**89**	**90**
A	D	D	B	B	C	*	A	D	C
91	**92**	**93**	**94**	**95**	**96**	**97**	**98**	**99**	**100**
C	B	C	C	A	C	C	B	A	A
101	**102**	**103**	**104**	**105**	**106**	**107**	**108**	**109**	**110**
C	B	D	C	D	A	A	C	B	B
111	**112**	**113**	**114**	**115**	**116**	**117**	**118**	**119**	**120**
B	A	C	B	A	A	B	B	*	A
121	**122**	**123**	**124**	**125**	**126**	**127**	**128**	**129**	**130**
C	B	D	B	C	A	C	D	D	B
131	**132**	**133**	**134**	**135**	**136**	**137**	**138**	**139**	**140**
D	A	A	C	C	*	A	B	B	C
141	**142**	**143**	**144**	**145**	**146**	**147**	**148**	**149**	**150**
B	*	B	D	D	A	A	A	A	C
151	**152**	**153**	**154**	**155**	**156**	**157**	**158**	**159**	**160**
D	B	B	C	C	A	A	A	D	C

SOME SELECTED EXPLANATORY ANSWERS

4. **ENCODED** are gene products and regulatory signals for:

- Replication of the viral genome.
- Assembly and packaging of the genome.
- Regulation and timing of the replication cycle.
- Modulation of host defenses.
- Spread to other cells and hosts.

Information **NOT** contained in viral genomes.

- No genes encoding the complete protein synthesis machinery.
- No genes encoding proteins involved in energy production or membrane biosynthesis.
- No classical centromeres or telomeres found in standard host chromosomes.
- Probably we haven't found them yet.

7. The excretory organs in insects are the Malpighian tubules, extensions of the digestive tract that branch off anterior to the hindgut. Urine is not formed by filtration in these tubules, because there is no pressure difference between the blood in the body cavity and the tubule. Instead, waste molecules and potassium (K^+) ions are secreted into the tubules by active transport. Secretion is the opposite of reabsorption—ions or molecules are transported from the body fluid into the tubule.

9. **Diploid :** In diploids (such as animals and plants), meiosis takes place only in specialized germ cells such as oocytes or spermatocytes.

10. **Structure of Microspore :** The wall of microspore is comprised of outer **exine** and inner **intine.** The exine is made of **sporopollenin** while intine is **pectocellulosic.** Sporopollenin is highly resistant to microbes and chemical decomposition. Therefore, pollen grains are well preserved during fossilization. In the exine, there are 1-3 germ pores in dicots and germinal furrows in monocots. The wall of pollen grain surrounds an uninucleate mass of protoplasm.

20. **Glutamate** is the major excitatory neurotransmitter and is distributed in all regions of the brain. Aspartate is closely related to **glutamate** and the two amino acids are often found together at axon terminals. Neurons synthesize **glutamate** and aspartate and are independent of dietary supply.

22. **Magnetic quantum number :** This is an quantum number which indicates the direction of an orbit in space in a magnetic field. It is represented by *m*. The value of magnetic quantum number *m* depend on the value of *l* and its value are from $-l$ to $+l$ including zero. Total Number of the value of $m = 2l + 1$.

27. For a zero order reaction, the rate = *k*. The concentration of products increases linearly with time. The unit of zero order rate constant is same as that of reaction rate, *i.e.*, mol lit^{-1} sec^{-1}.

34. Formaldehyde reacts with ammonia to give hexamethylene tetramine (HMTA), used as a source of formaldehyde in PF molding powders and in the pharmaceuticals industry.

36. $x^2 - y - 1 = 0$ and $x^2 + y - 1 = 0$

$y = x^2 - 1$ and $y = 1 - x^2$

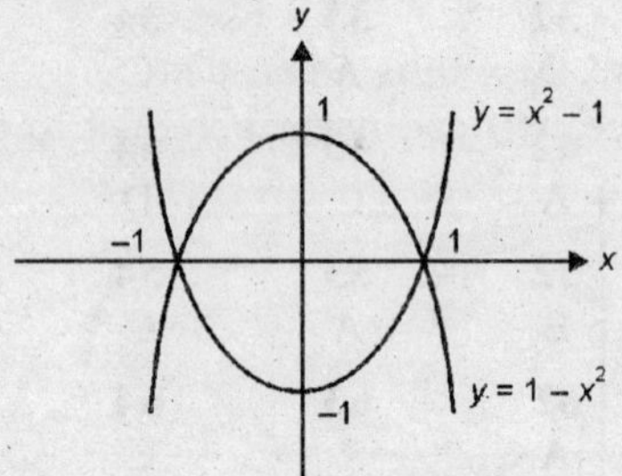

Hence, number of solution is unique.

39.

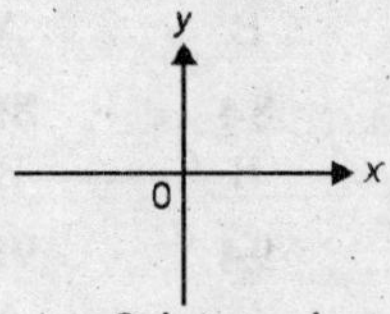

$y = 0$ is x-axis and $x = 0$ is y-axis, two axis intersect at origin only.

40.
$$A^2 = \begin{bmatrix} 1 & i \\ -i & 1 \end{bmatrix}\begin{bmatrix} 1 & i \\ -i & 1 \end{bmatrix}$$
$$= \begin{bmatrix} 1+1 & i+i \\ -i-i & 1+1 \end{bmatrix} = \begin{bmatrix} 2 & 2i \\ -2i & 2 \end{bmatrix}$$
$$A^2 - 2A = \begin{bmatrix} 2 & 2i \\ -2i & 2 \end{bmatrix} - 2\begin{bmatrix} 1 & i \\ -i & 1 \end{bmatrix}$$
$$= \begin{bmatrix} 2 & 2i \\ -2i & 2 \end{bmatrix} - \begin{bmatrix} 2 & 2i \\ -2i & 2 \end{bmatrix}$$
$$= \begin{bmatrix} 0 & 0 \\ 0 & 0 \end{bmatrix}.$$

41.
$$f(x) = x^3 - 1$$
$$f(x) = 0$$
$$\Rightarrow \quad x^3 - 1 = 0$$
$$x = (1)^{1/3}$$

From cube root of unity

$$x = 1, \frac{-1+\sqrt{3}i}{2}, \frac{-1-\sqrt{3}i}{2}$$

Hence, real root is $x = 1$.

43.
$$\frac{dx}{dt} = -x \Rightarrow \frac{dx}{x} = -dt$$

Integrating both sides, we get

$$\int \frac{dx}{x} = -\int dt$$

In $x = -t + c$

$$x = e^{-(t+c)}$$

$$x = K.e^{-t}$$

where, $K \neq 0$, is a constant

Hence, $x(t)$ can never be equal to 0.

46. 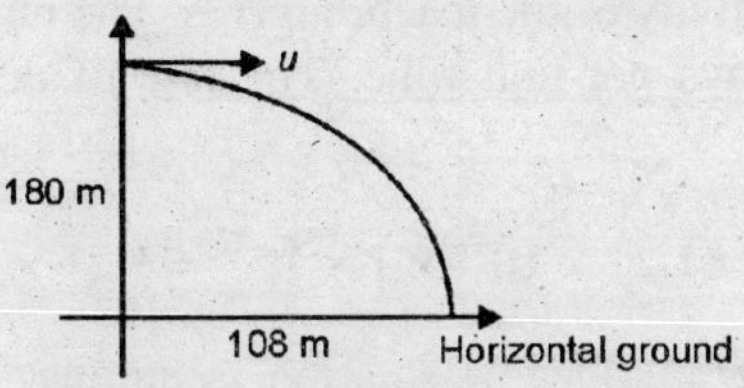

Let, the rock thrown horizontally with speed u m/sec.

From vertical displacement,

$$s_y = u_y.t + \frac{1}{2}g.t^2$$

$$180 = o.t + \frac{1}{2} \times 10 \times t^2$$

$$180 = 5t^2 \Rightarrow t = 6 \text{ sec}$$

From Horizontal displacement,

$$s_x = u_x.t$$

$$108 = u.6 \Rightarrow u = \frac{108}{6} = 18 \text{ m/sec.}$$

53. From conservation of momentum

$$m_g.v_g = m_b.v_b$$

$$\Rightarrow \quad (m_g.v_g)^2 = (m_b.v_b)^2$$

or $$\frac{\text{k.e. of gun}}{\text{k.e. of bullet}} = \frac{m_b}{m_g}.$$

54. The speed of light in a material can be related to its electrical and magnetic properties as

$$C = \frac{1}{\sqrt{\mu\varepsilon}}$$

where ε = electrical permittivity, and

μ = magnetic permeability.

Thus, $$n = \frac{c_0}{c} = \frac{\sqrt{\varepsilon\mu}}{\sqrt{\mu_0\varepsilon_0}} = \sqrt{\mu_r\varepsilon_r}$$

Since most substances are only slightly magnetic *i.e.*, $\mu_r \approx 1$.

Thus, $$n \equiv \sqrt{\varepsilon_r}$$

Thus, for transparent materials, index of refraction and dielectric constant are related.

56. The resonant frequency of LCR circuit

$$\nu = \frac{1}{2\pi}\sqrt{\frac{1}{LC}}$$

when, C changes to 2C

L changes to $\frac{L}{2}$

$$\therefore \quad \nu = \frac{1}{2\pi}\sqrt{\frac{1}{\frac{L}{2} \times 2C}}$$

$$= \frac{1}{2\pi}\sqrt{\frac{1}{LC}}.$$

75. When a neuron is at rest, it is not actively conducting a signal, there is an imbalance in the types of ions found inside and outside the cell walls of the neuron. This imbalance exists because openings in the axon, called *ion channels,* only allow some ions to pass into and out of the neuron.

79. Mycorrhizae, known as root fungi, form symbiotic associations with plant roots. In these associations, the fungi are actually integrated into the physical structure of the root. The fungi colonize the living root tissue during active plant growth.

83. Suspension Culture : Suspension culture contains a uniform suspension of separate cells in liquid medium. For the preparation of suspension culture, callus is transferred to liquid medium, which is agitated continuously which keeps the cells separate. Agitation can be achieved by rotary shaker system attached within the incubator at a rate of 50-150 rpm.

84. Meristem culture : In this way, using meristem tip which have no or very least quantity of viruses compare other parts of plant. Meristem tip of any plant is newly generated organ so it always free/less from infection compare to other parts of the plant.

92. Denaturing gradient gel electrophoresis (DGGE) separates DNA molecules differing in sequence by only a single base. As its name indicates, this technique combines gel electrophoresis with DNA denaturation. It is applicable to double-stranded DNA fragments of a few hundred bases, which are often generated by PCR.

95. **RNA editing** is a molecular process through which some cells can make discrete changes to specific nucleotide sequences within a RNA molecule after it has been generated by RNA polymerase. RNA editing is relatively rare, and common forms of RNA processing are not usually included as editing. Editing events may include the insertion, deletion, and base substitution of nucleotides within the edited RNA molecule.

102. The proper folding of proteins following synthesis is assisted by specific proteins called **chaperons,** which also assist in transport of proteins into organelles such as mitochondria.

114. Vasopressin (arginine vasopressin, antidiuretic hormone) is a non-opeptide released from the posterior pituitary gland in response to increased osmolarity and hypovolemia.

116. Class of inhibitors called *uncouplers* which dissociate the electron transport chain from oxidative phosphorylation, impairing the production of ATP. These uncouplers cause the membrane to become permeable to protons, allow oxidation to continue but uncouple oxidation from phosphorylation. This however, increases oxygen consumption in the absence of phosphorylation. 2, 4-dinitrophenol and valinomycin are two such uncouplers which inhibit oxidative phosphorylation.

121. Number of turn N

$$= \frac{E}{B.A.\omega} = \frac{170}{0.15 \times 2 \times 10^{-2} \times 2 \times 3.14 \times 60}$$

= 150 turns.

122. The half-life $t_{1/2}$ is the time elapsed before $\frac{N_0}{2}$ nuclei have decayed and $\frac{N_0}{2}$ remain active

$$t_{1/2} = \frac{0.693}{\lambda}$$

$$\lambda = \frac{0.693}{t_{1/2}} = \frac{0.693}{5730} = 1.21 \times 10^{-4}$$

The activity of the reaction gives. The number of decays per unit time. The rate of decay

$$-\frac{dN}{dt} = \lambda N$$

$$= 1.21 \times 10^{-4} \times 1 \times 10^{22} = 1.21 \times 10^{18}.$$

123.

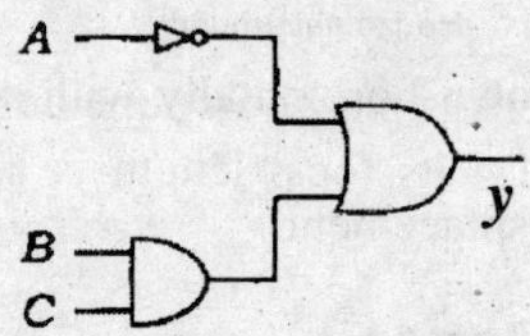

A	$\bar{A}$	B	C	B.C	$\bar{A}$+BC
1	0	0	0	0	0
1	0	1	0	0	0
1	0	0	1	0	0
0	1	1	1	1	1

124. *n*-plates form $(n - 1)$ capacitors in parallel.
∴ Net capacitance = $(n - 1)$ C.

126. Let, Correct temperature is T_c, then

$$\frac{T_c - 0}{100 - 0} = \frac{20 - 0.5}{100.8 - 0.5}$$

$$\frac{T_c}{100} = \frac{19.5}{100.3}$$

$$T_c = \frac{19.5}{100.3} \times 100 = 19.45°C.$$

127. Frequency of the reflected sound as heard by the driver

$$f' = \left(\frac{330}{330 + 30}\right) \times f = \frac{330}{360} \times 500 = 458 \text{ Hz.}$$

128. The optical power of a lens is defined as P_a

$$= \frac{\text{refractive index of the medium in which the lens is placed}}{\text{focal length of the lens in metre}}$$

For a lens placed in air,

$$P_a = \frac{\mu_a}{f_a} = \frac{1}{f_a(\text{in } m)}$$

Given, $P_a = -5D$.

Hence, $-5 = \dfrac{1}{f_a(\text{in } m)}$ which

gives $f_a = -\dfrac{1}{5} = -0.2$ m

The focal length of the lens placed in air is given by

$$\frac{1}{f_a} = \left(\frac{1.5-1}{1}\right)\left(\frac{1}{R_1} - \frac{1}{R_2}\right) \quad ...(i)$$

When the same lens is placed in a medium of refractive index 1.4 its focal length is given by

$$\frac{1}{f_m} = \left(\frac{1.5-1.4}{1.4}\right)\left(\frac{1}{R_1} - \frac{1}{R_2}\right) \quad ...(ii)$$

Dividing (*i*) by (*ii*), we have

$$\frac{f_m}{f_a} = \frac{(1.5-1)}{1} \times \frac{1.4}{(1.5-1.4)} = 7$$

or $f_m = 7f_a$

$= 7 \times (-0.2 \text{ m})$

$= 1.4$ m

$\therefore$ Power of the lens placed in medium of refractive index 1.4 is

$$P_m = \frac{1.6}{f_m(\text{in } m)}$$

$$= -\frac{1.4}{1.4} = -1D.$$

130. Equivalent Capacitance

$$C_{equ.} = \frac{1}{\frac{1}{C} + \frac{1}{C} + \frac{1}{C}}$$

$$= \frac{C}{3}$$

and Breakdown voltage

$$V' = \frac{q}{\frac{C}{3}} = \frac{3q}{C} = 3V.$$

131. $\vec{F} = q\vec{V} \times \vec{B}$

$$= \left(-2\times10^{-6}\right)\left(2\hat{i} + 3\hat{j}\right)\times 10^6 \times 2\hat{j}$$

$$= -4(2\hat{i}\times\hat{j} + 3\hat{j}\times\hat{j})$$

$$= -4(2\hat{k}) = -8\hat{k}$$

Hence, resultant force is 8 N in negative z – direction.

132. Moment of inertia of solid sphere about an axis passing through its centre

$$I_s = \frac{2}{5}MR^2$$

Moment of inertia of solid cylinder about an axis passing through its centre and parallel to its length.

$$I_c = \frac{1}{2}MR^2$$

$$\therefore \quad \frac{I_s}{I_c} = \frac{\frac{2}{5}MR^2}{\frac{1}{2}MR^2} = \frac{4}{5}.$$

134.
- The region from 4000 to 2500 cm^{-1} corresponds to absorptions caused by N–H, C–H and O–H single-bond stretching motions. N–H and O–H bonds absorb in the 3300 to 3600 cm^{-1} range; C–H bond stretching occurs near 3000 cm^{-1}.
- The region from 2500 to 2000 cm^{-1} is where triple-bond stretching occurs. Both $C \equiv N$ and $C \equiv C$ bonds absorb here.
- The region from 2000 to 1500 cm^{-1} is where double bonds (C = O, C = N, and C = C) absorb. Carbonyl groups generally absorb in the range 1670 to 1780 cm^{-1}, and alkene stretching normally occurs in the narrow range 1640 to 1680 cm^{-1}.
- The region below 1500 cm^{-1} is the fingerprint portion of the IR spectrum. A large number of absorptions due to a variety of C–C, C–O, C–N, and C–X single-bond vibrations occur here.

135. Interaction forces, and network of forces : Dispersions and gels show a yield point due to **intermolecular forces (*van der Waals* forces).** This includes dipole-dipole interactions between particles, and between particles and the surrounding dispersion agent. There are different kinds of interaction forces: Electrostatic interactions between permanent dipoles (*Keesom* forces), interactions due to induction between permanent and induced dipoles (*Debye* forces), and dispersion forces between mutually induced dipoles (*London* forces).

138. For a perfect gas,

$$\frac{P_1}{T_1} = \frac{P_2}{T_2}$$

$$P_1 = \frac{P_2}{T_2} \times T_1$$

$$= \frac{55}{180} \times 576$$

$$= 176 \text{ atm.}$$

140. Bond Order of Oxygen molecule (O_2) : Oxygen molecule will have 16 electrons (8 each from the atoms). According to the above formula, the pairs of electrons are distributed in the order of increasing energy levels are as follows.

Bonded pairs of electrons (BP) :

$\sigma 1s(2e)$ $\sigma 2s(2e)$ $\sigma 2px(2e)$ $\pi 2py(2e)$ $\pi 2pz(2e)$ Total 10 electrons.

Anti bond pairs of electrons (ABP) :

$\sigma 1s(2e)$ $\sigma 2s(2e)$ $\pi 2py(2e)$ Total 6 electrons.

Bond Order (B.O.)

$$= \frac{(BP - ABP)}{2} = \frac{10-6}{2} = 2$$

The **bond order** is 2 and so oxygen molecule will have a double bond and is stable molecule.

Bond Order of N_2

Total number of electrons in N_2 molecule is 7 + 7 = 14

As per the formula,

Bonded pair of electrons (BP):

$\sigma 1s(2e)$ $\sigma 2s(2e)$ $\pi 2py(2e)$ $\pi 2pz(2e)$ $\sigma 2px(2e)$ Total 10 electrons.

Anti bond pairs of electrons (ABP) :

$\sigma\ 1s^*(2e)\sigma 2s^*(2e)$. Total 4 electrons.

Bond order (B.O.)

$$= \frac{(BP - ABP)}{2} = \frac{10-4}{2} = 3$$

The bond order is 3 and so Nitrogen molecule will have a triple bond and is the most stable molecule.

Bond Order of Carbon molecule (C_2)

Total number of electrons in C_2 molecule is 6 + 6 = 12.

As per the formula,

Bonded pair of electrons (BP) :

$\sigma 1s(2e)$ $\sigma 2s(2e)$ $\sigma\ 2py(2e)$ $\pi 2pz(2e)$. Total 8 electrons.

Anti bond pairs of electrons (ABP):

$\sigma\ 1s^*(2e)$ $\sigma 2s^*(2e)$ Total 4 electrons.

Bond Order (B.O.)

$$= \frac{(BP - ABP)}{2} = \frac{8-4}{2} = 2$$

The **bond order** is 2 and so carbon molecule will have a double bond and is a stable molecule.

144. The energy required to excite an electron in an atom of hydrogen from n = 1 to n = 2 is given by the following relation (difference in energy E_2 and E_1)

$$E_2 = \frac{-1.312 \times 10^6 \times (1)^2}{(2)^2}$$

$$= -3.28 \times 10^5 \text{ J mol}^{-1}$$

$$E_2 = -1.312 \times 10^6 \text{ mol}^{-1}$$

$$\therefore \Delta E = E_2 - E_1$$

$$= [-3.28 \times 10^5] - [-1.312 \times 10^6] \text{ J mol}^{-1}$$

$$= (-3.28 \times 10^5 + 1.312 \times 10^6) \text{ J mol}^{-1}$$

$$= 9.84 \times 10^5 \text{ J mol}^{-1}.$$

148. The cell is

$Mg(s) \mid Mg^{2+}\ (aq)\ (0.130\ M) \parallel Ag^{+}\ (aq)\ (0.001\ M) \mid Ag(s)$

The Nernst equation is

$$E_{cell} = E^\circ - \frac{0.059}{n}\log\frac{[Mg^{2+}]}{[AG^{+}]^2}$$

$$E_{cell} = 3.17 - \frac{0.059}{2}\log\left[\frac{0.130}{(0.0001)^2}\right]$$

$$E_{cell} = 3.17 - \frac{0.059}{2}\times 7.114$$

$$E_{cell} = 3.17 - 0.21$$

$$E_{cell} = +\,2.96\ V.$$

149. The cathodic reaction is

$$Cu^{2+} + 2e^- \rightarrow Cu$$

$I = 1.5$ amperes

$t = 10 \times 60$

$= 600$ s

$\therefore\ Q = It = 1.5 \times 600$

$= 900$ C

The reaction states that 2×96487 C are required to deposit 63 g Cu at cathode

$\therefore$ 900 C will deposit $\dfrac{63\times 900}{2\times 96487}$

$= 0.2938$ g Cu at cathode.

152. The number of valence electrons in XeO_2F_2 is $8 + 2 \times 6 + 2 \times 7 = 34$.

These are distributed as shown in the following.

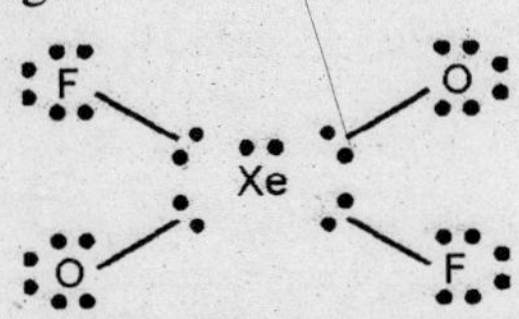

There are five paires of electrons around Xe. To accommodate these, Xe undergoes dsp^3 hybridization leading to the trigonal bipyramidal arrangement of five paires of electrons. The structure of XeO_2F_2 is

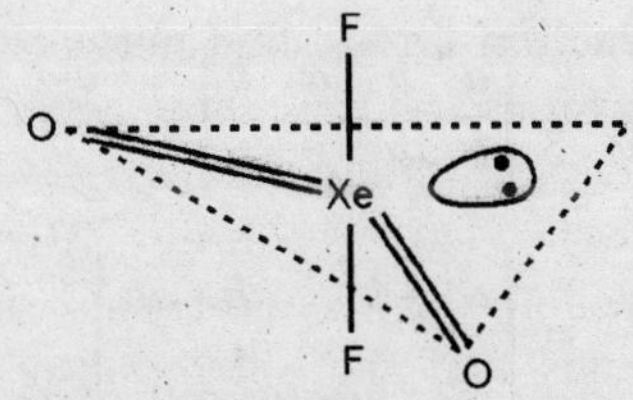

The shape of XeO_2F_2 is sea-saw or distorted trigonal bypyramidal.

153. $\tan A + \tan B = p$

$\tan A\ .\ \tan B = q$

$$\tan(A + B) = \frac{\tan A + \tan B}{1 - \tan A.\tan B}$$

$$= \frac{p}{1-q}.$$

154. Let, N = the number of subsets of the set which contain at most n elements

$$= {}^{2n+1}C_0 + {}^{2n+1}C_1 + {}^{2n+1}C_2 + \ldots + {}^{2n+1}C_n$$

We have,

$$2N = 2({}^{2n+1}C_0 + {}^{2n+1}C_1 + {}^{2n+1}C_2 + \ldots + {}^{2n+1}C_n)$$

$$= ({}^{2n+1}C_0 + {}^{2n+1}C_{2n+1}) + ({}^{2n+1}C_1 + {}^{2n+1}C_{2n}) + \ldots + ({}^{2n+1}C_n + {}^{2n+1}C_{n+1})$$

$$(\because {}^{n}C_r = {}^{n}C_{n-r})$$

$$= {}^{2n+1}C_0 + {}^{2n+1}C_1 + {}^{2n+1}C_2 + \ldots {}^{2n+1}C_{2n+1}$$

$$= 2^{2n+1} \Rightarrow N = 2^{2n}.$$

155. $x^2 + 4xy + 4y^2 + 5x + 10y + 4 = 0$

here, $a = 1, b = 4, g = \dfrac{5}{2}, f = 5, h = 2, c = 4$

$abc + 2fgh - af^2 - bg^2 - ch^2 = 0$

$$= 1\times 4\times 4 + 2\times 5\times\frac{5}{2}\times 2 - 1\times 5^2 - 4\times\frac{5^2}{4} - 4\times 2^2$$

$= 16 + 50 - 25 - 25 - 16 = 0$

Hence, given equation will represent a pair of st. lines.

Angle between two lines

$$\tan\theta = \frac{2\sqrt{h^2 - ab}}{a+b} = \frac{2\sqrt{2^2 - 1\times 4}}{1+4} = 0$$

Hence, two lines are parallel.

156. $A^2 = \begin{bmatrix} a & b \\ b & a \end{bmatrix}\begin{bmatrix} a & b \\ b & a \end{bmatrix}$

$$= \begin{bmatrix} a^2+b^2 & ab+ab \\ ab+ab & a^2+b^2 \end{bmatrix}$$

$$= \begin{bmatrix} a^2+b^2 & 2ab \\ 2ab & a^2+b^2 \end{bmatrix} = \begin{bmatrix} p & q \\ q & p \end{bmatrix}$$

$\therefore \quad p = a^2 + b^2$

$q = 2ab.$

158. As $f(x)$ is continuous at $x = [0, 3]$

$f(1) = (1)^2$

$= 2(1) + a$

or, $a + 2 = 1$

$\Rightarrow \quad a = -1$

$f(2) = 2(2) - 1 = b - 2$

$3 = b - 2$

$\Rightarrow \quad b = 5$

Hence, $a = -1, b = 5.$

160. $\lim\limits_{x\to 0} \dfrac{1-\cos^3 x}{x.\sin x.\cos x}$

From L'Hospital rule

$$= \lim_{x\to 0} \frac{3-\cos^2 x.\sin x}{\frac{1}{2}(\sin 2x + 2x\cos 2x)}$$

$$= \lim_{x\to 0} \frac{6.(-\sin^2 x.2.\cos x + \cos^3 x)}{(2\cos 2x + 2\cos 2x - 4x\sin 2x)}$$

$$= \frac{6}{2+2} = \frac{3}{2}.$$

Biotechnology AND Biochemistry

CHAPTER

1

Introduction

Biochemistry

In the year 1903, German biochemist Carl Neuberg first used the term 'biochemistry'. Biochemistry may be defined as the chemistry of living things.

The American Society of Biological chemist (1965) has been defined the term biochemist as "A biochemist is an investigator who utilizes chemical, physical or biological techniques to study the chemical nature and behaviour of living matter".

Biotechnology

Biotechnology is an important area of study. The complete definition of biotechnology is difficult, due to wide range of its usage.

Formentation, by microorganisms, formation of **cheese** and **curd** from milk, production of antibiotics, the process of baking and brewing and that of **vinegar** from molasses are example of **old biotechnology.**

Recombinant DNA Technology, Polymerase Chain Reaction (PCR), bioprocessing cell culture and fusion are the example of new biotechnology.

According to "U.S. National Science Foundation" Biotechnology may be defined as "the controlled use of biological agents such as micro-organisms or cellular components for beneficial use".

Biotechnology in India

In India at the 69th session of Indian Science Congress held at Mysore in 1982, Indian scientists emphasized the role of gene technology, consequently, Government of India constituted a **"National Biotechnology Board"** under the Department of Science and Technology to coordinate and promote research in this area. National Biotechnology Board decided to compile a list of biotechnologists (In India and abroad) working in the area given as under:

(*i*) Genetic engineering
(*ii*) Photosynthesis
(*iii*) Enzyme engineering
(*iv*) Immunotechnology
(*v*) Tissue culture
(*vi*) Alcohol fermentation

In the year 1986 Government of India was created separate Department of Biotechnology in the Ministry of Science and Technology. An International Centre for Genetic Engineering and Biotechnology (ICGEB) under the auspicies of United Nations (for developing countries) has also been established and it has two locations, one in India at New Delhi, and another in Italy at Triesta. Indian Centre operating in the year 1987.

International Institute of Biotechnology (IIB) having objective similar to those of ICGEB was established at Canterbury, Kent in U.K. which is also provides training (educational) through different methods like a course for the degree of Master of Biotechnology.

For the development of 'Biotechnology' in India, NBTB and DBT has been created both physical facilities *i.e.*, instrumentation and trained man power which are summarised as under:

(a) Plant Molecular Biology

For crop improvement through biotechnology six centres for plant molecular biology were established by DBT, Govt. of India:

(*i*) Madurai Kamraj University, Madurai.
(*ii*) Jawaharlal Nehrû University, New Delhi.
(*iii*) Tamil Nadu Agricultural University, Coimbatore.
(*iv*) Osmania University, Hyderabad.
(*v*) Bose Institute, Kolkata and
(*vi*) National Botanical Research Institute, Lucknow.

(b) Bioinformatics

In various institutions of the country, in the area of bioinformatics, 9 (nine) Distributed Information Centres (DIC's) were established in various institutions in the country.

The DIC's are established in the following institutions:

(*i*) Indian Institute of Science, Bangalore.
(*ii*) Madurai Kamraj University, Madurai.
(*iii*) Bose Institute, Kolkata.
(*iv*) Jawaharlal Nehru University, New Delhi.
(*v*) Poona University, Pune.
(*vi*) Indian Agricultural Research Institute, New Delhi.
(*vii*) Centre for Cellular and Molecular Biology, Hyderabad.
(*viii*) National Institute of Immunology, New Delhi and
(*ix*) Institute of Microbial Technology, Chandigarh.

To access the international databases in biotechnology, these centres are linked to each other and with the Central Information Network at DBT in New Delhi.

(c) Infrastructural facilities

DBT has set up the service oriented infrastructural facilities in order to develop research and development (R&D) which are given as under:

(*i*) Germplasm banks (for plants, animals, microbes and algae).
(*ii*) Animal houses
(*iii*) Production, import and distribution of enzymes, radio-labelled compounds and reagents.
(*iv*) DNA synthesizers.
(*v*) Genetic engineering, R&D units.
(*vi*) Bioprocess optimization and pilot plants.
(*vii*) Carbohydrate cell surface and Cellular Transport.
(*viii*) Protein and peptide sequencing.
(*ix*) NMR facility.
(*x*) Antibody development consortium.
(*xi*) Marine cyanobacterial germplasm collection.
(*xii*) Centre for reproductive biology and molecular endocrinology.

Research

In the field of biotechnology in India, some of the achievements are summarised as under:

(*i*) Successful use of embryo transfer and split embryos technology has become possible to give birth improved buffalo calves.
(*ii*) DNA finger printing using Bkm probe which have great importance in forensic medicine for identification of criminal.
(*iii*) Biopesticides and biofertilizers and being developed to meet the nitrogen demand and also to protect soil.
(*iv*) For farming of prawn and major carps aquaculture technology developed.
(*v*) For eradication of polio, oral polio vaccines developed.
(*vi*) For filaria, the diagnostic kit has been released.
(*vii*) Micropropagation of elite trees has been made possible at commercial scale for bamboos (at university of Delhi) and for teak, eucalyptus and sandal wood (one at TERI, New Delhi and other at NCL, Pune).

Application of Biotechnology

Applications of biotechnology can be summarised as under:

1. **Biotechnology in Medicine:** The insulin and interferon synthesized by bacteria have

already being released in market for the sale. In the field of medicines large number of vaccines for immunization against various diseases, monoclonal antibodies and DNA probes for diagnosis of diseases, other pharmaceutical drugs for treatment of diseases and human growth hormone are being released.

DNA finger printing and auto antibody finger printing techniques have great importance in forensic medicine for identification of criminals like rapists and murderers through the study of DNA/antibodies from blood and semen stains, tears, saliva, hair roots and urine etc.

2. **Biotechnology in Agriculture:** Another very important role of biotechnology in the area of agriculture which includes the following:
   (*i*) Tissue, plant cell and organ culture.
   (*ii*) Genetic engineering, in regeneration of plant to give transgenic plants carrying desirable traits like insect resistance, disease resistance and herbicide resistance.

   Also be used for increasing nitrogen fixing ability, storage proteins, hybrid crops and improved photosynthetic efficiency.
3. **Biotechnology in intellectual property rights:** It is also emphasised that protection of these rights may also affect the development of biotechnology. In recent years these have also been discussions on the protection of intellectual property rights (IPR) emanating from the use of biotechnology.
4. **Biotechnology in industrial microbiology:** Another very important area of biotechnology which receiving major attention of the biotechnologists. A variety of chemicals and pharmaceuticals drugs are being produced or will be in the process or produced to increase substantially the quantity and quality of these chemicals and drugs.
5. **Biotechnology in protein or enzyme engineering:** Use of biotechnology in the production of enzymes and storage proteins is of great importance protein engineer first prepares a completes aided protein model for a specific function and then prepares a synthetic gene that will produce this desired protein in a predictable manner. So, in future, proteins will be engineered in the desired manner.

   Biotechnology has also provided us a technique in the form of immobilized enzyme system, which allowed the production of substances. For *e.g.*, production of corn syrup with high fructose using an immobilized enzyme, glucose isomerase. Which is used as sweetening agent in the soft drink industry.
6. **Biotechnology in environment:** In the field of environment receiving major attention of biotechnologists. Environmental biotechnology, is conversed, both with implications and applications of biotechnology in the field of environment. Due to rapid growth of industrialization, urbanization and developments, there is a threat to clean environment. These are fear being expressed the implication of advances in biotechnology through release of harmful organisms developed through recombinant DNA technology. In this connection rules and laws have been framed from time to time to safeguards against the risks in which the recombinant DNA technology poses to the friendly and clean environment.

Different biotechnological techniques is also being used for some environmental problems like in the pollution control, restore of degraded land, depletion of natural resources and in conservation of biodiversity.

Through biotechnological techniques microbes are being developed to be used as biofertilizers, biopesticides, biosensors etc. and a recovery of metals. Genetic engineering and tissue culture, root nodulation and mycorrhizae (VAM fungi) are also being applied for reclamation of degraded soil.

Multiple Choice Questions

1. International Centre for Genetic Engineering and Biotechnology is situated at :
A. New Delhi
B. Trieste
C. Lucknow
D. Both A and B

2. Application of Biotechnology procedures in Medical procedures is known as :
A. Red Biotechnology
B. Green Biotechnology
C. Blue Biotechnology
D. White Biotechnology

3. What is biopatents?
A. Right to use biological entities
B. Right to use products
C. Right to use process
D. Right to use invention

4. In case of medicine, biotechnology can be used for which of the following?
A. Production of synthetic vaccines
B. Diagnosis of diseases
C. Production of rare drugs
D. All of these

5. Which of the following is an example of fortified food?
A. Addition of iodine to table salt (iodised salt)
B. Addition of calcium to fruit juice
C. Addition of Vitamin D to such foods as margarine
D. All of these

6. A germplasm bank of AMF (arbuscular mycorrhizal fungi) cultures was established at which of the following?
A. The Energy and Resources Institute, New Delhi
B. Indian Agricultural Research Institute, New Delhi
C. National Bureau of Plant Genetic Resources, New Delhi
D. Indian Institute of Pulses Research, Kanpur

7. Nutraceuticals differ from dietary supplements in respect to which of the following?
A. They include some products from nonfood sources
B. They must not only supplement the diet, but should also aid in prevention or treatment of diseases/disorders
C. They are represented for use as the conventional food
D. Both A and B

8. Which of the following bacteria generally used for genetic engineering?
A. *Pseudomonas*
B. *Agrobacterium*
C. *Bacillus*
D. *Clostridium*

9. Which is molecular scissors?
A. DNA polymerase
B. Ligase
C. Helicase
D. Restriction endonuclease

10. The technique of DNA finger printing was coined by :
A. Beadle and Tatum
B. Franklin
C. Alec Jeffreys
D. Stahl

11. What do you mean by DNA finger printing?
A. Molecular analysis of profiles of DNA samples
B. Techniques used for molecular analysis of different specimens of DNA
C. Analysis of DNA samples using imprinting devices
D. Techniques used for identification of finger prints of individuals

12. What is probiotics?
A. Safe antibiotics
B. New kind of food allergen
C. Cancer inducing microbes
D. Live microbial food supplement

13. The head quarter of Common Wealth Institute of Biological control is situated at :

A. Sydney B. Durban
C. Bangalore D. Trinidad

14. .... is commercially produced by micro-organism and used by microbes during their growth and therefore is called primary microbial product or primary metabolite.

A. Amino acid B. Antibiotics
C. Alcohol D. Organic acid

15. In PCR, the DNA is generally denatured by :

A. Temperature
B. Alkali treatment
C. Add treatment
D. None of these

16. Approximately how many Okazaki fragments are synthesized during one round of replication of the *E. coli* genome?

A. 4×10^6 B. 5000 to 8000
C. 2500 to 5000 D. 1000–1500

17. SDS and NaCl are used in the purification of genomic DNA from peripheral blood samples. Their functions in the process are respectively :

A. Salting out of DNA – denaturation of protein
B. Salting out of protein – denaturation of DNA
C. Denaturation of protein – salting out of DNA
D. Denaturation of protein – salting out of proteins

18. Genetic information transfer from DNA to RNA is known as :

A. Transformation
B. Transcriptase
C. Recombination
D. Transduction

19. Modern concept of chemotherapy was proposed by :

A. Joseph Lister
B. Elie Metchnikoff
C. Paul Ehrlich
D. R.J. Shepherd

20. National Center for Biotechnology Information (NCBI) was established on Nov. 4, 1988 as a Division of the

A. National Institute of Nutrition (NIN)
B. National Library of Medicine (NLM)
C. European Bioinformatics Institute
D. National Institute of Health (NIH)

21. GEM stands for :

A. Genetically engineered metabolism
B. Genetically engineered metazoa
C. Genetically engineered mustard
D. Genetically engineered microbes

22. Enzymatic procedure of DNA sequencing was developed by which of the following?

A. Gillbert and co-workers
B. Collins and co-workers
C. Sanger and co-workers
D. Maxam and co-workers

23. PCR can be used for which of the following?

A. Study of DNA polymorphism
B. Production of *c*DNA
C. Detection of presence of transgene
D. All of these

24. Which of the following ministeries established Department of Biotechnology in 1986?

A. Ministry of Science and Technology
B. Ministry of Agriculture
C. Ministry of Environment
D. Ministry of Human Resource Development

25. Bioaccumulation/biomagnification may occurs due to which of the following reasons?

A. Xenobiotic compounds are continuously taken up from the environment and accumulated in the lipid deposits of the body
B. Consumption of these organisms by other organisms leads to further accumulation of Xenobiotic compounds in the bodies of latter
C. The genobiotic compounds are recalcitrant and lipid soluble
D. All of these

26. Biofilms contain a complex community of organism, which degrade carbohydrates, proteins, lipids etc. into which of the following?

I. CO_2 II. PO_4^{2-}
III. NO_3^- IV. SO_4^{2-}

Options:

A. I, II, III, IV B. I, II, III
C. I, II, IV D. II, III, IV

27. Many of *Bacillus* species are pathogenic to Lepidopteran pests due to which of the following?

A. Endotoxin production
B. Exotoxin production
C. Antibiotic production
D. Chitinase production

28. Rabies virus affects which of the following organs?

A. Liver B. Spinal cord
C. Kidney D. Thymus

29. Which of the following statements are correct?

I. A reporter gene produces an enzyme that allows an easy identification or selection of the cells in which it is expressed.
II. A vector which is used for propagation of DNA insert in a suitable host is known as expression vector.
III. Process of producing clones is called cloning.
IV. A circular DNA molecule, which has autonomous replicating property in host cell it is called plasmid.

Options:

A. I, II, III, IV B. II, III, IV
C. I, II, IV D. I, III, IV

30. Which of the following viruses has acquired a gene from its host genome?

A. Adenoviruses
B. Epstein-Barr Virus
C. Acute transforming retroviruses
D. Replication competent tumour retro-viruses

31. In India, molecular markers can not be used to resolve issues related to which of the following?

A. Farmer's rights
B. Biodiversity
C. Plant breeder rights
D. Plant breeding rights

32. The action of chloramphenicol involves which of the following?

A. Inhibition of peptidy transferase
B. Inhibition of DNA replication
C. Prevention of initiator *t*RNA from entering the P site
D. Inhibition of transcription

33. Antisense RNA technology may be used to achieve which of the following?

A. Minimise the expression of a gene
B. Enhance the expression of a gene
C. Change the expression pattern of a gene
D. Stabilize the expression of a gene

34. The chief problem in the commercial exploitation of pharmaceutically valuable recombinant protein production in plants is which of the following?

A. Glycosylation of the proteins
B. Stability of the proteins
C. Folding of the proteins
D. Purification of the proteins

35. Penicillin acylase (amidase) is being used industrially for production of which of the following?

A. Secondary metabolites
B. Detergents
C. Synthetic penicillins
D. Mitigation of penicillin allergy

36. Which of the following techniques does not use radioactivity?

A. Southern blotting
B. RAPD
C. Western hybridization
D. RFLP

37. Stable integration of a gene into plant genome can be conclusively shown by which of the following?

A. Southern hybridization

B. Dot blot assay
C. PCR
D. Northern hybridization

38. In response of stress, which of the following amino acids accumulates?
A. Phenylalanine B. Lysine
C. Proline D. Histidine

39. A Barr body is a(n) :
A. Inactivated X chromosome
B. Inactivated Y chromosome
C. An abnormality seen in the cells of persons with down syndrome
D. Chromosome 21 trisomy complex

40. Cytogenetics is related to all the following except :
A. Karyotype
B. Chromosome behaviour
C. Pedigree analysis
D. Meiosis

41. Aminocentesis is ....
A. Cells are removed and grown in culture
B. Used to examine fetus for major abnormalities
C. Technique that can be used as early as eight weeks in pregnancy and yields
D. None of these

42. The lac repressor is a :
A. RNA molecule B. DNA molecule
C. Protein molecule D. Lipid molecule

43. Which is not a DNA-binding domain?
A. Intron-exon-intron
B. Zinc finger
C. Helix-turn-helix
D. Leucine zipper

44. Genes that are always transcribed are called :
A. Operator genes
B. Repressor genes
C. Transposons (Jumping genes)
D. Constitutive genes

45. What is the name for the transcripts that ultimately get converted to the form that is translated?
A. *m*RNA
B. Smaller nuclear RNA
C. Mature RNA
D. Heterogeneous nuclear RNA

46. A woman with mild symptoms of myotonic muscular dystrophy has a daughter with a very severe form of the disease. Which of the following statements is most correct?
A. The mother probably has a mutation with a larger number of repeats than the daughter
B. This is an example of anticipation
C. This must be an X-linked dominant condition
D. None of these

47. The bacteria used for the Ames test is :
A. *Thermus aquaticus*
B. *Salmonella typhimurium*
C. *Escherichia coli*
D. *Shigella dysenteriae*

48. What enzyme is responsible for proofreading repair after replication?
A. DNA polymerase B. RNA polymerase
C. Ligase D. Primase

49. Interleukins :
A. Halt the production of viruses
B. Used to treat individual with kidney disorders
C. Stimulate white cell production
D. Activate white blood cells which used to treat HIV and cancer

50. When investigating a crime scene detectives find a single hair on the floor. The sample is not enough to perform many tests. Which of the following methods could be used to make sufficient DNA to allow for comparison to the DNA of the prime suspect?
A. The hair root could be tissue cultured to make additional copies of the DNA
B. Polymerase chain reaction could be used to make many copies of DNA
C. Reverse transcriptase could be used to make DNA from RNA in the hair sample
D. The gene could be amplified by use of electrophoresis

ANSWERS

1	2	3	4	5	6	7	8	9	10
D	B	C	D	D	A	D	B	D	C
11	12	13	14	15	16	17	18	19	20
A	D	D	A	A	C	D	B	C	B
21	22	23	24	25	26	27	28	29	30
D	C	D	A	D	A	A	B	D	C
31	32	33	34	35	36	37	38	39	40
A	D	A	D	C	B	A	C	A	C
41	42	43	44	45	46	47	48	49	50
A	C	A	D	D	B	B	A	D	B

●●●

CHAPTER

2

Cells, Its Development and Division

CELL

Cell is a smallest unit of biological activity made of an organised mass of protoplasm surrounded by a protective and selectively permeable covering. Basically cells are of two types, prokaryotic and eukaryotic.

Cell theory was first modified in the light of Virchow's findings (1855, 1858) that cells develop from pre-existing cell-*Omnis cellula-e-cellula*. It is known as law of other modifications were carried out in the cell theory. The modern cell theory is called 'cell principle'.

Protoplasm Theory

It was proposed by Max Schultze (1861) which states that living matter of an organism in protoplasm and the cell is simply an accumulation of protoplasm limited by an outer membrane and containing a nucleus.

Organismal Theory

Sachs (1874) proposed that the whole organism functions as a single entity which is made of a continuous mass of protoplasm incompletely divided into cells.

Prokaryotic and Eukaryotic Cells

Basically cells are of two types, prokaryotic and eukaryotic.

A **prokaryotic** cell has one envelope system with no membrane lined internal organelles, except thylakoids, if present. A nucleus is absent, instead, a nucleoid made of naked and folded DNA lies freely in cytoplasm. Internal compartments are absent. Respiratory enzymes are located on the cell membrane. Cytoplasm does not show streaming. Sap vacuoles are absent. Gas vacuoles may occur-Cell wall, if present, contains peptidoglycan or muco-peptide. Flagella, if present are single stranded and made of flagellin.

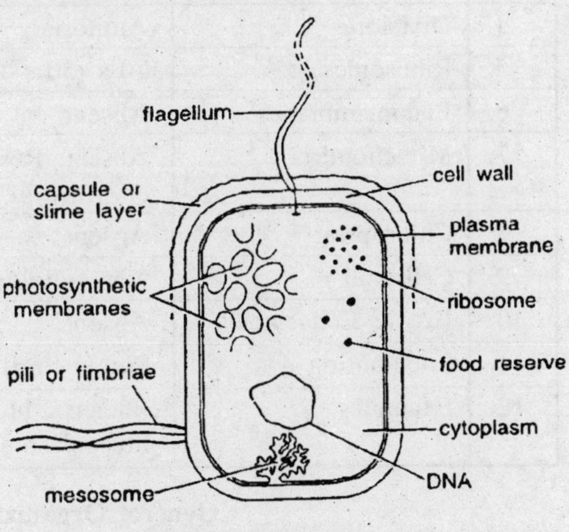

Fig. 1: *Prokaryotic cell.*

An **eukaryotic** cell has internal compartments and double envelope system, that is, besides cell membrane, the cell organelles are also covered by membranes. Mitosis and meiosis occurs in the life cycle. Cytoplasm show streaming, sap vacuoles occur. Flagella if present are double-stranded and made of tubulin.

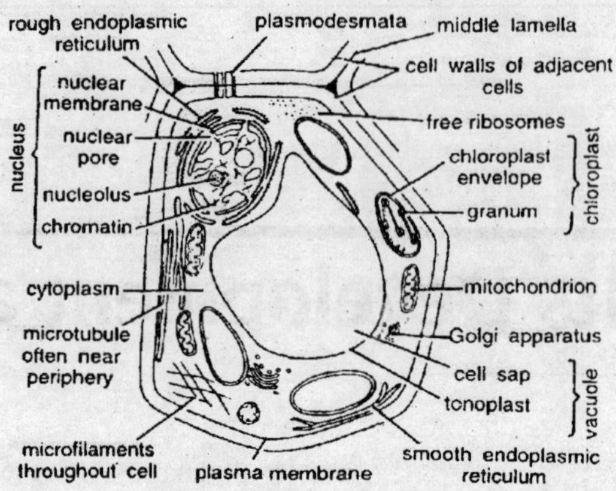

Fig. 2: *Eukaryotic cell.*

Comparison of Cell Organisation in Prokaryotic and Eukaryotic cells

	Cellular Component	*Prokaryotic Cell*	*Eukaryotic Cell*
1.	Nuclear Envelope	Absent	Present
2.	DNA	Naked	Combined with proteins
3.	Chromosome	Single	Multiple
4.	Division	Amitosis	Mitosis or meiosis
5.	Ribosomes	70 s (50 s + 30 s)	80 s (60 s + 40 s)
6.	Endomembranes	Absent	Present
7.	Mitochondria	Absent, Respiratory and photosynthetic enzymes in the plasma membrane	Present
8.	Chloroplast	Absent	Present in plant cells
9.	Cell wall	Non-cellulosic	Cellulosic, only in plants.
10.	Exo & Endocytosis	Absent	Present
11.	Locomotion	Single fibril flagellum	Cilia and flagella
12.	Examples	Bacteria, blue green algae mycoplasma (PPLO)	Protozoa, other algae, metaphyta & metazoa.

General Organization of the Eukaryotic Cell

	Main Component or components	*Subcomponent or subcomponents*	*Main Function*
1.	Cell Membrane	Cell wall, Cell coat, Plasma membrane	Protection cell, cell interaction permeability, endocytosis, exocytosis
2.	Nucleus	Chromatin & chromosomes, Nucleolus	Genetic information system Synthesis of Nucleoplasm, Ribosomes
3.	Cytoplasm matrix, cytosol, cytoskeleton	Soluble Enzymes, Microfilaments, Microtubules, Ribosomes	Glycolysis, Cell motility, Cell shape & Motility, Protein Synthesis.

	Main Component or components	*Subcomponent or subcomponents*	*Main Functions*
4.	Endomembrane System	Nuclear Envelope, Endoplasmic reticulum, rough & smooth Golgi complex	Nuclear permeability, Synthesis and transport of materials, Secretion
5.	Membrane organelles	Mitochondria, Chloroplast, Lysosomes, Peroxisomes	Cell respiration, Photosynthesis, Digestion, Peroxidation
6.	Microtubular organelles	Centrioles & spindle basal bodies, cilia and flagella	Cell division and Cell motility

Plant and Animal Cells

Both plant and animals posses eukaryotic cells. A plant cell is characterised by a cell wall, central vacuole, plastids and anastral spindle (centrioles are generally absent). Golgi apparatus consists of separate units called dictyosomes. Lysosomes are rare. Glyoxysomes and crystals may occur.

Animal cells are bathed by tissue fluid. They are covered by cell membranes only. Vacuoles are small. Plastids are absent. Spindle is amphiastral (centrioles present). Due to absence of wall, animal cells will burst if placed in hypotonic solution, Glyoxysomes and crystal are absent. Lysosomes are present.

A typical plant cell consists of cell wall and protoplast. Cell wall is absent in animal cells. Protoplast (Hanstein, 1880) is made of a plasmalemma, cytoplasm, nucleus and vacuoles.

The Functions and Structure of the Cellular Components

S. No.	*Cellular components*	*Functions*	*Structure*
1.	Plasma membrane	It controls entry and exit to and from the cell.	Fluid materials of proteins and lipids.
2.	Cytoplasm	It contains bulk of the cell material and enzymes responsible for maintaining the life of the cell.	Semifluid materials differentiated into several kinds of structure.
3.	Nucleus	It contains the genetic information which to passed on to daughter cells during the time of cell division.	Membrane surrounding fluid which contains nucleolid and DNA.
4.	Ribosomes	At the site of protein synthesis.	It contains RNA small granular particles.
5.	Endoplasmic reticulum	In the intracellular transport system.	A network of fluid filled tubular and vesicular structures. It is found in cytoplasm.
6.	Mitochondria	It is site for formation of ATP.	It is double walled rod like structure with internal cristae.
7.	Golgi apparatus	It works as storage and preparation for secretion of substances.	It contains small and large vacuoles.
8.	Lysosomes	The digestion of large molecules which enter the cell.	Fluid surrounded by lipoprotein membrane and which contains hydrolytic enzymes.
9.	Centrioles	Its involved in cell division.	Cylinder like structures.
10.	Microtubules	It helps in cellular support.	Like hollow tubes.
11.	Myofibrils	Developed formed, found in muscle cells.	Strands of proteins.
12.	Vacuoles	Storage and excretion of materials.	Droplets of fluid.

Cell Division

Cell division is the process by which cells replicate in order to replace cell loss, repair tissue damage and reproduce the organism.

Two types of cell division are encountered in the Eukaryotic cell *viz.* Mitotic and Meiotic divisions.

Mitosis

In mitotic division (Mitosis) two genetically identical daughter cells are produced from the original cell. The daughter cells, which are identical structurally and in genetic content, could be used to replace dead cells or to repair a damaged tissue. All cells of the body with the exception of the fully differentiated nerve cells undergo mitotic division.

Meiosis

In meiotic division (Meiosis), two genetically none identical daughter cells are produced from the mother cell. The product of meiotic division are involved in the duplication of the organism (Reproduction) hence this division only occurs in reproductive cells called Germ cells. In the human the product of meiotic division of the male germ cell is the Spermatozoon while that of the female germ cell is the Ovum. The union between the male and female germ cells in the process called fertilization results in the formation of a new organism.

Mitosis and Meiosis proceed in two phases *viz.* nuclear division (Karyokinesis) in which genetic materials are shared and cytoplasmic division (Cytokinesis) in which other organelles of the cell are shared.

Stages of Mitotic Division

A cell which is not actively dividing is said to be in interphase. As the cell prepares to divide, it replicates its DNA. This leads to elongation and thickening of the chromosomes. Subsequent to this, the cell proceeds to the first stage of mitosis.

Mitotic division proceeds through 5 stages which include:

(i) **Prophase**
Chromosomes condense, shorten thicken and become visible.

(ii) **Prometaphase**
Centrioles separate and migrate in opposite directions. Mitotic spindles (derived from Microtubules) are formed Nuclear envelope disintegrates (karyokinesis). Chromosomes are attached to mitotic spindle and are organized by the spindles and centrioles.

(iii) **Metaphase**
Chromosomes are arranged along the equatorial plane of the cell by mitotic spindles.

(iv) **Anaphase**
Chromosomes split at the centromere. Chromatids are pulled apart by mitotic spindles/centrioles in opposite directions.

(v) **Telophase**
Chromatids which are now new Chromosomes are assembles and enclosed in nuclear membrane on either side of the centre of the cell.
Cytoplasm is divided into two (Cytokinesis) leading to the formation of two daughter cells from one mother cell.

Stages of Mitosis

Cell Life Cycle

The lifespan of a cell entails two periods from the time the cell is formed to the time it undergoes cell division. The first period is referred to as the interphase period during which the cells carries out all its metabolic activities including growth. This period is also referred to as the metabolic phase of cell cycle. The second period is the period of cell division, which terminates the life of the cell but often results in the production of daughter cells. Somatic and germ cells proceed through the two periods of cell cycle. The interphase period is the longer of the two periods and it is further subdivided into phases.

Go Phase is the period of continuous differentiated function. Some cell undergoing this phase may have the option of coming out of it to undergo mitosis. Such cells are called Facultative dividers.

G1 Phase (Also called First Gap) is the period of specialization and execution of all special functions of the cell.

S Phase is the period of replication of DNA preparatory to mitotic division.

G2 Phase (Also call second Gap) is the period immediately preceding mitotic division.

Meiosis

Meiotic division occurs only in germ (reproductive) cells and consists of two successive phases, *viz.* Meiosis I and Meiosis II each of which further proceeds in several stages.

Meiosis I

This is also referred to as the reduction division in which the chromosome content of the daughter cell is reduced to half of that of the mother cell. As in mitotic division, meiosis I is preceded by replication of the DNA content of the dividing cell. The first phase of Meiosis I is the prophase. This goes through 5 stages as follows:

First Meiotic Prophase

- **Leptotene :** Chromosomes condense, thicken and begin to coil up and become visible.
- **Alternating :** Thick areas called chromomeres begin to appear on each chromosome.
- **Zygotene :** At this stage, homologous chromosomes pair up (Synapsis) to form bivalents. Homologous sex chromosome do not form bivalent but are only joined at the tips of their short arms.
- **Pachytene :** Chromosomes are much thicker and more pronounced. The chromatids of pairing chromosomes are also visible as a tetrad of four strands.
- **Diplotene :** Paired chromosomes begin to disengage and move apart. Areas at which exchange of materials would have occurred between homologous pairs (Chiasmata) are also becoming visible. At the end of this stage the Chiasmata terminalize *i.e.* Draw to the end of the chromosome arm.
- **Diakinesis :** Final terminalization of Chiasmata occur bringing prophase of first meiotic division to an end.

First Meiotic Metaphase

The nuclear membrane disintegrates and chromosomes migrate to the equatorial plane of the dividing cell.

First Meiotic Anaphase

Bivalent chromosomes disjoin and each member migrates randomly to opposite poles of the cell. This is the stage at which random and independent assortment of paternal and maternal chromosomes occurs.

First Meiotic Telophase

Chromosomes are assembles and enclosed in nuclear membrane on either side of the centre of the cell. The cytoplasm is divided into two (Cytokinesis) leading to the formation of two daughter cells from one mother cell. Each daughter cell now has the haploid number of chromosome (23 in the human).

Meiosis II

There is no intervening interphase and no DNA replication between Meiosis I and II. The two daughter cells proceed straight into Meiosis II. The stages of Meiosis II are similar to those of Mitosis, culminating in the production of two daughter cells from each of the two daughter cells of first Meiotic division. However, the chromosomes of the two daughter cells are not genetically identical.

The end result of Meiosis is the production 4 daughter cells, each with haploid number of unidentical chromosomes.

Multiple Choice Questions

1. The Cellular Organelles called 'suicide bags' are :

A. Lysosomes B. Nucleolus
C. Ribosomes D. Golgi bodies

2. Regarding microtubule assembly and disassembly during cell division, which will be the most appropriate answer?

A. Once formed, kinetochore microtubles depolymerize at the plus ends throughout mitosis.
B. Once formed, kinetochore microtubules-polymerise at the plus ends throughout mitosis
C. Kinetochore microtubules polymerise at their plus ends throughout the mitosis
D. Kinetochore microtubules polymerise at their minus ends up to cytokinesis, at which point they depolymerize -

3. Which of the following organelle does not carry out transcription?

A. Nucleus B. Peroxisome
C. Chloroplast D. Mitochondria

4. The correct order of chromatin organization from the following is :

I. Nucleosome
II. Solenoid
III. Double stranded DNA
IV. Chromosome

Choose the correct sequence.

A. I, IV, II, III B. III, I, II, IV
C. I, II, III, IV D. III, IV, I, II

5. The principlal pathway of transport of lysosomal hydrolases from the trans Golgi complex (pH 6.6) to the late endosome (pH 6.0) and the recycling of M6p (Mannose 6-Phosphate) receptors back to the Golgi depends on the pH difference between those two compartments. From what you know about M6p receptor binding and recycling and the pathways for delivery of material to lysosomes, predict what would happen if the pH in late endosomes was raised to 6.6.

A. M6p will bind to hydrolases but will not release the hydrolase in the late endosomes.
B. M6p will bind to hydrolases and will release the hydrolases in the late endosomes.
C. At higher endosomal pH, the receptor would not release the hydrolase and would not be recycled back to the *trn* Golgi network.
D. M6p will be degraded at higher pH

6. The diploid genome of a species comprises 6.4×10^9 bp and fit into a nucleus that 6 µm in diameter. If base pairs occur at intervals of 0.34 nm along the DNA helix, what is the total length of DNA in a resting cell?

A. 3.0 m B. 3.5 m
C. 2.2 m D. 4.0 m

7. Which one of the following sub-cellular structures is enclosed by a half unit membrane?

A. Vacuole B. Vesicle
C. Oleosome D. Chromoplast

8. Phosphorylation of serines as well as methylation and acetylation of lysines in histones tails affect the stability of chromatin structure above the nucleosome level and have important consequences for gene expression. The resulting changes in charge are expected to affect the ability of the tails to interact with DNA because :

A. DNA is negatively charged
B. DNA histone interaction is independent of net charge
C. Phosphorylation of serine increases DNA-histone interaction
D. Methylation and acetylation of lysine increases DNA-histone interaction

9. In the cell cycle, mitosis occur between :

A. G1 and S phase B. S and G1 phase
C. G1 and G2 phase D. S and G2 phase

10. If radioactive thymidine is used during S-phase of interphase. It will make chromosome radioactive :
A. Both end of chromosome
B. Near the centromere
C. Both euchromatin and heterochromatin region
D. Euchromatin region

11. Cells in G0 phase :
A. Can be stimulated to enter S-phase
B. Have the tetraploid amount of DNA
C. Occur in rapidly dividing tissue
D. Must accumulate division potential before entering M-phase

12. The major events of mitotic prophase include all of the following except :
A. Chromosome coiling
B. Nuclear envelope break down
C. DNA replication
D. Nucleolar disaggregation

13. Among phosphorylases and kinases :
A. Both require Pi as the source of Phosphate
B. Only phosphorylases require ATP as the source of Phosphate
C. Both require ATP as the source of Phosphate
D. Only kinases require ATP as the source of phosphate

14. Which of the following does not occur when cell enters M phase?
A. Chromatin condenses
B. Histone H1 is dephosphorylated
C. The Nuclear envelope, the endoplasmic reticulum and the Golgi breakdown
D. The spindle is formed

15. How many DNA molecules are present in the nucleus of human somatic cell in G2 stage of cell cycle?
A. 23 B. 46
C. 69 D. 92

16. Entry into M phase is not allowed if :
A. The cell is not big enough
B. If DNA replication is not complete
C. Mitotic cyclin is over expressed
D. Sufficient nutrients are not available

17. Cell cultured in colchicines undergo complete cell cycle, except for the fact that chromosomes don't migrate to the poles during anaphase. One can conclude that :
A. DNA replication does not necessary result in chromosome separation
B. The spindle is made after interphase
C. The centromere represent under replicated DNA
D. Colchicines does not affect the microtubules

18. Biochemical mechanisms that regulate progression through the cell cycle include :
A. Phosphorylation
B. Protein degradation
C. Dephosphorylation
D. All of these

19. G1-Phase of the cell cycle is usually absent in :
A. Early cleavage stage cells
B. HeLa cells in culture
C. Intestinal epithelial cells
D. Senescent fibroblast cells

20. In a dividing cell, DNA synthesis goes through only one round and is arrested before cytokinesis. The arrest is due to :
A. Specific cyclin protein along with CdC2 protein kinase
B. G protein with CdC2 protein kinase
C. CdC2 protein kinase
D. Specific cyclin protein

21. Cell cycle progression from G2 to M-phase is mainly controlled by :
A. Glycosyl transferases
B. Cyclins
C. Cyclic AMP
D. Cell Adhesion molecules

22. Cell cycle is controlled by :
A. Proteolysis of cyclin dependent kinase
B. Dephosphorylation of cyclin
C. Phosphorylation of cyclins
D. Proteolysis of cyclins

23. The basis for the blocking action of the alkaloid colchicine on cell division is :
A. G1 phase B. S phase
C. M phase D. G2 phase

24. Cyclin B breaks down :
A. During anaphase
B. During DNA replication
C. During G2/M transition
D. During cell wall formation

25. If haploid number of chromosomes in human is 23 and the amount of nuclear DNA in a sperm is 1C state how many chromosome a human cell has in the metaphase of mitosis and prophase I of meiosis respectively :
A. 46 and 23 B. 46 and 46
C. 23 and 46 D. None of these

26. If a sperm cell of a mouse contains 1.8 picogram of DNA per nucleus, the mass of DNA expressed in the nucleus of a kidney cell would be :–
A. 3.6 picogram B. 7.2 picogram
C. 0.9 picogram D. 1.8 picogram

27. In meiosis, replication of DNA occurs at :
A. Meiosis I
B. Meiosis II
C. Between Meiosis I and Meiosis II
D. In both Meiosis I and Meiosis II

28. Assembly of two subunits 40S and 60S of the ribosome is :
A. 100 S unit B. 80 S unit
C. 70 S unit D. 90 S unit

29. How many meiotic divisions are required to produce 12 pollengrains in *cyperaceae*?
A. 6 B. 2
C. 8 D. 12

30. The most convincing experiments that established the chromosomal basis of genes and gene-order involved :
A. Genetic crossing over parallel that of physical crossing over between homologous chromosomes
B. Non-disjunction of chromosomes
C. Karyotyping of chromosomes
D. Distributive pairing of chromosomes

31. When the centromere of a chromosome is located at one end of the chromosome, the position is described as :
A. Metacentric B. Submetacentric
C. Acrocentric D. Telocentric

32. Repeated chromatid replication and non-segregation of daughter chromatids leads to the formation of :
A. Pachytene chromosome
B. Polytene chromosome
C. Lampbrush chromosome
D. Leptotene chromosome

33. An organism has $2n = 12$ chromosomes which form 6 bivalents at meiosis. A mutant strain of this organism was found to have 5-bivalents and one trivalent (-association of three chromosomes). The mutant strain could be :
A. Monosomic B. Trisomic
C. Autotetraploid D. Allotriploid

34. A cell undergoing meiosis produces four daughter cells, two of which are aneuploids, while two are haploid. This can occur due to :
A. Non-disjunction during first meiotic division only
B. Non-disjunction during second meiotic division only
C. Non-disjunction during either first or second meiotic divisions
D. Non-disjunction during both first and second meiotic divisions

35. In contrast with plant cells, the most distinctive feature of cell division in animal cells is :
A. Control of cell cycle transitions by protein kinases
B. Enzymes responsible for DNA replication
C. Ubiquitin dependent pathway for protein degradation
D. Pattern of chromosome movement

36. What kind of aneuploid gametes will be generated if meiotic non-disjunction occurs at first division? (*n* represents the haploid number of chromomes).

A. Only $n + 1$ and n
B. Only $n - 1$ and n
C. Both $n + 1$ and $n - 1$
D. Either $n + 1$ or $n - 1$

37. While studying the cell with the electron microscope, a scientist notes the following : numerous ribosomes, a well developed endoplasmic reticulum, chloroplasts, and a cell wall :

A. A fungus
B. An animal
C. A bacterium
D. A plant

38. Regarding mitosis and cytokinesis, one difference between higher plants and animals is that in plants :

A. The spindles contains cellulose microfibril in addition to microtubules, whereas animal spindles do not contain microfibrils
B. Sister chromatids are identical, whereas in animals they differ from one another
C. A cell plate begins to form at telophase, whereas in animals a cleavage furrow is initiated at that stage
D. Chromosomes become attached to the spindle at prophase, whereas in animals chromosomes do not become attacked until anaphase

39. In contrast with plant cells, the most distinctive feature of cell division in animal cell is :

A. Pattern of chromosome movement
B. Control of cell cycle
C. Enzymes responsible for DNA replication
D. Ubiquitin-dependent pathway for protein degradation

40. Chloroplast distribution in a photosynthesizing cell is governed by blue light sensing phototropin 2 (PHOT2). When the cells are irradiated with high intensity blue light, the chloroplast?

A. Move to the side walls
B. Aggregate in the middle of the cell
C. Are sparsely distributed
D. Aggregate in small cluster

41. The membrane lipid molecules assemble spontaneously into bilayers when placed in water and form a closed—spherical structure known as :

A. Lysosome
B. Endosome
C. Peroxisome
D. Liposome

42. Intra cellular transport and cytoskeletal organization of a cell is regulated by nucleotide exchange of different small molecular weight GTPases of Ras super family. Over expression of which of the following GTPase modulates the actin-cytoskeleton of HeLa cells?

A. Ran is GDP bound form
B. Ran is GTP bound form
C. Rho in GTP bound form
D. Rho in GDP bound form

43. Aneuploid females with only one X chromosome is a characteristics of individuals with :

A. *cri du chat* syndrome
B. Klinefelter syndrome
C. Down syndrome
D. Turner syndrome

44. Transport of water across aquaporins is regulated by the presence of which of the following sequence of three highly conserved amino acids?

A. Ala-Asn-Pro
B. Pro-Asn-Ala
C. Asn-Pro-Ala
D. None of these

45. Which type of the cyclins have/has essential functions in S-phase of cell cycle?

A. A-type
B. B-type
C. D-type
D. None of these

46. Integrin molecules link extracellular matrix (ECM) to the actin cytoskeleton of cell. Integrin binds to which of the following ECM macromolecules?

A. Fibronectin
B. Laminin
C. Vitronectin
D. Collagen

47. Major disadvantage of using liposome as a targeted drug delivery vehicle is that :

A. It's drug entrapment efficiency is very low

B. It is very unstable and has low shelf-life
C. It gets intercalated in cell membranes
D. It gets internalized by phagocytosis inside lysosomes

48. Phosphatidyl serine, an important component of biological membrane, is located in :
A. The middle of the bilayer
B. The outer leaflet but flipflops to inner leaflet under specific conditions
C. The inner leaflet but flipflops to outer leaflet under specific conditions
D. None of these

49. In the area of desmosome the adjacent cells have :
I. Inter cellular thickening material
II. Cilia
III. Transmembrane linker
IV. Protein sheath
A. I and IV are correct
B. II and IV are correct
C. I, II and III are correct
D. I and III are correct

50. You are following the intracellular sorting of an integral plasma membrane protein in a living cell, in culture. You have decided to probe this protein by metabolic labelling technique with 35S-methionine (Pulse-chase technique). After one cycle of division, the cells were treated with a potent inhibitor of protein biosynthesis and processed for subcellular fractionation. In which of the following fraction will you expect the presence of this protein upon immuno-precipitation with a specific antibody?
A. Only cytoplasm
B. Only plasma membrane
C. Both endoplasmic reticulum and plasma membrane
D. Only secretory vesicles and endoplasmic reticulum

ANSWERS

1	2	3	4	5	6	7	8	9	10
B	C	B	B	C	C	A	C	B	C
11	**12**	**13**	**14**	**15**	**16**	**17**	**18**	**19**	**20**
A	C	D	B	D	B	A	D	D	A
21	**22**	**23**	**24**	**25**	**26**	**27**	**28**	**29**	**30**
B	D	C	A	B	A	A	B	D	A
31	**32**	**33**	**34**	**35**	**36**	**37**	**38**	**39**	**40**
D	B	B	B	A	C	D	C	A	A
41	**42**	**43**	**44**	**45**	**46**	**47**	**48**	**49**	**50**
D	C	D	C	A	A	D	C	D	C

●●●

CHAPTER

3

Acids and Bases

Introduction

Acids are generally a class of substances that taste sour, such as vinegar, which is a dilute solution of acetic acid. Bases or alkaline substances, are characterized by their bitter taste and slippery feel.

The first precise definition of an acid and base was given by Svante Arrhenius, and is referred to as Arrhenius Theory.

Arrhenius Theory

In 1830s it was known that all acids contain hydrogen, but not all hydrogen containing compounds were acids. In 1889 Swedish chemist Svante Arrhenius through his theory connected acidic properties with the presence of hydrogen (H^+) ions and basic porperties with presence of hydroxyl (OH^-) ions. Arrhenius defined an acid as a substance that ionizes in water to give hydroxide ions.

- If a solution contains more H^+ than OH^- ions, then it is acidic.
- If a solution contains more OH^+ than H^+ ions, then it is basic.
- If H^+ and OH^- ions are equal, then it is a neutral solution.

HCl is a strong acid, and is very soluble in water. It dissociates into its component ions in the following manner:

$$HCl(g) \rightarrow H^+(aq.) + Cl^-(aq.)$$

The hydrogen ion interacts strongly with a lone pair of electrons on the oxygen of a water molecule. The resulting ion, H_3O^+ is called the hydronium ion.

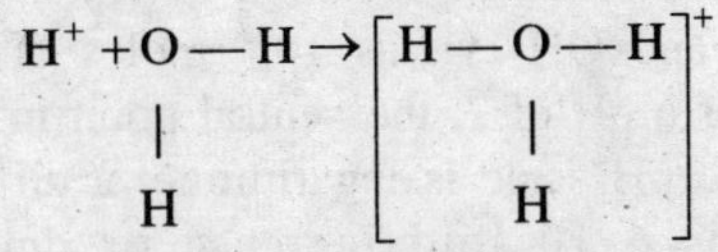

Acidic solutions are formed when an acid transfers a proton to water. The reaction of HCl with water can be written in either of the following ways:

$$HCl(aq.) + H_2O(l) \rightleftharpoons H_3O^+(aq.) + Cl^-(aq.)$$

$$HCl(aq.) \rightleftharpoons H^+(aq.) + Cl^-(aq.)$$

The Bronsted-Lowry Concept of Acids and Bases

According to Bronsted and Lowry acids are substances that are capable of donating a proton, and bases are substances capable of accepting a proton. So, in the example above, HCl acts as a Bronsted Acid by donating a proton in water, and water in turn acts as a Bronsted Base by accepting a proton from HCl. Thus, water can act as an acid or a base.

e.g., $NH_3(aq.) + H_2O(l) \rightleftharpoons NH_4^+(aq.) + OH^-(aq.)$

Here, H_2O acts as a Bronsted acid by donating a proton to NH_3, which acts as a Bronsted base. Using the Arrhenius definition, we say that the resulting solution is basic because it contains OH^- ions, the we say that the NH_3 molecule is basic (a proton acceptor).

pH Scale

The pH scale is used as a measure of acidity or the concentration of H^+ ions in the aqueous medium. The symbol of pH is an abbreviation for *"pondus hydrogenii"* of French meaning "Hydrogen power".

It is defined as the negative logarithms (10^{-n}) of the hydronium ions (H_3O^+) or hydrogen ions (H^+) concentration expressed in moles per liter of water.

Dr. Sorensen has been credited as the founder of the modern pH concept.

$$pH = -\log([H_3O^+]) \text{ or } ([H^+])$$

Since normal dissociation of water molecules into H^+ and OH^- yields 10^{-7} moles of H^+, pure water has a pH of 7, the neutral point in the scale. Since the pH scale is logarithmic, a difference in 1 unit is a 10 fold increase or decrease in concentration of hydrogen ions.

$$pH = -\log [H^+]$$

$$pH = -\log 10^{-7} = 7$$

From this it is obvious that at 25°C

$pH < 7$, solution is acidic

$pH = 7$, solution is neutral

$pH > 7$, solution is basic

Notice that the pH of a solution measures the concentration of dissociated protons, and not the total concentration of acid in a solution.

The negative log scale is useful for measuring other minute quantities, for example to measure $[OH^-]$:

$$pOH = -\log [OH^-]$$

Knowing this, we obtain the following useful expression:

$$pH + pOH = -\log K_W = 14.00.$$

Buffers

In any biological system, one of the most important parameters of an aqueous solution is the concentration of protons ($[H^+]$) or the pH. Although the $[H^+]$ is quite low, typically 10^{-6} to 10^{-8} M, it must be maintained within this range for life to exist. If an aqueous solution is able to maintain its pH near constant it can be called a buffer. A buffer is a solution of a weak acid or a weak base and its corresponding salt that keeps pH constant under diverse conditions. Weak acids and bases are those, which dissociate very little in solution (as opposed to strong acids and bases like HCl and NaOH which dissociate completely). The weak acid or base is in equilibrium with its dissociated salts. By the law of mass action, an increase in the concentration of either reactant or product forces the reaction in the opposite direction, thereby maintaining the original equilibrium. An acidic buffer (pH below 7) contains a weak acid and a salt of the acid (conjugate base). A basic buffer (pH above 7) contains a weak base and its conjugate acid (salt of the base).

Strong Acids and Bases

A strong acid under goes complete dissociation into ions in aqueous solution.

For e.g., HCl when added to water under goes ionization as follows

$$HCl + H_2O \rightarrow H_3O^+ + Cl^-$$

NaOH is a strong base because it completely dissociates into Na^+ and OH^- ions in water. When there is complete ionization like this, that solution cannot act as buffer.

Weak Acids and Bases

A buffer is a solution of a weak acid or a weak base and its corresponding salt that keeps pH constant under diverse conditions. Weak acids and bases are those, which dissociate very little in solution (as opposed to strong acids and bases like HCl and NaOH which dissociate completely). In water, a weak acid and bases dissociate partially. The reaction can be represented as follows.

$$HA \rightleftharpoons A^- + H^+$$

The actual reaction in water is

$$HA + H_2O \rightleftharpoons A^- + H_3O^+$$

Since, the concentration of water is constant, it can be represented more precisely by the first equation.

The weak acid or base is in equilibrium with its dissociated salts. By the law of mass action, an increase in the concentration of either reactant or product forces the reaction in the opposite direction, thereby maintaining the original equilibrium.

Weak acids and bases can therefore serve as reservoirs that can react with any extra hydrogen or hydroxide ions that may enter the medium. For example, if excess acid is added to a buffer solution of a weak acid and its salt (left), the introduced hydrogen ions initially increase the pH (middle). But as the weak acid's anions (negatively charged ions) take up the extra H^+ the original pH becomes reinstated (right). If a base is added, the free protons combine to form additional salt. In both cases, the ratio of [buffer acid] to [buffer salt] remains constant; and so does the pH.

The law of mass action can explain the above reversible reaction and the dissociation constant for this reaction is given by

$$K_a = \frac{[A^-][H^+]}{[HA]}$$

According to Handerson and Hasselgach equation

$$pH = pK_a + -\log\frac{[A^-]}{[HA]}$$

So, if the pK_a of a weak acid or base is known, it is easy to make a buffer of the desired pH by taking appropriate proportion of acid and base. At equilibrium, when the concentration of the acid ([HA]) is equal to its conjugated base ($[A^-]$).

$$pH = pK_a.$$

Multiple Choice Questions

1. The pH of gastric juice of infants is :
A. 3.0 B. 4.0
C. 5.0 D. 4.50

2. Major buffering system in blood is :
A. Protein and H protein
B. K_2HPO_4 and KH_2PO_4
C. $NaHCO_3$ and H_2CO_3
D. Haemoglobin and H Haemoglobin

3. In the reaction $HADH^+ + A^-$, the conjugate acid and conjugate base are :
A. HA and H^+ B. HA and H^-
C. $H^+ + A^-$ D. A^+ and H^+

4. The pH of a buffer is determined by pH = pK_a + log [salt]/[acid] and the name of this equation is :
A. Gibbs-Donnam
B. Henderson-Joules
C. Henderson-Hasselbach
D. All of the above

5. The sedimentation velocity of a particle depends on :
A. Its mass only B. Mass and density
C. Shape of particle D. All of the above

6. During diffusion, the movement of solute particle occurs as :
A. From lower concentration to higher
B. From higher concentration to lower
C. All of the above
D. None of the above

7. Vant Hoff's law is concerned with :
A. Strong electrolytes B. Ionic solutions
C. Osmotic pressure D. Weak electrolytes

8. In dialysis protein molecules are separated on the basis of :
A. Molecular size
B. Net molecular charge
C. Dialysis coefficient
D. Pressure across semi permeable membrane

9. The osmotic pressure of a solution increase with the rise in :
A. Humidity B. Temperature
C. Cold D. All of the above

10. Adsorption decrease with rise in :
A. Temperature B. Humidity
C. Cold D. Dryness

11. The sedimentation coefficient "S" is defined as :
A. The velocity divided by the centrifugal force
B. The velocity divided by centrifugal field
C. The centrifugal force divided by the effective mass of the particle
D. The centrifugal field divided by the velocity

12. The molecular weight of a solute particle is :
A. Inversely proportional to osmotic pressure
B. Directly proportional to osmotic pressure
C. Independent of osmotic pressure
D. Inversely proportional to square of osmotic pressure

13. The viscosity of a solution increase due to the pressure of :
A. Small sized particle
B. Soluble particles
C. Suspended particle
D. All of the above

14. Surface tension of solution is increased by :
A. Acetic acid
B. Concentrated H_2SO_4
C. Bile salts
D. Bile acid

15. The phenomenon of osmosis is opposite to :
A. Adsorption B. Effusion
C. Diffusion D. Dialysis

16. In case of non-homogenous solution diffusion is expressed as ds/dt=DA.dc/dx by :
A. Ficks law B. Peterson law
C. Faraday's law D. Dicks law

17. The average pH of urine is :
A. 5.6 B. 6.0
C. 6.4 D. 6.8

18. The osmotic pressure of a solution relating to solute molecule depends on the :
A. Size B. Shape
C. Number D. Volume

19. If a cell immersed in a concentrated solution, it follows the phenomenon of :
A. Turger pressure B. Plasmolysis
C. Hemolysis D. Paralysis

20. The intracellular fluid of red cells and the red cell membrane in 0.92 per cent NaCl solution are :
A. Osmosis B. Adsorption
C. Diffusion D. Absorption

21. The Purgative action of Epsom ($MgSO_4$) salt follows :
A. Osmosis B. Diffusion
C. Effusion D. Imbibition

22. Hemolysis is caused by the dilution of RBC by :
A. Osmosis B. Diffusion
C. Effusion D. Imbibition

23. The process of adsorption is applied in the purification of :
A. Enzymes B. Vitamin
C. Coenzymes D. Hormones

24. Bile salt make emulsion with fat for the action of :
A. Amylose B. Lipase
C. Pepsin D. Trypsin

25. The size of each colloidal particle in nm is :
A. 4 to 40 B. 6 to 60
C. 8 to 80 D. 10 to 100

26. The sperm colloidal protein particles may be precipitated by the addition of large amount of :
A. Ammonium sulphate
B. Calcium sulphate
C. Barium sulphate
D. Magnesium sulphate

27. Water is not expelled by squeezing in :
A. Imbibition B. Precipitation
C. Combination D. Dilution

28. Fatty acids can be transported into and out of mitochondria through :
A. Active transport
B. Passive diffusion
C. Simple diffusion
D. Non facilitated transport

29. The absorption of intact protein from the gut in the fetal and new born animal takes place by :
A. Pinocytosis B. Passive diffusion
C. Simple diffusion D. Active transport

30. A person was admitted in a coma. Analysis of the arterial blood gave the following values: PCO_2 16 mm Hg, HCO_3 – 5 mmol/l and pH 7.1. What is the underlying acid-base disorder?
A. Metabolic Acidosis
B. Metabolic Alkalosis
C. Respiratory Acidosis
D. Respiratory Alkalosis

31. In a man undergoing surgery, it was necessary to aspirate the contents of the upper gastrointestinal tract. After surgery, the following values were obtained from an arterial blood sample: pH 7.55, PCO_2 52 mm Hg and HCO_3 – 40 mmol/l. What is the underlying disorder?
A. Metabolic Acidosis
B. Metabolic Alkalosis
C. Respiratory Acidosis
D. Respiratory Alkalosis

32. A young woman is found comatose, having taken an unknown number of sleeping pills an unknown time before. An arterial blood sample yields the following values: pH – 6.90, HCO_3 – 13 meq/liter, $PaCO_2$ 68 mmHg. This patient's acid-base status is most accurately described as :
A. Uncompensated metabolic acidosis
B. Uncompensated respiratory acidosis
C. Simultaneous respiratory and metabolic acidosis
D. Respiratory acidosis with partial renal compensation

33. A student is nervous for a big exam and is breathing rapidly, what do you expect out of the following :
A. Metabolic Acidosis
B. Metabolic Alkalosis
C. Respiratory Acidosis
D. Respiratory Alkalosis

34. A 45-year-old female with renal failure, missed her dialysis and was feeling sick, what could be the reason?
A. Metabolic Acidosis
B. Metabolic Alkalosis
C. Respiratory Acidosis
D. Respiratory Alkalosis

35. An 80-year-old man had a bad cold. After two weeks he said, "It went in to my chest, I am feeling tightness in my chest, I am coughing, suffocated and unable to breathe!" What could be the possible reason?
A. Metabolic Acidosis
B. Metabolic Alkalosis
C. Respiratory Acidosis
D. Respiratory Alkalosis

36. A post operative surgical patient had a naso gastric tube in for three days. The nurse caring for the patient stated that there was much drainage from the tube that is why she felt so sick. What could be the reason?
A. Metabolic Acidosis
B. Metabolic Alkalosis
C. Respiratory Acidosis
D. Respiratory Alkalosis

37. The pH of the body fluids is stabilized by buffer systems. Which of the following compounds is the most effective buffer system at physiological pH?
A. Bicarbonate buffer B. Phosphate buffer
C. Protein buffer D. All of the above

38. Which of the following laboratory results below indicates compensated metabolic alkalosis?
A. Low pCO_2, normal bicarbonate and, high pH
B. Low pCO_2, low bicarbonate, low pH
C. High pCO_2, normal bicarbonate and, low pH
D. High pCO_2, high bicarbonate and high pH

39. The greatest buffering capacity at physiological pH would be provided by a protein rich in which of the following amino acids?
A. Lysine B. Histidine
C. Aspartic acid D. Leucine

40. Which of the following is most appropriate for a female suffering from Insulin dependent diabetes mellitus with a pH of 7.2, HCO_3 – 17 mmol/L and pCO_2 – 20 mm HG?
A. Metabolic Acidosis
B. Metabolic Alkalosis
C. Respiratory Acidosis
D. Respiratory Alkalosis

41. Causes of metabolic alkalosis include all the following except.
A. Mineralocorticoid deficiency
B. Hypokalemia
C. Thiazide diuretic therapy
D. Recurrent vomiting

42. Renal Glutaminase activity is increased in :
A. Metabolic Acidosis
B. Respiratory Acidosis
C. Both of the above
D. None of the above

43. Causes of lactic acidosis include all except :
A. Acute Myocardial infarction
B. Hypoxia
C. Circulatory failure
D. Infections

44. Which out of the following conditions will not cause respiratory alkalosis?
A. Fever
B. Anxiety
C. Laryngeal obstruction
D. Salicylate toxicity

45. All are true about metabolic alkalosis except one :
A. Associated with hyperkalemia
B. Associated with decreased ionic calcium concentration
C. Can be caused due to Primary hyperaldosteronism
D. Can be caused due to Renin secreting tumor

46. Choose the incorrect statement out of the following :
A. Deoxy haemoglobin is a weak base
B. Oxyhaemoglobin is a relatively strong acid
C. The buffering capacity of haemoglobin is lesser than plasma protein
D. The buffering capacity of haemoglobin is due to histidine residues.

47. Carbonic anhydrase is present at all places except :
A. Gastric parietal cells
B. Red blood cells
C. Renal tubular cells
D. Plasma

48. All are true for renal handling of acids in metabolic acidosis except :
A. Hydrogen ion secretion is increased
B. Bicarbonate reabsorption is decreased
C. Urinary acidity is increased
D. Urinary ammonia is increased

49. Choose the incorrect statement about anion gap out of the following :
A. In lactic acidosis anion gap is increased
B. Anion gap is decreased in Hypercalcemia
C. Anion gap is decreased in Lithium toxicity
D. Anion gap is decreased in ketoacidosis

50. Excessive citrate in transfused blood can cause which of the following abnormalities?
A. Metabolic alkalosis
B. Metabolic acidosis
C. Respiratory alkalosis
D. Respiratory acidosis

ANSWERS

1	2	3	4	5	6	7	8	9	10
C	C	B	C	D	B	C	A	B	A
11	**12**	**13**	**14**	**15**	**16**	**17**	**18**	**19**	**20**
B	A	C	B	C	A	B	C	B	C
21	**22**	**23**	**24**	**25**	**26**	**27**	**28**	**29**	**30**
A	A	A	B	D	A	A	B	A	A
31	**32**	**33**	**34**	**35**	**36**	**37**	**38**	**39**	**40**
B	C	D	A	C	B	A	D	B	A
41	**42**	**43**	**44**	**45**	**46**	**47**	**48**	**49**	**50**
A	C	D	C	A	C	D	B	D	A

●●●

CHAPTER

4

Bacteria and Viruses

Bacteria

Anton Van Leeuwenhoek (1683) of Holland observed sewage water, saliva and excretion peeled off from the teeth, under the crude microscope prepared by him. In these he found numerous tiny animalcules. These were bacteria. Linnaeus (1758) placed them in genus vermes.

Ehrenberg (1829) first of all called them *Bacteria*. Louis Pasteur (1812-95) worked on fermentation and reported that it takes place by bacteria. Anton Van Leeuwenhoek is called **Father of Bacteriology**.

Bacteria are prokaryotic and single celled microorganisms that differ from all other organisms (the eukaryotes) in lacking a true nucleus and organelles such as mitochondria, chloroplasts and lysosomes. Their genetic material consists of a single loop of double stranded DNA, whereas, the genetic material of eukaryotes consists of multiple chromosomes, which are complex structures of DNA and protein.

Bacteria are cosmopolitan and they occur in water, soil, air, animals and plants. *i.e.,* they are found in all habitat. They are also found in snow as well as hot water streams at 78°C. Some bacteria are harmful and some beneficial.

Structure of Bacterial Cell

- Bacterial cell is surrounded by a cell wall made up of polysaccharides, lipids and proteins.
- Cell wall contains a substance, a mucopeptide in which molecules of *N-acetylglucosamine* and *N-acetylmuramic acid* join each others.

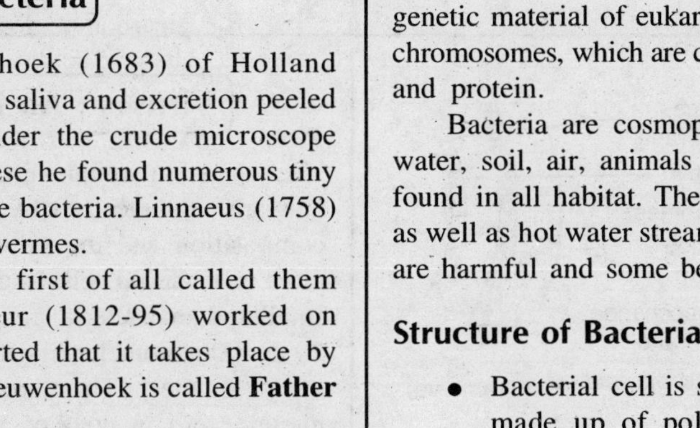

Fig. 1: *Electron microscopic structure of Bacterial cell. (Capsulated forms are usually non-flagellated).*

Table 1 : *Differences between Cell Wall Composition of Gram-Positive and Gram-Negative Bacteria*

	Gram Positive	***Gram Negative***
Cell wall	About 5 times thicker than Gram negative bacteria (25-30 mm)	Thin (10-15 mm)
Teichoic acid	Present	Absent
Lipopolysaccharide	Absent	Present (50%)
Amino sugar content (Peptidoglycan)	10-30%	1-10%
Amino acids	3-4	14-18
Lipid content	0-2%	10-20%
Effect of pencilin	Destroys it	Does not destroy it so are pathogenic

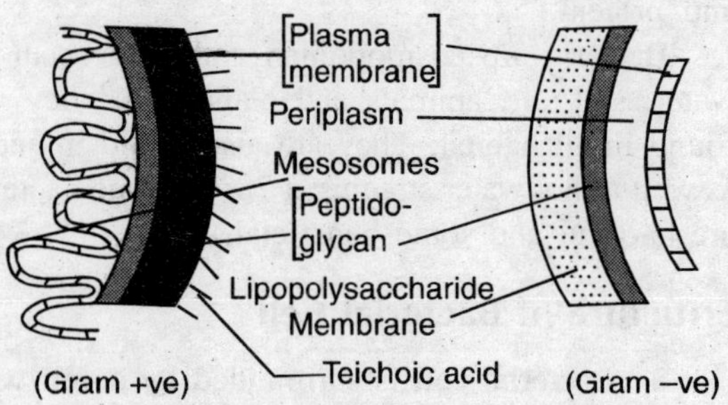

Fig. 2 : *Comparative structure of Gram +ve and Gram –ve bacterial cell wall.*

PLASMIDS

In addition to the chromosome an *E. coli* cell often possesses *extrachromosomal genetic material* each called a *plasmid*. Plasmid may contain 0.5 to 2.0% of the total DNA of the cell. Plasmids maintain a distinct existence from the main chromosome and replicate independently of it.

In *E. coli*, there are three classes of plasmids :

1. *F (i.e., Fertility) factors :* They promote sexual conjugation. All *E. coli* cells do not have this. Those which contain F factor are called F^+ and those that lack it are called F^-. F^+ forms sex pilli.
2. *R (i.e., Resistance) factors* : R factors offers to bacteria resistance to certain drugs.
3. *Col (i.e., Colicinergic) factors* : Gives to bacteria capacity to secrete colicins which have an antibiotic function.

Viruses : An Introduction

Viruses are a unique group of infectious agents that are characterized by their small size, simple composition and the need to grow in an animal, plant or bacterial cell. In general, viruses are much smaller than bacteria, ranging in size from about 20 to 400 nanometer (nm) in diameter (1 nanometer = 10^{-9} meter). They are composed of a core of nucleic acid, a coat of protein (capsid) and, in some cases, lipid (fatty acid) and carbohydrate material. Viruses vary in their detailed morphology (form and structure) and in their specificity for different types of host cells. Viruses were first recognized by their property of being able to pass through the pores of porcelain filters and by the pathological changes they produced in cells. They can be collected by ultracentrifugation and visualized under the electron microscope.

Historical Aspects

1. The term virus originated from a Latin word *virion*—means poison.
2. *Louis Pasteur* (1879-1895): Discovered and developed vaccine (artificial immunity) against rabies and anthrax.
3. *F. Loeffer* (1897) : Dicovered first animal virus, foot & mouth disease virus.
4. *Mayer* (1886) : Described the tobacco mosaic disease but could not find causal agent.

5. The Russian biologist *Ivanowsky* (1892) first demonstrated their occurrence in tobacco leaves suffering from mosaic disease.
6. *Popper* (1898) discovered poliomyelitis virus.
7. *Beijerinck* (1898), Leoffter and Frosch (1898) etc. established the existence of viruses by discovering several similar diseases in plants and animals.
8. Of human diseases, *yellow fever* was the first to be discovered viral disease.
9. *Francis Peyton Rous* (1911) discovered that certain viruses cause *Cancer* in some animals.
10. *Twort* (1915) and *D'Herelle* (1917) discovered an interesting virus and named it *Bacteriophage* because it infects bacteria.
11. *W.M. Stanley*, an American microbiologist and Nobel Prize winner (1946) succeeded in isolating Tobacco Mosaic Virus (TMV) from infected leaf-juice and prepared its pure crystals (1935).
12. Acquired Immune Deficiency Syndrome (AIDS) was discovered in early 1980s.
13. *Dr. Robert Galo* of America and *Dr. Paolo Lusso* of Italy first identified AIDS virus in 1984 and named it HIV (Human Immuno deficiency Virus). The research was conducted at the National Cancer Institute Laboratory in Bethesda, Maryland.
14. *F.C. Bawden* and *N.W. Pirie* (1937-1938) discovered that TMV contained RNA also in addition to protein *i.e.,* it was nucleoprotein.
15. *J. Enders* (1949) made the first successful cultivation of polio virus in tissue culture.
16. *A.D. Harshey* and *M. Chase* (1952): Discovered that bacteriophage DNA enters host but its protein does not, so infective part of a virus is DNA.
17. *A. Lwoff* and *Wollman* (1953) discovered temperate phage or prophages.
18. *J. Salk* and *Sapine* (1957) discovered vaccination against polio.
19. *R.L. Sinsheimer* (1959) discovered $\phi \times 174$ bacteriophage with single stranded DNA.
20. *R.S. Schafferman* and *M.E. Morris* (1963) discovered cyanophages *i.e.,* viruses attacking cyanobacteria.
21. *D. Schidolovski* and *R. Ahmad* (1973) discovered that virus caused cancer in primates.
22. *Lu Montagnier et al.* (1983) discovered AIDS virus.

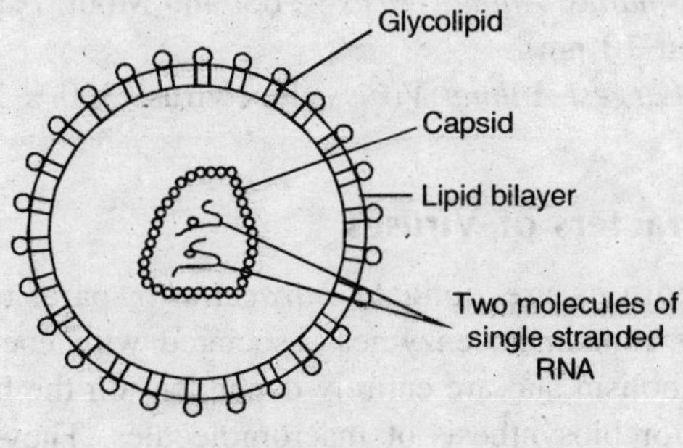

Fig. 3 : *AIDS Virus*

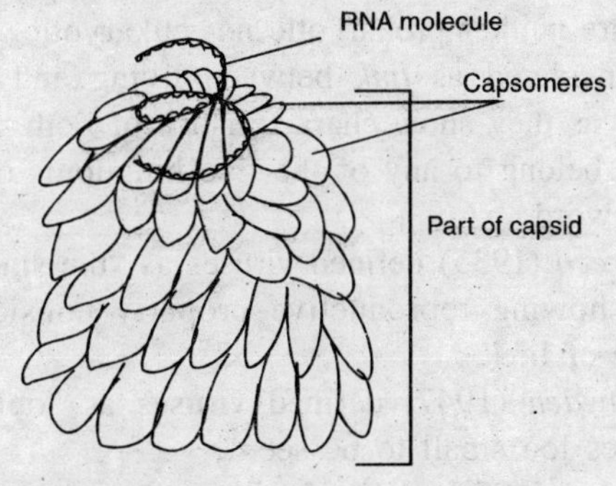

Fig. 4: *Tobacco Mosaic Virus (TMV)*

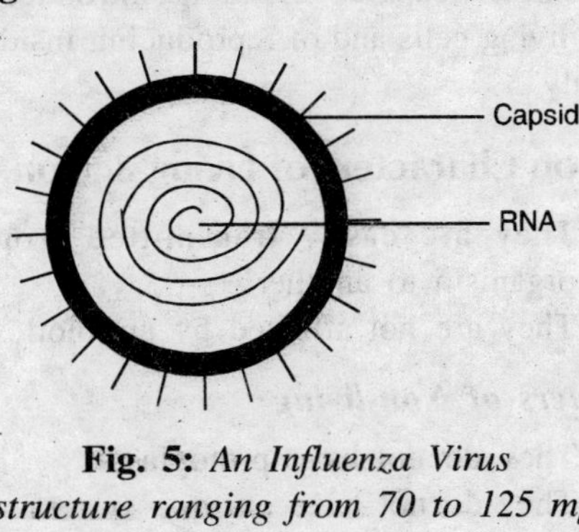

Fig. 5: *An Influenza Virus*
(structure ranging from 70 to 125 mμ)

Shape and Size of Virus

Shape : Virus may be spherical, cuboidal, cylindrical, polyhedral or tadpole like in shape.

Size : Plant viruses range in size from 8 nm – 450 nm. Animal viruses range from 10 nm to 30 nm.

Smallest Plant Virus : Satellite tobacco necrosis virus.

Longest Plant Virus : Citrus tristeza (2000 × 12 nm).

Smallest Animal Virus : Foot and Mouth Virus (about 10 nm).

Largest Animal Virus : Pox virus (300 × 230 nm).

Characters of Viruses

All viruses are obligate *intracellular* parasities. They contain no enzymes associated with energy metabolism and are entirely dependent on the host cell for biosynthesis of macromolecules. They do not have the capacity to live on non-living food substrate. They are non-cellular infection entities. They are neither prokaryotic nor eukaryotic. They are considered as *link* between living and non-livings as they show characteristics of both. They do not belong to any of the five kingdoms of the living word.

Green (1935) defined viruses as "the smallest units showing reproductive property considered typical of life".

Bawden (1947) defined viruses as "obligate parasites too small to be seen".

Lusia (1953) defined viruses as "submicroscopic entities capable of being introduced into specific living cells and of reproducing inside such cells only".

Common Characters of Living & Non-living

1. They are easily transmitted from one organism to another.
2. They are not affected by antibiotics.

Characters of Non-living

1. They do not have protoplasm.
2. They do not have enzyme system.
3. They do not respire, grow or move in extracellular state.
4. They can be crystallized & purified.
5. Attempts to culture viruses in different types of culture media (outside the cells) have failed.

Characters of Living Beings

1. They replicate, although inside the living cells.
2. Nucleic acids present in their body are capable of synthesizing protein for their coat, although they are ribosomes of the host for the purpose.
3. Nucleic acids show similar gene mutations and recombination as chromosomes of living organisms.
4. They cause diseases like bacteria & fungi.
5. They can adapt to their changing environment.

Classification

According to Holmes (1948), viruses are classified on the basis of their hosts into three kinds :

(*a*) **Animal viruses (Zoophagineae) :** They infect man, fowl, pigeon, parrot, dog, cow, etc., and arthropods insects. They usually have DNA or may also have RNA (*e.g.*, retroviruses).

(*b*) **Plant viruses (Phytophagineae) :** They infect angiosperms, like potato, tobacco, sugarcane, cucurbits, beans and higher plants. They have RNA and lack lipoprotein coat over capsid.

(*c*) **Bacterial viruses (Phagineae) :** They are commonly called bacteriophages or phages *i.e.,* eaters of bacteria. They have DNA.

Viruses and Nucleic Acids

With reference to the number of strands, four types of nucleic acids have been found in viruses :

1. Double stranded DNA
2. Single stranded DNA
3. Double stranded RNA
4. Single stranded RNA

Table 2. *Types of Nucleic Acid and Number of Strands in various Viruses*

DNA virus	*Strands*	*RNA virus*	*Strands*
Polyoma virus	DNA (2)		
Pox virus	DNA (2)	Reovirus	RNA (2)
Coliphage T2	DNA (2)	Wound Tumour	RNA (2)
Coliphage T3	DNA (2)	Virus	
Coliphage T4	DNA (2)	Retrovirus (Rous Sarc- oma virus)	RNA (2)
Colipage T6	DNA (2)	Rice dwarf virus	RNA (2)
Coliphage T7	DNA (2)	ϕ 6 phages	RNA (2)
Herpes virus	DNA (2)	TMV	RNA (1)
Adenovirus	DNA (2)	Poliomyelitis	RNA (1)
Papilloma virus	DNA (2)	Togavirus	RNA (1)
Caulimo virus	DNA (2)	Influenza	RNA (1)
Cauliflower mosaic virus	DNA (2)	Mumps	RNA (1)
λ phages	DNA (2)	Rabies Tobacco necrosis virus ϕ_2 $R_{17}MS_2$ aB phages	RNA (1) RNA (1) RNA (2) RNA (1) RNA (1)
$\phi \times 174$ *E. coli* phage	DNA (1)		
S 13 *E. coli* phage	DNA (1)	Bacterio- phage MS-2	RNA (1)
Parvovirus	DNA (1)	Avian Leukemia	RNA (1) RNA (1)

Viroids

- Viroids are even more streamlined and simple than viruses.
- They consist of only highly wound genetic material that is untwisted.
- Viroids causes several exotic sounding plant diseases, including 'Coconut cadang cadang' Tomato bunchy top and Cucumber pale fruit disease.
- Viroids nearly destroyed the *Chrysanthemum* industry in the United States in early 1950s.
- They are infectious to plants only.
- They consists of only very small RNA genome (only 300 nucleotides) without a protein coat.
- They were discovered by Diener & Raymer (1973).

Prions

- Made up of protein only without any protein coat.
- Prions are thinner in size than smallest known virus.
- In 1982, San Francisco biologist S.B. Prusiner described prions and implicated them in several disorders.
- Prions are thought to be an abnormal form of a glycoprotein found in the cells of many mammals.
- The build up of prions usually occurs in the brain, causing neurodegenerative disorders such as scrapie in sheep and goats, mad cow disease, in English bovines and creutzfeldt Jakob Syndrome and Gerstmann-Straussler-Scheinker syndrome in humans, Kunu (Laughing death in human) and also it causes deadly Alzhemer's disease.
- In 1976 D.C. Gajdusek was awarded Nobel Prize for research on prion based diseases.
- This is smallest proteinaceous infection.
- In 1985 it was shown that prion protein is encoded by a gene with the cell's own chromosome.

Virusoids

- These are viroids located inside protein coat of a true virus.
- RNA is circular or linear and non-infectious.

Multiple Choice Questions

1. Vinegar is obtained from sugar by :
   A. *Lactobacillus* B. *Salmonella*
   C. *Acetobacter* D. *Nitrosomonas*
2. Bacteria bearing flagella all over the body are called :
   A. Amphitrichous B. Cephalotrichous
   C. Peritrichous D. Lophotrichous
3. Methane producing bacteria belong to :
   A. Cyanobacteria B. Archaebacteria
   C. Eubacteria D. Eukarya
4. The bacterium associated with denitrification :
   A. *Azatobacter* B. *Rhizobium*
   C. *Pseudomonas* D. *Nitrosomonas*
5. In Gram staining safranin act as :
   A. Primary stain B. Mordant
   C. Counter stain D. Decoloriser
6. If you discovered a bacterial cell that had no restriction enzyme, which of the following would you expect to happen?
   A. The cell would be unable to replicate its DNA
   B. The cell would create incomplete plasmids
   C. The cell would be easily infected and lysed by viruses
   D. The cell would become a parasite
7. The direct method of study of bacteriophage is :
   A. Lysogenic conversion
   B. Serological testing
   C. Electron microscopy
   D. Plaque study
8. Which genus of bacteria listed below grows best at 70-100°C?
   A. *Pyrococcus* B. *Halobacterium*
   C. *Methanosarcina* D. *Escherichia*
9. X174 has
   A. Double stranded RNA
   B. Single stranded RNA
   C. Single stranded DNA
   D. Double stranded DNA
10. M13 is a :
    A. Plasmid vector B. Cosmid vector
    C. Viral vector D. Phagemid vector
11. On exposure to desiccation, which of the following bacteria are least likely to experience rapid water loss?
    A. Isolated rods B. Rods in chain
    C. Cocci in chain D. Cocci in clusters
12. Occurrence of 2 or more flagella at both of the poles are called :
    A. Amphitrichous B. Lophotrichous
    C. Cephalotrichous D. Peritrichous
13. Occurrence of multiple flagella in the same area are called :
    A. Amphitrichous B. Lophotrichous
    C. Cephalotrichous D. Peritrichous
14. An example for filamentous type of bacteria is :
    A. *Diplococcus* B. *Acetobacter*
    C. *Beggiota* D. *Sarcinae*
15. Gram negative bacteria, *Klebsiella pneumoniae,* up on infecting humans results in severe septic shock after a few hours of infection. Which of the following is not true for this type of infection?
    A. Cell wall endotoxins cause overproduction of cytokines
    B. Septic shock can be treated by anti TNFα antibodies
    C. Recombinant bacterial proteins can be used for the treatment of septic shock
    D. Recombinant TNFα receptor antagonist can be used for the treatment of Septic shock
16. Free living nitrogen fixers can survive in different ecological niches. Identify the incorrect combination for the following list :
    A. *Azotobacter* – acidic soil
    B. *Deraxia* – alkaline soil
    C. *Beijerinckia* – acid soil
    D. *Frankia* – neutral soil

17. Microbial leaching involves the process of dissolution of metals from ore breaking rocks using microorganisms. Which one of the following bacteria helps in leaching copper from its ore?

A. *Acidithiobacillus ferroxidans*
B. *Pseudomonas putida*
C. *Deinococcus radiodurans*
D. *Rhodopseudomonas capsulata*

18. A rapidly growing bacterial species such as *E. coli* exhibits a typical phase of growth cycle in liquid nutrient broth (lag phase → log phase → stationary phase → death phase). If a bacterial culture has a starting density of 10^3 cells/ml has a lag time of 10 minutes, what will be the cell density be at (cells/ml) 30 minutes?

A. 6.0×10^3 B. 2.0×10^3
C. 3.0×10^3 D. 4.0×10^3

19. Assertion (A) : Quorum sensing is the ability of bacteria to communicate and coordinate behaviour.

Reason (R) : Bacteria those use Quorum sensing to produce and secrete a signaling chemical called as "inducer"

A. Both (A) and (R) are true and (R) is the correct explanation
B. Both (A) and (R) are true, but (R) is not the correct explanation
C. (A) is true, but (R) is false
D. (A) is false, but (R) is true

20. A bacterial population has a plasmid with copy number '*n*'. It was observed that on an average in one out of 2(*n*-1) cell divisions, there was spontaneous plasmid curing. It was inferred from the observation that :

(*a*) Each cell division does not have equal probability of plasmid curing
(*b*) There is no evidence for any mechanism of plasmid segregation in the two daughter cells
(*c*) Plasmid distribution to daughter cells is random
(*d*) Each plasmid has an equal chance of being in either of the two daughter cells

Which of the combinations of above statements is true?

A. (*a*) and (*b*) B. (*b*) and (*d*)
C. Only (*a*) D. (*b*), (*c*) and (*d*)

21. Identify the proteobacteria based on the key given below:

(*i*) Cause disease in humans (*ii*)
(*i*) Do not cause disease in humans (*iii*)
(*ii*) An obligate intracellular parasite (A)
(*ii*) Not an obligate intracellular parasite (B)
(*iii*) Live in insects (C)
(*iii*) Do not live in insects (*iv*)
(*iv*) Chemoautotrophic (D)
(*iv*) Notmoautotrophic (*v*)
(*v*) Plant pathogen (E)
(*v*) Not a plant pathogen (*vi*)
(*vi*) Fix nitrogen (*vii*)
(*vi*) Do not fix nitrogen (F)
(*vii*) Associated with legumes (G)
(*vii*) Not associated with legumes (H)

A. A–*Rickettsia*; B–*Brucella*; C–*Wolbachia*; D–*Nitrobacter*; E–*Agrobacterium*; F–*Acetobacter*; G–*Rhizobium*; H–*Azospirillum*
B. A–*Rickettsia*; B–*Wolbachia*; C–*Brucella*; D–*Nitrobacter*; E–*Acetobacter*; F–*Agrobacterium*; G–*Rhizobium*; H–*Azospirillum*
C. A–*Rickettsia*; B–*Brucella*; C–*Wolbachia*; D–*Nitrobacter*; E–*Agrobacterium*; F–*Acetobacter*; G–*Azospirillum*; H–*Rhizobium*
D. A–*Rickettsia*; B–*Brucella*; C–*Wolbachia*; D–*Nitrobacter*; E–*Acetobacter*; F–*Agrobacterium*; G–*Azospirillum*; H–*Rhizobium*

22. Plasmids are self replicating small circular DNA elements in bacterial cells that can be said to have a stable symbiotic existence with the host cell. They often carry genes useful to the host. Which of the following is a potential threat to the evolution and stability of the symbiotic coexistence?

A. 'Copy-up' mutations that increase the rate of plasmid replication per host Cell cycle.

B. Reversible integration of plasmid DNA into the host DNA.

C. Transfer of plasmids to new cells by conjugation.

D. Spontaneous curing of plasmids in a small proportion of host cells.

23. In the phylogenetic tree, branch-lengths are drawn proportional to the number of changes along a lineage. The following inferences were made from this tree.

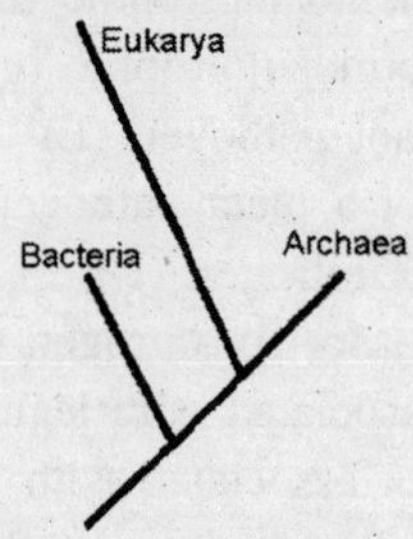

(*a*) Bacteria are more closely related to Eukarya than to Archaea.

(*b*) Bacteria and Archaea are more similar to each other than either is to Eukarya.

(*c*) Archaea and Eukarya diverged from each other after their common ancestor diverged from bacteria.

Which of the following represents a combination of correct inferences?

A. (*a*), (*b*) and (*c*) B. (*a*) and (*b*) only

C. (*b*) and (*c*) only D. (*a*) and (*c*) only

24. Major stimulus for spore formation in bacteria is :

A. Nutrition limitation

B. Heat stress

C. Cold stress

D. pH stress

25. *Mycobacterium tuberculosis* is an intra-cellular bacterium. It prefers to infect :

A. Macrophages B. B-cells

C. T-cells D. Neutrophils

26. Which of the following organism is widely used as a biocontrol agent in organic farming?

A. *Rhizobium tropici*

B. *Trichoderma viridis*

C. *Fusarium oxysporum*

D. *Nostoc muscorum*

27. Which of the following statements is NOT true during the infection of plant cells with *Agrobacterium?*

A. The protein products of virulence genes *virA* and *virG* perceive Acetosyringone.

B. The *VirB* protein forms a connection between *Agrobacterium* and the plant cell and facilitates T-DNA transfer into the plant.

C. The T-DNA is excised and bound to *Vir*D2 protein.

D. The T-DNA, after becoming coated with *Vir*F binds to phosphorylated VIPI.

28. The scatter plot of growth rate and growth yield for 100 random environmental isolates of bacteria is shown below: Which of the following can be inferred from the data?

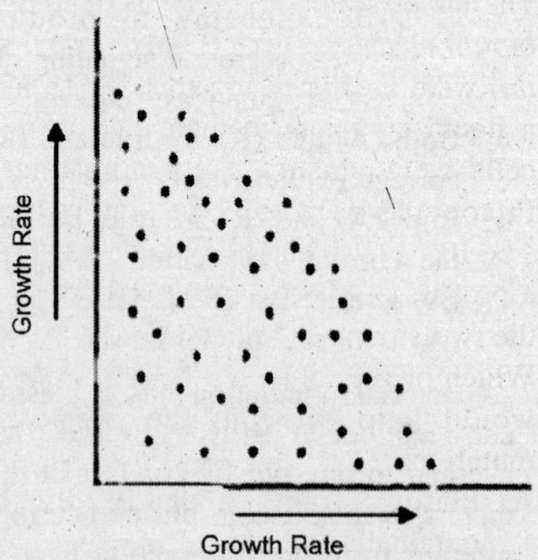

A. The two parameters are not related

B. Growth rate is inversely proportional to growth yield

C. Growth yield is negatively correlated with growth rate

D. High growth rate cannot be accompanied by high growth yield

29. Three *E. coli* mutants are isolated which require compound 'A' for their growth. The compounds B, C and D are known to be involved in the biosynthetic pathway to A. In order to determine the pathway, the mutants were grown in a medium supplemented with one of the compounds, A to D. The results obtained are summarized below:

Mutant	*Medium Supplemented with Compound*			
	A	B	C	D
1	+	0	0	0
2	+	0	0	+
3	+	0	+	+

'+' indicates growth; '0' indicates lack of growth. Which of the following equations represents the biosynthetic pathway of A?

A. B → C → D → A
B. C → D → B → A
C. B → D → C → A
D. A → C → D → B

30. The auxotrophic strains of *E. coli:* A *(met-bio-thr+leu+thi+)* and B *(met + bio + trhleu-thi)* were incubated together for 18 hours in a liquid complete medium and then ~ 108 cells were plated on a minimal medium. Prototrophs were observed at a frequency of 1×10^{-7} cells. This may have happened by a process of genetic recombination between the two strains or by mutation of the strains. Which of the following control experiments would help rule out the possibility of mutations?

A. Plating strains A and B directly on minimal medium
B. Growing the mixture of strains A and B for 18 hours and then plating on complete medium
C. Growing strains A and B individually in a liquid complete medium for 18 hours and then plating them on a minimal medium
D. Growing the obtained prototrophs in a liquid complete medium for 18 hours and then plating them on a minimal medium

31. The rates of mutation in *E. coli* from *lac*– to *lac*+ are determined using a medium containing lactose, as the only source of energy. The principle of spontanity of mutations can be said to be violated if :

A. The rate of mutation increases during starvation
B. In the presence of lactose, the rate of mutation from *lac*– to *lac*+ is increased. But overall rate of mutation is not.
C. The rate of mutation in *lac* gene is always greater than that in other genes.
D. The rate of mutation in *lac* gene is always less than that in other genes.

32. The presence of *Salmonella* in tap water is indicative of contamination with :

A. Industrial effluents B. Human excreta
C. Agricultural waste D. Kitchen waste

33. A fixed smear of a bacterial culture is subjected to the following solutions in the order listed below and appeared red.

P. Carbolfuchsin (heated)
Q. Acid-alcohol
R. Methylene blue

Bacteria stained by this method can be identified as :

A. Non-acid fast *E. coli*
B. Acid-fast *Mycobacterium* sp.
C. Gram-positive *E. coli*
D. Gram-negative α

34. In the global nitrogen cycle, the following microbial organisms are involved in three important process-denitrification, nitrification and nitrogen fixation.

P. *Rhizobium* Q. *Nitrosomonas*
R. *Nitrobacter* S. *Pseudomonas*
T. *Azotobacter*

Which of the following is the correctly matched pair of process and its causative species?

A. Denitrification–(Q); nitrogen fixation–(R) and (T); nitrification–(S)
B. Denitrification–(S); nitrogen fixation–(P) and (T); nitrification–(R)
C. Denitrification–(T); nitrogen fixation–(P) and (S); nitrification–(S)
D. Denitrification–(Q); nitrogen fixation–(P) and (S); nitrification–(R)

35. A culture of an *E. coli* strain that is lysogenic for phage lambda is grown at 32°C. Induction of the prophage from the host chromosome will occur when the culture is exposed to :
A. 40°C
B. Ultra violet radiation
C. Infrared radiation
D. Wild type *E. coli* culture

36. Intracellular pathogens like *Mycobacteria, Salmonella, Listeria* survive in macrophages by modulating host cellular machinery. In order to study the fate of these intracellular pathogens in macrophages, cells were labelled with lysotracker Red and infected with GFP-labelled organisms. After 2 hours at 37°C, cells were fixed, stained with anti-transferring receptor antibody and probed with secondary antibody conjugated-blue dyes. Cells were viewed under confocal microscope.
Observation: GFP-labelled, *Mycobacteria, Salmonella* and *Listeria* were localized in the same compartment labelled with blue dyes, whereas GFP *Leishmania* colocalize with red labelled compartment. Which of the following statement is true based on these observations?
A. *Mycobacteria, Salmonella* and *Listeria* reside in the lysosomes
B. *Leishmania* reside in lysosome like compartment
C. *Leishmania* reside in a compartment which bears characteristics of early Endocytic compartment
D. *Mycobacteria, Salmonella* and *Listeria* lyse the phagosomal membrane and reside in cytosol.

37. Industrial products in which bacteria are employed for production are shown in the following table:

(I) List of products	**(II) Microorganism**
P. 2,3-Butane diol	1. *Leuconostoc*
Q. Dextran	2. *Brevibacterium*
R. Glutamic acid	3. *Bacillus polymyxa*
S. Cobalamine	4. *Propionibacterium*

The correct combination in :
A. P-3, Q-1; R-2, S-4
B. P-1, Q-2; R-3, S-4
C. P-3, Q-2; R-4, S-1
D. P-2, Q-3, ; R-4, S-1

38. Who separated the protein coat virus from RNA core by chemical means?
A. Frankel Conrat B. Nishimura
C. Twort D. Schreider

39. One of the following is an incorrect statement :
A. Herps disease is due to *Myxomyces* sp.
B. Virus with nucleic acid and protein coat is called virion
C. Virus with nucleic acid but without protein coat is called viroid
D. Tadpole shaped becteriophages are of different forms

40. The spike disease in sandalwood tree is caused by :
A. Bacteria B. Fungi
C. Virus D. Algae

41. The nucleoid of Retrovirus is :
A. SS-RNA haploid B. SS-RNA diploid
C. DS-RNA haploid D. DS-RNA diploid

42. Which one of the following enzyme is present in the bacteriophage?
A. Protease
B. Lysozyme
C. Urease
D. Succinic dehydrogenase

43. Which of the following disease is caused by viroids?
A. Potato spindle tuber
B. Wart disease of potato
C. Early blight of potato
D. Potato leaf roll

44. The common viral disease of papaya :
A. Leaf roll B. Big bud
C. Leaf msaic D. Leaf curl

45. Which part of the plant cannot be infected by virus :
A. Shoot apex B. Pith
C. Phloem D. Cortex

46. Viruses that persist in the cell and cause recurrent disease are called :
A. Oncogenic viruses B. Latent viruses
C. Cytopathic viruses D. Resistant viruses

47. Assertion (A) : Interferons are produced by the host against viral infection.
Reason (R) : Interferons are essential for viral replication.
A. Both (A) and (R) are true and (R) is the correct explanation
B. Both (A) and (R) are true, but (R) is the false explanation
C. (A) is true, but (R) is false
D. (A) is false, but (R) is true

48. The sequence of the steps involved in the viral replication :
I. Attachment
II. Penetration
III. Genome replication and expression
IV. Uncoating
Arrange the above steps chronologically
A. I, II, III, IV B. III, II, I, IV
C. II, III, IV, I D. I, II, IV, III

49. Viruses contain the following :
I. Genome II. Capsid
III. Capsule IV. Enzyme
A. I, II and III are correct
B. I, II, III and IV are correct
C. I, II and IV are correct
D. II, III and IV are correct

50. Why lysogenic cycle is more beneficial to a virus than lytic cycle under certain circumstances?
A. The lysogenic cycle prevents local extinction of the host while still retaining its infection potential
B. By integrating with the bacterial chromosomes, the genetic instructions for the virus become refreshed after one or more replication events during binary fission
C. Lysogenic infection cycles do not harm their host cells, so they can produce virus particles indefinitely
D. Lysogency cause more mutations to occur in the virus, creating more variants upon which natural selection can operate

ANSWERS

1	2	3	4	5	6	7	8	9	10
C	C	B	C	C	C	C	A	C	C
11	**12**	**13**	**14**	**15**	**16**	**17**	**18**	**19**	**20**
D	A	B	C	C	A	A	D	A	D
21	**22**	**23**	**24**	**25**	**26**	**27**	**28**	**29**	**30**
A	A	C	A	A	B	D	D	A	C
31	**32**	**33**	**34**	**35**	**36**	**37**	**38**	**39**	**40**
B	B	B	B	B	B	A	A	A	C
41	**42**	**43**	**44**	**45**	**46**	**47**	**48**	**49**	**50**
B	B	A	D	A	B	C	D	C	A

●●●

CHAPTER

5

Nucleic Acids

Introduction

A macromolecule composed of phosphoric acid, pentose sugar and organic bases. The two nucleic acids, deoxyribonucleic acid (DNA) and ribonucleic acid (RNA), are made up of long chains of molecules called (q.v.). They were first isolated a part of a protein complex in 1871, and were separated from the protein moiety in 1889.

Nucleic Acid Structure

Nucleic acids are formed of C, H, O, N and P. Nucleic acids were first discovered by Friedrich Miescher (1868-69) as *nuclein* and named *nucleic acid* by Friedrich Altman (1889). "Nucleic acids is a biologically-occurring polynucleotides in which the nucleotides residues are linked in a specific sequence by a phosphodiester bonds. Nucleic acids are of two types, DNA and RNA.

(A) DNA *(Deoxyribose nucleic acid)*

DNA is helically coiled macromolecule which is made up of two antiparallel polydeoxyribonucleotide chains held together by hydrogen bonds. The two chains are plectonemically (interlocked) coiled around a common axis. Maximum amount of DNA occurs in chromatin of nucleus in eukaryotes and nucleoid of prokaryotes. A small amount (1-5%) occurs in plastids and mitochondria. DNA has a diameter of 20 Å. One turn of spiral has a distance of 34 Å. It contains 10 nucleotides in each chain so that the distance between adjacent nucleotides is 3.4 Å. Each DNA chain has a backbone of alternate deoxyribose sugar ($C_5H_{10}O_4$) and phosphate. Nitrogen bases are attached to sugar molecules at right angles to main strand. Phosphate is connected to sugar of its nucleotide at 5′ carbon, sugar is attached to nitrogen base at 1′ carbon while nitrogen base is attached by N-atom at 3′ (pyrimidine) or 9′ (purine) position. Phosphate is attached to next sugar. The direction is 3′ → 5′ in one chain and 5′ → 3′ in the other. Nitrogen bases are of four types, viz., adenine, guanine (purines), cytosine and thymine (pyrimidine). Adenine of one chain lies opposite to thymine (pyrimidine) of the other. They are held together by two hydrogen bonds. Similarly, cytosine of one chain lies opposite to guanine of the other. They are held together by three hydrogen bonds. Because of this, the two chains of DNA are complementary. Hydrogen bonds are easily broken at high temperature (80°-90°C). The phenomenon is called *denaturation* or *melting*. *Renaturation* or *annealing* occurs on cooling.

Structure of DNA was discovered by James D. Watson and Francis H.C. Crick (1953). Prior to them, Chargaff (1950) had discovered certain structural pecularities called **Chargaff's rules**; which are

(*a*) Purines and pyrimidines occur in equal amounts.

(*b*) Phosphate and deoxyribose sugar occur in equal number.

(*c*) Molar amount of adenine is equal to that to thymine and cytosine equal to guanine.

(*d*) Base ratio A + T/C + G is specific for a species. It is rarely one.

Purines have 9 membered double ring in which nitrogen is at 1′, 3′, 7′ & 9′ positions and pyrimidines are six membered single ring in which nitrogen is at 1′ & 3′ positions.

DNA duplex of Watson and Crick is right handed and is called *B-DNA*. The plane of base pairs are perpendicular to axis of the helix. Four more models have been proposed.

A-DNA a single turn of right handed helix contains 11 base pairs which are tilted 20° away from perpendicular to helix axis.

C-DNA : Right handed C-DNA has 9 base pairs.

D-DNA : Right handed D-DNA has 8 base pairs per turn of spiral.

Z-DNA : It in left handed double helix which contains zig-zag back and 12 base pairs per turn of spiral which is 49 Å in length.

DNA can form its duplicate in the process of replication (*autocatalytic function*). It is semiconservative where both the strand function as templates for synthesis of new chains. DNA also produces RNA in the process called *transcription* (heterocatalytic function).

Genetic information is contained in the base sequence of DNA strands with three adjacent nitrogen bases functioning as a *codon*. *Cistron* is a linear sequence of codons which determines the synthesis of a polypeptide.

DNA has been synthesized *in vitro* by Kornberg-1959.

Linear and Circular DNA : Linear DNA has both ends free. It occurs in eukaryotic nuclei, certain prokaryotes (PPLO, blue green algae) and some automonous cell organelle occassionally. Circular DNA doesn't have free ends as the ends closely held covalently. It occurs in bacteria randomly as in autonomous cell organelles (plastids, mitochondria). Both are coiled variously to form compact masses.

(B) RNA *(Ribose Nucleic Acid)*

It is a single stranded chain of ribonucleotides. Four types of ribonucleotides are found. AMP, GMP, CMP, UMP. Backbone of the strand is produced by alternate ribose and phosphate molecules. Nitrogen bases (uracil, adenine, guanine & cytosine) are attached to ribose sugar molecules ($C_5H_{10}O_5$). RNA are of various types according to the location where they are found.

(*a*) *Ribosomal RNA or Insoluble RNA (r RNA):* It constitutes 70-80% of total RNA of the cell which is present inside ribosomes as 28 s, 5.8 s, 5 s (larger subunit) and 18 s (smaller subunit). 5.8 s is absent in prokaryotes. 5 s is transcribed separately others are transcribed as hn RNA with 5′ → 3′ and sequence of 18 s-5.8 s-28 s. 3′ end of 18 s RNA has nucleotides complementary to those of cap region of *m*RNA. 5 s *r*RNA provides binding to *t*RNA. Some *r*RNAs (28 s and 23 s) are also the longest of all RNAs.

(*b*) *Messenger RNA or Template RNA (m RNA):* It is a long RNA which constitutes only 5% of the total. In brings coded information from DNA to cytoplasm for polypeptide synthesis. *m*RNA has codon for initiation and termination of polypeptides. It has methylated cap region at 5′ end. Cap is followed by *initiation codon* AUG or GUG coding region, *termination codon* UAA, UAG or UGA & poly A part at 3′ end for attachment to ribosome.

(*c*) *Transfer RNA (t RNA = s RNA or Soluble RNA or adaptor RNA) :* It constitutes about 15% of the total RNA. *t*RNA is the smallest (4 s). It has an *anticodon* and *aminoacid* carrying points. It is folded to form clover leaf or L-shaped structure. *t*RNA carries amino acids to exposed codon of *m*RNA. Many nitrogen bases are modified so that it becomes clover like (two dimensional, Holley-1964) or L-shaped (Klug-1974). It has G at 5′ end and CCA at 3′ end. CCA of 3′ end is meant for attaching to a specific amino acid (AA-binding site). On the opposite site lies an anticodon that is complementary to a specific codon of *m*RNA. The two are called *recognition sites*. There are two loops. TΨC having *pseudo uridine* for attaching to ribosomes and DHU (dihydro-uridine) for binding to AA-*t*RNA synthetase or AA activating enzyme. A variable extra arm or loop is often present between TΨC loop and anticodon. Its function is not yet known.

(*d*) *Genetic or Genomic RNA* : It is found in riboviruses (*e.g.*, TMV, Rheovirus and HIV).

(*e*) *Small Stable RNA (ssRNA)*. 'Found in eukaryotes' nucleus and cytoplasm. Small nuclear RNA are significantly involved in *m*RNA processing and gene regulation. U_1, U_2, U_4, U_5 & U_6 *Sn*RNAs are involved is intron removal and processing of hn RNA into *m*RNA. *Sn*RNA are 90-300 nucleotides, 100,000-1,00,000 copies/cell.

Genetic Code

The genetic language consists of only 4 letters "GACU". These four letters can be combined and to form 64 genetic words, each consisting of 3 letters. Each triplet code for particular amino acids. Genetic code is the relationship of amino acid sequence in a polypeptide and nucleotide/base sequence in DNA/*m* RNA. Crick *et al.* (1961) found that deletion or addition of three consecutive nitrogen bases caused less disturbance than one or two bases in the DNA of T_4 bacteriophage. Nirenberg and Mathaei (1961) prepared polymers of individual nucleotides and found polyuridylic acid (poly u) stimulated peptide sequence polyphenylalanine, poly A polylysine and poly C polyproline. Poly G did not function as it is formed of triple stranded structure. They further found that the number of amino acids in a peptide was one third of the number of nucleotides in the polymer. Khorana (1964) observed that nucleotide copolymers UGUGUGUG..... stimulated peptide with alternate cysteine and valine. Ochoa (1965) employed other copolymers and, therefore, cracked a number of codons. DNA and *m*RNA codons are complementary. AAA of DNA becomes UUU on *m*RNA.

(*i*) *Triplet Code* : Three adjacent nitrogen bases constitute a codon and specify one amino acid.

(*ii*) *Commaless* : There are no pauses so that genetic code reads continuously. Any change in nucleotide sequence will make the code read differently.

(*iii*) *Number* : Possible number of triplets is 64 (4^3) but amino acids involved are only 20.

(*iv*) *Non-overlapping Codons* : Adjacent codons are independent with no base being member of two codons.

(*v*) *Universality of Codons* : A codon specifies the same amino acid in all organisms from virus to human being.

(*vi*) *Nonambiguous Codons* : One codon specifies only one amino acid and not any other.

(*vii*) *Degeneracy of Codon* : Tryptophan is specified by only UGG and methionine by only AUG. All other amino acids are specified by 2-6 codons. *e.g.*, phenyl alanine by two UUC & UUU and arginine by six viz. AGA, AGU, CGA, CGC, CGG and CGU.

(*viii*) *Colinearity* : The sequence of codons of DNA/*m*RNA correspond to the sequence of amino acids in a polypetide.

(*ix*) *Start Signal* : The first codon starting a polypeptide is either AUG (methionine specific) or GUG (valine specific).

(*x*) *Stop Signal* : Termination of polypeptide synthesis occurs by one of the three nonsense or termination codon—UAA, UAG and UGA.

(*xi*) *Related Codon* : Amino acids with similar properties have usually related codons, *e.g.*, aromatic amino acids tryptophan (UGG), phenyl alanine (UUC, UUU) and tyrosine (UAC, UAU).

(*xii*) *Wobble Hypothesis* : (Crick, 1968). The first two bases of a codon are more important in determining the amino acid while the third one may differ without affecting the coding (Wobble-loose-pairing). *e.g.*, CCA, CCG, CCC and CCU all specify amino acid proline.

Central Dogma

Central dogma of molecular biology (Crick, 1958) proposes a *unidirectional* or one way *flow of information* from DNA to RNA (Transcription) and from RNA to polypeptide (Translation).

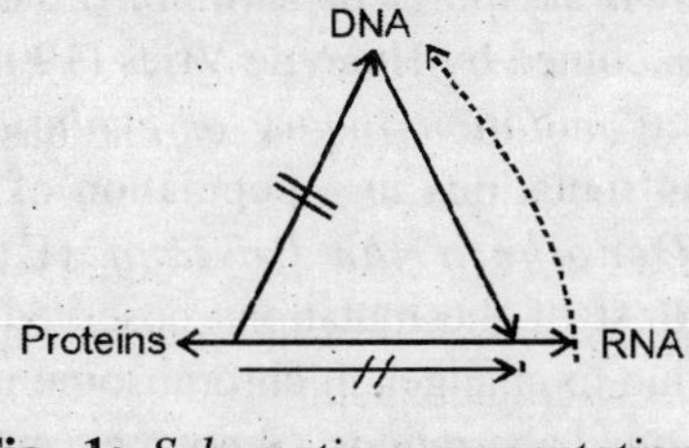

Fig. 1: *Schematic representation of the central dogma.*

$$\text{DNA} \xrightarrow{\text{Transcription}} \text{RNA} \xrightarrow{\text{Translation}} \text{Protein (Polypeptide)}$$

Since DNA is formed from DNA, the central dogma is changed to

$$\text{DNA} \xrightarrow{\text{Replication}} \text{DNA} \xrightarrow{\text{Transcription}} \text{RNA} \xrightarrow{\text{Translation}} \text{Protein}$$

Commoner (1968) suggested a *circular flow* of *information*. Temin (1970) found that retroviruses perform *central dogma reversee* that involves reverse transcription or teminism (forming DNA from RNA).

$$\text{DNA} \underset{\text{Reverse Transcription}}{\overset{\text{Transcription}}{\rightleftharpoons}} \text{RNA} \xrightarrow{\text{Translation}} \text{Polypeptide}$$

Transcription or RNA synthesis occurs over DNA. Translation or protein synthesis occurs over ribosomes. The two are separated in time and space. This protects DNA from respiratory enzymes and RNAs from nucleases.

As DNA contains all the hereditary information, it is called master copy of genetic information. It replicates or forms its carbon copies when new master copies are required. Normally it forms *working copies* in the form of *m*RNAs.

Replication (Semi Conservative Replication of DNA)

Replication is the formation of exact replica carbon copy. It is an autocatalytic function of DNA. Replication can occur by three methods.

(*i*) *Conservative :* A new structure (replica) is formed over the template of old structure (conserved).

(*ii*) *Disruptive :* Old structure undergoes fragmentation. Fragments synthesise complementary structures, both of which assemble randomly to form two replicas.

(*iii*) *Semi conservative :* In each replica, one half is parent structure and one half new structure. Watson and Crick (1953) suggested that DNA replication is semi-conservative. During replication, the two strands of DNA separate, function as templates and help in synthesis of their complementary chains/strands so that in each daughter DNA duplex, one strand in old and one new. Semiconservative replication of chromosome was found out by Taylor (1957) in *Vicia faba* (Broad Bean) using tritiated thymidine. Semiconservative replication of DNA was proved by the work of Messelson and Stahl (1958).

Classes of Proteins involved in Replication

Protein	*Function*
1. DNA polymerase →	*Deoxynucleotide polymerization*
2. Helicases →	*Processive unwinding of DNA*
3. Topoisomerases →	*Relieve torsional strain that results from helicase-induced unwinding*
4. DNA primase →	*Initiate synthesis of RNA primers*
5. Single strand Binding proteins (SSB) →	*Prevent premature reannealing of dsDNA*
6. DNA ligase →	*Seals the single strand nick between the nascent chain and Okazaki fragments on lagging strand.*

Transcription

It is a heterocatalytic function of DNA which involves transfer of coded information from DNA through the synthesis of RNA over the template of DNA. Only one of the strand of DNA called *sense* strand (master strand) is involved. The segment of DNA involved in transcription is **cistron**. It has a *promoter* region where initiation begins and a terminator region where *transcription* ends. Enzyme involved in transcription is RNA polymerase. It is single in prokaryotes.

There are three types of RNA polymerase in eukaryotes, I for 28 s, 18 s & 5.8 s RNA. II for *m*RNA and *Sn* RNA and III for *t*RNA, 5 *s*RNA & ScRNA.

RNA polymerase has polypeptides—σ, α, β, ρ, β′ and ω. σ or *sigma factor* recognises the promoter region while the remaining or *core enzyme* takes part in transcription. A rho (ρ) or ρ factor is needed for termination of transcription.

Steps involved in transcription

(*a*) *Activation of Ribonucleotides.*
(*b*) *Cistron*
(*c*) *Base Pairing*
(*d*) *Strand Formation*
(*e*) *Separation of RNA*
(*f*) *Duplex Formation*
(*g*) *Post-Transcriptional Processing*

Translation

The letters A, G, T & C correspond to the nucleotide found in DNA. They are organized into three letter code words called codons and the collection of these codons make up the genetic code. A linear array of codons (a gene) specifies the synthesis of various RNA molecules, most of which are involved in some aspect of protein synthesis. Protein synthesis occurs in three major steps : **initiation, elongation** and **termination**.

This process resembles DNA replication and transcription in the general features and in the fact that it, too, follows a 5′ and 3′ polarity.

Mutations

There are sudden inheritable/transmissible discontinuous variations which appear in organisms due to permanent change in their genotypes. They are also known as *sports* or *saltations*. The term mutation was coined by Hugo de Vries (1901) who also proposed *mutation theory of evolution*. He observed 834 mutations in a population of 54343 plants of *Oenothera lamarckiana* (Evening Primrose). Most of the mutations observed by de Vries were due to changes in chromosome number and chromosomal aberrations. Genotype can change due to changes in chromosome numbers, chromosome structure and gene mutations.

Changes in Chromosome Number/ Genomatic Mutations

They are of two types, *euploidy* and *eneuploidy*. Euploidy is the condition in which chromosome number is exact multiple of a genome, *e.g.*, monoploidy, diploidy, polyploidy.

Polyploidy

It is the phenomenon of having more than two genomes or sets of chromosomes, *e.g.*, 3n (triploid), 4n (tetraploid), 5n (pentaploid), etc. Polyploids with odd number of genomes are sexually sterile *e.g.*, Banana, Pineapple. They are propagated vegetatively. Other polyploids also become genetically distinct because they do not cross-breed freely with diploids. Polyploidy can develop due to:

(*i*) Failure of spindle development
(*ii*) Failure of meiosis I
(*iii*) Fusion between unreduced gametes, unreduced and normal gametes, one egg with more than one sperm, etc.

Polyploidy is induced by

(*i*) Decaptiation
(*ii*) High or low temperatures
(*iii*) Mechanical injury
(*iv*) Infection
(*v*) Irradiation

(*vi*) Chemicals like colchicine, granosan, sulphanilamide, veratrine sulphate etc.

Colchicine ($C_{22}H_{25}O_6N$) is an alkaloid obtained from Autumn Crocus (*Colchicum autumnale*). It is a mitotic poison (Blakeslee, 1937) which inhibits spindle formation without affecting replication of chromosomes.

Polyploidy is of three types-

1. Autopolyploidy
2. Allopolyploidy
3. Autoallopolyploidy.

Monosomic

It is an aneuploid in which a chromosome is deficient of its homologue causing genetic imbalance resulting in weaker than a normal man. Monosomic is formed by the union of deficient gamete (n–1) and normal gamete. A sex chromosome monosomic in human beings is TURNER's SYNDROME (44 + X). In double monosomic, two different chromosomes are deficient of their homologous ($2n$ – 1 – 1).

Nullisomic

It is an aneuploid in which a pair of homologous chromosomes is deficient ($2n$ – 2) resulting in absence of certain genes. Nullisomy is lethal in diploid organisms but can survive in polyploidy. It is produced by the union of two gametes deficient in the same chromosome (n – 1) × (n – 1).

Trisomic

Trisomic is an aneuploid or heteroploid which has one chromosome represented in triplicate ($2n$ + 1). It is followed by the union of a normal gamete (n) with a gamete having an extra chromosome (n + 1). Trisomics show gigas effect for certain genes and abnormality for others. **'Down's syndrome'** of human beings is due to trisomy of chromsome 21 while **Klinfelter's Syndrome'** is due to additional X-chromosome in otherwise male individual (XY). All the possible trisomics have been studied in *Datura*.

Tetrasomic

It is an aneuploid in which one chromosome is represented in quadruplicate. Tetrasomic is formed by union of two gametes each having the similar extra chromosome, (n + 1) × (n + 1). Two chromosomes are represented in quadruplicate in *double tetrasomic* (2n + 2 + 2) and three in *triple tetrasomic* (2n + 2 + 2 + 2). Tetrasomic super females (44 + XXXX) are infertile and have mental abnormality. Otherwise, tetrasomics are most stable of all the aneuploids showing gigas effect of many genes.

Pentasomic

An aneuploid with one chromosome represented five times ($2n$ + 3) *e.g.,* a rare super female in human beings (44 + XXXXX).

Mixed Aneuploid

It is an aneuploid with both hypoploidy (deficiency) and hyperploidy (addition), *e.g.,* $2n$ + 1A – 1B.

Chromosomal Aberrations

They are changes involving chromosome morphology which result in changes in number and sequence of genes without altering ploidy or gene structure. Chromosomal aberrations involve breaking of chromosome segments, their loss or union with same (intrachromosomal aberration) or different chromosomes (interchromosomal aberration).

Gene Mutations

They are sudden stable changes in the structure of gene or cistron due to change in nucleotide type and/or nucleotide sequence. Most of gene mutations involve a change in single nucleotide. They are called **'point mutations'.**

Spontaneous Mutations

They are mutations which develop at random naturally, automatically or spontaneously in an organism due to internal reasons without any relation to any external/environmental factor. The frequency of spontaneous mutations is different for different organisms and their different genes, *e.g.,* 1 in one million in *Drosophila*, 1 in 10 million cell generations in bacteria & 1 in 50,000 in human beings.

Induced Mutations

They are mutations that are produced in response to specific external factors or chemicals. Muller (1927) was the first scientist to produce induced mutations in *Drosophila* with the help of X-rays (upto 150 times the spontaneous sex-linked lethal mutations). He got nobel prize in 1946.

Same Sense, Mis-Sense and Non-Sense Mutations

Due to **wobble** position, a change in one nucleotide of a codon does not change amino-acid specificity. This is called *same sense* (or silent) mutation, *e.g.,* AGA = AGG = AGC. When a nucleotide change in one codon causes the change of one amino acid of a polypeptide, it is called *mis-sense mutation.*

If a mutation causes early termination of polypeptide synthesis, it is termed as *non-sense mutation.*

Hardy Weinberg Genetic Equilibrium

In 1908, an English mathematician, G.H. Hardy and a German physician, W. Weinberg, independently discovered, the principle concerned with the frequency of alleles in a population, called the ***Hardy-Weinberg equilibrium principles.*** It states that at equilibrium, both allele and genotype frequencies remain constant from generation to generation. This occurs among diploid, sexually reproducing organisms with nonoverlapping generations in which mating is random and no selection or other factors are present for changing the allele frequencies or as long as the following assumptions are met :

1. The population size is very large
2. Random mating is occurring
3. No mutation takes place
4. No genes are input from other sources
5. (No migration takes place)
6. No selections occurs.

Multiple Choice Questions

1. A nucleoside consists of :
A. Nitrogenous base
B. Purine or pyrimidine base + sugar
C. Purine or pyrimidine base + phosphorus
D. Purine + pyrimidine base + sugar + phosphorus

2. A nucleotide consists of :
A. A nitrogenous base like choline
B. Purine + pyrimidine base + sugar + phosphorus
C. Purine or pyrimidine base + sugar
D. Purine or pyrimidine base + phosphorus

3. A purine nucleotide is :
A. AMP B. UMP
C. CMP D. TMP

4. A pyrimidine nucleotide is :
A. GMP B. AMP
C. CMP D. IMP

5. Adenine is :
A. 6-Amino purine
B. 2-Amino-6-oxypurine
C. 2-Oxy-4-aminopyrimidine
D. 2,4-Dioxypyrimidine

6. 2,4-Dioxypyrimidine is :
A. Thymine B. Cystosine
C. Uracil D. Guanine

7. The chemical name of guanine is :
A. 2,4-Dioxy-5-methylpyrimidine
B. 2-Amino-6-oxypurine
C. 2-Oxy-4-aminopyrimidine
D. 2,4-Dioxypyrimidine

8. Nucleotides and nucleic acids concentration are often also expressed in terms of :
A. Ng B. Mg
C. Meq D. OD at 260 nm

9. The pyrimidine nucleotide acting as the high energy intermediate is :
A. ATP B. UTP
C. UDPG D. CMP

10. The carbon of the pentose in ester linkage with the phosphate in a nucleotide structure is :

A. C1 B. C3
C. C4 D. C5

11. Uracil and ribose form :

A. Uridine B. Cytidine
C. Guanosine D. Adenosine

12. The most abundant free nucleotide in mammalian cells is :

A. ATP B. NAD
C. GTP D. FAD

13. The mean intracellular concentration of ATP in mammalian cell is about :

A. 1 mM B. 2 mM
C. 0.1 mM D. 0.2 mM

14. The nucleic acid base found in mRNA but not in DNA is :

A. Adenine B. Cytosine
C. Guanine D. Uracil

15. In RNA molecule 'Caps' :

A. Allow *t*RNA to be processed
B. Are unique to eukaryotic *m*RNA
C. Occur at the 3′ end of *t*RNA
D. Allow correct translation of prokaryotic *m*RNA

16. In contrast to eukaryotic *m*RNA, prokaryotic *m*RNA :

A. Can be polycistronic
B. Is synthesized with introns
C. Can only be monocistronic
D. Has a poly A tail

17. The size of small stable RNA ranges from :

A. 0-40 nucleotides
B. 40-80 nucleotides
C. 90-300 nucleotides
D. More than 320 nucleotides

18. The number of small stable RNAs per cell ranges from :

A. 10-50,000
B. 50,000-1,00,000
C. 1,00,000-10,00,000
D. More than 10 lakhs

19. Molecular weight of heterogenous nuclear RNA (hnRNA) is :

A. More than 107
B. 105 to 106
C. 104 to 105
D. Less than 104

20. In RNA molecule guanine content does not necessarily equal its cytosine content nor does its adenine content necessarily equal its uracil content since it is a

A. Single strand molecule
B. Double stranded molecule
C. Double stranded helical molecule
D. Polymer of purine and pyrimidine ribonucleotides

21. The nitrogenous base present in the RNA molecule is :

A. Thymine B. Uracil
C. Xanthine D. Hypoxanthine

22. RNA does not contain :

A. Uracil B. Adenine
C. Thymine D. Ribose

23. The sugar moiety present in RNA is :

A. Ribulose B. Arabinose
C. Ribose D. Deoxyribose

24. In RNA molecule :

A. Guanine content equals cytosine
B. Adenine content equals uracil
C. Adenine content equals guanine
D. Guanine content does not necessarily equal its cytosine content

25. Methylated purines and pyrimidines are characteristically present in :

A. *m*RNA B. *hn*RNA
C. *t*RNA D. *r*RNA

26. Thymine is present in :

A. *t*RNA
B. Ribosomal RNA
C. Mammalian *m*RNA
D. Prokaryotic *m*RNA

27. The approximate number of nucleotides in *t*RNA molecule is :

A. 25 B. 50
C. 75 D. 100

28. In every cell, the number of *t*RNA molecules is at least :
A. 10 B. 20
C. 30 D. 40

29. The structure of *t*RNA appears like a :
A. Helix B. Hair pin
C. Clover leaf D. Coil

30. Although each specific *t*RNA differs from the others in its sequence of nucleotides, all *t*RNA molecules contain a base paired stem that terminates in the sequence CCA at :
A. 32 Termini B. 52 Termini
C. Anticodon arm D. 3252 Termini

31. Transfer RNAs are classified on the basis of the number of base pairs in :
A. Acceptor arm B. Anticodon arm
C. D arm D. Extra arm

32. In *t*RNA molecule D arm is named for the presence of the base :
A. Uridine B. Pseudouridine
C. Dihydrouridine D. Thymidine

33. The acceptor arm in the *t*RNA molecule has :
A. 5 Base pairs B. 7 Base pairs
C. 10 Base pairs D. 20 Base pairs

34. In *t*RNA molecule, the anticodon arm possesses :
A. 5 Base pairs B. 7 Base pairs
C. 8 Base pairs D. 10 Base pairs

35. The TΨC arm in the *t*RNA molecule possesses the sequence :
A. T, pseudouridine and C
B. T, uridine and C
C. T, dihydrouridine and C
D. T, adenine and C

36. Double helical structure model of the DNA was proposed by :
A. Pauling and Corey
B. Peter Mitchell
C. Watson and Crick
D. King and Wooten

37. DNA does not contain :
A. Thymine B. Adenine
C. Uracil D. Deoxyribose

38. The sugar moiety present in DNA is :
A. Deoxyribose B. Ribose
C. Lyxose D. Ribulose

39. DNA rich in A-T pairs have :
A. 1 Hydrogen bond
B. 2 Hydrogen bonds
C. 3 Hydrogen bonds
D. 4 Hydrogen bonds

40. In DNA molecule :
A. Guanine content does not equal cytosine content
B. Adenine content does not equal thymine content
C. Adenine content equals uracil content
D. Guanine content equals cytosine content

41. DNA rich in G-C pairs have :
A. 1 Hydrogen bond
B. 2 Hydrogen bonds
C. 3 Hydrogen bonds
D. 4 Hydrogen bonds

42. The fact that DNA bears the genetic information of an organism implies that :
A. Base composition should be identical from species to species
B. DNA base composition should charge with age
C. DNA from different tissues in the same organism should usually have the same base composition
D. DNA base composition is altered with nutritional state of an organism

43. The width (helical diameter) of the double helix in β-form DNA in nm is :
A. 1 B. 2
C. 3 D. 4

44. The number of base pair in a single turn of β-form DNA about the axis of the molecule is :
A. 4 B. 8
C. 10 D. 12

45. The distance spanned by one turn of β-form DNA is :
A. 1.0 nm B. 2.0 nm
C. 3.0 nm D. 3.4 nm

46. In a DNA molecule the thymine concentration is 30 per cent, the guanosine concentration will be :
A. 10 per cent B. 20 per cent
C. 30 per cent D. 40 per cent

47. In a DNA molecule, the guanosine content is 40 per cent, the adenine content will be :
A. 10 per cent B. 20 per cent
C. 30 per cent D. 40 per cent

48. An increased melting temperature of duplex DNA results from a high content of :
A. Adenine + Guanine
B. Thymine + Cytosine
C. Cytosine + Guanine
D. Cytosine + Adenine

49. A synthetic nucleotide analogue, 4-hydroxypyrazolopyrimidine is used in the treatment of :
A. Acute nephritis
B. Gout
C. Cystic fibrosis of lung
D. Multiple myeloma

50. A synthetic nucleotide analogue, used in the chemotherapy of cancer and viral infections is :
A. Arabinosyl cytosine
B. 6-Mercaptopurine
C. 4-Hydroxypyrazolopyrimidine
D. 6-Thioguanine

ANSWERS

1	2	3	4	5	6	7	8	9	10
B	B	A	C	A	C	B	D	C	D
11	**12**	**13**	**14**	**15**	**16**	**17**	**18**	**19**	**20**
A	A	A	D	B	A	C	C	A	A
21	**22**	**23**	**24**	**25**	**26**	**27**	**28**	**29**	**30**
B	C	C	D	C	A	C	B	C	A
31	**32**	**33**	**34**	**35**	**36**	**37**	**38**	**39**	**40**
D	A	B	A	A	C	C	A	B	D
41	**42**	**43**	**44**	**45**	**46**	**47**	**48**	**49**	**50**
C	C	B	C	D	B	A	C	B	A

●●●

CHAPTER

6

Carbohydrates and Its Metabolism

Introduction

Carbohydrates are carbon compounds that contain hydrogen and oxygen in the ratio 2 : 1. Chemically carbohydrates are aldehyde or ketone derivatives of polyhydric alcohols; these are known as aldoses or ketoses. Hydrated carbons are also called carbohydrates.

Classification of Carbohydrates

A. Monosaccharides

• Glyceraldehyde	• Lyxoketose	• Talose
• Dihydroxyacetone	• Ribulose	• Atrose
• Erythrose	• Xylulose	• Allose
• Threose	• Fructose	• Glucoheptose
• Xyloketose	• Sorbose	• Mannoheptose
• Arabinose	• Glucose	• Sedoheptulose
• Xylose	• Mannose	• Glucooctase
• Ribose	• Galactose	• Mannooctase
• Lyxose	• Gulose	• Mannononose
	• Idose	• Glucodecose

B. Disaccharides

I. Reducing sugars

- Maltose α±1, 4 Glucose + Glucose
- Lactose β±1, 4 Glucose + Galactose
- Cellobiose β±1, 4 Glucose + Glucose
- Gentiobiose β±1, 6 Glucose + Glucose
- Melibiose α±1, 4 Glucose + Galactose
- Isomaltose α±1, 6 Glucose + Glucose

II. Non-reducing sugars

- Sucrose α±1, β-2 Glucose + Fructose
- Trehalose α-1, β-2 Glucose + Glucose

C. Trisaccharides

I. Reducing sugars

- Mannotriose — Galactose + Galactose + Glucose
- Robinose — Galactose + Rhamnose + Rhamnose
- Rhamninose — Galactose + Rhamnose + Rhamnose

II. Non-Reducing sugars

- Raffinose — Fructose + Glucose + Galactose
- Gentobiose — Fructose + Glucose + Glucose
- Melezitose — Glucose + Fructose + Glucose

D. Oligosaccharides

- Stachylose — Fructose + Glucose + Galactose + Galactose
- Verbascose — Fructose + Glucose + Galactose + Galactose + Galactose
- Ajugose — Fructose + Glucose + Galactose + Galactose + Galactose + Galactose

E. Polysaccharides

I. Homopolysaccharides

- Starches
- Dextrins
- Glycogens
- Celluloses
- Inulin

- Chitin
- Dextrans

II. Heteropolysaccharides

- Agar
- Mucilages
- Gumarabic
- Gumacacia
- Pectins
- Alginic acids

III. Mucopolysaccharides

- Hyaluronic acid
- Heparin
- Chondroitin sulphates
- Sialic acids

Which saccharides present in honey?

Following different saccharides are present in honey:

Maltose, Sucrose, Gentiobiose, Isokestose, Kojibiose, Maltulose, Erlose, Melezitose, Turanose, Nigerose, Panose, Isomaltotetrose, Isomaltose, α, β-trehalose, Maltotriose etc.

Relative sweetness of different sugars

Sugar	*Relative sweetness*
Sucrose	100
Fructose	170
Glucose	70
Galactose	32
Maltose	30
Lactose	16
Lactulose	55
Saccharin	40000-50000

Sweetness of polyols

Sugar polyol	*Relative sweetness*
Gucitol	50
Lactitol	40
Maltitol	68
Mannitol	40
Xylitol	90
Sorbitol	63
Galactitol	58

Different Sugar Alcohols Present in Higher Plants

- Pentitol: Ribitol
- Hexitols: Sorbitol, D-Mannitol, Dulcitol, Allitol, L-Iditol, Polygalitol, Styracitol and Hamamalitol
- Heptitols: Volemitol, Perseitol
- Octitol: D-Erythro-D-galacto-octitol

Properties of Amylose and Amylopectin

Property	*Amylose*	*Amylopectin*
Molecular configuration	Essentially linear	Highly branched
Weight-average (mol. wt.)	Ca. 10^6	Ca. 10^8
X-ray diffraction	Crystalline	Amorphous
Action of β-amylase and Z-enzyme	Complete hydrolysis	Residual dextrin of high mol. wt.
Complex formation with iodine and polar substances	Readily forms complexes	Very limited complexes formation
Stability in aqueous solution	Unstable, tends to retrograde	Stable

Starch Degrading Enzymes

Enzyme	*Reaction catalyzed*
α-amylase	Hydrolytic cleavage of α-1, 4 bonds in a glucan
β-amylase	Liberation of maltose from the non-reducing end of a glucan
α-glucosidase	Hydrolytic cleavage of α-1-4, or α-1-6, bonds in oligo-saccharides
Debranching enzyme	Hydrolysis of α-1 ± 6 bonds
D-enzyme	G donor + G acceptor ↔ G donor + acceptor-1, + Glucose

Enzyme	*Reaction catalyzed*
Glucosyl-glucan transferase	Transfer of glycosyl, maltosyl or maltotriosyl residues to glucose
Phosphorylase	Glucose 1 ± P + α-glucan ↔ α-glucose 1 ± P + Glucose
Maltose phosphorylase	Maltose + Orthophosphate ↔ Glucose ±1-P + glucose

Biochemical Functions of Carbohydrates

- Source of energy
- Conversion of solar energy into chemical energy
- Central to metabolism
- Utilization of fats
- Sweetness of food
- Appetizing and flavouring
- Bulk of food
- Peristaltatic movement of stomach
- Movement of joints
- Structural support
- Storage of food material
- Transport between cells and organs
- Accumulation of fats
- Helps in pollination
- Helps in transamination
- Helps for fat burns
- Responsible for ketone body formation
- It exerts sparing effect of protein
- It acts as drugs
- It helps in protein synthesis
- Responsible for colour development to the products
- The ability to bind water and control water activity in foods is one of the most important properties of carbohydrates
- It helps in the sugar-flavourant formation
- Burnt carbohydrates gives colour to the food products
- It also acts as a lubricant at joints

Carbohydrates Metabolism

Since all digestible forms of carbohydrates are eventually transformed into glucose, it is important to consider how glucose is able to provide energy in the form of adenosine triphosphate (ATP) to various cells and tissues. Glucose is metabolized in three stages:

Glycolysis
The Krebs Cycle
Oxidative phosphorylation

During exercise, hormonal levels shift and this disruption of homeostasis alters the metabolism of glucose and other energy-bearing molecules. Therefore, in this SparkNote the metabolism of carbohydrates will be considered in the context of exercise strategies and hypotheses.

1. Glycolysis

The breakdown of glucose to provide energy begins with glycolysis. To begin with, glucose enters the cytosol of the cell, or the fluid inside the cell not including cellular organelles. Next, glucose is converted into two, three-carbon molecules of pyruvate through a series of ten different reactions. A specific enzyme catalyzes each reaction along the way and a total of two ATP are generated per glucose molecule. Since ADP is converted to ATP during the breakdown of the substrate glucose, the process is known as substrate-level phosphorylation. During the sixth reaction, glyceraldehyde 3-phosphate is oxidized to 1,3 bisphosphoglycerate while reducing nicotinamide adenosine dinucleotide (NAD) to NADH, the reduced form of the compound. NADH is then shuttled to the mitochondria of the cell where it is used in the electron transport chain to generate ATP via oxidative phosphorylation, which will be described later.

The most important enzyme in glycolysis is called phosphofructokinase (PFK) and catalyzes the third reaction in the sequence. Since this reaction is so favorable under physiologic conditions, it is known as the "committed step" in glycolysis. In other words, glucose will be completely degraded to pyruvate after this reaction has taken place. With this in mind, PFK seems as if it would be an excellent site of control for glucose metabolism. In fact, this is exactly the case. When ATP or energy is plentiful in the cell, PFK is inhibited and the breakdown of glucose for energy slows down. Therefore, PFK can regulate the degradation of glucose to match the energy needs of the cell. This type of regulation is a recurring theme in biochemistry.

2. Krebs Cycle and Oxidative Phosphorylation/Electron Transport Chain

There are many compounds that are formed and recycled during the Krebs Cycle (Citric Acid Cycle). These include oxidized forms of nictotinamide adenine dinucleotide (NAD^+) and flavin adenine dinucleotide (FAD) and their reduced counterparts: NADH and $FADH_2$. NAD^+ and FAD are electron acceptors and become reduced while the substrates in the Krebs Cycle become oxidized and surrender their electrons.

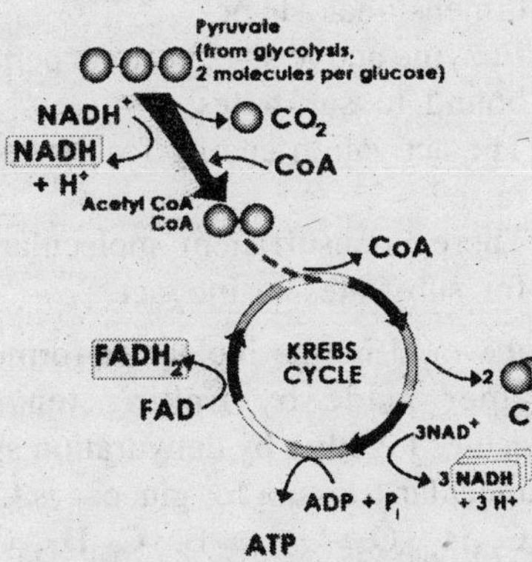

Fig. 1: *The Krebs Cycle*

The Krebs Cycle begins when the pyruvate formed in the cytoplasm of the cell during glycolysis is transferred to the mitochondria, where most of the energy inherent in glucose is extracted. In the mitochondria, pyruvate is converted to acetyl CoA by the enzyme pyruvate carboxylase. In general, Acetyl-CoA condenses with a four carbon compound called oxaloacetate to form a six carbon acid. This six-carbon compound is degraded to a five and four carbon compound, releasing two molecules of carbon dioxide. At the same time, two molecules of NADH are formed. Finally, the C-4 carbon skeleton undergoes three additional reactions in which guanosine triphosphate (GTP), $FADH_2$ and NADH are formed, thereby regenerating oxaloacetate. $FADH_2$ and NADH are passed on to the electron transport chain (see below) that is embedded in the inner mitochondria membrane. GTP is a high-energy compound that is used to regenerate ATP from ADP. Therefore, the main purpose of the Krebs Cycle is to provide high-energy electrons in the form of $FADH_2$ and NADH to be passed onward to the electron transport chain.

The high-energy electrons contained in NADH and $FADH_2$ are passed on to a series of enzyme complexes in the mitochondrial membrane.

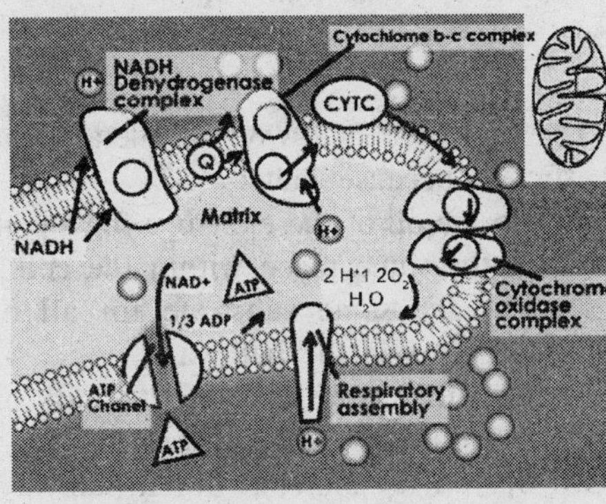

Fig. 2.: *Electron Transport Chain*

Multiple Choice Questions

1. Metabolic reactions that break down complex molecules into smaller compounds, thereby releasing usable energy for the cell, are best described as :
   A. Biosynthetic B. Catabolic
   C. Catalytic D. Photosynthetic
2. The ultimate source of energy that sustains living systems is :
   A. Glucose B. Oxygen
   C. Sunlight D. Carbon dioxide
3. Which of the following is not a disaccharide?
   A. Amylose B. Cellobiose
   C. Lactose D. None of these
4. What is the cause of the genetic disease known as Galactosemia?
   A. Deficiency in lactase

B. Absence of galactose 1-P uridyl transferase
C. Absence of lactose synthetase
D. Non functioning of semnase

5. What is the consensus N-glycosylation site in a protein sequence?
A. Asn-Xaa-(Ser or Thr)
B. (Ser or Thr)-Asn-Ala
C. Thr-(Asn or Gln)-Ala
D. None of these

6. Which of the following statements about the energy needs of cells is false?
A. Without a continuous input of energy, cell disorder will increase
B. The laws of thermodynamics force cells to acquire energy
C. Many cellular reactions have an associated activation energy
D. The most usable energy for cells comes from the rapid combustion of glucose

7. The production or breakdown of is often coupled with the metabolic reactions of biosynthesis and catabolism.
A. Aspirin B. DNA
C. ATP D. CO_2

8. The multistep pathways of metabolism are efficient because they :
A. Locate all of the enzymes for a pathway in the same place within the cell
B. Use the same substrate for all of the enzymes in the pathway
C. Use the same enzyme for all of the substrates in the pathway
D. Spread the enzymes for a pathway into several different organelles

9. Which of the following is not involved in the biosynthesis of DNA?
A. Energy from ATP
B. Mononucleotides
C. Carbonic anhydrase
D. Enzymes

10. Which of the following would be considered a part of metabolism?
A. Biosynthetic pathways that build DNA
B. Catabolic pathways that breakdown complex carbohydrates
C. The capture of light energy for use in making glucose
D. All of the above

11. A common way that cells capture the energy released during the breakdown of large molecules is to add electrons to smaller, specialized molecules that can accept them. This process of electron acceptance is otherwise known as :
A. Biosynthesis B. Metabolism
C. Reduction D. Catalysis

12. When living organisms are cooled below some critical body temperature, the metabolic reactions within their cells cease to function properly. This malfunction occurs because :
A. Their enzymes lose the proper three-dimensional shape
B. Enzyme active sites become permanently bound to substrates
C. The activation energy for the reaction is raised
D. There is insufficient molecular motion for substrates to interact

13. What would be the molecular formula for a polymer made by linking ten glucose molecules together by dehydration synthesis, if molecular formula for glucose is $C_6H_{12}O_6$?
A. $C_{60}H_{100}O_{50}$ B. $C_{60}H_{120}O_{60}$
C. $C_{60}H_{102}O_{51}$ D. $(C_6H_{12}O_6)_{10}$

14. Boat and chair conformations are found :
A. In pyranose sugars
B. In any sugar without axial -OH groups
C. In any sugar without equatorial -OH groups
D. only in D-glucopyranose

15. In glucose the orientation of the -H and -OH groups around the carbon atom 5 adjacent to the terminal primary alcohol carbon determines :
A. D or L series
B. Dextro or levorotatory
C. Epimer and anomers
D. Epimers

16. The carbohydrate of the blood group substances is :

A. Sucrose B. Fucose
C. Arabinose D. Maltose

17. Erythromycin contains :
A. Dimethyl amino sugar
B. Trimethyl amino sugar
C. Sterol and sugar
D. Glycerol and sugar

18. A sugar alcohol is :
A. Mannitol B. Trehalose
C. Xylulose D. Arabinose

19. The major sugar of insect hemolymph is :
A. Glycogen B. Pectin
C. Trehalose D. Sucrose

20. The sugar found in DNA is :
A. Xylose B. Ribose
C. Deoxyribose D. Ribulose

21. The sugar found in RNA is :
A. Ribose B. Deoxyribose
C. Ribulose D. Erythrose

22. The sugar found in milk is :
A. Galactose B. Glucose
C. Fructose D. Lactose

23. Invert sugar is :
A. Lactose
B. Sucrose
C. Hydrolytic products of sucrose
D. Fructose

24. Sucrose consists of :
A. Glucose + glucose
B. Glucose + fructose
C. Glucose + galactose
D. Glucose + mannose

25. The monosaccharide units are linked by $1 \rightarrow 4$ glycosidic linkage in :
A. Maltose B. Sucrose
C. Cellulose D. Cellobiose

26. Which of the following is a non-reducing sugar?
A. Isomaltose B. Maltose
C. Lactose D. Trehalose

27. Which of the following is a reducing sugar?
A. Sucrose B. Trehalose
C. Isomaltose D. Agar

28. A dissaccharide formed by 1,1-glycosidic linkage between their monosaccharide units is :
A. Lactose B. Maltose
C. Trehalose D. Sucrose

29. Which of the following interconversions of monosaccharides does not require a nucleotide linked sugar intermediate
A. Galactose-1 phosphate to glucose 1-phosphate
B. Glucose-6-phosphate to mannose-6-phosphate
C. Glucose phosphate UDP Glucose
D. D-glucuronic acid to UDP glucoronic acid

30. Mutarotation refers to change in :
A. pH
B. Optical rotation
C. Conductance
D. Chemical properties

31. A polysaccharide which is often called animal starch is :
A. Glycogen B. Starch
C. Inulin D. Dextrin

32. The homopolysaccharide used for intra-venous infusion as plasma substitute is :
A. Agar B. Inulin
C. Pectin D. Starch

33. The polysaccharide used in assessing the glomerular filtration rate (GFR) is :
A. Glycogen B. Agar
C. Inulin D. Hyaluronic acid

34. The constituent unit of inulin is :
A. Glucose B. Fructose
C. Mannose D. Galactose

35. The polysaccharide found in the exoskeleton of invertebrates is :
A. Pectin
B. Chitin
C. Cellulose
D. Chondroitin sulphate

36. Which of the following is a heteroglycan?
A. Dextrins B. Agar
C. Inulin D. Chitin

37. The glycosaminoglycan which does not contain uronic acid is :
A. Dermatan sulphate
B. Chondroitin sulphate
C. Keratan sulphate
D. Heparan sulphate

38. The glycosaminoglycan which does not contain uronic acid is :
A. Hyaluronic acid
B. Heparin
C. Chondroitin sulphate
D. Dermatan sulphate

39. Keratan sulphate is found in abundance in :
A. Heart muscle B. Liver
C. Adrenal cortex D. Cornea

40. Repeating units of hyaluronic acid are :
A. N-acetyl glucosamine and D-glucuronic acid
B. N-acetyl galactosamine and D-glucuronic acid
C. N-acetyl glucosamine and galactose
D. N-acetyl galactosamine and L-iduronic acid

41. The approximate number of branches in amylopectin is :
A. 10 B. 20
C. 40 D. 80

42. In amylopectin the intervals of glucose units of each branch is :
A. 10-20 B. 24-30
C. 30-40 D. 40-50

43. A polymer of glucose synthesized by the action of leuconostoc mesenteroids in a sucrose medium is :
A. Dextrans B. Dextrin
C. Limit dextrin D. Inulin

44. Glucose on reduction with sodium amalgam forms :
A. Dulcitol
B. Sorbitol
C. Mannitol
D. Mannitol and sorbitol

45. Glucose on oxidation does not give :
A. Glycoside
B. Glucosaccharic acid
C. Gluconic acid
D. Glucuronic acid

46. Oxidation of galactose with conc HNO_3 yields :
A. Mucic acid B. Glucuronic acid
C. Saccharic acid D. Gluconic acid

47. A positive Benedict's test is not given by :
A. Sucrose B. Lactose
C. Maltose D. Glucose

48. Starch is a :
A. Polysaccharide B. Monosaccharide
C. Disaccharide D. None of these

49. A positive Seliwanoff's test is obtained with :
A. Glucose B. Fructose
C. Lactose D. Maltose

50. Osazones are not formed with the :
A. Glucose B. Fructose
C. Sucrose D. Lactose

ANSWERS

1	2	3	4	5	6	7	8	9	10
B	C	A	B	A	D	C	A	C	D
11	**12**	**13**	**14**	**15**	**16**	**17**	**18**	**19**	**20**
C	D	C	A	A	B	A	A	C	C
21	**22**	**23**	**24**	**25**	**26**	**27**	**28**	**29**	**30**
A	D	C	B	A	D	C	C	B	B
31	**32**	**33**	**34**	**35**	**36**	**37**	**38**	**39**	**40**
D	A	C	B	B	B	C	B	D	A
41	**42**	**43**	**44**	**45**	**46**	**47**	**48**	**49**	**50**
D	B	A	B	A	A	A	A	B	C

●●●

CHAPTER

7

Proteins and Its Metabolism

Introduction

The term protein was first suggested by Geradus Johannes Mulder in 1838. Protein occupy a central position in the architecture and functioning of living matter.

Amino Acids

Amino acids are organic acids formed of C, H, O, N and in some S. The amino acids are regarded as 'building blocks of proteins'. They possess an amino group attached to α-carbon, hence called α-amino acids. The α-carbon also bears one hydrogen and a hydrocarbon which is case of glycine (simplest and smallest) is a second hydrogen, $-NH_2.CHR.COOH$. Proline and hydroxyl proline have –NH (imino group) instead of $-NH_2$. They are therefore, *imino acids* instead of amino acids. R-hydrocarbon can be polar (*e.g.,* glutamic acid, serine) or non-polar (*e.g.,* alanine). Amino acids may be neutral (*e.g.,* glycine, alanine, valine, leucine, isoleucine, commonly monoamino monobasic), acidic (monoamino dicarboxylic, *e.g.,* arginine, lysine), S-containing (*viz.,* cystine, cysteine, methionine), alcoholic (*e.g.,* serine, threonine), aromatic (*e.g.,* phenylalanine, tyrosine, tryptophan) or heterocyclic (*e.g.,* histidine, proline, hydroxyproline, tryptophan). There are 20 protein amino acids. Except for glycine, all other protein amino acids are laevorotatory. Three important nonprotein amino acids are *ornithine*, citrulline (urea cycle) and diamino-pinetic acid. Animals cannot synthesize ten protein amino acids. They are called *essential amino acids* (viz. Arginine, histidine, lysine, methionine, leucine, isoleucine, valine, tryptophan and phenylalanine).

The amino acids which commonly occur in proteins are summarised in the following tables:

Table-1 : *Amino acids commonly occur in Proteins*

Sr. No.	Name	Symbol	Mol. Wt.
1.	Aspartic acid	Asp	133
2.	Glutanic acid	Gln	147
3.	Lysine	Lys	146
4.	Arginine	Arg	174
5.	Histidine	His	154
6.	Tyrosine	Tyr	181
7.	Cysteine	Cys	121
8.	Aspargine	Asu	132
9.	Glutamine	Glu	146
10.	Serine	Ser	105
11.	Threonine	Thr	119
12.	Glycine	Gly	75
13.	Alanine	Ala	89
14.	Valine	Val	117
15.	Leucine	Leu	131
16.	Isoleucine	Ile	131
17.	Methionine	Met	149
18.	Proline	Pro	115
19.	Phenylalanine	Phe	165
20.	Tryptophan	Try	204

Essential and Non-essential Amino Acids are summarised in table-2:

Table-2 : *Essential and Non-essential Amino acids*

S. No.	*Essential Amino Acids*	*S. No.*	*Non-essential Amino Acids*
1.	Isoleucine	1.	Alanine
2.	Leucine	2.	Arginine
3.	Lysine	3.	Aspartic Acid
4.	Methionine	4.	Aspargine
5.	Phenylalanine	5.	Cysteine
6.	Threonine	6.	Glutamine
7.	Tryptophan	7.	Glutanic acid
8.	Valine	8.	Glycine
		9.	Histidine
		10.	Proline & hydroxy proline
		11.	Serine
		12.	Tyrosine

Proteins

The major macromolecular constituent of cells, consisting of carbon, hydrogen, oxygen, nitrogen and usually sulphur and phosphorus a linear polymer of amino acids linked together by peptide bond in specific sequence. They are amino acid polymers formed of one or more polypeptides (monomeric proteins and oligomeric proteins). The minimum molecular weight of a protein is 4500 (Adreno corticotrophin hormone with 39 amino acids), 5733 (bovine insulin, two polypeptides, 51 amino acids A 21, B 30 in bovine and 53 with 22 + 31 amino acids in human) and 6000 (bacterial ferredoxin). Haemoglobin has 4 polypeptides with 574 amino acid residues and molecular weight of 64,500. Glutamate dehydrogenase has a molecular weight of about 1 million with 40 polypeptides and 8300 amino acids.

Proteins exhibit four levels of structural organisation, primary, secondary, tertiary and quaternary. A polypeptide is formed of a linear row of amino acids joined by peptide bonds. This constitutes the primary structure. Secondary structure is of two main types, α-helix and β-pleated. In α-helix, a single amino acid chain is coiled spirally by establishing hydrogen bonds between first and fourth amino acids. In β-pleated arrangement a sheet is produced by holding of two or more polypeptide chains in parallel or antiparallel manner by means of hydrogen bonds. The polypeptide chains are folded variously in tertiary structure. The bonds formed during tertiary structure can easily be broken by high temperature, high energy radiations and salts of heavy metals.

The process is called *denaturation*, Quaternary structure is present in case of Oligomeric proteins. In shape, proteins can be fibrous (elastin, collagen, actin, myosin, keratin) or *globular* (histone, haemoglobin, egg albumin, serum globulin).

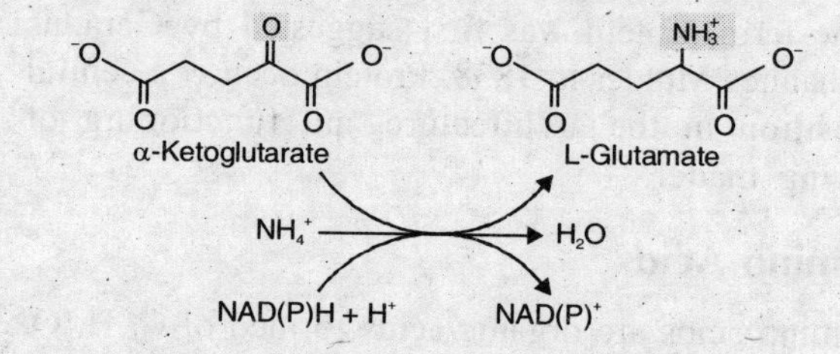

Fig. 1: *The glutamate dehydrogenase reaction. Reductive amination of α-ketoglutarate by NH_4^+ proceeds at the expense of NAD(P)H.*

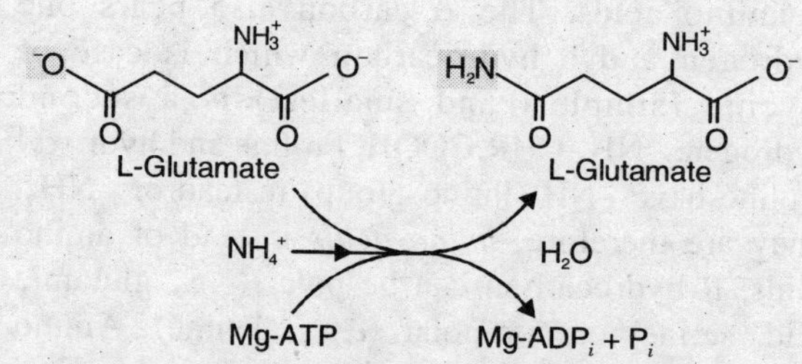

Fig. 2 : *The glutamine synthetase reaction.*

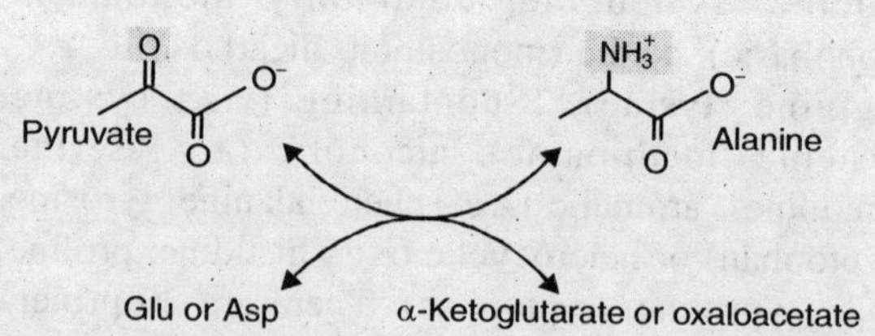

Fig. 3 : *Formation of alanine by transamination of pyruvate. The amino donor may be glutamate or aspartate. The other product thus is α-ketoglutarate or oxaloacetate. If oxaloacetate rather then pyruvate is the ox acid, acceptance of the amino group of glutamate form aspartate.*

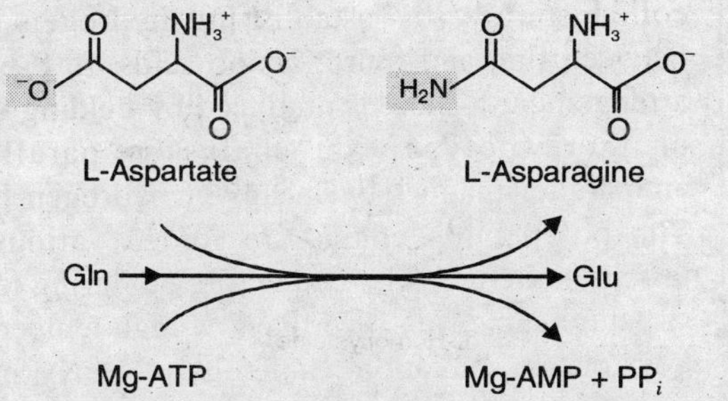

Fig. 4 : *The aspargine synthetase reaction. Note similarities to and differences from the glutamine synthetase reaction.*

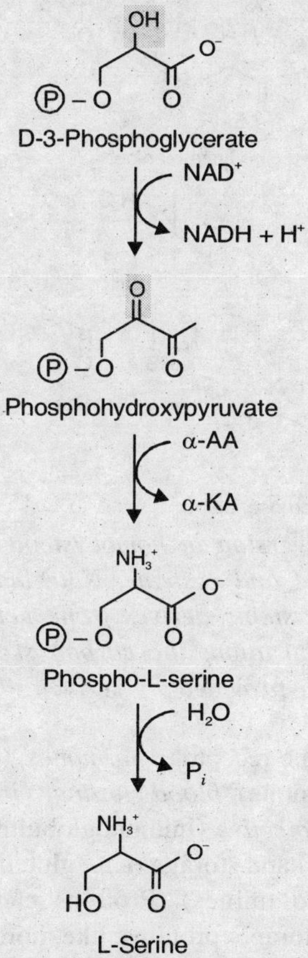

Fig. 5 : *Serine biosynthesis. (α-AA, α-amino acids; α-KA, α-keto acids.)*

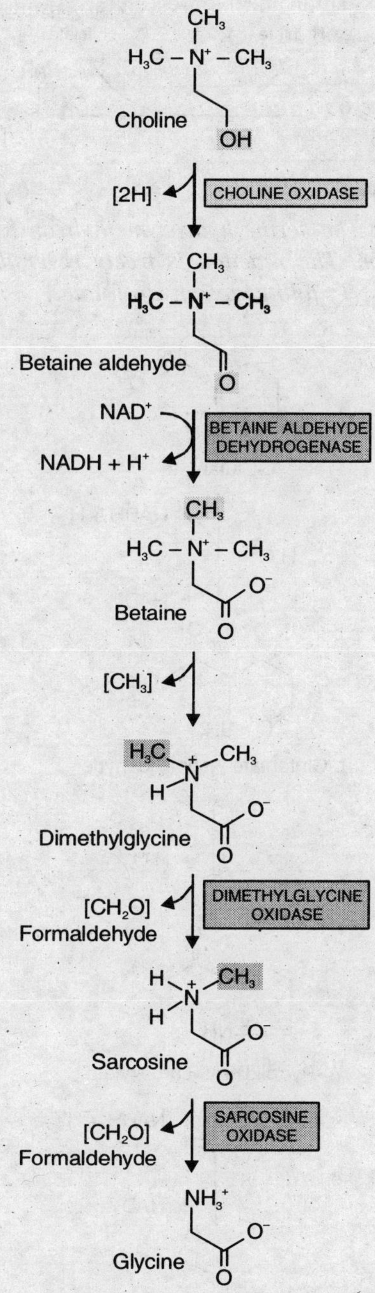

Fig. 6 : *Formation of glycine from choline.*

In function they can be *structural* (*e.g.,* Elastin in ligaments, collagen in tendons and cartilage, tubulin in microtubules, chondrin in cartilage, ossein in bone, fibroin in silk, keratin in hair skin and nails.) *contractive* (*e.g.,* actin, myosin), *carrier/transport proteins* (haemoglobin for oxygen, lipoprotein of plasma for lipids from liver to other organs), visual pigments (*e.g.,* rhodopsin), *receptor*

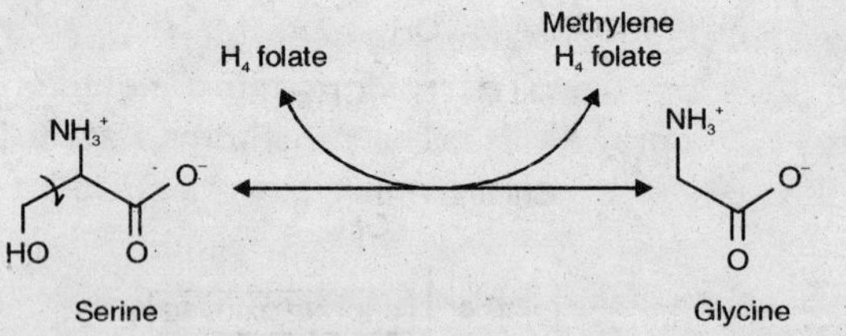

Fig. 7 : *The serine hydroxymethyltransferase reaction. The reaction is freely reversible. (H_4 folate, tetrahydrofolate.)*

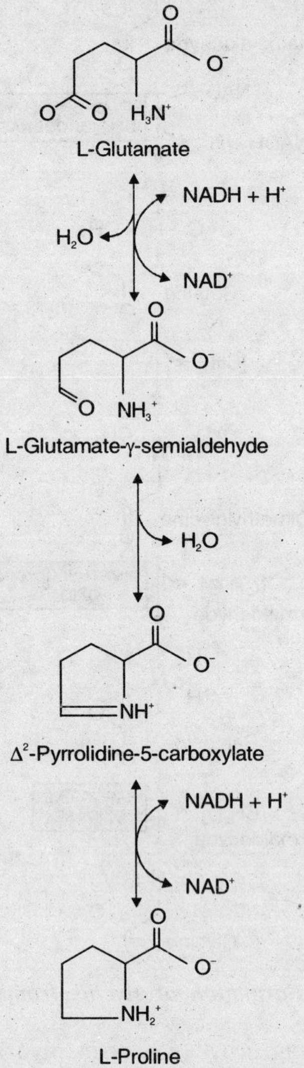

Fig. 8 : *Biosynthesis of proline from glutamate by reversal of the reactions of proline catabolism.*

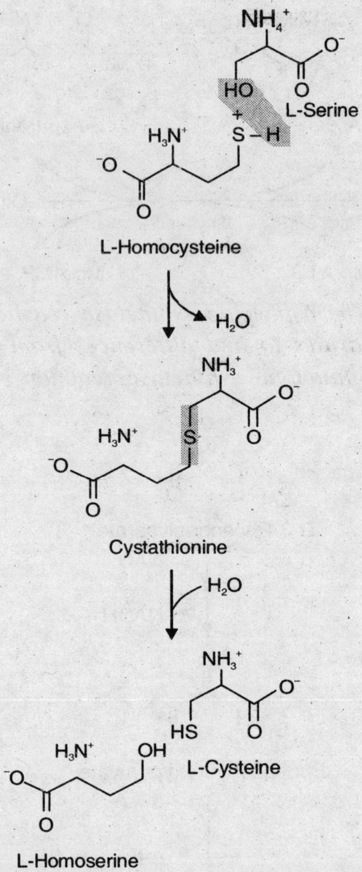

Fig. 9 : *Conversion of homocysteine and serine to homoserine and cysteine. Note that while the sulphur of cysteine derives from methionine by transsulphuration, the carbon skeleton is provided by serine.*

(*e.g.*, adrenaline receptor), *hormones*, (*e.g.*, insulin, GH, parathormone), *blood clotting* (*viz.* fibrinogen, thrombin), *protective* (immunoglobulin) enzymatic (*e.g.*, amylase) and storage (*e.g.*, glutelin, albumins, globulins, prolamines). Proteins can be *simple* (*e.g.*, most storage proteins like hordein, zeatin, glutenin of wheat, oryzenin) or *conjugated* (with additional groups). Conjugated proteins include nucleoproteins (with DNA, RNA), chromoproteins (flavoproteins with FMN or FAD, haemocyanin and cytochrome with Fe, rhodopsin with vitamin A, haemocyanin and cytochrome oxidase with Cu, Chloroplastin with chlorophyll, biliproteins with

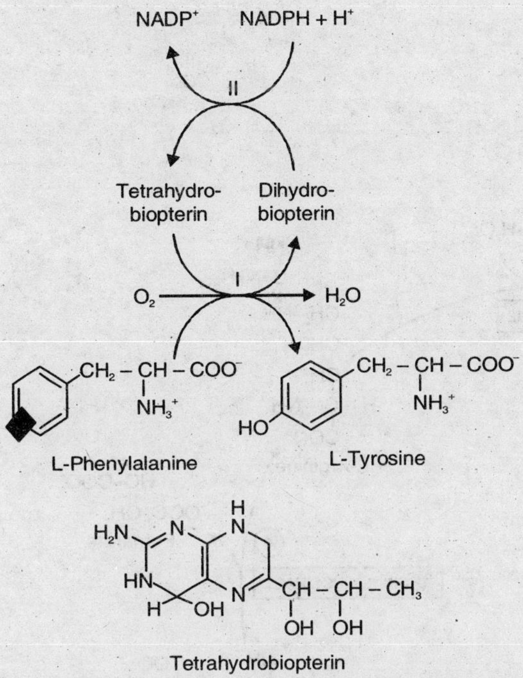

Fig. 10 : *The phenyl hydroxylase reaction. Two distinct enzymatic activities are involved. Activity II catalyzes reduction of dihydrobiopterin by NADPH, and activity I the reduction of O_2 to H_2O and phenylalanine to tyrosine. This reaction is associated with several defects of phenylalanine metabolism.*

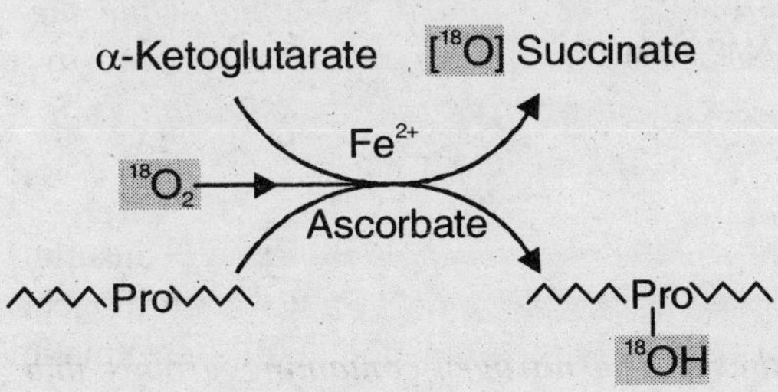

Fig. 11 : *The prolyl hydroxylase reaction. The substrate is a proline-rich peptide. During the course of the reaction, molecular oxygen is incorporated into both succinate and proline (shown by the use of heavy oxygen, $^{18}O_2$). Lysyl hydroxylase catalyzes an analogous reaction.*

phycobilins), metaprotein (Fe in Ferritin), lipoprotein (with lipid like lipovitellin of egg yolk), phosphoprotein (phosphate as in casein, ovovitellin of egg yolk), glycoprotein, mucoprotein (*e.g.*, mucin of saliva) etc. Derived proteins include proteoses, peptones as also a protein called monellin. Proteins present in meat, egg and milk are called *complete proteins* as they contain all the essential amino acids. Amongst plant foods, only soybean protein is nearly complete. The *most abundant* protein is *rubisco* (RuBP carboxylase) present in chloroplasts.

All vertebrates, including humans, can form the 12 nutritionally non-essential amino acids from amphibolic intermediates or from other dietary amino acids. Vertebrates cannot, however, biosynthesize the 10 nutritionally essential amino acids. Vertebrates biosynthesize amino acids from amphibolic intermediates via metabolic pathways that involve 5 or fewer enzyme-catalyzed reactions. The parent amphibolic intermediates and the amino acids to which they give rise are citric acid cycle intermediates α-ketoglutarate (Glu, Gln, Pro, Hyp) oxaloacetate (Asp, Asn) and the glycolytic intermediate 3 phosphoglycerate (Ser, Gly). Three other amino acids (Cys, Tyr, Hyl) are formed from nutritionally essential amino acids. Serine provides the carbon skeleton and homocysteine, the sulphur for cysteine biosynthesis, while phenylalanine hydroxylase converts phenylalanine to tyrosine. Since no codon or *t*RNA dictates the insertion of Hyp or Hyl into peptides, neither dietary hydroxyproline nor dietary hydroxylysine is incorporated into proteins. There hydroxylated amino acids arise via post-translational hydroxylation by mixed-function oxidases of peptidyl Pro or Lys. By contrast, selenocysteine an essential active site residue in several mammalian enzymes, arises by cotranslational insertion of a previously modified *t*RNA.

Catabolism of Protein

Protein turnover, the continual synthesis and catabolism of proteins, occurs in all life forms. Each day, human subjects degrade 1-2% of their body protein, principally from skeletal muscle. Rates of protein degradation vary widely between protein and rates of degradation may vary in different physiologic states. The half-lives of proteins, the

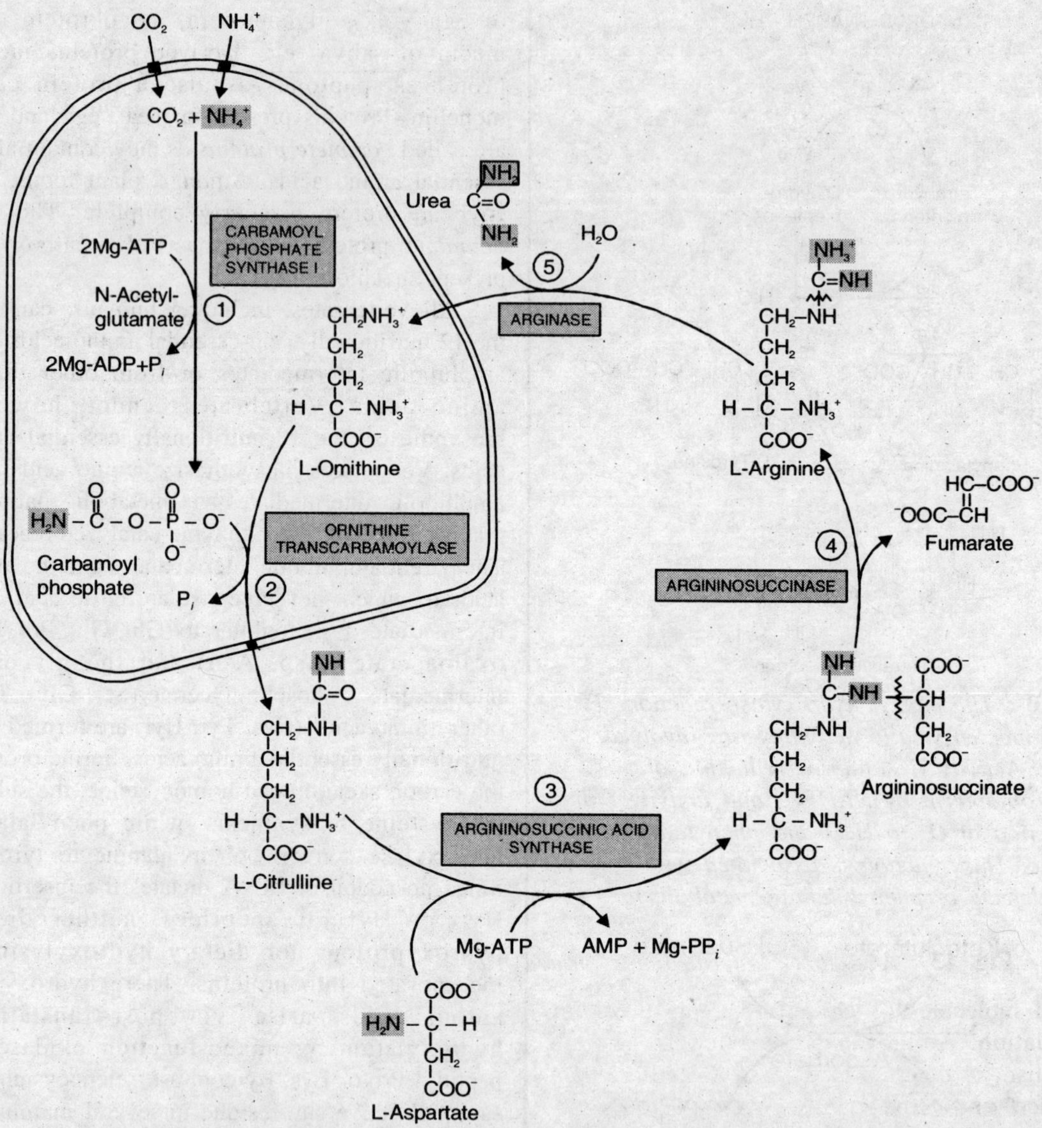

Fig. 12 : Reactions and intermediates of urea biosynthesis. The nitrogen-containing groups that contribute to the formation of urea are shaded. Reaction (1) and (2) occur in the matrix of liver mitochondria and reaction (3), (4) and (5) in liver cytosol. CO_2 (as bicarbonate), ammonium ion, ornithine and citrulline enter the mitochondrial matrix via specific carriers (•) present in the inner membrane of liver mitochondria.

time required to degrade half of the existing protein, vary from 30 minutes to over 150 hours. Extremely short half-lives often characterize enzymes that perform key-regulatory roles. Proteases and peptidases degrade proteins by both ATP-dependent and ATP independent pathways. For circulating glycoproteins, liver cell-surface asialoglycoprotein receptors bind and internalize asialoglycoproteins destined for degradation by lysosomal proteases. For many intracellular proteins, attachment of

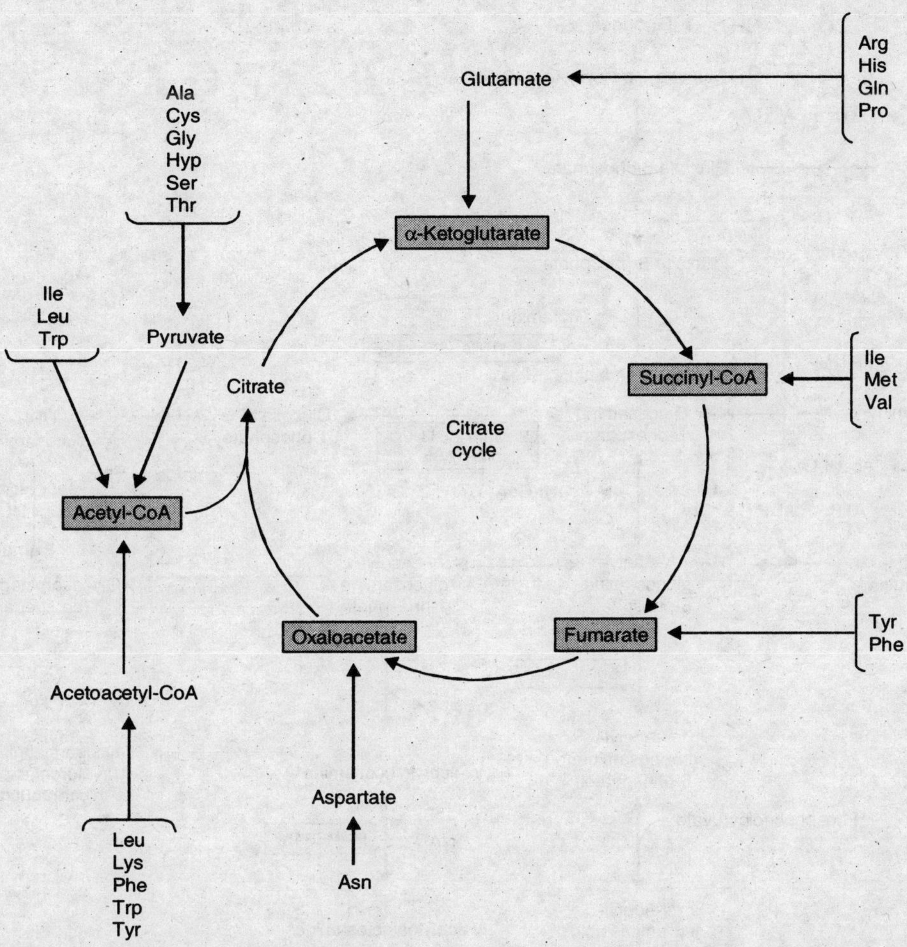

Fig. 13 : *Amphibolic intermediates formed from the carbon skeleton of amino acids.*

several molecules of ubiquitin targets them for degradation. Amino acids liberated by protein catabolism or ingested in excess of need are then degraded, not stored.

Ammonia is highly toxic to all animals. While fish excrete ammonia directly, the need to conserve water precludes this as a major route of nitrogen elimination in birds or land animals. Birds detoxify ammonia by converting it to uric acid. Humans and other higher vertebrates convert ammonia to urea. In either case, the initial reaction in amino-acid catabolism is removal of the α-amino group by transamination, a reaction that requires pyridoxal phosphate. Transamination thus channels α-amino acid nitrogen into glutamate.

L-amino acid oxidase also deaminates α-amino acids, although this is of less certain physiologic significance. L-Glutamate dehydrogenase occupies a central position in nitrogen metabolism. Ammonia intoxication is life threatening, and glutamine synthase converts ammonia to non-toxic glutamine for transport to the liver. Liver glutaminase then releases ammonia from glutamine for use in urea synthesis. Exchange between organs maintains circulating levels of amino acids, and a complex set of interorgan exchanges characterizes the postabsorptive state.

Urea, the major end product of nitrogen catabolism in humans, is synthesized from ammonia, carbon dioxide, and the amide nitrogen of aspartate.

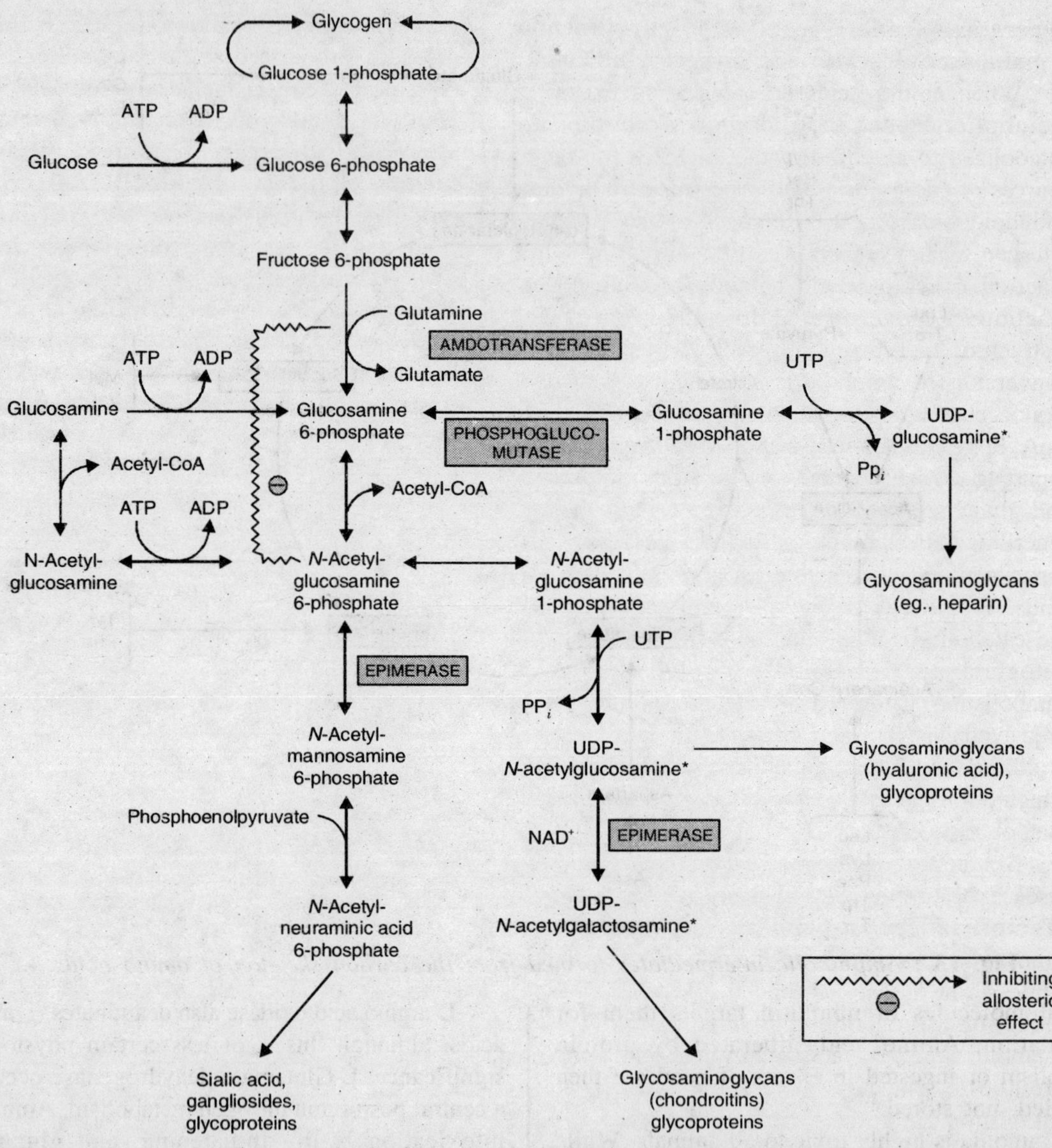

Fig. 14 : *Summary of the interrelationships in metabolism of amino sugars. (At asterisk* : Analogous to UDPGlc.) Other purine or pyrimidine nucleotides may be similarly linked to sugars or amino sugars. Examples are thymidine diphosphate (TDP)-glucosamine and TDP-N-acetylglucosamine.*

The reaction takes place in part in the mitochondrial matrix and in part in the cytosol. Synthesis of carbamoyl phosphate from ammonium ion and CO_2 takes place in liver mitochondria, as does condensation of carbamoyl phosphate with ornithine to form citrulline. The subsequent reactions are cytosolic. The final reaction, catalyzed by arginase, cleaves arginine to urea and ornithine and completes the cycle. Regulation of urea biosynthesis involves both, changes in enzyme levels and allosteric regulation of carbamoyl phosphate synthase activity by N-acetylglutamate. Inborn errors of metabolisim are associated with each reaction of the urea cycle. There include

hyperammonemia types 1 & 2, **citrullinemia, argininosuccinicaciduria** & **hyperargininemia**.

When amino acids are present in excess of metabolic needs, their carbon skeletons are catabolized to amphibolic intermediates for use as sources of energy or as substrates for carbohydrate and lipid biosynthesis. Removal of the α-amino nitrogen by transmination is most often the initial reaction of amino acid catabolism. Subsequent reactions remove any additional nitrogens and restructure the remaining hydrocarbon skeleton for conversion to amphibolic intermediates such as oxaloacetate, α-ketoglutarate, pyruvate and acetyl CoA. For example, deamination of asparagine forms aspartate, which becomes oxaloacetate. Glutamine and glutamate form α-ketoglutarate by analogous reactions. No known metabolic disorders are associated with the catabolism of these four amino acids. Two **hyperprolinemias** are associated with the catabolic pathway that converts proline to α-ketoglutarate. The initial reaction of arginine catabolism is catalyzed by arginase, an enzyme of urea synthesis. Transamination of the δ-nitrogen of ornithine then forms glutamate-γ-semialdehyde, at which point the catabolism of arginine merges with that of histidine.

Six amino acids form pyruvate. Metabolic diseases associated with glycine catabolism include **glycinuria** and primary **hyperoxaluria.** Transamination of alanine forms pyruvate directly, while serine is catabolized by glycine. Following reduction of cystine to cysteine, two sets of catabolic reactions convert cysteine to pyruvate. Metabolic diseases associated with cysteine catabolism include **cystine-lysinuria,** cystine storage diseases, and the **homocystinurias.** Threonine catabolism merges with that of glycine after threonine aldolase cleaves threonine to glycine and acetaldehyde. Oxidation of acetaldehyde forms acetate and ultimately acetyl CoA. **Hydroxy prolinemia** is associated with defective ability to degrade 4-hydroxyproline to pyruvate and glyoxylate.

Twelve amino acids form acetyl CoA. Following transamination, the carbon skeleton of tyrosine is degraded to fumarate and acetoacetate via a protracted reaction sequence. Metabolic diseases associated with enzymatic defects in tyrosine catabolism include **tyrosinosis, Richner-Hanhart syndrome, neonatal tyrosinemia**, and **alkaptonuria.** Hydroxylation of phenylalanine forms tyrosine. Metabolic disorders of phenylalanine catabolism include **phenylketon-uria** (PKU) and several **hyperpherylalaninemias.**

No nitrogen of lysine undergoes transamination. The α-amino nitrogen is removed indirectly via saccharopine. Metabolic diseases of lysine catabolism include periodic and persistant forms of **hyperlysinemia/ammonemia.** The abnormal tryptophan metabolite **xantherenate** accumulates in vitamin B_6 deficiency. Methionine, isoleucine and valine are catabolized to succinyl CoA. The catabolism of leucine, valine and isoleucine presents many analogies to fatty acid catabolism. Oxidative decarboxylation of branched α-keto acids is analogous to conversion of pyruvate to acetyl CoA, and dehydrogenation of branched acyl CoA thioesters is analogous to a reaction of fatty acid catabolism. Three reactions are sepecific to leucine catabolism, four to valine catabolism, and three to isoleucine catabolism. Metabolic disorders of branched-chain amino acid catabolism include **hypervalinemia, maple syrup urine disease,** intermittent **branched chain ketonuria, isovaleric acidemia,** and **methylmalonic aciduria.**

In addition to fulfilling specific structural and functional roles in proteins and polypeptides, amino acids participate in a wide variety of biosynthetic processes. For example, glycine participates in the biosynthesis of heme, purines, and creatine and is conjugated to bile acids and to the urinary metabolites of many drugs.

In addition to its roles in phospholipid and sphingosine biosynthesis, serine provides carbon 2 & 8 of purines and the methyl group of thymine. The methionine derivative S-adenosylmethionine, the methyl group donor for many biosynthetic processes, also participates directly in spermine and spermidine biosynthesis. Glutumate and ornithine form the neurotransmitter γ-**aminobutyrate (GABA).** Both the thioethanolamine of coenzyme A and the taurine of taurocholic and other taurine-linked bile acids arise from cysteine.

Decarboxylation of histidine forms histamine, while several dipeptides are derived from histidine and the nonprotein amino acid and cytosine catabolite β-alanine. Arginine serves as the formanidine donor for creatine biosynthesis and, via ornithine, participates in polymine biosynthesis. Arginine also serves as the nitrogen source for synthesis of nitric acid (NO). Important tryptophan metabolites include serotonin in turn serves as a precursor of melatonin. Tyrosine forms both epinephrine and nor-epinephrine and its iodination forms thyroid hormone.

Coordination of Metabolism

The basic strategy of metabolism is to form ATP, reducing power and building blocks for biosynthesis. Central themes of it are as follows :

(*i*) **ATP is the Universal currency of Energy:** The high phosphoryl transfer potential of ATP enables it to serve as the energy source in muscle contraction, active transport, signal amplification and biosynthesis. The hydrolysis of an ATP molecule changes the equilibrium ratio of products to reactants in a coupled reaction by a factor of about 10^6. Hence, a thermodynamically unfavourable reaction sequence can be made highly favourable by coupling it to the hydrolysis of a sufficient number of ATP molecules. For example, three molecules of ATP are consumed in the conversion of mevalonate into isopentenyl pyrophosphate, the activated five-carbon unit in the synthesis of cholesterol.

(*ii*) **ATP is generated by the oxidation of fuel molecules such as glucose, fatty acids and amino acids :** The common intermediate in most of these oxidations is Acetyl CoA. The carbons of the acetyl unit are completely oxidized to CO_2 by the citric acid cycle with the concomitant formation of NADH and $FADH_2$. These electron carriers then transfer their high-potential electrons to the respiratory chain. The subsequent flow of electrons to O_2 leads to the pumping of protons across the inner mitochondrial membranes. This proton gradient is then used to synthesize ATP. Glycolysis also generates ATP, but the amount formed is much smaller than in oxidative phosphorylation. The oxidation of glucose to pyruvate yields only 2ATP; whereas, 30 (or 32) ATPs are formed when glucose is completely oxidized to CO_2. However, glycolysis can proceed rapidly for a short time under anaerobic conditions; in contrast with oxidative phosphorylation which requires a continuous supply of O_2.

(*iii*) **NADPH is the major electron donor in reductive biosynthesis :** The most biosynthesis the products are more reduced than the precursors; and so reductive power is needed as well as ATP. The high-potential electrons required to drive these reactions are usually provided by NADPH. For example, in fatty acid biosynthesis, the keto group of an added two carbon unit is reduced to a methylene group by the input of four electrons from two molecules of NADPH. The activation of O_2 by mixed function of P_{450} oxygenases in hydroxylation reactions also illustrates the pervasive role of NADPH as a reductant. The pentose phosphate pathway supplies much of the required NADPH. This reductant is also formed by the tandem operation of the mitochondrial citrate-pyruvate shuttle and the malate enzyme in the cytosol.

(*iv*) **Biomolecules are constructed from a relatively small set of building blocks :** The highly diverse molecules of life are synthesized from a much smaller number of precursors. The metabolic pathways that generate ATP and NADPH also provide building blocks for the biosynthesis of more complex molecules. For example, dihydroxy acetone phosphate formed in glycolysis gives rise to the glycerol backbone of phosphatidyl choline and other phosphoglycerides. Phosphoenol-pyruvate, another glycolytic intermediate provides part of the carbon skeleton of the aromatic amino acids. Acetyl CoA, the common intermediate in the breakdown of most fuels, supplies a two carbon unit in a wide variety of biosynthesis. Succinyl CoA, formed in the citric acid cycle, is one of the precursors of porphyrins. Ribose-5 phosphate, which is formed in addition to NADPH by the pentose phosphate pathway, is the source of the sugar unit of nucleotides. Many biosynthesis also require one carbon units. Tetrahydrofolate is a source of these units at several oxidation levels. The formations of these derivatives and of S–adenosyl-methionine, the major donor of methyl groups, is closely tied to amino acid metabolism. Thus, the central metabolic pathways have ANABOLIC as well as CATABOLIC roles.

(*v*) **Biosynthetic and degradative pathways are almost always distinct :** For example, the pathway for the synthesis of fatty acids is different from that of their degradation. Likewise, glycogen is synthesized and degraded by different sets of reactions. This separations enables both biosynthetic and degradative pathways to be thermodynamically favourable at all times. A biosynthetic pathway is made exergonic by coupling it to the hydrolysis of sufficient number of ATP. For example, four more ~(P) are spent in converting pyruvate into glucose in gluconeogenesis than are gained in converting glucose into pyruvate in glucolysis. The four additional ~(P) assure that gluconeogenesis like glycolysis, is highly exergonic under all cellular conditions. The essential point is that the rate of metabolic pathways are governed more by the activities of key enzymes than by mass action. The separation of biosynthetic and degradative pathways contributes greatly to the effectiveness of metabolic control.

Multiple Choice Questions

1. All proteins contain the :
A. Same 20 amino acids
B. Different amino acids
C. 300 Amino acids occurring in nature
D. Only a few amino acids

2. Proteins contain :
A. Only L-amino acids
B. Only D-amino acids
C. DL-Amino acids
D. Both A and B

3. The optically inactive amino acid is :
A. Glycine B. Serine
C. Threonine D. Valine

4. At neutral pH, a mixture of amino acids in solution would be predominantly :
A. Dipolar ions
B. Nonpolar molecules
C. Positive and monovalent
D. Hydrophobic

5. The true statement about solutions of amino acids at physiological pH is :
A. All amino acids contain both positive and negative charges
B. All amino acids contain positively charged side chains
C. Some amino acids contain only positive charge
D. All amino acids contain negatively charged side chains

6. pH (isoelectric pH) of alanine is :
A. 6.02 B. 6.6
C. 6.8 D. 7.2

7. Since the pK values for aspartic acid are 2.0, 3.9 and 10.0, it follows that the isoelectric (pH) is :
A. 3.0 B. 3.9
C. 5.9 D. 6.0

8. Sulphur containing amino acid is :
A. Methionine B. Leucine
C. Valine D. Asparagine

9. An example of sulphur containing amino acid is :
A. 2-Amino-3-mercaptopropanoic acid
B. 2-Amino-3-methylbutanoic acid
C. 2-Amino-3-hydroxypropanoic acid
D. Amino acetic acid

10. All the following are sulphur containing amino acids found in proteins except :
A. Cysteine B. Cystine
C. Methionine D. Threonine

11. An aromatic amino acid is :
A. Lysine B. Tyrosine
C. Taurine D. Arginine

12. The functions of plasma albumin are :
A. Osmosis B. Transport
C. Immunity D. Both A and B

13. Amino acid with side chain containing basic groups is :
A. 2-Amino 5-guanidovaleric acid
B. 2-Pyrrolidine carboxylic acid
C. 2-Amino 3-mercaptopropanoic acid
D. 2-Amino propanoic acid

14. An example of L-amino acid not present in proteins but essential in mammalian metabolism is :
A. 3-Amino 3-hydroxypropanoic acid
B. 2-Amino 3-hydroxybutanoic acid
C. 2-Amino 4-mercaptobutanoic acid
D. 2-Amino 3-mercaptopropanoic acid

15. An essential amino acid in man is :
A. Aspartate B. Tyrosine
C. Methionine D. Serine

16. Non essential amino acids :
A. Are not components of tissue proteins
B. May be synthesized in the body from essential amino acids
C. Have no role in the metabolism
D. May be synthesized in the body in diseased states

17. Which one of the following is semi-essential amino acid for humans?
A. Valine B. Arginine
C. Lysine D. Tyrosine

18. An example of polar amino acid is :
A. Alanine B. Leucine
C. Arginine D. Valine

19. The amino acid with a nonpolar side chain is :
A. Serine B. Valine
C. Asparagine D. Threonine

20. A ketogenic amino acid is :
A. Valine B. Cysteine
C. Leucine D. Threonine

21. An amino acid that does not form an α-helix is :
A. Valine B. Proline
C. Tyrosine D. Tryptophan

22. An amino acid not found in proteins is :
A. L-Alanine B. Proline
C. Lysine D. Histidine

23. In mammalian tissues serine can be a biosynthetic precursor of :
A. Methionine B. Glycine
C. Tryptophan D. Phenylalanine

24. A vasodilating compound is produced by the decarboxylation of the amino acid :
A. Arginine B. Aspartic acid
C. Glutamine D. Histidine

25. Biuret reaction is specific for :
A. –CONH-linkages
B. $-CSNH_2$ group
C. $-(NH)NH_2$ group
D. All of these

26. Sakaguchi's reaction is specific for :
A. Tyrosine B. Proline
C. Arginine D. Cysteine

27. Million-Nasse's reaction is specific for the amino acid :
A. Tryptophan
B. Tyrosine
C. Phenylalanine
D. Arginine

28. Ninhydrin with evolution of CO_2 forms a blue complex with :
A. Peptide bond B. L-Amino acids
C. Serotonin D. Histamine

29. The most of the ultraviolet absorption of proteins above 240 nm is due to their content of :
A. Tryptophan B. Aspartate
C. Glutamate D. Alanine

30. Which of the following is a dipeptide?
A. Anserine B. Glutathione
C. Glucagon D. L-Lipoprotein

31. Which of the following is a tripeptide?
A. Anserine B. Oxytocin
C. Glutathione D. Kallidin

32. A peptide which acts as potent smooth muscle hypotensive agent is :
A. Glutathione B. Bradykinin
C. Tryocidine D. Gramicidin-s

33. A tripeptide functioning as an important reducing agent in the tissues is :
A. Bradykinin B. Kallidin
C. Tyrocidin D. Glutathione

34. An example of metalloprotein is :
A. Casein B. Ceruloplasmin
C. Gelatin D. Salmine

35. Carbonic anhydrase is an example of :
A. Lipoprotein B. Phosphoprotein
C. Metalloprotein D. Chromoprotein

36. An example of chromoprotein is :
A. Haemoglobin B. Sturine
C. Nuclein D. Gliadin

37. An example of scleroprotein is :
A. Zein B. Keratin
C. Glutenin D. Ovoglobulin

38. Casein, the milk protein is :
A. Nucleoprotein B. Chromoprotein
C. Phosphoprotein D. Glycoprotein

39. An example of phosphoprotein present in egg yolk is :
A. Ovoalbumin B. Ovoglobulin
C. Ovovitellin D. Avidin

40. A simple protein found in the nucleoproteins of the sperm is :
A. Prolamine B. Protamine
C. Glutelin D. Globulin

41. Histones are :
A. Identical to protamine
B. Proteins rich in lysine and arginine
C. Proteins with high molecular weight
D. Insoluble in water and very dilute acids

42. The protein present in hair is :
A. Keratin B. Elastin
C. Myosin D. Tropocollagen

43. The amino acid from which synthesis of the protein of hair keratin takes place is :
A. Alanine B. Methionine
C. Proline D. Hydroxyproline

44. In one molecule of albumin the number of amino acids is :
A. 510 B. 590
C. 610 D. 650

45. Plasma proteins which contain more than 4 per cent hexosamine are :
A. Microglobulins B. Glycoproteins
C. Mucoproteins D. Orosomucoids

46. After releasing O_2 at the tissues, hemoglobin transports :
A. CO_2 and protons to the lungs
B. O_2 to the lungs
C. CO_2 and protons to the tissue
D. Nutrients

47. Ehlers-Danlos syndrome characterized by hypermobile joints and skin abnormalities is due to :
A. Abnormality in gene for procollagen
B. Deficiency of lysyl oxidase
C. Deficiency of prolyl hydroxylase
D. Deficiency of lysyl hydroxylase

48. Proteins are soluble in :
A. Anhydrous acetone
B. Aqueous alcohol
C. Anhydrous alcohol
D. Benzene

49. A cereal protein soluble in 70 per cent alcohol but insoluble in water or salt solution is :

A. Glutelin B. Protamine

C. Albumin D. Gliadin

50. Many globular proteins are stable in solution inspite they lack in :

A. Disulphide bonds B. Hydrogen bonds

C. Salt bonds D. Non polar bonds

ANSWERS

1	2	3	4	5	6	7	8	9	10
A	A	A	A	A	A	A	A	A	D
11	**12**	**13**	**14**	**15**	**16**	**17**	**18**	**19**	**20**
B	A	A	C	C	B	B	C	B	C
21	**22**	**23**	**24**	**25**	**26**	**27**	**28**	**29**	**30**
B	A	B	D	A	C	B	B	A	A
31	**32**	**33**	**34**	**35**	**36**	**37**	**38**	**39**	**40**
C	B	D	B	C	A	B	C	C	B
41	**42**	**43**	**44**	**45**	**46**	**47**	**48**	**49**	**50**
B	A	B	C	C	A	A	B	D	A

●●●

CHAPTER

8

Lipids and Its Metabolism

Introduction

The term 'lipid' was first used by Bloor in 1943 (German-Biochemist). The compounds in this group are characterized by their relative insolubility in water (hydrophobic in nature) and solubility in organic solvents like chloroform, ether and benzene. The cause of this general property of lipids and related compounds is the predominance of long hydrocarbon chains or benzene rings. There structures are non-polar and hydrophobic. In many lipids, there chains may be attached at one end to a polar group, rendering it capable of binding water by hydrogen bonds.

Classification: On the basis of chemical composition, Bloor in 1943 has proposed the following classification of lipids.

A. Simple lipids or Homolipids
B. Compound lipids or Heterolipids
C. Derived lipids.

In 1976, Conn and Stumpt classified lipids traditionally into six classes.

(*i*) Acyl glycerols
(*ii*) Waxes
(*iii*) Phospholipids
(*iv*) Sphingolipids
(*v*) Glycolipids
(*vi*) Terpenoid lipid (including carotenoids and steroids)

(A) Simple lipids or Homolipids

Simple lipids are alcohol esters of fatty acids.

(*a*) *Neutral Fats :* (glycerides). Often called triglycerides; they are triesters of fatty acids and glycerols. Neutral fats accumulate in adipose tissue.

(*b*) *Waxes :* Having a higher melting point than natural fats, they are esters of fatty acids with alcohols other than glycerol, such as in bees wax.

(B) Compound Lipids or Heterolipids

(*a*) ***Phospholipids and Biological Membranes:*** Compound lipids are more complex and upon hydrolysis yield other substances in addition to fatty acids and alcohol. Phospholipids have only two fatty acids attached to the glycerol molecule. The third hydroxyl group of glycerol is esterified to phosphoric acid instead of to a fatty acid. This phosphate is also bound to a second alcohol molecule, which can be choline, ethanolamine, inositol, or serine, depending on the type of phospholipid. From this formula it is possible to derive the main phospholipids such as 'phosphatidyl choline (*lecithin*), phosphatidyl ethanolamine (*cephalin*), phosphatidyl serine and phosphatidyl inositol'.

Phospholipids have two long hydrophobic fatty acid 'tails' and a hydrophilic (polar) phosphate-containing "head". Phospholipids are thus 'amphipathic' molecules (*i.e.,* they contain a hydrophilic and a hydrophobic region), and their configuration accounts for many of the properties of biological membranes. Such membranes are bilayers of phospholipids with the hydrophilic heads (phosphate-containing regions) positioned at the water interface and the long hydrophobic tails arranged in the interior, when phospholipids are mixed with water, the bilayer arrangement-polar head outside; nonpolar tails inside is adopted spontaneously.

(*b*) *Glycolipids* and *sphingolipids* are characterized by the fact that glycerol is replaced by the amino alcohol *sphingosine*. To their groups belong the sphingomyelins, mainly in the myelin sheath of nerves, the *cerebrosides*, which are characterized by the presence of galactose or glucose in the molecule, the sulphatides, which contain sulphuric acid esterified to galactose, and *gangliosides*.

(*c*) *Steroids* are lipids that are derived from a complex ring structure. One of the best known is cholesterol, which is present in cell membranes. Steroids have different functions depending on the groups attached to the basic structure. Several hormones such as estrogen, progesterone, corticosterone and vitamin A are steroids.

(*d*) Other complex lipids are sulpholipids and aminolipids. Lipoproteins may also be placed in this category.

(C) Derived Lipids

The substances derived from simple and compound lipids by hydrolysis. These include fatty aldehydes, ketone bodies, hydrocarbons, lipid soluble vitamins, hormones, steroids, terpenes and carotenoids. Due to uncharged nature glycerides and cholesterol esters, are also called **neutral lipids.**

(D) Biosynthesis of Fatty Acids

(*a*) The synthesis of long chain fatty acids (lipogenesis) is carried out by two enzyme systems present in the cytosol of the cell: Acetyl CoA carboxylase and fatty acid synthase.

(*b*) The pathway converts acetyl CoA to palmitate and requires NADPH, ATP, Mn^{+2}, biotin, pantothenic acid and HCO_3^- as cofactors.

(*c*) Acetyl CoA carboxylase is required to convert acetyl CoA to malonyl CoA. In turn, fatty acid synthase, a multienzyme complex of one polypeptide chain with seven separate enzymatic activities, catalyzes the assembly of palmitic from one acetyl CoA and seven malonyl CoA molecules.

(*d*) Lipogenesis is regulated at the acetyl CoA carboxylase step by allosteric modifiers covalent modification, induction and repression of enzyme synthesis. Citrate activates the enzyme, and long chain acetyl CoA inhibits its activity. Insulin activates acetyl CoA carboxylase in the short term by dephosphorylation and in long term by induction of synthesis. Glucagon and epinephrine have opposite actions to insulin.

(*e*) Lengthening of long chain fatty acids takes place in the endoplasmic reticulum catalyzed by a microsomal elongase enzyme system.

(E) Oxidation of Fatty Acids (KETOGENESIS)

(*a*) Fatty acid oxidation in mitochondria leads to the generation of large quantities of ATP by a process called β-oxidation that cleaves acetyl CoA units sequentially from fatty acyl chains. The acetyl CoA is oxidized in the citric acid cycle, generating further ATP.

(*b*) Oxidation of odd-numbered carbon fatty acids yield acetyl CoA plus one molecule of propionyl CoA, which is glucogenic.

(*c*) Peroxisomes are capable of oxidizing very long chain fatty acids but only as far as octanoyl CoA, which must then be transferred to mitochondria for further oxidation.

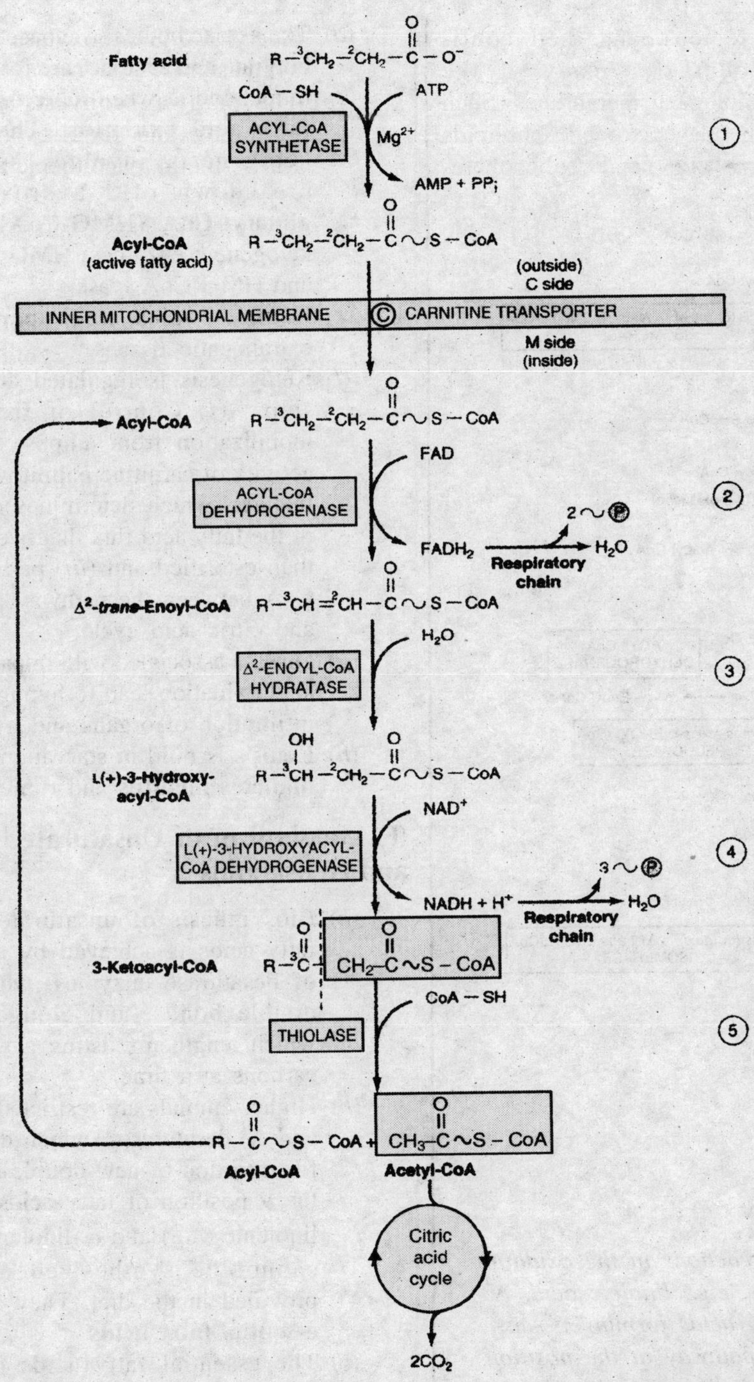

Fig. 1 : *β-Oxidation of fatty acids. Long-chain acyl-CoA is cycled through reactions 2-5, acetyl-CoA being split off, each cycle, by thiolase (reaction 5). When the acyl radical is only four carbon atoms in length, two acetyl-CoA molecules are formed in reaction 5.*

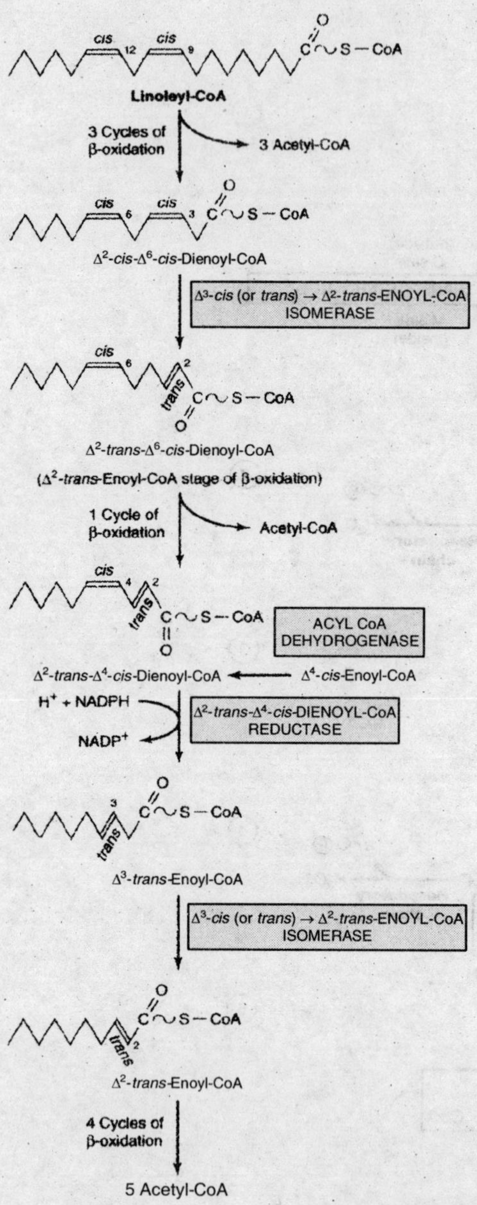

Fig. 2 : *Sequence of reactions in the oxidation unsaturated fatty acids. e.g., linoleic acid.* Δ^4-*cis-fatty acid or fatty acids forming* Δ^4-*cis-enoyl-CoA enter the pathway at the position shown. NADPH for the dienoyl-CoA reductase step is supplied by intramitochondrial sources so as glutamate dehydrogenase, isocitrate dehydrogenase and NAD(P)H transhydrogenase.*

(*d*) The ketone bodies (acetoacetate, 3-hydroxy butyrate and acetone) are formed in hepatic mitochondria when there is a high rate of fatty acid oxidation. The pathway of ketogenesis involves synthesis and breakdown of 3-hydroxy-3 methyl-glutaryl-CoA (HMG-CoA) by two key ketogeneic enzymes, HMG CoA synthase and HMG-CoA lyase.

(*e*) Ketone bodies are important fuels in extrahepatic tissues.

(*f*) Ketogenesis is regulated at three crucial steps: (*i*) Control of free fatty acid mobilization from adipose tissue (*ii*) the activity of carnitine palmitoyl transferase-I in liver, which determines the proportion of the fatty acid flux that is oxidized rather than esterified and (*iii*) partition of acetyl CoA between the pathway of ketogenesis and citric acid cycle.

(*g*) Disease associated with impairment of fatty acid oxidation lead to hypoglycemia, fatty infiltration of organs and hypoketonemia.

(*h*) Ketosis is mild in starvation but severe in **diabetes mellitus** and ruminant ketosis.

(F) Metabolism of Unsaturated Fatty Acids and Elcosanoids

(*a*) Biosynthesis of unsaturated long-chain fatty acids is achieved by a combination of desaturase enzymes, which introduce double bonds, and elongase enzymes, which lengthen existing acyl chains by 2 carbons at a time.

(*b*) Higher animals are restricted to Δ^4, Δ^5, Δ^6 and Δ^9 desaturases, which does not allow for insertion of new double bonds beyond the 9 position of fatty acids. As a result, linolenic (ω_6) and α-linolenic (ω_3) acids cannot be synthesized and must be provided in the diet. They are called the **essential fatty acids**.

(*c*) The essential fatty acids give rise C_{20} (eicosanoic) fatty acids from which are synthesized every important groups of physiologically and pharmacologically active compounds, the **eicosanoids.** They

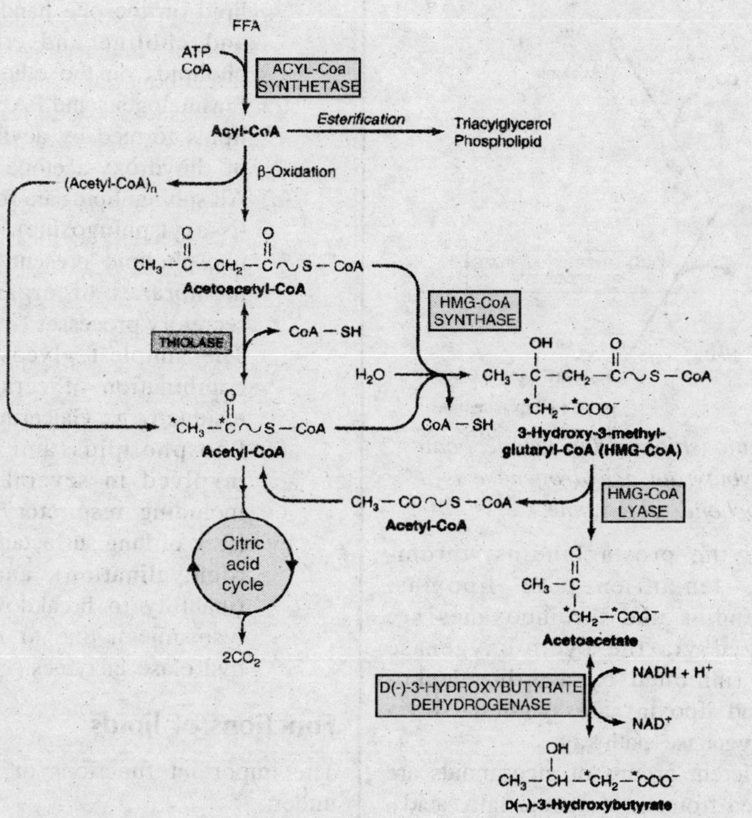

Fig. 3 : *Pathways of ketogenesis in the liver. (FFA, free fatty acids; HMG, 3-hydroxy-3-methylglutaryl.)*

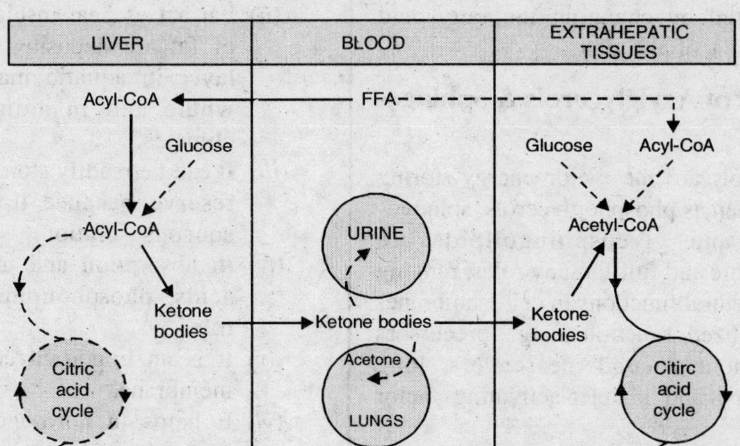

Fig. 4 : *Formation, utilization, and excretion of ketone bodies. (The main pathway is indicated by the solid arrows.)*

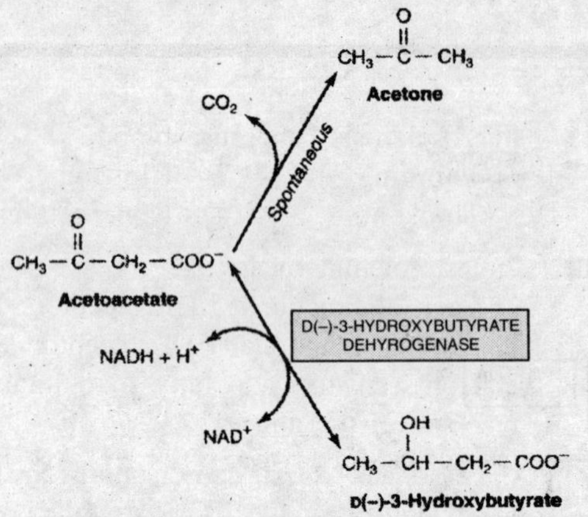

Fig. 5 : *Interrelationships of the ketone bodies. D(–)-3-Hydroxybutyrate dehydrogenase is a mitochondrial enzyme.*

comprise the prostaglandins, thromboxanes, leukotrienes & lipoxins. Prostaglandins and thromboxanes are synthesized via the cyclo-oxygenase pathway (inhibited by aspirin), leukotrienes and lipoxins are synthesized via the lipoxygenase pathway.

(*d*) Since different groups of eicosanoids are synthesized from the essential fatty acids, the balance between the physiologic effects of the various eicosanoids may be manipulated by changing the fatty acid composition of the diet.

(G) Metabolism of Acylglycerols & Sphingolipids

(*a*) Triglycerols are the major energy-storing lipids, whereas phosphoglycerols, sphingomyelin, and glycosphingolipids are amphipathic and fulfill many roles, ranging from structural functions in cell membranes to specialized functions, *e.g.*, precursors for hormone second messengers, long surfactants, and platelet-activating factor (PAF).

(*b*) Triacylglycerols and some phosphoglycerols are synthesized by progressive acylation of glycerol 3-phosphate. The pathway bifurcates at phosphatidate, forming inositol phospholipids and cardiolipid on the one hand and triacylglycerol and choline and ethanolamine phospholipids on the other.

(*c*) Plasmalogens and PAF are either phospholipids formed by acylation and alkylation of dihydroxy-acetone phosphate.

(*d*) All sphingolipids are formed from ceramide (N-acylsphingosine). Spingomyelin is a phospholipid present characteristically in membranes of organelles involved in secretory processes (*e.g.*, Golgi apparatus). The simplest glycosphingolipids are a combination of ceramide plus a sugar residue (*e.g.*, Galcerin myelin).

(*e*) Phospholipids and sphingolipids are involved in several disease processes including respiratory distress syndrome (lack of lung surfactant), multiple sclerosis (demyelination), and sphingolipidoses (inability to breakdown sphingolipids in lysosomes), due to inherited defects in hydrolase enzymes.

Functions of lipids

The important functions of lipids are given as under:

(*i*) One gram of lipid produces 9.3 kilocalories of heat.

(*ii*) Fat act as heat insulator. Much quantities of fat are deposited in the subcutaneous layer in aquatic mammals for example whale and in animals living in cold climates.

(*iii*) It can be readily stored in the body as food reserve because lipid are insoluble in aqueous solution.

(*iv*) In absorption and transportation of fatty acids, phospholipids play an important role.

(*v*) It is an important constituent of the cell membrane.

(*vi*) It helps in hormone synthesis. The sex hormones adrenocorticoids, cholic acids and also vitamin D are all synthesized from a steroidal lipid.

(*vii*) It acts as carriers of natural fat soluble vitamins (vitamins A, D and E).

Multiple Choice Questions

1. Synthesis of fatty acid takes place when :
A. Fatty acid are plentiful
B. Carbohydrate is plentiful
C. Carbohydrate and energy are plentiful
D. None of these

2. Triacylglycerols are :
A. Soluble in water
B. Insoluble in water
C. Soluble in water at elevated temperature
D. Partially soluble in water

3. Cholestrol is the precursor of :
A. Steroid hormones
B. Vitamin A
C. Bile salts
D. Both A and C

4. How many classes of steroid hormones are there?
A. 3 B. 2
C. 4 D. 5

5. Which of the following is called milk ejection hormone?
A. Prolactin B. Vasopressin
C. Oxytocin D. All of these

6. How many types of lipoproteins are there?
A. 2 B. 6
C. 8 D. 5

7. HDLs are synthesized in :
A. Blood B. Liver
C. Intestine D. Pancreas

8. VLDLs are synthesized in :
A. Blood B. Liver
C. Intestine D. Pancreas

9. Chylomicrons are synthesized in :
A. Blood B. Liver
C. Intestine D. Pancreas

10. What is the major protein constituent of high-density lipoprotein (HDL)?
A. Apolipoprotein A-1
B. Apolipoprotein C-1
C. Apolipoprotein E
D. None of these

11. Atherosclerosis can cause blood :
A. Thinning B. Clotting
C. Thickening D. None of these

12. Cholesterolemia means :
A. Lack of functional LDL receptors
B. Lack of functional HDL receptor
C. High sensitivity to fatty food intake
D. None of the above

13. Arachidonate has 20 carbon atoms with :
A. 3 double bonds B. 2 double bonds
C. 4 double bonds D. 8 double bonds

14. How many ATPs are formed during complete oxidation of Palmitate?
A. 35 B. 96
C. 129 D. 131

15. Fatty acid synthesis takes place in :
A. Mitochondria
B. Cell membrane
C. Cytosol
D. Endoplasmic reticulum

16. Citric acid cycle occurs in :
A. Cytoplasm
B. Mitochondria
C. Endoplasmic reticulum
D. Golgi bodies

17. How many ATPs are produced during citric acid cycle?
A. 10 B. 13
C. 12 D. 8

18. In what form does the product of glycolysis enter the TCA cycle?
A. Acetyl CoA B. Pyruvate
C. NADH D. Glucose

19. Why the TCA cycle is the central pathway of metabolism of the cell?
A. It occurs in the center of the cell
B. Its intermediates are commonly used by other metabolic reactions
C. All other metabolic pathways depend upon it
D. None of the above

20. The first intermediate in TCA cycle is :
A. Succinate B. Fumerate
C. Citrate D. Malate

21. In eukaryotes, electron transport occurs in :
A. Membranes of mitochondria
B. Endoplasmic reticulum
C. Cytoplasm
D. All of the above

22. Important function of cholesterol is to :
A. Modulate fluidity
B. Enhance blood circulation
C. Prevent bile salts formation
D. None of these

23. Cholesterol can be synthesized *de novo* in :
A. Pancreas
B. Intestine
C. Liver
D. Cell membrane

24. Oxidation of a molecule involves :
A. Gain of electron
B. Loss of electron
C. Gain of proton
D. Loss of proton

25. A positive redox potential means substance has :
A. Lower affinity for electron
B. Higher affinity for electron
C. Lower affinity for proton
D. Higher affinity for proton

26. Which one of the following is not the intermediate of Kreb's cycle?
A. Isocitrate B. Succinate
C. Fumarate D. Stearate

27. Which of the following enzyme does not take part in the TCA cycle?
A. Citrate synthase
B. Iso-citrate dehydrogenase
C. Pyruvate dehydrogenase
D. Malate dehydrogenase

28. Standard redox potential for a substance is measured under standard condition and is expressed as :
A. Mili-Ampere B. Volt
C. Without unit D. Ohm

29. Which one is not the main protein in electron transport chain?
A. NADH dehydrogenase
B. Cytochrome bcl complex
C. Cytochrome oxidase
D. Citrate synthease

30. Given the following redox couples and standard oxidation-reduction potentials : E'0 (V)
2 cytochrome c ox +2 e- ¬ >2 cytochrome c red +0.254V
2 cytochrome a.3ox + 2 e- ¬ >2 cytochrome a 3 red +0.385V
which way would electrons move between these couples under standard conditions?
A. From cytochrome c red to cytochrome a 3 ox
B. From cytochrome a 3 red to cytochrome c ox
C. From cytochrome c red to cytochrome c ox
D. From cytochrome a 3 red to cytochrome a 3 ox

31. How many molecules of ATPs are synthesized per NADH oxidation?
A. 2 B. 1
C. 3 D. 4

32. To stop ATP synthesis which chemical is generally used?
A. DNSA
B. 2,4 dinitrophenol
C. DDT
D. None of the chemical can stop ATP synthesis

33. Malate-asparatate shuttle operates in :
A. Lungs and liver
B. Heart and liver
C. Pancreas and liver
D. None of these

34. The enzymes of the TCA cycle in a eukaryotic cell are located in the :
A. Nucleus
B. Mitochondria
C. Plasma membrane
D. Lysosomal bodies

35. Which of the following is involved in energy production?
A. Generation of proton gradients across membranes
B. Transport of electrons on organic molecules
C. Conversion of compounds with high energy to those of low energy
D. All of the above

36. Most multi-cellular organisms obtain energy for the synthesis of ATP during oxidative phosphorylation from :
A. High energy phosphate compounds
B. A proton gradient across the inner mitochondrial membrane
C. A proton gradient across the cell membrane
D. A proton gradient across the outer mitochondrial membrane

37. The catabolism of sugars and fatty acids is similar because :
A. Both of these compounds are funnelled through the TCA/citric acid cycle
B. Both of these compounds generate redox energy during catabolism
C. Both of these compounds generate chemical energy during catabolism
D. All of the above

38. Most of the enzymes of the citric acid cycle in a eukaryotic cell are located in the :
A. Inner mitochondrial membrane
B. Cytosol
C. Mitochondrial matrix
D. Intermembrane space

39. The end product of glycolysis is pyruvate, which enters the citric acid cycle after being converted to :
A. Acetic acid B. Acetyl-CoA
C. Acetaldehyde D. None of these

40. During cellular respiration, most of the ATP made, is generated by :
A. Oxidative phosphorylation
B. Photophosphorylation
C. Substrate-level phosphorylation
D. Glycolysis

41. The $FADH_2$ and NADH produced by the oxidation of one acetyl-CoA results in the synthesis of about :
A. 3 ATPs B. 6 ATPs
C. 11 ATPs D. 15 ATPs

42. Energy that is released from glucose during respiration but not transferred to ATP bonds can be detected as :
A. CO_2 B. AMP
C. ADP D. Heat

43. Glycosphingolipids are a combination of :
A. Ceramide with one or more sugar residues
B. Glycerol with galactose
C. Sphingosine with galactose
D. Sphingosine with phosphoric acid

44. An important finding in Histidinemia is :
A. Impairment of conversion of α-glutamate to α-ketoglutarate
B. Speech defect
C. Decreased urinary histidine level
D. Patients can not be treated by diet

45. An important feature of maple syrup urine disease is :
A. Patient can not be treated by dietary regulation
B. Without treatment death of patient may occur by the end of second year of life
C. Blood levels of leucine, isoleucine and serine are increased
D. Excessive brain damage

46. An inborn error, maple syrup urine disease is due to deficiency of the enzyme :
A. Isovaleryl-CoA hydrogenase
B. Phenylalnine hydroxylase
C. Adenosyl transferase
D. α-Ketoacid decarboxylase

47. Maple syrup urine disease becomes evident in extra uterine life by the end of :
A. First week B. Second week
C. Third week D. Fourth week

48. Alkaptonuria occurs due to deficiency of the enzyme:
A. Maleylacetoacetate isomerase

B. Homogentisate oxidase
C. p-Hydroxyphenylpyruvate hydroxylase
D. Fumarylacetoacetate hydrolase

49. An important finding in glycinuria is :
A. Excess excretion of oxalate in the urine
B. Deficiency of enzyme glycinase
C. Significantly increased serum glycine level
D. Defect in renal tubular reabsorption of glycine

50. Increased urinary indole acetic acid is diagnostic of :
A. Maple syrup urine disease
B. Hartnup disease
C. Homocystinuria
D. Phenylketonuria

ANSWERS

1	2	3	4	5	6	7	8	9	10
C	B	D	D	C	D	A	B	C	A
11	**12**	**13**	**14**	**15**	**16**	**17**	**18**	**19**	**20**
B	A	C	C	C	B	C	A	B	C
21	**22**	**23**	**24**	**25**	**26**	**27**	**28**	**29**	**30**
A	B	C	B	B	D	C	B	D	A
31	**32**	**33**	**34**	**35**	**36**	**37**	**38**	**39**	**40**
C	B	B	B	B	B	A	C	B	A
41	**42**	**43**	**44**	**45**	**46**	**47**	**48**	**49**	**50**
C	D	A	B	D	D	A	B	D	B

●●●

CHAPTER

9

Vitamins

Introduction

Vitamins are required for normal growth and maintenance of all animal life. Vitamins are important for their regulatory and protective functions. These are required in very small amounts. But it is necessary to provide vitamins in the diet because many of them cannot be manufactured by the body. The lack of vitamins result in definite deficiency disorders which are specific for each particular vitamin.

Vitamins are conveniently classified into the groups on the basis of their solubility *i.e.,* fat-soluble and water soluble vitamins. Fat-soluble vitamins include A, D, E and K and water-soluble vitamins include the B-group and vitamin C. Foods differ greatly in the amount and kinds of the vitamin they supply.

Functions

1. Vitamin A is necessary for normal growth and development of body.
2. Vitamin A is needed for proper tooth structure.
3. It also helps in the integrity of the epithelium, especially the mucous membranes, which line the eyes, mouth, alimentary canal, salivary glands, respiratory and genito-urinary tracts.
4. Vitamin A maintains normal reproductive functions in males.
5. Vitamin A is essential for the maintenance of normal vision in dim light.
6. Vitamin A is also known as anti-infective vitamin as it provides resistance power to the body by maintaining normal structure of the organs.
7. Vitamin D is responsible for the proper absorption of calcium and phosphorus from the digestive tract and their deposition in bones. These vitamins are also responsible for the formation of strong and rigid bones.
8. One important function of vitamin E is its antioxidative properties. This property helps in reducing the oxidation of vitamin A and C and sulphur containing amino acids.
9. Vitamin K is essential for the formation of prothrombin by liver. Prothrombin is a normal constituent of the blood and helps clotting of blood on contact with air. Deficiency of vitamin K prolongs the clotting time.
10. Thiamine is necessary for catalyzing the oxidation of carbohydrates in the body. This reaction releases energy in the system. Thiamine helps in the normal functioning of the nervous system and the heart. It is essential for the proper growth.
11. Riboflavin is essential for the health of skin mucous membrane structures of alimentary canal and eyes.
12. Niacin functions in the body as a component of two important co-enzymes involved in respiration and the breakdown of glucose to produce energy.

13. Nicotinic acid is essential for the normal functioning of the skin, intestinal and nervous systems.
14. Conversion of tryptophan to niacin requires pyridoxine (B_5). Therefore if pyridoxine is deficient in diet this conversion step is not materialized and niacin is not formed.
15. Ascorbic acid (Vit. C) performs a number of important functions in the body. It is a part of the cementing material which holds the body cells in place. It helps the body to build resistance to infection. It also helps in absorption of calcium and iron. It is also essential for rapid healing of wounds as it helps in the formation of connective tissue. Vit. C is essential for cholesterol metabolism.

RDA

A daily requirement of an adult for vitamin A is of 750 mcg of retinol or 3000 mcg of β-carotene per day derived from foods of either animal or vegetable origin. The allowance for infants is 300 mcg to 400 mcg (about 1200 I.U.), the need increases gradually as the child grows to adolescence. No increased allowance during pregnancy is recommended but the allowance is increased to 1380 mcg or 4600 mcg of β-carotene during lactation. This is in conformity with FAO/WHO recommendations.

For Vitamin D: About 200 International units are stated to be the daily it of child. The amount of Thiamine required is dependent on the calorie intake because Thiamine is involved in carbohydrate utilization. The recommended Thiamine allowance is 0.5 mg per 1000 calories.

For Riboflavin: 0.60 mg of riboflavin per 1000 calories is recommended and the daily allowances are calculated from the calorie allowance for each stage of life. Thus the recommended allowance varies from 0.7 mg for an infant to 1.7 mg for an adolescent. The need for riboflavin increases during pregnancy and lactation and also with increased activity and calorie intake.

Niacin is required 6.6 mg per 1000 calorie. The total niacin required daily on the basis of calorie requirement can range from 8 mg to 26 mg depending on the age and occupation of the individual.

One of the amino acids, tryptophan has been shown to be a precursor of niacin in the body. It has been estimated that about 60 mg of tryptophan can give rise to 1 mg of niacin.

The recommended daily allowance is 20 to 40 mg for children and 40 mg for an adult. An intake of 80 mg per day is recommended for a nursing mother.

Sources of Vitamin

Vitamin A is present in animal foods only. Liver is the richest source of Vit. A. Other sources include butter, ghee, milk, curd and egg yolk. Refined oils and vanaspati are good sources of vitamin A if these are fortified with vitamin A to the extent of 750 mcg per 100 g. Vitamin A is not present in vegetable foods but these foods contain the pigment, β-carotene, which is precursor of vitamin A and is therefore also known as 'Pro-vitamin A'. Leafy vegetables such as spinach, amaranth, coriander, drumstick leaves as well as ripe fruits such as mango, papaya and yellow pumpkin are good sources of β-carotene.

Foods are not good sources of vitamin D. It is found in small quantities in liver, egg yolk, milk and milk fat (butter and ghee), obtained from animals fed on pastures exposed to sunlight.

Almost all food contains some amount of Thiamine, except sugars, fats and oils. Plant sources include pulses, nuts, oilseeds and whole grain cereals. Parboiled rice and fresh peas are good sources of Thiamine. Green leafy vegetables and animal foods such as milk, eggs, fish and meat are fair sources.

The main sources of Thiamine in Indian diet are cereals but refining of cereals reduces Thiamine content with most of the Thiamine being lost in the polishing. Parboiling of rice helps to conserve Thiamine. If cooking water is not discarded, a minimum mineral quantity of Thiamine is lost.

Milk is the richest source of riboflavin. There is no riboflavin in butter and ghee because the vitamin is water-soluble and remains in the water extract during the removal of butter from milk or curds. Liver and kidney of animals and birds are also good sources of riboflavin.

Meat is a rich source of niacin, the higher concentrations found in organ meat such as liver groundnuts are the best source of niacin. Cereals are the major source of niacin in the Indian diet, unrefined and parboiled cereals retain more niacin than refined ones.

Amla is one of the richest source of ascorbic acid. Citrus fruits such as oranges, sweet lime, grape fruit are also excellent sources of ascorbic acid. Guava, cashew apple and drumsticks have a high ascorbic acid content. Sprouted pulses such as Bengal gram and green gram are good sources and have proved to be of value during drought and famine conditions when other sources of ascorbic acids, such as fresh fruits and vegetables are not available.

Multiple Choice Questions

1. Vitamins are essential because the organism :
A. Can't synthesize these compounds at all
B. Can synthesize these compounds partially
C. Can't synthesize these compounds in the adequate amounts
D. None of the above

2. The vitamin riboflavin, which occurs as a yellow pigment in egg yolk and milk become :
A. Colourless on reduction with Zn in acid and regained its colour on re-oxidation
B. Colourless on oxidation and regained its colour on reduction with Zn in acid
C. More deep in colour on reduction with Zn in acid and regained its colour on re-oxidation
D. More deep in colour on oxidation and regained its colour on reduction with Zn in acid

3. A deficiency of niacin causes :
A. Pellagra B. Scurvy
C. Cataract D. Anemia

4. Which of these is a symptom of vitamin A deficiency?
A. Osteoporosis
B. Impaired taste perception
C. Blindness
D. Impaired blood clotting

5. Which of these is a vitamin A precursor?
A. Cobalamin B. Pyridoxine
C. β-Carotene D. Thiamine

6. The vitamin riboflavin part of the ________ molecule.
A. Ferredoxin
B. FAD
C. Pyridoxal phosphate
D. Pyrophosphate

7. Vitamin niacin is part of the ____ molecule.
A. Ferredoxin
B. Pyridoxal phosphate
C. Pyrophosphate
D. NAD

8. Main function of insulin hormone is to :
A. Increase glycogen in liver
B. Decrease glycogen in liver
C. Increase blood sugar
D. Decrease blood sugar

9. An example of a digestive hormone is :
A. Lipase B. Pepsin
C. Amylase D. Gastrin

10. Which of these hormones is a catecholamine?
A. Follitropin
B. Norepinephrine
C. Tetraiodothyronine
D. Tetrahydrofolate

11. Which of these molecules is vitamin H?
A. Biotin B. Carnitine
C. Folic acid D. None of these

12. The occurrence of metals such as iron or molybdenum in some flavoproteins can :
A. Stabilize the semiquinone

B. De-stabilize the semiquinone
C. Form chelation
D. All of these

13. In one iron-metalloflavoprotein, the iron is present as a :
A. Heme-protein B. Non heme type
C. Both A and B D. Flavin moiety

14. Dihydroorotate dehydrogenase contains :
A. 4 flavins and 4 atoms of iron per molecule
B. 2 flavins and 4 atoms of iron per molecule
C. 2 flavins and 2 atoms of iron per molecule
D. 4 flavins and 2 atoms of iron per molecule

15. Lipoic acid exists in :
A. Oxidized form
B. Reduced form
C. Oxidized and reduced form both
D. None of these

16. Lipoic acid is a co-factor of the :
A. A-ketoglutaric dehydrogenase
B. Pyruvic dehydrogenase
C. Di-hydroorotate dehydrogenase
D. Both A and B

17. Biotin occurs mainly in combined forms bound to protein through :
A. E-N-lysine moiety
B. E-S-lysine moiety
C. S-N-biotinyl-L-lysine
D. C-N-lipoyl-L-lysine

18. A deficiency of thiamin produces the disease known as :
A. Beri-beri B. Scurvy
C. Cataract D. Anemia

19. Which of the following compounds/(s) belong/(s) to the vitamin B_6 group?
A. Pyridoxal B. Pyridoxine
C. Pyridoxamine D. All of these

20. An enzyme, L-folate reductase reduces folk acid to :
A. Hydrofolic acid
B. Dihydrofolic acid
C. Trihydrofolic acid
D. Tetrahydrofolic acid

21. Which of the following are reduced coenzymes?
A. NAD II and $FADH_2$
B. NAD^+ and FAD
C. ATP and GTP
D. Coenzyme A and ubiquinone

22. In the co-enzyme B_{12} the position occupied by a cyanide ion in vitamin B_{12} is bonded directly to the __________ of the ribose of adenosine.
A. Adenine
B. 5-6 dimethylbenzimidazole
C. Hydroxycobalamin
D. Cyanocobalamin

23. Vitamin B_{12} is useful in the prevention and treatment of :
A. Pernicious anemia
B. Scurvy
C. Cataract
D. Beri-beri

24. The reductant, NADH, transfers the electrons via a flavoproteins to the specific disulphide (S-S) protein to form a dithiol (SH,SH) protein which converts vitamin :
A. B_{12} (Co^{2+}) to B_{12} (Co)
B. B_{12} (Co) to B_{12} (Co^{2+})
C. B_{12} (Co^{2+}) to B_{12} (Co^+)
D. B_{12} (Co^+) to B_{12} (Co^{2+})

25. Acyl carrier protein (ACP) plays an important role in the biosynthesis of :
A. Fatty acids B. Amino acids
C. Sugars D. Carbohydrates

26. *E. coli* ACP has its molecular weight as around :
A. 9000 B. 19000
C. 39000 D. 90000

27. Vitamin-C is considered as a :
A. Water soluble
B. Fat soluble
C. Fat and water soluble
D. None of these

28. Who discovered vitamin C (ascorbic acid)?
A. Paul Berg
B. Linus Pauling
C. Albert Szent-Gyorgyi
D. Kerry Mullis

29. The absence of ascorbic acid in the human diet gives rise to :
A. Rickets B. Pernicious anemia
C. Cataract D. Beri-beri

30. Ascorbic acid acts as an :
A. Reducing agent
B. Oxidizing agent
C. Oxidizing and reducing agent both
D. None of the above

31. β-carotene together with α-carotene, γ-carotene and cryptoxanthine are synthesized by :
A. Plants
B. Animal
C. Plants and animals both
D. None of these

32. An early sign of retinol deficiencies in man is :
A. Night blindness B. Keratinization
C. Xeropthalmia D. None of these

33. The symptoms of retinol excess are :
A. Bone fragility B. Nausea
C. Weakness D. All of these

34. Vitamin-D deficiency can cause :
A. Rickets
B. Pernicious anemia
C. Cataract
D. Beri-beri

35. The most prominent role that tocopherol has in *in-vitro* systems is as a strong :
A. Antioxidants B. Reducing agent
C. Oxidizing agent D. All of these

36. Selenium is an essential component of the enzyme glutathione peroxidase which :
A. Scavenges toxic hydoperoxycompounds in tissues
B. Reduces toxic hydoperoxycompounds in tissues
C. Oxidizes toxic hydoperoxycompounds in tissues
D. None of the above

37. Vitamin K_1 was first isolated from alfalalfa and has the phytyl side chain consisting of :
A. Four isoprene units
B. Six isoprene units
C. Nine isoprene units
D. Four isoprene units

38. A deficiency of Vitamin K results in a decreased level of :
A. Prothrombin B. Thrombin
C. Fibrin D. Fibrinogen

39. Models of end-linked Osaka VI Fibrinogen dimers, a bilayer dimer is linked at :
A. Both ends by one disulphide bond
B. Either end via two disulphide bonds
C. Either end via a single disulphide bond
D. Both ends by two disulphide bonds

40. The disease Beri-beri is due to a dietary deficiency in :
A. Vitamin B_1 (thiamine)
B. Vitamin B_2 (riboflavin)
C. Vitamin B_6 (pyridoxine)
D. Vitamin B_{12}

41. What metal ion is specifically bound by vitamin B_{12}?
A. Cobalt B. Copper
C. Zinc D. Iron

42. What compound of raw egg white causes a syndrome similar to vitamin B deficiency?
A. Avidin B. Betabindin
C. Ovalbumin D. Albumin

43. Vitamin B_{12} (Cobalamin) is only synthesized by :
A. Fishes B. Micro-organisms
C. Plants D. Animals

44. The prosthetic group biotin is a carrier of which type of molecule?
A. Activated carbon dioxide (CO_2)
B. Ammonia
C. Methyl group
D. Sulfhydryl group

45. A fat-soluble vitamin that regulates blood clotting is :

A. Vitamin A B. Vitamin K
C. Vitamin C D. Niacin

46. What transports copper from the intestinal cells to the liver?

A. Ceruloplasmin B. Secretin
C. Acrolein D. Albumin

47. What condition is caused by iodine deficiency during pregnancy and is characterized by stunted growth, deafness and mental retardation?

A. Cretinism
B. Keshan disease
C. Multiple sclerosis
D. Crohn's disease

48. Vitamin A or retinal is a :

A. Steroid
B. Polyisoprenoid compound containing a cyclohexenyl ring
C. Benzoquinone derivative
D. 6-Hydroxychromane

49. Retinal is reduced to retinol in intestinal mucosa by a specific retinaldehyde reductase utilising :

A. NADPH + H^+ B. FAD
C. NAD D. NADH + H^+

50. Retinol and retinal are interconverted requiring dehydrogenase or reductase in the presence of :

A. NAD or NADP B. NADH + H^+
C. NADPH D. FAD

ANSWERS

1	2	3	4	5	6	7	8	9	10
A	A	A	C	B	D	D	D	D	B
11	**12**	**13**	**14**	**15**	**16**	**17**	**18**	**19**	**20**
A	A	A	A	C	C	A	A	D	B
21	**22**	**23**	**24**	**25**	**26**	**27**	**28**	**29**	**30**
A	A	A	C	A	A	A	C	B	A
31	**32**	**33**	**34**	**35**	**36**	**37**	**38**	**39**	**40**
A	A	D	A	A	A	A	A	D	D
41	**42**	**43**	**44**	**45**	**46**	**47**	**48**	**49**	**50**
A	A	A	A	B	A	A	B	A	A

●●●

CHAPTER

10

Enzymes and Co-enzymes

Introduction

Enzymes (Kuhne, 1878) or biocatalysts are proteinaceous substances that are capable of catalysing chemical reactions of biological systems without themselves undergoing any change. Buchner (1877) found that yeast extract could cause fermentation of sugar. He coined the name of *zymase* for the complex of biocatalysts extracted from yeast. In 1903, Buchner isolated the first enzyme and this will be denoted zymase. He was awarded Nobel Prize for this. Protein nature of enzyme was first found out by Sumner. Sumner (1926) also purified the enzyme *Urease* and obtained it in crystalline form. For this Sumner was awarded Nobel Prize in 1946. Soon Northrop crystallised pepsin and trypsin.

For naming enzymes, previously enzymes were divided into two categories, **hydrolysing** and **desmolysing**. Most of the hydrolysing or digestive enzymes are located in lysosomes.

Depending upon the substrate, **hydrolysing enzymes** are of the following types :

(*i*) *Carbohydrase lactase* act on lactose to form glucose and galactose, sucrase/ invertase act on sucrose to form glucose, cellulase act on cellulose to produce glucose.

(*ii*) *Lipases* act on fats to form fatty acids and glycerol.

(*iii*) *Proteases* act on proteins like pepsin from proteins to peptones, trypsin from proteins to peptones and peptides, carboxypeptidases from larger peptides to smaller ones, aminopeptidases from peptides to smaller peptides and amino acids, dipeptidases.

(*iv*) *Esterases* hydrolyse ester linkages *e.g.*, lipases.

(*v*) *Phosphorylases*.

(*vi*) *Nucleases.* Restriction endonucleases are used in genetic engineering for breaking DNA at specific sites.

Desmolysing enzymes (other than hydrolysing) include oxidases, dehydrogenases, transaminases, carboxylases etc.

Presently enzymes are classfied into six categories :

I. Oxidoreductases : Oxidases, reductases, dehydrogenases, peroxidases, ogygenases.

II. Transferases : Group transfer *e.g.*, amino group from one amino acid to an organic acid.

III. Hydrolases : Hydrolyse, *e.g.*, amylase, sucrase, lactase, maltase.

IV. Lyases : Cleavage without hydrolysis, addition to double bond or breaking, *e.g.*, aldolase.

V. Isomerases : Rearrangement of molecular structure, *e.g. isomerases* (aldose to ketose or *vice versa*), *epimerases* (change of main constituent in its position) and *mutases* (position change of a side group).

VI. Ligases/Synthetases : Bonding of two molecules/atoms with energy from ATP, *e.g.*, Pyruvate carboxylase, Glutamine synthetase.

Enzymes are named by adding the suffix-*ase* after the substrate (*e.g.,* lipase, maltase, amylase) or chemical reaction (succinate dehydrogenase). Some old names also persist. *e.g.,* pepsin, trypsin.

Lock and Key Hypothesis of Enzyme Action Mechanism

(Template Hypothesis, Emil Fischer, 1898). The active sites of the enzyme have a specific geometric shape wherein the substrate molecules fit in just like a key in a particular lock. The substrate molecules undergo change forming products. The products are released and the active sites of enzyme become free to pick up more substrate molecules.

Induced Fit Theory (Koshland, 1958)

The active site of enzyme posseses two groups, **buttressing** or **supporting** and **catalytic**. As the substrate gets associated with buttressing region, the active site changes configuration so that the catalytic group comes to lie in opposite area where substrate bonds are to be changed.

Enzymes differ from catalysts is being proteinaceous, having high molecular weight, high efficiency, narrow range of pH, narrow range of effective temperature, susceptibility to poisons and short wave radiations.

Optimum pH varies for enzymes to enzymes, *e.g.,* 2 for pepsin, 8.5 for trypsin. Most intracellular enzymes function at near neutral pH. Enzymes are *thermolabile*. High temperature denatures enzymes. Lower temperature inactivates them. Within limits, Q_{10} is 2-3. *Time factor* (initial rise followed by fall) also operates. Enzymes inhibition can be due to protein poisons (*e.g.,* cyanide, lead, nickel), end products (allosteric, feedback or retro-inhibition, *e.g.,* excess of glucose phosphate inhibits hexokinase or glucokinase activity, excess of isoleucine inhibits threonine deaminase) and specific enzyme inhibitors.

Enzyme Inhibitors

They are natural or synthetic chemicals that bind (reversibly or irreversibly) to specific enzymes bringing about change in their activity.

I. Irreversible Inhibitors

They combine covalently with amino acids of functional groups or active sites, *e.g.,* TPCK for chymotrypsin. The inhibitors are not released by dilution. The effect is similar to reversible noncompetitive inhibitors and, therefore, often included under **noncompetitive inhibition.**

II. Reversible Inhibitors

They form a temporary association with enzyme that blocks enzyme activity. Inhibition by reversible inhibitors is of three types :

(*i*) *Competitive Inhibition or Substrate Analogue Inhibition:* Their inhibitor combines reversibly to the free enzyme at or near the active site. It may be substrate analogue (*e.g.,* malonate inhibits succinate dehydrogenase activity because its structure is similar to succinate) or a different chemical like product of the reaction (*i.e.,* feed back inhibition). The effect is reduced or nullified by increased substrate concentration.

(*ii*) *Non-competitive Inhibition :* Here no competition occurs between the substrate and the inhibitor. The inhibitor combines reversibly with enzyme or enzyme substrate complex to form EIS. The effect is not reversed by increased substrate concentration.

(*iii*) *Uncompetitive Inhibition :* The inhibitor combines reversibly with enzyme substrate complex (ES) and not the free enzyme. Therefore, the effect is less at low substrate concentrations as compared to effect of noncompetitive inhibitor which is equal at all substrate concentrations.

Enzyme Related Hereditary Diseases are caused by deficient production of enzymes due to defective heredity, *e.g.,* **albinism** (deficiency of tyrosinase), **galactosemia** (defective galactose I-P uridyl transferase), **phenylketonuria** (deficient phenylalanine hydroxylase), **methaemoglobinemia** (defective methaemoglobin reductase), **fructosuria** (defective fructokinase).

ELISA—Enzyme Linked Immuno Sorbent Assay

It is a technique which can detect and even quantitate extremely small amount of a protein, antibody or antigen with the help of enzymes. The enzyme catalysed reaction is allowed to proceed in specialized well of ELISA plate or tray. The reaction is quenched with denaturing reagents and the extent of product determined spectrophotometrically or by chromogenic substance. AIDS is detected by doing ELISA.

Enzymes Kinetics

Kuhne (1878) : Coined the term enzyme.

Buchner (1897, 1903) : Isolated enzymes (including Zymase from yeast) for the first time.

Duclaux (1883) : Provided a system for naming enzymes by using the syffix-*ase*.

Sumner (1926) : Reported that enzymes are proteinaceous in nature.

Cech et al. (1981) : Found ribozyme in RNA.

It is found that plants contain substances, which greatly influence the rates of reactions of metabolism without themselves being used up. Such substances were designated by Kunhe (1867) as ENZYMES; the term actually meaning in *yeastes*. Enzymes are also known as plant *catalysts*, *cell ferments* or *organic catalysts*.

The existence of biological catalysts was predicted by a Swedish chemist Jon Jakob Berzelius in 1835. **Haldane** defined enzymes as "soluble colloidal", organic catalysts produced by living organisms. The chemistry of the enzyme was known for the first time in 1926, when J.B. Sumner obtained a reasonably pure crystaline form of *urease* from Jack bean meal. Since then several other enzymes have been purified and crystallized and all have proved to be protein.

Nomenclature of Enzymes

The *trivial* names as ending in *'Sin'* were used previously, *e.g., pepsin, trypsin* etc. These are also named after the type of reaction catalyzed by them, *e.g.,* carboxylase, oxidases etc. The use of the suffix-*ase* after the name of substrate on which an enzyme acts upon was also in vogue *e.g.,* lipase, sucrase etc. To call a group of enzymes by one synthetic name, the International commission on enzymes prepared a nomenclature in 1961. According to which there are six classes :

(*i*) *Oxidoreductases :* Which causes oxidation reduction reactions *e.g.,* malate dehydrogenase, catalase, etc.

(*ii*) *Transferases :* Which catalyse the transfer of one group of atoms from one molecule to another *e.g.,* aminoacyl transferase etc.

(*iii*) *Hydrolases :* Which catalyse the hydrolysis of a complex molecule into two components *e.g.,* amylase, sucrase.

(*iv*) *Lyases :* Which work as the remover of a group of atoms from their substrates. *e.g.,* aldolase.

(*v*) *Isomerase :* Which cause rearrangement of intramolecular atoms in their substrate *e.g.,* Phosphofructo isomerase.

(*vi*) *Ligases :* Which catalyse the linkage of two molecules *e.g.,* DNA-ligase.

The name of the enzymes are generally designated by the suffix '—*ase* after the name of the substrate or a certain group of compounds on which enzymes act upon; *e.g., protease, inulase, sucrase, lipase, esterase* etc.

Structure of Enzyme

Most of the enzymes are made up of two portions—a coenzyme and an apoenzyme.

$$\text{Coenzyme (prosthetic group)} + \text{apoenzyme} \rightleftharpoons \text{(holo) enzyme}$$

(*i*) *Specificity :* Enzymes are substrate specific. A single enzyme will generally catalyse only a single substrate or a group of closely related substrates. Some or few enzymes may however catalyse two or more totally unrelated reactions.

(*ii*) *Enzyme Optima :* Enzymes generally work best under a certain narrowly defined conditions referred to as *optima*. These include appropriate pH, temperature etc.

(*iii*) Enzymes can react with acidic and alkaline substances. As large part of the enzyme is made of proteins which are amphoteric in nature.

(iv) *Large Size :* The size of the enzyme molecule is far larger than the substrate molecules. Hence, a few molecules of enzymes may provide reacting surface to a large number of substrate molecules.

(v) *Enzyme Inhibition :* Enzymes may become less effective or even non-functional by certain chemical agents known as inhibitors. Enzyme inhibition may be reversible or irreversible. A number of insecticides and drugs are irreversible inhibitors.

Factors Affecting Enzyme Activity

(i) *Substrate Concentration :* Enzyme activity increases with increase in concentration of the substrate to a maximum and then it levels off. Probably at this concentration of substrate, amount of enzyme becomes limiting. Some enzymes are inhibited even by high substrate concentration. This phenomenon is called 'substrate inhibition'.

Two types of graphs are obtained with substrate concentration and enzyme activity plot. Straight line or Michaelis type of curve is obtained with the enzymes which are made up of one unit. The enzymes which are made up of several subunits and show interaction amongst subunits (called **allosteric enzymes**), exhibit sigmoidal type of curve.

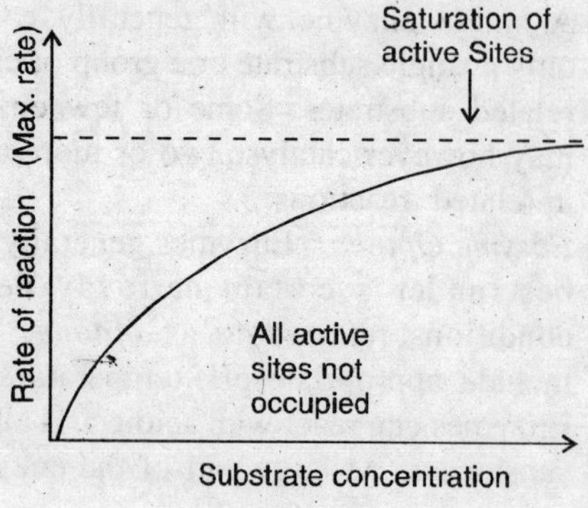

(ii) *Enzyme Concentration :* In general, the rate of reaction will increase with increasing enzyme concentration, provided substrate concentration is maintained at a high level and other conditions such as pH and temperature are kept constant. Increase in enzyme concentration increases the rate of reaction by providing more active sites to the substrates.

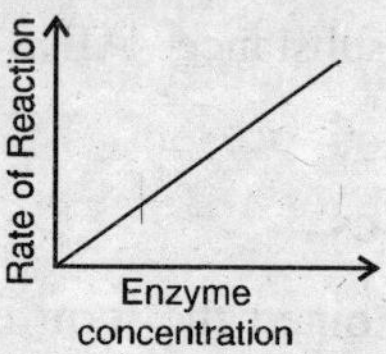

(iii) *Temperature :* In most of the enzymatic reactions, rise of 10°C in the temperature doubles the rate of reaction between 5-40°C. At higher temperature Q_{10} drops to 1° or lower. Enzymes are denatured irreversibly at 50°C to 60°C, because of their proteinaceous nature. But a few enzymes continue to be active above 60°C also *e.g.*, the enzymes of dry tissues, such as seeds and spores are more resistant to high temperature.

(iv) *pH :* Enzymes normally function most effectively at particular hydrogen ion concentrations. Each one of them shows reduced activity outside its own optimum pH range. Certain enzymes like *pepsin* have very low pH requirements, whereas enzymes like trypsin are active at high pH. Enzymes like urease are active only near neutrality.

The hydrogen ion concentration may alter the ionization and solubility of the enzymes, the substrates, inhibitors and activators. It may also change the enzymes susceptibility to heat. This may explain why the optimum pH of an enzyme varies with the temperature.

(v) *Redox Potential :* Enzymes are sensitive to the redox potential of the cell also. Oxidising and reducing enzymes are involved in changing the redox potential.

Many other enzymes are affected by redox potential because of the presence of readily oxidisable -SH group in their molecule.

Enzyme Kinetics

The rate at which a reaction proceeds depends on the concentration of substrate, product and active enzyme. Product is removed as soon as it is formed. So the rate of reaction will be limited by the concentration of either the enzyme or the substrate. If enzyme is present in excess, so that the concentration of the substrate A determines the rate at which A is converted to the product P.

$$A \xrightarrow{k} P$$

The rate of conversion can be expressed as

$$\frac{-d[A]}{dt} = k[A]$$

In which [A] is the *instantaneous concentration* of the substrate, k is the *rate constant* of the reaction, and $\frac{-d[A]}{dt}$ is the rate at which A is converted to P. The disappearance of A and the appearance of P are plotted as function of time in following figure:

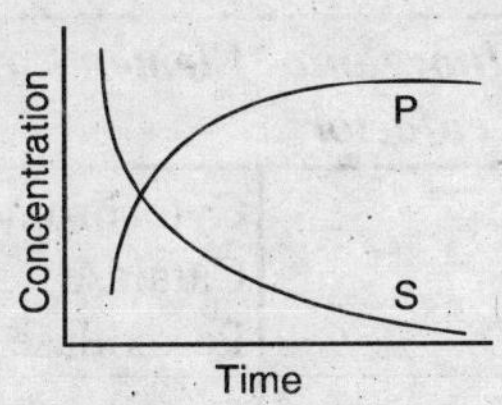

where S = Substrate
P = Product

Note that as [A] decreases exponentially, [P] increases exponentially. An exponential time function is always generated when the rate of change of a quality $\frac{d[A]}{dt}$ (in this example) is proportional to the instantaneous value of that quantity [A] in this example. The kinetics of such a reaction is said to be *first order*. The rate constant of a first order reaction has the dimensions of reciprocal time–that is "per second", or S^{-1}. The rate constant can be inverted to yield to *time constant*, which has the dimensions of time. Thus, a first order reaction with a rate constant of 10/S has a time constant of $\frac{1}{10}$S.

I[st] Order Reaction

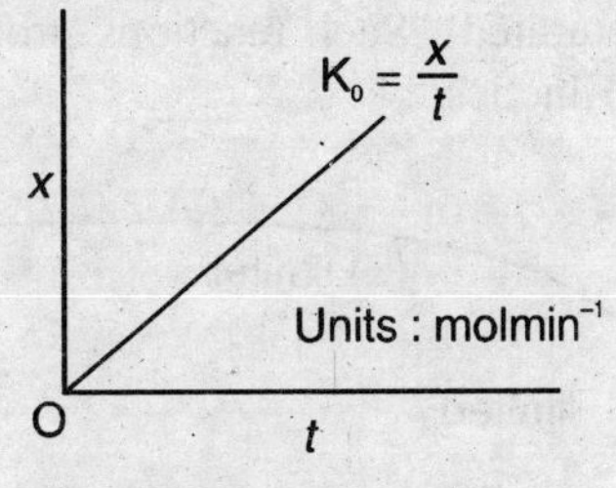

Zero order Reaction

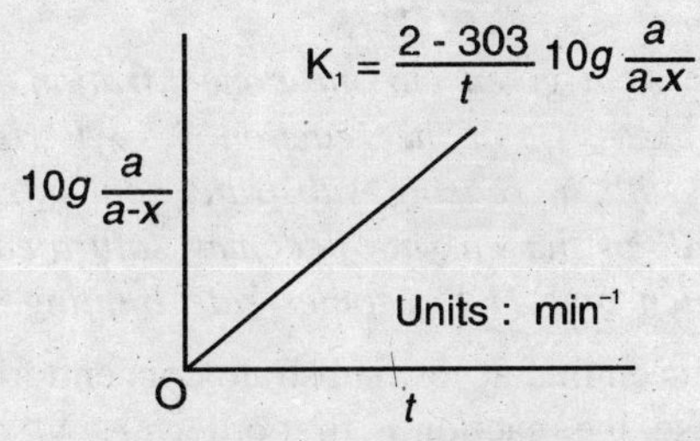

where x = the amount of substrate S reacting.
t = time
a = initial amount of A at time zero.

Note : First order plot is semilog, and therefore the straightline represents an exponential time course.

In a reaction with two substrates A and B, in the presence of enzyme,

$$A + B \xrightarrow{k} P$$

the rate of disappearance of A will be proportional to the product [A] [B]. Thus,

$$\frac{d[A]}{dt} = k[A][B]$$

The reaction will proceed with *second-order kinetics*.

It is noteworthy that the order of reaction is determined by the number of substrate species participating as reactant, but instant by the number of species present in rate limiting concentrations. Thus, if, B were present in great excess over A, the reaction A + B → P would become first order, since its rate would be limited by only one substrate concentration.

The rate of reaction is independent of substrate concentrations when the enzyme is present in limiting concentrations and all the enzyme molecules are complexed with substrate (*i.e.*, the enzyme is saturated). Such reactions proceed with *Zero Order* Kinetics

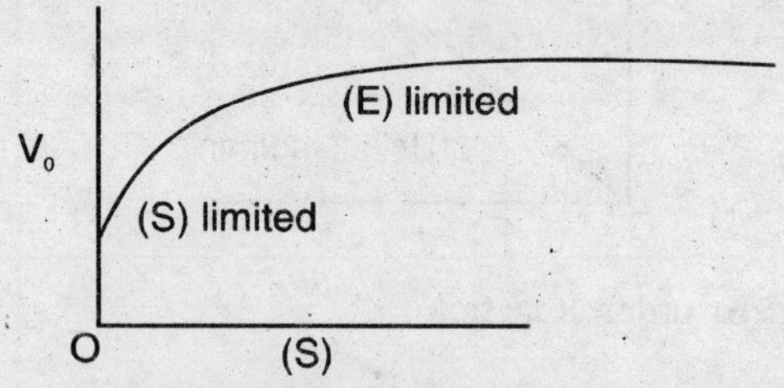

Fig. 1: *At a given enzyme concentration, the* initial rate V_0, *of the reaction* $S \rightarrow P$ *rises linearly with increasing substrate concentration untill all of the enzyme becomes saturated at which time [E] becomes rate limiting.*

It is seen that at low substrate concentration in above case, the reaction is first order *i.e.*, $V_0 \alpha [S]$. At higher substrate concentrations, however, the reaction becomes zero order, as all the enzyme is saturated with substrate, and the concentration of enzyme, not substrate, limits V_0. **In the living cells, all orders of reaction, as well as mixed order reactions, occur.**

Enzyme Inhibition

Enzyme inhibitors are of two types : *reversible* and *irreversible.* While the reversible can be removed by dialysis and the irreversible ones cannot be removed because they probably bring about alternation or destruction of the active sites of the enzymes. The reversible inhibitors are of three types:

(*a*) Competitive
(*b*) Non-competitive
(*c*) Uncompetitive

Co-enzymes

Organic molecules acting as cofactor of some enzyme. *e.g.*, NAD^+, $NADP^+$ etc. with the exception of a small group of catalytic RNA molecules, enzymes are proteins. Their catalytic activity depends upon the integrity of their native protein conformation. If an enzyme is denatured or dissociated into subunits, catalytic activity is usually lost. If an enzyme is broken down into its component amino acids, its catalytic activity is always destroyed. Thus, the primary, secondary, tertiary and quaternary structures of protein enzymes are essential to their catalytic activity. Some enzymes require no chemical groups other than their amino acid residues for activity. Others require an additional chemical component called a COFACTOR. The cofactor may be either one or more inorganic ions such as Fe^{+2}, Mg^{+2} or a complex organic or metallo-organic molecule called a *co-enzyme*. Some enzymes require both, a co-enzyme and one or more metal ions for activity. A co-enzyme or metal ion that is covalently bound to the enzyme protein is called a *prosthetic group*. A complete catalytically active enzyme together with its co-enzyme and/or metal ions is called a *HOLO ENZYME*. The protein part of such an enzyme is called **apoenzyme or apoprotein.**

Co-enzymes functions as transient carriers of specific functional groups.

List of some Inorganic Elements required by enzymes as cofactors.

(*i*) Fe^{++} or Fe^{+++}	Cytochrome oxidase Catalase Peroxidase
(*ii*) Cu^{++}	Cytochrome oxidase
(*iii*) Zn^{++}	Carbonic anhydrase Alcohol dehydrogenase
(*iv*) Mg^{++}	Hexokinase Glucose-6-phosphatase Pyruvate Kinase
(*v*) Mn^{++}	Arginase Ribonucleotide reductase
(*vi*) K^+	Pyruvate Kinase
(*vii*) Ni^{++}	Urease
(*viii*) Mo	Dinitrogenase
(*ix*) Se	Glutathione peroxidase.

Table 1. *Some Co-enzymes Serving as Transient Carriers of Specific Atoms or Functional Group*

Co-enzyme	*Examples of some chemical group transferred*	*Dietary precursor in mammals*
(*i*) Thiamine pyrophosphate	Aldehydes	Thiamine (Vitamin B_1)
(*ii*) Flavin adenine dinucleotide	Electrons	Riboflavin (Vitamin B_2)
(*iii*) Nicotinamide adenine dinucleotide	Hydride ion $(:H^-)$	Nicotinic acid (Niacin)
(*iv*) Coenzyme A	Acyl groups	Pantothenic acid plus other molecules
(*v*) Pyridoxal phosphate	Amino group	Pyridoxine (Vitamin B_6)
(*vi*) 5^1-Deoxyadenosyl cobalamine (co-enzyme B_{12})	H atoms and alkyl groups	(Vitamin B_{12})
(*vii*) Biocytin	CO_2	Biotin
(*viii*) Tetrahydrofolate	One carbon groups	Folate
(*ix*) Lipoate acid	Electrons & acyl groups	Not required in diet

Interrelationship Between Vitamins & Co-enzymes

Vitamins are organic substances which cannot be synthesized by some organisms and must, therefore, be supplied to them regularly, but in small quantities—in order to ensure a normal metabolism. Certain vitamins, notably water soluble vitamins and particularly the B group vitamins are co-enzymes or essential constituents of co-enzymes. It must be noted that same compounds often serve as co-enzymes for all living organisms, but are vitamins only for some of them. In animals and man a prolonged deficiency in a vitamin called *a vitaminosis* produces a set of pathological symptoms due to the inability of the organisms to catalyse some reactions (because the corresponding co-enzyme is missing).

Allosteric Compounds

These are special types of molecules which regulate (promote or inhibit) enzyme activity. They act on the enzyme at the site other than active site called the *allosteric site* (allo = other). The promoter or inhibitor (allosteric compound) causes change in enzyme's shape and structure and determines whether or not the enzyme can react with appropriate substrate. The effect of alosteric compound is reversible, *i.e.*, when they are withdrawn the enzyme resumes the original activity.

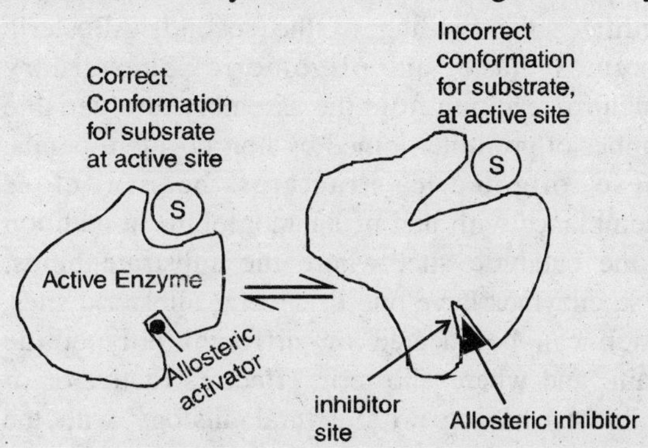

In case of allosteric activators, the affinity of the enzyme for the substrate increases ($K_m' < K_m$). Even the shape of the curve can change from the sigmoid form to the hyperbolic form. When an allosteric inhibitor binds to the allosteric site, an allosteric transition (also reversible) takes place, causing a change in the active site, which now takes a conformation less favourable to the binding of the substrate. If v is plotted as a function of [S], the curve is again a sigmoid, "but a decrease of the affinity of the enzyme for the substrate is observed ($K_m'' > K_m$). In fact, *this is only one of the various possible mechanisms, because there are several types of inhibitions*. The activity of numerous enzymes is controlled, not by the formation of complexes between enzymes and regulatory molecules, but by *covalent modifications*, of the enzyme. Of such modifications, the best known is the phosphorylation/dephosphorylation of hydroxylated amino acids, this modification catalyzed in one direction by a kinase, and in other by a phosphatase, plays a very important role in numerous mechanisms, like transmission of nerve impulse or metabolism of glycogen.

Allosteric enzymes differ from other enzymes in that the plot *v* as a function of [S], obtained, when studying the kinetics of the reactions catalyzed by allosteric enzymes, does not give a branch of an equilateral hyperbola corresponding to the Michaelis-Menten equation, but a **sigmoid curve (S-shaped).** We come across such a curve while studying the haemoglobin—oxyhaemoglobin interconversion. In the case of allosteric enzymes, it also reflects a *co-operative effect*, *i.e.,* the fact that the binding of the first substrate molecule facilitates the binding of the second. Allosteric enzymes have an oligomeric quarternary structure, resulting from the assembly of a variable number of promoter joined by non-covalent bonds. These oligomeric structures bear a close resemblance with that of haemoglobin. In addition to the catalytic site, where the substrate binds, these enzymes have one or several allosteric sites, which can be located on different polypeptide chain, and where allosteric effectors (activator or inhibitors) having no structural analogy with the substrate can bind.

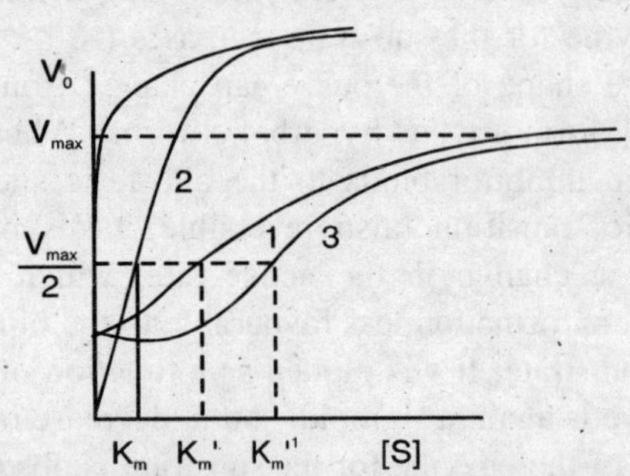

Curve 1 → in the absence of allosteric effector
Curve 2 → in the presence of allosteric activator
Curve 3 → in the presence of an allosteric inhibitor.

Isoenzymes or Isozymes

Isoenzymes are two or more slightly different molecular forms of the same enzyme and, therefore, they differ in their physical and chemical properties.

Isoenzymes of an enzyme differ from one another in electrical charges and hence, they can be separated by electrophoresis. Such multiple forms of enzymes are called isoenzymes. They are widespread in nature. Over a 100 enzymes are now known to be of isoenzymic nature.

Isoenzymes are useful in separation of taxa, *i.e.,* chaemotaxonomic studies and catalysis of a reversible reaction, one form of enzyme can catalyse reaction in one direction, while other form of the same enzyme can catalyse reaction is opposite direction.

Multiple Enzyme System

These are groups of individual enzymes which are physically correlated to function together. Thus in such "enzyme complexes", all the enzymes of the group are functional only when they are together. Individual enzymes are however inactive.

The Ribozymes

Thomas Cech (1983) found that the precursor of a ribosomal RNA is *Tetrahymena* (a ciliated protozoan) undergoes self splicing.

The intron in this precursor RNA molecule is precisely removed by a catalytic activity of the RNA itself. This liberated intron, then loses a short 5^1 terminal sequence to form a 395-nucleotide RNA molecule that catalyzes the transformation of other RNA molecules. This intron-derived RNA catalyzes the cleavage and joining of RNA chains at specific sites without itself being consumed. Hence, it is a true enzyme-RIBOZYME.

Catalytic Antibodies (Abzymes)

In 1969, William Jencks proposed that antibodies, specific for the transition state of a chemical reaction; should have catalytic power. This incisive prediction was realized in 1986 when the laboratories of Richard Lerner and Peter Schultz found that the *catalytic antibodies could be produced by using transition state analogues as immunogens.*

The preparation of an antibody that catalyze the insertion of a metal-ion into a porphyrin, nicely illustrates this experimental approach. Ferrochelatase, the final enzyme in the bio-synthetic pathway for the production of haeme, catalyzes the insertion of Fe^{++} into protoporphyrin IX. The nearly planar porphyrin must be bent for iron to enter. N-methyloprotoporphyrin is a potent inhibitor of ferrochelatase. This compound resembles the transition state because, N-alkylation forces the porphyrin to be bent. Moreover, it was known that N-alkylporphyrin chelate metal ions 10^4 times than their unalkylated counterparts. Bending increases the exposure of the pyrrole nitrogen lone pairs of electrons to solvent, which facilitates chelation. Indeed, an effective catalyst was produced using an N-alkylporphyrin as the immunogen. The resulting antibody distorted a planar porphyrin to facilitate the entry of a metal ion. Eighty porphyrin molecules were metallated per hour per antibody molecule, a rate only tenfold less than that of ferrochelatase. The uncatalysed reaction was 2500 times as slow as the antibody catalyzed reaction. Antibodies catalyzing many other kinds of chemical reactions—exemplified by ester and amide hydrolysis, amide bond formation, transesterification, photoinduced cleavage, photoinduced dimerization, decarboxylation and oxidization—have been produced using similar strategies. The power of transition state analogues is now evident since

(*i*) they provide insight into catalytic mechanisms.

(*ii*) they can serve as potent and specific inhibitors of enzymes.

(*iii*) they can be used as immunogens.

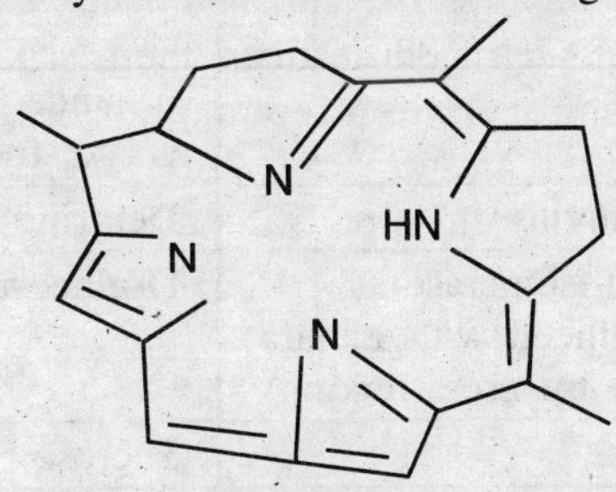

(A)

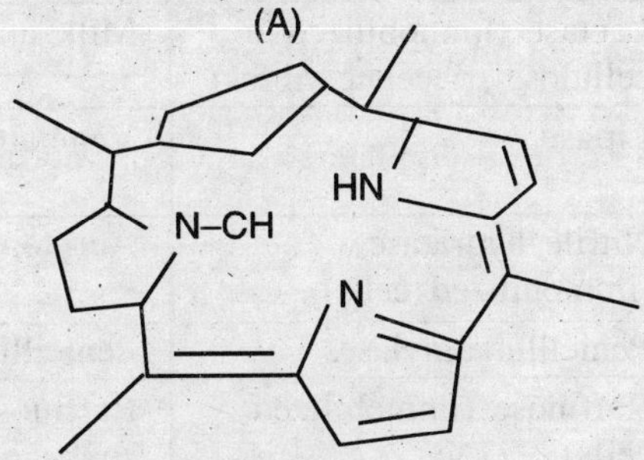

(B)

Fig. 2 *(A) The insertion of a metal ion into a porphyrin proceeds through a transition state in which the porphyrin is bent.*

Fig. 2 *(B) N-methylmesoporphyrin, a bent porphyrin is an effective immunogen, because it resembles the transition state of the reaction.*

Uses of Immobilized Enzymes

The advantages offered by immobilized enzyme systems include the following:

(*i*) use of continuous processes which reduces the plant size to nearly one-half,

(*ii*) easy, cheap, off-site fabrication of plants,

(*iii*) greatly increased productivity/unit of enzyme used, and

(*iv*) often process advantages.

Table 1. *List of some important applications of immobilized enzymes*

Enzyme	*Substrate*	*Enzyme reactor used**	*Product*
Aminoacylase (immobilized on anion exchange resins)	N-acyl-DL-amino acids	PBR	L-amino acids
Aspartate ammonialyse	Fumaric acid + NH_4^+	PBR	L-aspartic acid

Enzyme	*Substrate*	*Enzyme reactor used**	*Product*
Cyanidase	Cyanide present in industrial waste; food or feed		Formic acid
Glucoamylase	Dextrins produced by α-amylase		D-glucose
Glucose isomerase (immobilized with glutaraldehyde by cross-linking)	D-glucose in glucose syrup	PBR	High fructose corn syrup
Invertase	Sucrose	PBR	Invert sugar
Lactase (immobilized in cellulose triacetate fibres)	Milk and whey	STR	Lactose-free milk and whey
Lipase	Vegetable oils, *e.g.*, palm oil		Cocoa butter substitute
Nitrile hydratase (immobilized cells)	Acrylonitrile		Acrylamide
Penicillin amylases	Penicillin G and penicillin V	STR, PBR	Penicillins
Raffinose (immobilized cells)	Raffinose in beet juice, soybean milk	STR	Raffinose-free solutions
Thermolysin	L-aspartic acid and L-phenylalanine methyl ester		Aspartame

* *PBR, packed bad reactor; STR, stirred tank reactor.*

Uses of Enzymes in Solution

The various uses of enzymes in solution are briefly described below.

Detergents

Detergents represents the largest industrial application of enzymes amounting to 25-30% of the total sales of enzymes. The enzymes used in detergent must be cost effective, safe to use and be able to perform the task in the presence of anionic and non-ionic detergents, soaps, oxidants etc. at pH between 8 and 10.5. Enzymes constitute only 0.4-0.8 crude enzyme by weight (about 1% by cost) of detergents. The chief enzymes used are proteases, α-amylase and, sometimes, cellulase.

Leather Industry

Alkaline proteases (0.1-1% w/w) are used to remove hair from hides; this is safer and more pleasant than the traditional method using sodium sulphide. Dehaired hides are processed or bated often using pancreatic enzymes to increase their suppleness and softness in appearance. Bating is necessary for the production of soft leather clothing.

Wool Industries

Wool fibres are covered with overlapping scales pointing towards the tip; this favours movement of the fiber away from the tip and considered to lead to shrinkage.

Glucose from Cellulose

A typical waste cellulosic material is < 50% cellulose, the remainder being mostly lignin and pentosan in roughly equal quantities. Therefore, a mixture of enzymes is needed for its degradation. The concerned enzymes are relatively unstable, have low activity against native lignocellulose and show both substrate and product inhibition. Therefore, the cost of conversion of cellulose to glucose using the various available cellulolytic enzymes is high and uneconomical.

Food, Dairy, Juice and Beverages Industries

Several processes in the production of food, beverages etc. utilize enzymes.

Maltose Syrup

High maltose syrup are used to produce hard candy and frozen deserts, while high conversion syrups are used for soft candy, in soft drinks, brewing and baking. High maltose syrups (40-50 DE, 45-60% maltose, 2-7% glucose) are produced from liquefied starch (DE 11) by using fungal α-amylase (2000 U/kg) at 55°C for 48 hr. Pullulunase could be used for obtaining higher maltose contents.

The production of high conversion syrups (60-70 DE, 30-37% maltose, 35-43% glucose, 10% maltotriose, 15% other oligosaccharides) depends on the use of fungal α-amylase and glucoamylase.

Bread Whitening Lipoxygenase is used for whitening of bread.

Uses in Medicine

Enzyme applications in medicine are as extensive as a industry; some of these are listed in Table 3 Pancreatic enzymes have been used in digestive disorders since nineteenth century. Most enzymes are used extracellularly for *(i)* topical applications, *e.g.*, collagenase, *(ii)* removal of toxic substances, *e.g.*, rhodonase, *(iii)* disorders within blood circulation system, *e.g.*, streptokinase, urokinase etc.

Table 3. *Selected Applications of Enzymes in Treatment of Diseases*

Enzyme	*Reaction Catalyzed*	*Used for*
α-Amylase	Starch hydrolysis	Digestive disorders
Asparaginase	Asparagine → aspartate	Leukemia
Collagenase	Collagen hydrolysis	Skin ulcers
Ficin	Protein hydrolysis	Deworming
Glutaminase	L-Glutamine → L-Glutamate	Leukemia
Lysozyme	Bacterial cell wall hydrolysis	Used as antibiotic
Lipase	Lipid hydrolysis	Digestive disorders
Papain	Protein hydrolysis	Deworming
Proteases	Protein hydrolysis	Digestive disorders
Rhodonase	Formation of cyanide sulphide	Cyanide poisoning
Ribonuclease	RNA digestion	As antiviral agent
β-Lactamase	Penicillin → Penicilloate	Penicillin allergy
Streptodekase	Plasminogen → plasmin	Blood clots
Uricase	Urate → allantion	Gout
Urokinase	Plasminogen → plasmin	Blood clots

Enzyme Engineering

Improvement in the activity and usefulness of existing enzymes or creation of new enzyme activity by making suitable changes in its amino acid sequences is called *enzyme engineering.* When this approach is used to modify the properties of any protein, whether enzyme or non-enzyme, it is termed as *protein engineering.* Since enzymes are proteins, enzymes engineering is a part of the larger activity of protein engineering. Enzymes engineering utilizes recombinant DNA technology to introduce the desired changes in the amino acid sequences of enzymes.

Objectives of Enzyme Engineering

The chief objective of enzyme engineering is to produce an enzyme that is more useful for industrial and/or other applications. The various properties of an enzyme that may be modified to achieve this objectives are as follows :

(*i*) improved kinetic properties,
(*ii*) elimination of allosteric regulation,
(*iii*) enhanced substrate and reaction specificity,
(*iv*) increase a thermostability,
(*v*) alternation in optimal pH,
(*vi*) suitability for use in organic solvent,
(*vii*) increased/decreased optimal temperature etc.

Principles of Enzyme Engineering

The structure and function of an enzyme molecule, for that matter of any protein molecule, are chiefly determined by its amino acid sequence, *i.e.,* its primary structure. Therefore, any change in the properties of an enzyme is always reflected in its primary structure. Conversely, a change in the amino acid sequence should alter the properties of the enzyme. But this is not always the case because the enzymatic properties etc. are changed only when amino acid changes are introduced in certain critical regions for the various functions of an enzyme, and to be able to predict the effect of specific amino acid changes in these areas on the various functions. However, the present knowledge of the relationships between amino acid sequence, and three-dimensional structure and properties of enzymes, obtained from a large database, is only partially operative. It allows an explanation of the changes in structure and function on the basis of the changes in amino acid sequence, but it does not allow a dependable prediction of the influences of specific amino acid changes on the structure and functions of enzymes.

It may, however, be reasonable to anticipate that as more elaborate databases and improved softwares become available, it should become possible to predict with far greater confidence the structural and functional changes in enzymes produced by the specific changes in their amino acid sequences. The effectiveness of enzyme engineering will be greatly enhanced then, and their this activity may have a tremendous influence on enzyme technology.

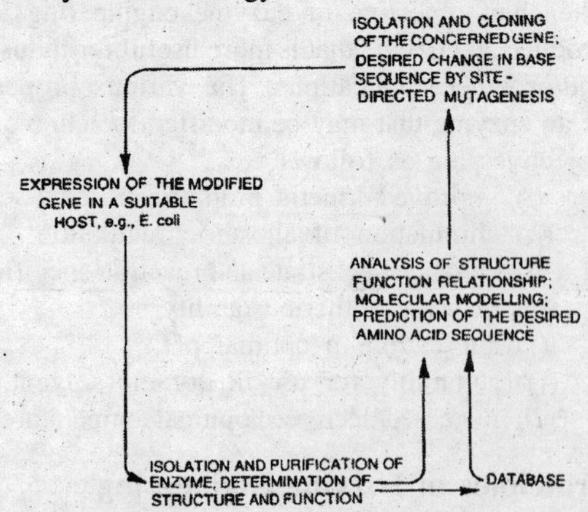

Fig. 3: *A summary of the various steps involves in enzyme engineering.*

Synzymes

These are generally synthetic polymers, sometimes proteins, which have enzymatic activities; they are called *synzymes* or *artificial enzymes.* Synzymes must possess two functional sites; one substrate-binding site and a catalytically effective site. Creation of binding site is far more easier than that of catalytic site. However, if the synzyme has a binding site for the transition state of the substrate, the binding site itself serves as the catalytic site as well, *e.g.,* in case of enzymes. Synzymes also follow the Michaelis-Menten kineties just as natural enzymes do.

Some synzymes are simply derivatised proteins.

Solvent Engineering

Solvent engineering seeks to usefully modify properties of enzymes simply by using appropriate organic solvents to carry out the enzyme-catalyzed reactions. This is in contrast to enzyme engineering (Section 17) which aims to modify the properties of an enzyme by making appropriate changes in its amino acid sequence. A brief comparison between the two approaches is presented in the Table 4.

Table 4. *Comparison of Enzyme Engineering and Solvent Engineering*

Feature of enzyme	*Enzyme engineering*	*Solvent engineering*
Amino acid sequence	Changed	Unchanged
Recombinant DNA technology	Must	Not used
Site-specific mutagenesis	Must	Not used
Catalytic activity	Modified	Modified
Enzyme specificity	Modified	Modified
Thermostability	Increased	Increased
pH dependence	Modified	Modified
Allosteric function	Modified	Unchanged
Creation of new enzyme activities	Possible	Possible
Stability against denaturants	Achieved	Not achieved
Stereospecificity	Modified	Modified
Use of organic solvent for enzyme action	Not necessary	Essential

Multiple Choice Questions

1. The compound which has the lowest density is :
A. Chylomicron B. α-lipoprotein
C. β-lipoprotein D. Pre α-lipoprotein

2. Non steroidal anti inflammatory drugs, such as aspirin act by inhibiting the activity of the enzyme :
A. Lipoxygenase
B. Cyclooxygenase
C. Phospholipase A2
D. Lipoprotein lipase

3. From arachidonate, synthesis of prostaglandins is catalysed by :
A. Cyclooxygenase
B. Lipoxygenase
C. Thromboxane synthase
D. Isomerase

4. A Holoenzyme is :
A. Functional unit B. Apo enzyme
C. Co-enzyme D. All of these

5. Gaucher's disease is due to the deficiency of the enzyme :
A. α-fucosidase
B. α-galactosidase
C. β-glucosidase
D. Sphingomyelinase

6. Neimann-Pick disease is due to the deficiency of the enzyme :
A. Hexosaminidase A and B
B. Ceramidase
C. Ceramide lactosidase
D. Sphingomyelinase

7. Krabbe's disease is due to the deficiency of the enzyme :
A. Ceramide lactosidase
B. Ceramidase
C. β-galactosidase
D. GM1 β-galactosidase

8. Fabry's disease is due to the deficiency of the enzyme :
A. Ceramide trihexosidase
B. Galactocerebrosidase
C. Phytanic acid oxidase
D. Sphingomyelinase

9. Farber's disease is due to the deficiency of the enzyme :
A. β-galactosidase
B. Ceramidase
C. β-glucocerebrosidase
D. Arylsulphatase A

10. A synthetic nucleotide analogue, used in organ transplantation as a suppressor of immunologic rejection of grafts is :
A. Theophylline
B. Cytarabine
C. 4-Hydroxypyrazolopyrimidine
D. 6-Mercaptopurine

11. Example of an extracellular enzyme is :
A. Lactate dehydrogenase
B. Cytochrome oxidase
C. Pancreatic lipase
D. Hexokinase

12. Enzymes, which are produced in inactive form in the living cells, are called :
A. Papain B. Lysozymes
C. Apoenzymes D. Proenzymes

13. An example of ligases is :
A. Succinate thiokinase
B. Alanine racemase
C. Fumarase
D. Aldolase

14. An example of lyases is :
A. Glutamine synthetase
B. Fumarase
C. Cholinesterase
D. Amylase

15. Activation or inactivation of certain key regulatory enzymes is accomplished by covalent modification of the amino acid :
A. Tyrosine B. Phenylalanine
C. Lysine D. Serine

16. The enzyme which can add water to a carbon-carbon double bond or remove water to create a double bond without breaking the bond is :
A. Hydratase B. Hydroxylase
C. Hydrolase D. Esterase

17. Fischer's 'lock and key' model of the enzyme action implies that :
A. The active site is complementary in shape to that of substance only after interaction
B. The active site is complementary in shape to that of substance
C. Substrates change conformation prior to active site interaction
D. The active site is flexible and adjusts to substrate

18. From the Lineweaver-Burk plot of Michaelis-Menten equation, K_m and V_{max} can be determined when V is the reaction velocity at substrate concentration S, the X-axis experimental data are expressed as :
A. 1/V B. V
C. 1/S D. S

19. A sigmoidal plot of substrate concentration ([S]) verses reaction velocity (V) may indicate
A. Michaelis-Menten kinetics
B. Co-operative binding
C. Competitive inhibition
D. Non-competitive inhibition

20. The K_m of the enzyme giving the kinetic data as below is :
A. – 0.50 B. – 0.25
C. + 0.25 D. + 0.33

21. The kinetic effect of purely competitive inhibitor of an enzyme :
A. Increases K_m without affecting V_{max}
B. Decreases K_m without affecting V_{max}
C. Increases V_{max} without affecting K_m
D. Decreases V_{max} without affecting K_m

22. If curve X in the graph (below) represents no inhibition for the reaction of the enzyme with its substrates, the curve representing the competitive inhibition, of the same reaction is :
A. A B. B
C. C D. D

23. An inducer is absent in the type of enzyme :
A. Allosteric enzyme
B. Constitutive enzyme
C. Co-operative enzyme
D. Isoenzymic enzyme

24. A demonstrable inducer is absent in :
A. Allosteric enzyme
B. Constitutive enzyme
C. Inhibited enzyme
D. Co-operative enzyme

25. In reversible non-competitive enzyme activity inhibition :
A. V_{max} is increased
B. K_m is increased
C. K_m is decreased
D. Concentration of active enzyme is reduced

26. In reversible non-competitive enzyme activity inhibition :
A. Inhibitor bears structural resemblance to substrate
B. Inhibitor lowers the maximum velocity attainable with a given amount of enzyme
C. K_m is increased
D. K_m is decreased

27. In competitive enzyme activity inhibition :
A. The structure of inhibitor generally resembles that of the substrate
B. Inhibitor decreases apparent K_m
C. K_m remains unaffective
D. Inhibitor decreases V_{max} without affecting K_m

28. In enzyme kinetics V_{max} reflects :
A. The amount of an active enzyme
B. Substrate concentration
C. Half the substrate concentration
D. Enzyme substrate complex

29. In enzyme kinetics K_m implies :
A. The substrate concentration that gives one half V_{max}
B. The dissocation constant for the enzyme substrate complex
C. Concentration of enzyme
D. Half of the substrate concentration required to achieve V_{max}

30. In competitive enzyme activity inhibition :
A. Apparent K_m is decreased
B. Apparent K_m is increased
C. V_{max} is increased
D. V_{max} is decreased

31. In non competitive enzyme activity inhibition, inhibitor :
A. Increases K_m
B. Decreases K_m
C. Does not effect K_m
D. Increases K_m

32. An enzyme catalyzing oxidoreduction, using oxygen as hydrogen acceptor is :
A. Cytochrome oxidase
B. Lactate dehydrogenase
C. Malate dehydrogenase
D. Succinate dehydrogenase

33. The enzyme using some other substance, not oxygen as hydrogen acceptor is :
A. Tyrosinase
B. Succinate dehydrogenase
C. Uricase
D. Cytochrome oxidase

34. An enzyme which uses hydrogen acceptor as substrate is :
A. Xanthine oxidase
B. Aldehyde oxidase
C. Catalase
D. Tryptophan oxygenase.

35. Enzyme involved in joining together two substrates is :
A. Glutamine synthetase
B. Aldolase
C. Gunaine deaminase
D. Arginase

36. The pH optima of most of the enzymes is :
A. Between 2 and 4
B. Between 5 and 9
C. Between 8 and 12
D. Above 12

37. Co-enzymes are :
A. Heat stable, dialyzable, non protein organic molecules
B. Soluble, colloidal, protein molecules
C. Structural analogue of enzymes
D. Different forms of enzymes

38. An example of hydrogen transferring co-enzyme is :
A. CoA B. NAD^+
C. Biotin D. TPP

39. An example of group transferring co-enzyme is :
A. NAD^+ B. $NADP^+$
C. FAD D. CoA

40. Cocarboxylase is :
A. Thiamine pyrophosphate
B. Pyridoxal phosphate
C. Biotin
D. CoA

41. A co-enzyme containing non aromatic hetero ring is :
A. ATP B. NAD
C. FMN D. Biotin

42. A co-enzyme containing aromatic hetero ring is :
A. TPP B. Lipoic acid
C. Co-enzyme Q D. Biotin

43. Isoenzymes are :
A. Chemically, immunologically and electrophoretically different forms of an enzyme
B. Different forms of an enzyme similar in all properties
C. Catalysing different reactions
D. Having the same quaternary structures like the enzymes

44. Isoenzymes can be characterized by :
A. Proteins lacking enzymatic activity that are necessary for the activation of enzymes
B. Proteolytic enzymes activated by hydrolysis
C. Enzymes with identical primary structure
D. Similar enzymes that catalyse different reaction

45. The isoenzymes of LDH :
A. Differ only in a single amino acid
B. Differ in catalytic activity
C. Exist in 5 forms depending on M and H monomer contents
D. Occur as monomers

46. The normal value of CPK in serum varies between :
A. 4-60 IU/L B. 60-250 IU/L
C. 4-17 IU/L D. > 350 IU/L

47. Factors affecting enzyme activity :
A. Concentration B. pH
C. Temperature D. All of these

48. The normal serum GOT activity ranges from :
A. 3.0-15.0 IU/L B. 4.0-17.0 IU/L
C. 4.0-60.0 IU/L D. 0.9-4.0 IU/L

49. The normal GPT activity ranges from :
A. 60.0-250.0 IU/L B. 4.0-17.0 IU/L
C. 3.0-15.0 IU/L D. 0.1-14.0 IU/L

50. The normal serum acid phosphatase activity ranges from :
A. 5.0-13.0 KA units/100 ml
B. 1.0-5.0 KA units/100 ml
C. 13.0-18.0 KA units/100 ml
D. 0.2-0.8 KA units/100 ml

ANSWERS

1	2	3	4	5	6	7	8	9	10
A	B	A	D	C	D	C	A	B	D
11	**12**	**13**	**14**	**15**	**16**	**17**	**18**	**19**	**20**
C	D	A	B	D	A	B	C	B	D
21	**22**	**23**	**24**	**25**	**26**	**27**	**28**	**29**	**30**
A	A	B	B	D	B	A	A	A	B
31	**32**	**33**	**34**	**35**	**36**	**37**	**38**	**39**	**40**
C	A	B	C	A	B	A	B	D	C
41	**42**	**43**	**44**	**45**	**46**	**47**	**48**	**49**	**50**
D	A	A	B	C	A	D	B	C	B

●●●

CHAPTER

11

Recombinant DNA Technology

Recombinant DNA (*r*DNA) is a form of artificial DNA that is created by combining two or more sequences that would not normally occur together through the process of gene splicing. In terms of genetic modification, it is created through the introduction of relevant DNA into an existing organismal DNA, such as plasmids of bacteria, to code for or alter different traits for specific purpose, such as antibiotic resistance.

It differ from genetic recombination in that it does not occur through natural processes within the cell, but is engineered. A recombinant protein is a protein that is derived from *r*DNA.

Cloning in biology is the process of producing similar populations of genetically identical-individuals that occurs in nature when organisms such as bacteria, insects or plants reproduce asexually. Cloning in biotechnology refers to processes used to create copies of DNA fragments, cells or organisms.

Molecular cloning refers to the process of making multiple molecules. Cloning is commonly used to amplify DNA fragments containing whole genes, but it can also be used to amplify any DNA sequence such as promoters, non-coding sequences and randomly fragmented DNA.

Recombinant Proteins

These are the proteins produced by the transferred gene or transgene; they themselves are of commercial value, *e.g.*, insulin, interferon etc. produced in bacteria etc.

Metabolic Engineering

When metabolic activities of an organism are modified by introducing into it transgenes which affect enzymatic, transport and/or regulatory functions of its cells, it is known as *metabolic engineering*. It is imperative for metabolic engineering that the biosynthetic pathways to be modified and their regulatory controls are well known, and the genes involved are identified and cloned.

1. A transgene may be added which encodes an enzyme to modify a metabolite produced by the organism to yield a new product of interest. For example, *Acremonium chrysogenum* produces cephalosporin C. The gene encoding D-amino acid oxidase from *Fusarium* was introduced into *A. chrysogenum.* This enzyme converts cephalosporin C into 7-aminocephalosporanic acid, a precursor of several semisynthetic antibiotics.
2. The enzyme encoded by transgene may enable a better utilization of the substrate or even of previously inaccessible components of the substrate. For example, normal yeasts are unable to utilize cyclodextrins present in the malt; this increase the caloric content of beer. Transgenic yeasts capable of utilizing cyclodextrins are now commercially used to produce low calorie beer with 1% more alcohol content.

3. All the genes of an entirely new biosynthetic pathway may be transferred to generate new products. For example, two genes are involved in the conversion of acetyl-CoA to polyhydroxybutyrate (PHB) which is used to produce biodegradable plastic. The two genes were transferred into *E. coli* from *Alcaligenes eutrophus*. Transgenic *E. coli*, under appropriate conditions, accumulate PHB to upto 50% of their dry weight.
4. Several gene transfers have enhanced growth rates of the organisms, reduced their nutrient requirements and enabled their growth to higher cell densities. For example, transfer of the gene encoding glutamate dehydrogenase from *E. coli* to glutamate synthate (GOGAT) deficient mutants of Methylophilus methylotrophus increased the efficiency of carbon conversion from 4% to 7%. *M. methylotrophus* is used to produce microbial biomass from methanol, which is used as animal feed. Glutamate dehydrogenase utilizes ammonium to produce glutamic acid but uses little energy, while glutamate synthase pathway (in which GOGAT is involved uses up one additional ATP molecule for each molecule of glutamic acid produced).
5. In some cases, conversion of an intermediate product to the end product is slow due to a low activity of the rate limiting enzyme. In such cases, the activity of rate limiting enzyme can be increased by increasing its dosage. For example, in case of *C. acremonium* the enzyme (encoded by the gene *cefEF*) that converts pencillin N intermediate in the cephalsporin C biosynthesis is rate limiting. The dosage of *cefEF* was increased leading to a 15% higher cephalosporin C yield. (The pencillin content was reduced by a factor of 15).

The above listed examples are only a fraction of the possibilities offered by the recombinant-DNA technology. The powers of this technology shall increase with the refinements of techniques for genetic modification of industrially important micro-organisms and the knowledge of various details of their metabolic pathways.

Process Development

Determination of the optimum conditions for fermentation, *e.g.*, medium requirements, aeration, agitation, temperature, pH, duration, etc., using a microbe producing a metabolite of interest is called *process development*. Considerations of cost, labour, time and space make it mandatory to carry out the process development in 3 distinct steps :

(*i*) laboratory process development using flasks,

(*ii*) scaling up using small to medium fermenters, and

(*iii*) production scale fermentation experiments.

Shake-Flask Fermentation

Culture flasks (usually Erlenmeyer) of 250 or 500 ml or larger are used for growing micro-organisms; these are shaken generally by a gyratory shaker at 200-250 r.p.m. Shake cultures are usually applied to aerobic process. In general, filamentous micro-organisms are grown for the production of secondary metabolites, which begins 1-3 days after inoculation and continues 3-4 days thereafter. In all such cases, the shake-cultures are used for *(i)* strain improvement as well as *(ii)* for determination of the optimum conditions for the fermentation process. *(iii)* In many industrial process, they are also used for the initial stages of inoculum development.

Micropropagation

Propagation or multiplication of heterozygous genotypes is of necessity by vegetative propagation as it involves only mitotic cell division. Progeny obtained by vegetative propagation (or asexual reproduction) of a single plant or individual constitute a *clone*. By definition, members of a single clone have the same genotype. Further, different varieties of vegetatively propagated plants are distinct clones. Conventionally, vegetative reproduction is achieved by cuttings, budding, grafting, etc. Tissue culture also enables rapid clonal propagation of plants, this is usually called *micropropagation*. This is the only process of plant biotechnology which is being utilized by industries in India for commercial production of mainly ornamental plants and fruit tree, *e.g.,* banana. The ornamental plants are produced almost entirely for export generally under contract with foreign companies, but fruits like banana and plantation crops like cardamom are mainly for the domestic market.

The chief objective of clonal or vegetative propagation is to produce progeny plants which are identical in genotype with their parent plant. This is achieved by the following process which are employed for micropropagation of different plant species :

(*i*) proliferation of axillary buds,

(*ii*) induction of adventitious buds, bulbs, protocorms, etc., and

(*iii*) somatic embryogenesis.

Enhancing Biomass Yields

Virtually all high value biochemicals from cultured plant cells are secondary metabolites, which are usually produced in differentiated cells or organised tissues. Therefore, most such biochemicals are not produced by rapidly growing cell cultures, and the culture conditions favouring growth suppress biochemical production (and vice-versa). Therefore, the production strategy should consist of two distinct phases :

(*i*) growth phase for cell biomass accumulation, and

(*ii*) *production phase* for biosynthesis and accumulation of the biochemicals.

Ordinarily, some growth will take place during the production phase as well, which should be exploited. But in at least some cases, culture growth and biochemical production occur together, *e.g.,* berberine production in *Thalictrum minus*.

Biomass accumulation can be improved by using optimum culture conditions of which nutrient medium and inoculum size are particularly important. Experience with *L. erythrohizon* cell cultures which yield shikonin (the first commercial example) suggests the following. The different standard medium formulations may differ considerable in terms of culture growth; for example, *L. erythrorhizon* suspension cultures yielded only 6.8 g cell dry weight/l on White's medium, while the yield on LS (Linsmaier and Skoog) medium was 16.8 g/l. Further improvements in culture growth may be obtained by appropriate modifications of the most suitable standard recipe. A modification of the LS medium (called M-5 medium) resulted in a 15% improvement in cell dry weight yields from *L. erythrohizon* cultures.

Multiple Choice Questions

1. The deliberate modifications of an organism's genetic information by directly changing its nucleic acid content is a subject matter of :

A. Genetic engineering
B. Population genetics
C. Microbiology
D. Protein engineering

2. Enzymes that recognize and cleave specific 4 to 8 base pair sequences of DNA are :

A. DNA ligase
B. Helicases
C. Restriction endonucleases
D. DNA gyrase

3. What is the normal role of restriction endonucleases in bacterial cells?

A. To degrade the bacterial chromosome into small pieces during replication
B. To degrade invading phage DNA
C. To produce RNA primers for replication
D. All of the above

4. Which type of restriction endonuclease cuts the DNA within the recognition site?

A. Type I B. Type II
C. Type III D. All of these

5. Bacterial cells protect their own DNA from degradation by restriction endonucleases by :
A. Methylating the DNA at the sites that the enzyme recognizes
B. Deleting all recognition sites from the genome
C. Not producing any restriction endonucleases
D. Having anti restriction endonucleases

6. Vectors are :
A. Molecules that degrade nucleic acids
B. Molecules that help in replication
C. Molecules that are able to covalently bond to and carry foreign DNA into cells
D. Molecules that protect host cells from invasion by foreign DNA

7. Charged molecules are separated based on varying rates of migration through a solid matrix when subjected to an electric field. This technique is known as :
A. Photoreactivation
B. Gel electrophoresis
C. Autoradiography
D. Blotting

8. The Southern blotting technique depends on :
A. Similarities between the sequences of probe DNA and experimental DNA
B. Similarities between the sequences of probe RNA and experimental RNA
C. Similarities between the sequences of probe protein and experimental protein
D. The molecular mass of proteins

9. A short molecule containing 2-20 nucleotide is :
A. Plasmid B. Vector
C. Oligonucleotide D. Mononucleotide

10. A molecular technique in which DNA sequences between two oligonucleotide primers can be amplified is known as :
A. Southern blotting
B. Northern blotting
C. Polymerase chain reaction
D. DNA replication

11. The advantage of using DNA polymerases from thermophilic organisms in PCR is that :
A. The DNA polymerases of these bacteria are much faster than those from other organisms
B. The DNA polymerases of these bacteria can withstand the high temperatures needed to denature the DNA strands
C. The DNA polymerases of these bacteria never make mistakes while replicating DNA
D. All of the above

12. Which of the following enzymes is used to covalently bond foreign DNA to a vector plasmid?
A. DNA polymerase
B. Restriction endonuclease
C. DNA ligase
D. DNA helicase

13. A genomic library is :
A. A database where the sequence of an organism's genome is stored
B. A collection of many clones possessing different DNA fragments from the same organisms bound to vectors
C. A book that describes how to isolate DNA from a particular organism
D. A place where the information of the genetic organization of organisms are kept

14. Which of the following is obtained using processed *m*RNA molecules as a template?
A. *r*DNA
B. *m*DNA
C. *c*DNA
D. *t*DNA

15. For gene probes to be useful they must :
A. Be large enough to contain gene-specific sequences
B. Be labelled in some manner to allow detection
C. Both A and B
D. None of the above

16. Which of the following is not commonly used as vector?
A. Artificial chromosome
B. Cosmid
C. Fungi
D. Plasmid

17. In genetic engineering, a chimera is :
A. An enzyme that links DNA molecules
B. A plasmid that contains foreign DNA
C. A virus that infects bacteria
D. A fungi

18. Which of the following vector can maintain the largest fragment of foreign DNA?
A. YAC B. Cosmid
C. Plasmid D. Phage

19. An animal, that has gained new genetic information from the acquisition of foreign DNA, is considered as :
A. An enzyme that links DNA molecules
B. A transgenic animal
C. A vector
D. A chimera

20. Electroporation is :
A. The process of separating charged molecules through a gel maintained in an electric field
B. The process of combining foreign DNA to an electrically charged vector molecule
C. The application of high voltage pulses
D. The process of multiplication of the cells

21. The piece of equipment, that introduces DNA into cells via DNA-coated micro projectiles is known as :
A. Laser
B. Inoculating needle
C. Gene gun
D. DNA probe

22. Problems in obtaining large amounts of proteins encoded by recombinant genes can often be overcome by using :
A. BACS
B. Expression vectors
C. YACS
D. All of these

23. *Agrobacterium tumefaciens* is :
A. A disease in humans that causes loss of sight
B. A bacterium that can be used to introduce DNA into plants
C. A fungi that is used to produce antibiotics in large amounts
D. A bacterium that can be used to introduce DNA into plants

24. The first therapeutic product formed by means of recombinant DNA technology is :
A. Vaccine for foot and mouth disease
B. Insulin
C. Hepatitis B vaccine
D. Human growth hormone

25. Human insulin formed by recombinant DNA technology is known as :
A. H insulin B. R insulin
C. Humulin D. Huinsulin

26. How is human insulin synthesized using recombinant DNA technology?
A. By using chemically synthesized DNA sequences for the two chains separately
B. By isolating DNA from the islets of Langerhans of pancreas
C. By using *c*DNA for insulin
D. By using chemically synthesized DNA sequences for the complete insulin protein

27. Chemically synthesized DNA sequences for the two chains are separately inserted into the plasmid *pBR* 322 by the side of :
A. β-Galactosidase B. Galactokinase
C. Acid phosphatase D. Glucokinase

28. The two chains of insulin can be separated by detaching it from β-galactosidase using :
A. Hydrogen chloride
B. Cyanogen bromide
C. Peptidase
D. Protease

29. The two chains are joined together to constitute native insulin using :
A. Sodium dissulphonate and sodium sulphite

B. Sodium dissulphonate and sodium sulphate
C. Sodium sulphate and sodium sulphite
D. Sodium sulphate only

30. Human growth hormone (hGH) is secreted by :
A. Pitutary gland B. Hypothalamus
C. Pancreas D. None of these

31. The recombinant hGH lacks :
A. Terminal glutamine
B. Terminal lysine
C. Terminal methionine
D. Terminal arginine

32. Recombinant vaccine for Hepatitis B virus has been synthesized against which of the following antigen?
A. Viral core antigen (HBcAg)
B. Viral surface antigen (HBcAg)
C. E antigen (HBeAg)
D. All of the above

33. Recombinant vaccine for HBV was produced by cloning viral genes in :
A. Plasmids
B. Cosmids
C. Autonomously replicating plasmid of yeast
D. Phagemids

34. The promoter used for cloning HBV surface antigen is :
A. Galactokinase
B. Alcohol dehydrogenase I
C. β-galactosidase
D. Polyhedrin

35. In India first genetically engineered vaccine against HBV was developed by :
A. Ranbaxy Pvt Ltd
B. Shantha Biotechnics Pvt Ltd
C. Dabur Pvt Ltd
D. Glaxo Pvt Ltd

36. Foot and mouth disease (FMD) of animals is caused by :
A. RNA virus B. DNA virus
C. Bacteria D. Protozoa

37. How many nucleotides are there in Ss RNA molecule of picorna virus causing FMD?
A. 1000 B. 5000
C. 8000 D. 10000

38. Which one of the following capsid protein of picorna virus is immunogenic in nature?
A. VP1 B. VP2
C. VP3 D. VP4

39. The gene coding for VPI is cloned in :
A. PMB 9 B. PBR 322
C. PUC 18 D. PUC 19

40. Which of the following statements is incorrect regarding the different forms of the DNA double helix?
A. Z-DNA is left-handed and assumes a zigzag conformation.
B. B-DNA is the most tightly wound form with 12 bp per turn.
C. A-DNA is more compact than B-DNA.
D. B-DNA is the most common DNA in living cells.

41. A DNA sequence refers to the _____ .
A. Rules pertaining to base pairing
B. Linear order of nucleotides along a DNA strand
C. Orientation of the base relative to the sugar
D. Three-dimensional structure of a nucleotide

42. A polymer of DNA 20 nucleotides in length has _____ different sequence possibilities.
A. 20^4 B. 20
C. 4^{20} D. 10

43. _____ adds new nucleotides to the _____ end of the elongating strand during DNA replication.
A. Polynucleotide phosphorylase; 32
B. DNA polymerase; 32
C. Replicase; 52
D. Sequenase; 52

44. Template-dependent DNA synthesis makes sure that _____.
A. The product made retains the B-form

B. Gene expression is always active
C. The DNA strands are not antiparallel
D. The double helix is an exact copy of the original pattern of nucleotides

45. The function of genes may be deduced from the _____.
A. Structure of the helix
B. Sequence of nucleotides
C. Base-pairing rules
D. Contigs

46. Long, uninterrupted stretches of DNA sequence making up _____ are determined by searching for overlaps of short DNA fragments.
A. Vectors
B. Primers
C. Contigs
D. Oligonucleotides

47. Sonication is a method employing _____.
A. Enzymes to cleave DNA for cloning purposes
B. High frequency sound waves to fragment DNA
C. Computers to locate overlaps in DNA sequence
D. Chemicals to treat bacteria to facilitate uptake of recombinant vectors

48. Cloning vectors are _____.
A. Exclusively linear DNA molecules
B. Able to replicate when introduced into *Escherichia coli*
C. Enzymes used to fragment DNA
D. Modified deoxynucleotides used for the chain termination method of DNA sequencing

49. _____ is used to join together short DNA fragments and cloning vectors.
A. DNA ligase
B. DNA polymerase
C. DNA recombinase
D. DNA replicase

50. A cloning vector that contains ligated DNA fragments is _____.
A. Competent
B. Recombinant
C. Primed
D. Polymerized

ANSWERS

1	2	3	4	5	6	7	8	9	10
A	C	B	B	A	C	B	C	C	C
11	**12**	**13**	**14**	**15**	**16**	**17**	**18**	**19**	**20**
B	C	B	C	C	C	C	A	B	C
21	**22**	**23**	**24**	**25**	**26**	**27**	**28**	**29**	**30**
A	D	B	B	C	A	A	B	A	A
31	**32**	**33**	**34**	**35**	**36**	**37**	**38**	**39**	**40**
C	B	C	B	B	A	C	A	B	B
41	**42**	**43**	**44**	**45**	**46**	**47**	**48**	**49**	**50**
B	C	B	D	B	C	B	B	A	B

• • •

CHAPTER

12

Plant Tissue Culture

Introduction

Plant tissue culture is a practice used to propagate plants under sterile conditions, often to produce clone of plants. Plant tissue culture is used widely in plant science, it also has a great commercial applications.

Tissue culture is a technique whereby small pieces of viable tissue (called 'explants') are isolated from parent plants and grown in a defined nutritional and controlled environment for prolonged period under aseptic conditions. For successful plant tissue culture it is best to proceed with an explant rich in undermined cells, *e.g.,* those from the cortex or meristem, because such cells are capable of rapid proliferation. The explants to proceed with tissue culture technique in plants are usually taken from leaves, buds, root apex, shoot-apex, nodal segments or germinating seeds. These explants are transferred onto the suitable culture media under specific conditions where they grow into an undifferentiated mass of cells called 'callus'. To develop a callus from the explant it is essential that the nutrient medium used contains phytohormones such as auxins (*e.g.,* IAA), cytokinins.

These hormones vary for different tissue explants from different parts of the same plant and for the same explant from different genera of plants. Thus there is no 'ideal' medium, but, however, the typical medium used in plant tissue culture technology is Murashige-Scoog (MS) medium.

(A) Technique

There are four main stages involved in plant tissue culture technology. These are:

(a) Initiation of Culture

The most important factor in tissue culture technology is the maintenance of aseptic conditions. To meet this condition, the plant material which usually contains bacterial or fungal spores has first to be surface sterilized by the use of chlorine water or mercuric chloride or some other surface-sterilizing agents to eliminate micro-organisms. This is necessary because the culture medium on which these tissue pieces are inoculated contains mineral salts, vitamins, sugars, phytohormones and other growth regulators on which microbial contaminants would grow at a faster rate than plant tissue. After inoculation, the cultures are incubated under controlled conditions at 25°C in light.

(b) Multiplication or Sub-Culture Stage

After 2-3 weeks, the explants show visible growth by forming either callus or differentiated organs like shoots, roots, or complete plant depending upon the composition of the medium. If one desires to obtain large number of plantlets from the callus, he sub-cultures the callus. What is done is that the callus produced from the explant is taken out, cut into small pieces and each piece is transferred to fresh medium in a separate tube where the callus pieces grow into big mass of callus. This is called sub-culture stage. Sub-culturing may be repeated if required.

(c) Development and Differentiation of Sub-Culture

The sub-cultured callus is now induced to proceed for further development and finally, differentiation. If the concentration of phytohormones in the medium are altered then the callus can be induced to differentiate. A very high auxin to cytokinin ratio induces root formation and a high cytokinin-to-auxin ratio induces shoot formation. This is referred to as *'organogenesis'*. Alternatively, a structure called *embryoid* may develop if media concentration is so altered, this process is called *'somatic embryogenesis'*. If the hormonal conditions are correctly balanced then an entire plantlet can be induced to grow on culture medium, this process is called *'regeneration'*.

(d) From Test-Tubes to Field

In the last stage, which is followed only when full-grown plants are to be obtained, the test-tube rooted plantlets are first subjected to *"aclamatization"* so that they can adjust later in the field conditions, and then transferred to the field. The plantlet is takes out from the rooting medium, is washed slowly with running water for an hour or more to ensure that no piece of agar is left on the surface of the plantlet.

If agar remains their it would invite microbes to grow and destroy the plantlet. Now the plantlet is put to low minimal salt medium (LMSM) for 24-48 hours and then transferred to a pot that contains autoclave—sterilized mixture of clay-soil, core sand leaf moults in 1 : 1 : 1 proportion. The plantlet—containing pot is covered generally with transparent polythene (beakers may also be used) to maintain the humidity.

This condition is maintained for about 15-30 days. At this stage the plantlet is fully aclamatized and is able to withstand open environment conditions. The plantlet then is transferred to soil and the techniques employed are similar to normal agricultural practices.

(B) Applications

(a) Improvement of crops and ornamentals

Plant tissue culture technology bears tremendous potential for the improvement of crops and ornamentals. Some major areas of this application are :

(i) Large Scale Plant Regeneration : Clonal Propagation (Micropropagation)

One of the most important aspects of plant tissue culture technology is its use in clonal propagation of selected species of crops and ornamentals. The clonal propagation via tissue culture technique (*i.e., in vitro* clonal propagation) is popularly called *"micropropagation"*. In most cases, *in vitro* clonal propagation has been achieved using shoot apex tissue or axillary bud tissues. The apical shoot tip or axillary bud is surface sterilized and placed onto a culture medium that promotes formation of additional shoot (the medium is called shooting medium). Subsequently, multiple shoots are separated and transferred onto such a medium that induces root formation (the medium is called *rooting medium*). Plantlets are then aclamatized and transferred to the green-house or field.

Interest in *in vitro* clonal propagation of plants originated from the success of G. Morel (1960) with orchids. At present, however, *in vitro* clonal propagation is the only commercially viable method of propagation of orchids and almost all economically important orchids are clonable *in vitro*. Certain species of ornamentals or parental breeding lines are often difficult or time consuming to propagate using conventional sexual methods. In either case, mass clonal propagation using tissue culture technique could be commercially useful. Shoot apex clonal propagation has already been used to commercially propagate a large number of marketable ornamentals. For instance, ornamental plants such as *Gerbera, Cordyline, Dracaena, Hemerocallis* are now commercially propagated via cloning.

[*Clonal propagation*, which is popularly called 'micropropagation' represents the technique in which vegetative tissue are used to produce plants that are genetically identical to their parents. A *clone* is however, the genetically alike progeny of plant derived from its vegetative tissue.]

Clonal propagation is, however, being used commercially for large-scale plant propagation in certain crop species (*e.g.*, potato) by the production of *minitubers*.

When the axillary shoot cultures of potato are cultivated in the presence of appropriate levels of phytohormones such as cytokinin and gibberellin, they result in large number of small tubers called 'minitubers'. The later are transferred directly to the field and generate normal plants. One can see from the following table that the method of minituber-oriented plant propagation is advantageous.

Traditional	***Micropropagation***
Year I	
100 g tuber	100 g tuber
↓	↓
1 mature plant	1 preconditioned plant
↓	↓
1600 g tuber	10 nodal cultures
	↓
	Shoot multiplication
	↓
	Dormant 650000 minitubers
Year II	**Ist season**
16 mature plants	617500 plants
↓	↓
16 × 600 g tubers = 25.6 kg.	617500 × 500 g tubers = 308750 kg.

Chart : *Comparison of traditional and tissue culture method for propagating potato tubers. The reduced number of plants obtained from the minitubers is an allowance for 5% loss on establishment in soil.*

The significant advantages offered by *in vitro* clonal propagation over the conventional methods of clonal propagation (methods of vegetative propagation) *in vivo* are :

1. Over a million plants can be regenerated from a small even microscopic, piece of plant tissue within a year. Contrary to it, none of the *in vivo* methods of clonal propagation can result in such a prolific rate of multiplication.
2. *In vitro* clonal propagation can continue throughout the year irrespective of the season.
3. An enhanced rate of *in vitro* clonal propagation reduces considerably the period between the selection and the release of a new culture.
4. *In vitro* clonal propagation provides reliable and economic methods of maintaining pathogen free plants in a state that can achieve rapid multiplication when needed, irrespective of the time of the year.

(ii) Non-Clonal Production

Contrary to the shoot apex propagation which faithfully produces clones, there are sizable number of evidences which suggest that the regeneration of plants from callus, leaf explants, protoplasts does not result in production of clones, the regenerated plants bear genetic variability. Regeneration of plants from long-term callus culture is associated with *in vitro* chromosome instability and recovery of aneuploid plants. Regeneration of plants directly from leaf explants has resulted in morphological variation. It has been reported that the regenerates of potato and sugarcane varied in response to diseases. Plants regenerated from callus of sugarcane were examined in detail and have been found resistant to eye-spot diseases. Fiji disease and downy mildew for which the parent was susceptible. These regenerate-lines are being incorporated into a conventional breeding programme. In this way the large scale plant regeneration using callus, leaf explants etc. may lead to the production of genetically variable crop species which may prove novel agriculturally. However, no new crop varieties have yet been produced using the culture induced genetic variability although this method will probably produced new varieties shortly.

(b) Production of Virus Free Plants

Virus infection has been a major problem of crop and ornamental plants. Such infections greatly reduce the yield, vigour and quality of plants and their products. For instance, virus infection of ornamental plants reduces the size and number of blooms produced and infection of fruit crops reduce yield by upto 90%. The traditional method of virus elimination is by heat-treatment, but, only a few viruses can be eliminated by this method. It has been observed in a large number of species, however, that the concentration of infective virus is low in the apical meristems of a plant. It is so because the lack of vascular differentiation in the meristem prevents intercellular movement of viruses and the active metabolism of mitotic cells precludes viral infection.

It is considered, therefore, that if a small enough piece of apical meristem is exised and cultured, it is often possible to obtain virus-free plant regenerates. Genetic stability as measured by phenotypic and chromosomal stability is usually preserved in plant regenerates recovered from apical meristem cultures.

G. Morel and C. Martin (1952) succeeded for the first time in obtaining virus-free *Dahlia* plants by culturing apical meristem from infected plants.

(c) Forestry

One of the most important aspects of tissue culture technology is rapid regeneration of forest plants. This have been successfully demonstrated in case of orchids, herbaceous plants and some fruit tree species.

However, realization of the importance of tissue culture in accelerating improvement programmes for forest tree spp. is a recent origin. A number of reports on methods of producing plants by tissue culture of a whole range of forest tree belonging to angiosperms such as teak, eucalyptus, sandalwood, populus, birch, tamarind, rubber, oil palm, and gymnosperms such as *Picea abies*, *Pinus radiata*, *Sequoia sempervirens*, *Thuja plicata* have recently been published.

(d) Production of Secondary Metabolites

Plant tissues cultured *in vitro* produce some or all of the primary and secondary metabolites produced by mature plants. The primary metabolites consists of proteins, carbohydrates, fats, vitamins etc. which are essential for growth, while secondary metabolities like alkaloids, steroids, phenolics, flavonoids may be produced as defence mechanisms by the plant. Secondary metabolites have a wide range of applications in the pharmaceuticals, chemical and food industries. Commercial production of these high cost secondary metabolites by plant tissue cultures is attracting greater attention now than before. This is evident from the large number of commercial companies taking keen interest to exploit the potential.

Table 1. *A list of suggested plant products of industrial potential, which can be obtained using plant tissue culture system is given :*

Product	*Plant*	*Source*
(1) Pharmaceuticals Ajmalicine, Vincristine, Serpentine, Vinblastine	*Cathasanthus roseus*	Root
Atropine, Hyoscyamine	*Atropa belladona*	
Codeine, Morphine, Thebaine	*Papaver somniferum* and *P. bracteatum*	Capsule, Seedlings
Anthraquinones	*Cassia tora*, *Morinda citrifolia*	Fruit
Berberine	*Coptis japonica*	Rhizome
Digitoxin-Digoxin	*Digitalis purpurea*	Leaves
Reserpine	*Digitalis lanata*	Leaves

Product	*Plant*	*Source*
Visnagin	*Rauwalfia serpentina, Ammi visnaga*	Roots
(2) Chemicals		
Diosgenin	*Disocorea deltoidea*	Tubers
Essential oils	*Mentha* spp.	Leaves
(3) Flavours		
Angelica	*Angelica sylvestric*	Roots
Capsicum	*Capsicum annum*	Fruit
Celery	*Apium graveolens*	Leaves
Garlic	*Alium sativum*	Bulbs
Ginger oil	*Zingiber officinale*	Roots

Product	*Plant*	*Source*
Hops	*Humulus lupulus*	Flowers
Onion	*Alium cepa*	Bulbs
(4) Insecticides		
Pyrethrins	*Chrysanthemum Cineraria folium*	Flowers
(5) Colourants		
Anthocyanin, Carotenoids	*Dacus carota*	Roots
Betanine	*Beta vulgaris*	Roots
(6) Non-Nutritive *Sweeteners*		
Thaumatin	*Thamatococcus danielli*	Fruits
Stevioside	*Stevia rebaudiana.*	

Representative Indian Institutions

High cost secondary metabolitic production via plant tissue culture technique have attracted the Indian Scientist's attention and the work is in progress in various institutions. Following are some representative institutions which are actively engaged in such activities :

Institutions	*Metabolities*
Bhabha Atomic Research Centre, Bombay	Fermentative production of Ajmalicine, Vinblastin, Vincristine
Central Food Technological Research Institute, Mysore	Capsaicin (pungent spice), Pyrethrines, (bioinsecticide for grain storage), Saffron (flavour & colour), Garlic (flavour) and Carum (spice)
Central Institute of Medicinal and Aromatic plants, Lucknow	Artemesin, Essential oils
M.S. University of Baroda	Atropine, Nicotine, Phenolics, Solasodine, Phyto-ecdysones
Regional Research Laboratory, Jammu-Tawi	Diosgenin, Indole Alkaloids, Morphine alkaloids, Scopoletin and Visnagin
University of Rajasthan, Jaipur	Atropine, Berberine, Diosgenin, Opium alkaloids, Hyoscyamine, precursor biotransformation
University of Jodhpur, Jodhpur	Ephedrine, Indole alkaloids

Advantage and Disadvantage of Tissue Culture

A variety of research in biotechnology depend on tissue culture. Tissue culture offers a number of advantages for studies in different areas, but there are certain disadvantages, so that tissue culture technique should not be used indiscrimately. The advantages and disadvantages of tissue culture are summarised as under:

S. No.	*Advantages*	*Disadvantages*
1.	Controlled and defined physiological conditions like constitution of medium etc.	It is about 10 times more expensive for same quantity of animal tissues, therefore reasons for its use should be compelling.
2.	Controlled physico-chemical environment like temperature, pH, osmotic pressure, carbon dioxide, oxygen etc.	For the interpretation and regulation of cell culture behaviours expertise is needed.
3.	It is economical, because smaller quantities of chemicals/reagents are needed in comparison to *vivo.*	Chromosome constitution is unstable and aneuploids.
4.	There is homogeneity of cell types which is achieved through serial passages.	
5.	Moral, legal and ethical questions of animal-experimentation are avoided.	

Multiple Choice Questions

1. The tumor phenotype, which can be maintained indefinitely in tissue culture, results from the expression of genes on the :

A. T-DNA B. *c*-DNA
C. *r*-DNA D. *m*-RNA

2. Which of the following have been used for cryo-preservation of germplasm with 90 per cent tissue survival and 10 per cent plant regeneration that are used in international exchange of germplasm?

A. Meristem culture
B. Embryo culture
C. Both A and B
D. Embryo cultures supplemented with GA3 and coconut milk

3. Establishment of embryogenic cell suspensions with high regeneration potential is a pre-requisite for successful :

A. Protoplast culture
B. Detection of nucleic acid sequence
C. Restriction of fragment length polymorphisms
D. Mircoinjection

4. Which of the following have been used for cryo-preservation of germplasm with 90 per cent tissue survival and 10 per cent plant regeneration that are used in international exchange of germplasm?

A. Meristem tip cultures
B. Embryo cultures
C. Embryo cultures supplemented
D. Both A and B with GA3 and coconut milk

5. Establishment of embryogenic cell suspensions with high regeneration potential is a pre-requisite for successful :

A. Protoplast culture
B. Detection of nucleic acid sequence

C. Restriction of fragment length polymorphisms
D. Microinjection

6. Random amplified polymorphic DNA (RAPD) is a method that :
A. Stimulate production of sense RNA to compensate
B. Activate the expression of all genes in a biochemical pathway
C. Eliminate the expression of all genes in a biochemical pathway
D. Reveals intra-specific variation and diversity between species

7. Which of the following transfers a set of genes encoded in a region called T-DNA of the Ti-plasmid into cells at wound positions?
A. *A. tumefaciens*
B. *L. peruvianum*
C. *L. birsutum*
D. *A. sterilis*

8. The intervention in the communication system of nematodes offer several possibilities of bio control. Which of the following(s) play(s) a very important role in the host recognition?
A. Protein
B. Lipid
C. Carbohydrate moiety
D. Minerals

9. Wheat variety Rendezvous, with resistance to eye spot *(Pseudo-cercosporella herpotrichoides)* derived from *Aegilops ventricosa,* has a chloroplast genome derived from :
A. *Aegilops*
B. *Triticum*
C. Both (A) and (B)
D. S alleles

10. Human serum albumin in case of transgenic plant is secreted in :
A. Transgenic leaf tissues
B. Transgenic root tissues
C. Is not secreted out of plant
D. Transgenic stem tissues

11. Protoplast fusion technique generally involves:
A. Hybridization between two species
B. Isolation and fusion
C. Isolation, fusion and culturing
D. Isolation only

12. Cytoplast (somatic cell hybridization) is the process in which :
A. Two protoplast along with their nuclei fuse
B. A normal protoplast along with an enucleated protoplast fuse
C. Both (A) and (B)
D. Two enucleated protoplasts fuse

13. α and β–cyclo dextrins are used for :
A. Pharmaceutical delivery system
B. Flavour and odour enhancement in foods
C. Removal of undesired compounds like caffeine from foods
D. All of the above

14. Microinjection technique of introducing NA is a :
A. Physical method
B. Biological method
C. Combination of physical and biological method
D. Chemical method

15. MS medium, is found essential for initial morphogenesis of the immature embryos and to enhance callus development, when supplemented with :
A. GA_3
B. GA_3 and coconut milk
C. Adenine
D. Adenine and kinetin

16. Plant tissue culture technique is used to select somaclones that are :
A. Resistant to herbicide
B. Not resistant to herbicide
C. Resistant to insecticides
D. Not resistant to insecticides

17. Which of the following mutant/(s) has/have increased amount of lysine and/or methionine in corn seeds?
A. Opaque-2 B. Sugary-1
C. Flory-2 D. All of these

18. In transgenic plants of *Arabidipsis thaliana* which of the following gene is transferred for the synthesis of PHB (polyhydroxy-butyrate)?
A. Polyhydroxybutyrate synthease
B. Acetyl CoA reductase
C. Both A and B
D. Pyrophosphorylase

19. Cultivated tomato species *(Lycopersicon esculentum)* has a high degree of resistance towards :
A. *Meloidogyne* spp. nematodes
B. *L. peruviaanum*
C. Both (A) and (B)
D. *Solarium melongena*

20. *c*DNA clone prepared to stigma *m*RNA of *Brassica oleracea* can be used as a probe on DNA digests of segregating populations to recognize :
A. *Aegilops* B. *Triticum*
C. Both (A) and (B) D. S alleles

21. Advantage of microprojectile method over microinjection method for gene transfer in plants include :
A. Intact cells are used
B. Method is universal in its application irrespective of all shape, size, type and presence or absence of cell wall
C. Gene can be transferred to many cells simultaneously
D. All of the above

22. Liposomes mediated gene transfer in plants involves :
A. Plasmid DNA enclosed in a lipid bag
B. Fusion of liposomes with protoplast
C. Use of polyethylene glycol (PEG)
D. All of the above

23. Which of the following plant signal molecules regulate the expression of vir B, C, D and E in case of tobacco?
A. Acetosyringone
B. α-hydroxy syringone
C. Both (A) and (B)
D. None of these

24. Which of the following is true about T-DNA?
A. Integration of T-DNA can occur at many different, apparently random, sites in the plant nuclear DNA
B. Integration of T-DNA occurs only at one specific sites in the plant nuclear DNA
C. Integration of T-DNA occurs at two specific sites in the plant nuclear DNA
D. Integration of T-DNA occurs at one site that may be random in the plant nuclear DNA

25. Which of the following is not true about the direct repeats flanking T-DNA?
A. They are conserved between nopaline and octopine Ti-plasmids
B. These repeats are transferred intact to the plant genome
C. These are important in integration mechanism
D. All of the above

26. Which of the following is not true for microinjection technique that involves transfer of DNA into protoplast?
A. It is carried out with the help of micromanipulator
B. The recipient cells are immobilized on artificial support or artificially bound to substrate
C. It employs needle with diameter greater than cell diameter
D. All of the above

27. Virulent strains of Agrobacterium contain large Ti-plasmids, which are responsible for the DNA transfer and subsequent disease symptoms. It has been shown that Ti-plasmids contain :
A. One set of sequence necessary for gene transfer
B. Two sets of sequence necessary for gene transfer
C. Three sets of sequence necessary for gene transfer
D. Four sets of sequence necessary for gene transfer

28. T-DNA transfer and processing into plant genome requires products of which of the following genes?
A. *Vir* A, B
B. *Vir* G, C
C. *Vir* D, E
D. All of these

29. Because of large size of Ti-plasmid, intermediate vectors (IV) are developed in which T-DNA has been subcloned into :
A. PBR 322 based plasmid vector
B. pRK 2013
C. pRN 3
D. All of these

30. In response to the activating signal molecule, an endonuclease is produced that causes nicks in the T-DNA. It is encoded by :
A. *Vir* A
B. *Vir* B
C. *Vir* C
D. *Vir* D

31. Microinjection involves :
A. Injection of large amount of DNA
B. Injection of DNA into bigger cells
C. Injection with needle having diameter greater than cell diameter
D. All of the above

32. Which of the following are used as selection marker for the cells transformed with *Agrobacterium*?
A. Neomycin phosphotransferase
B. Streptomycin phosphotransferase
C. Hygromycin phosphotransferase
D. Any of the above

33. Fermentation medium for oxytetracyclin (terramycin) consist of :
A. CSL, starch, $(NH_4)_2SO_4$, sodium chloride and $CaCO_3$
B. CSL, $(NH_4)_2SO_4$, sodium chloride and $CaCO_3$
C. CSL, starch, $(NH_4)_2SO_4$, ammonium chloride and $CaCO_3$
D. CSL, $(NH_4)_2SO_4$, ammonium chloride and $CaCO_3$

34. Media composition for the production of streptomycin is :
A. Soybean meal, glucose, peptone, malt extract, sodium chloride
B. Soybean meal, glucose, peptone, malt extract, ammonium chloride
C. Soybean meal, glucose, peptone, malt extract, calcium carbonate
D. Soybean meal, glucose, peptone, malt extract, ammonium sulphate

35. Media used for the production of chlorotetracyclin consists of :
A. Sugar, corn steep liquor, $CaCO_3$, $(NH_4)_2SO_4$ and NH_4Cl
B. Corn steep liquor, $CaCO_3$ and $(NH_4)_2SO_4$
C. Sugar, $CaCO_3$ and $(NH_4)_2SO_4$
D. Corn steep liquor, $CaCO_3$, $(NH_4)_2SO_4$ and NH_4Cl

36. Which of the following event occurs during second phase of growth of *P. chrysogenum?*
A. Synthesis of penicillin is high
B. Mycelial mass increases
C. pH increase
D. Both (A) and (B)

37. Inoculum preparation for the fermentation medium for penicillin takes place on :
A. Wheat seeds B. Barley seeds
C. Rice seeds D. All of these

38. Antibiotics are typically produced in fed batch reactors because :
A. Antibiotic yields are generally higher when cells enter the stationary phase
B. The precursors are often toxic to the cells
C. Antibiotic yields are generally higher when cell growth slows
D. All of the above

39. Which of the following event occurs during third phase of growth of *P. chrysogenum*?
A. Concentration of antibiotic increases in the medium
B. Autolysis of the medium starts
C. Slight rise in pH due to liberation of ammonia
D. All of the above

40. Which of the following changes occur during first phase of growth of *Penicillium chrysogenum?*
A. Growth of mycelia occurs
B. Ammonia is liberated in the medium
C. Lactic acid present in the corn steep liquor is utilized at the maximum rate
D. All of the above

41. Nitrogen in the plant cell culture media is provided by either ammonia or nitrate salt. In the media :
A. Utilization of ammonium cause culture pH to drop while utilization of nitrate cause culture pH to rise
B. Utilization of nitrate cause culture pH to drop while utilization of ammonium cause culture pH to rise
C. Utilization of both ammonium and nitrate result in rise in pH
D. Utilization of both ammonium and nitrate result in drop in pH

42. Which of the following is not true about nurse or conditioned medium?
A. It is liquid removed from the suspension of fast growing cells
B. It contains uncharacterized growth factor released by growing cells
C. It is used in the culture of regenerating protoplast
D. It is removed aseptically from the culture and is autoclaved before use

43. Very high sugar concentration (40-100 g/l) have been used :
A. In specialized secondary metabolite production
B. To adjust the osmotic potential of the media in short term treatment for regeneration
C. Both A and B
D. None of these

44. What is 'nurse' or conditioned medium?
A. It is the media full of growth factors used for the growth of cells
B. It is the medium added to nurse the callus culture
C. Both A and B
D. It is the liquid medium removed from the suspension of fast growing cells

45. Neutralized activated charcoal is occasionally added to young regenerating cultures to :
A. Remove toxic phenolics produced by the stressed plant cell
B. Help to remove plants growth regulators introduced at an earlier stage
C. Both A and B
D. Maintain the pH of the medium

46. Protoplasts can be produced from suspension cultures, callus tissues or intact tissues by enzymatic treatment with :
A. Cellulotic enzymes
B. Pectolytic enzymes
C. Both cellulotic and pectolytic enzymes
D. Proteolytic enzymes

47. Which of the following is considered as the disadvantage of conventional plant tissue culture for clonal propagation?
A. Multiplication of sexually derived sterile hybrids
B. Less multiplication of disease free plants
C. Storage and transportation of propagates
D. Both B and C

48. What is meant by 'Organ culture'?
A. Maintenance alive of a whole organ, after removal from the organism by partial immersion in a nutrient fluid
B. Introduction of a new organ in an animal body with a view to create genetic mutation in the progenies of that animal
C. Cultivation of organs in a laboratory through the synthesis of tissues
D. The aspects of culture in community which are mainly dedicated by the need of a specified organ of the human body

49. In a callus culture :
A. Increasing level of cytokinin to a callus induces shoot formation and increasing level of auxin promote root formation

B. Increasing level of auxin to a callus induces shoot formation and increasing level of cytokinin promote root formation
C. Auxins and cytokinins are not required
D. Only auxin is required for root and shoot formation

50. Which breeding method uses a chemical to strip the cell wall of plant cells of two sexually incompatible species?
A. Mass selection
B. Protoplast fusion
C. Transformation
D. Transpiration

ANSWERS

1	2	3	4	5	6	7	8	9	10
A	A	A	A	A	D	A	C	A	A
11	**12**	**13**	**14**	**15**	**16**	**17**	**18**	**19**	**20**
C	C	D	A	B	A	D	C	B	D
21	**22**	**23**	**24**	**25**	**26**	**27**	**28**	**29**	**30**
D	D	C	A	B	C	B	D	A	D
31	**32**	**33**	**34**	**35**	**36**	**37**	**38**	**39**	**40**
C	D	A	A	A	D	B	D	D	D
41	**42**	**43**	**44**	**45**	**46**	**47**	**48**	**49**	**50**
A	D	C	D	C	C	C	A	A	B

●●●

CHAPTER

13

Genomics

Introduction

It deals with the systematic use of genome information to provide answers in biology, medicine and industry. Genomics has the potential of offering new therapeutic methods for the treatment of some diseases, as well as new diagnostic methods. Major tools and methods related to genomics are bioinformatics, genetic analysis, measurement of gene expression, and determination of gene function.

Functional Genomics

(*i*) Expressed genes can be identified in a *c*DNA library.
(*ii*) A microarray can identify genes that are transcribed.
(*iii*) The coordinate regulation of many genes is revealed by a DNA microarray analysis.

Proteomics

(*i*) The proteome is much larger than genome.
(*ii*) 2D gel electrophoresis is used to separate a mixture of cellular proteins.
(*iii*) Mass spectrometry is used to identify proteins.
(*iv*) Protein microarrays can be used to study.

Bioinformatics

(*i*) Sequence files are analyzed by computer programmes.
(*ii*) The scientific community has collected sequence files and stored then in large computer databases.
(*iii*) Different computational strategies can identify functional genetic sequences.
(*iv*) Several computer-based approaches can identify structural genes within a nucleotide sequence.
(*v*) Computer programmes can identify-homologous sequences.
(*vi*) A simple dotmatrix can compare the degree of similarity between two-sequences.
(*vii*) A databased can be searched to identify homologous sequences.
(*viii*) Genetic sequences can be used to predict the structure of RNA and proteins.

Multiple Choice Questions

1. Researchers used _______ to orient and overlap sequence data to create complete sequence maps of entire chromosomes.
A. Sequence tagged sites (STSs)
B. DNA microarrays
C. Fluorescent single-molecule detection
D. TIGR

2. Which of these would support the hypothesis that a disease is caused by a certain mutation?

A. Finding the mutant gene expressed in cells affected by the disease using ESTs
B. Finding the mutant gene expressed in the genome of affected patients using DNA microarrays
C. Finding the mutant gene in animals affected by the disease using comparative genomics
D. All of these

3. The K_m (Michaelis constant) of an enzyme for a substrate is defined operationally as :
A. Half the substrate concentration at which the reaction rate is maximal
B. The substrate concentration at which the reaction rate is half maximal
C. The dissociation constant of the enzyme substrate complex
D. The dissociation constant of the enzyme product complex

4. Eukaryotic cells with DNA damage often cease progression through the cell cycle until the damage is repaired. This type of control over the cell cycle is referred to as :
A. Proteosome control
B. Damage control
C. Checkpoint control
D. Anticyclin control

5. A microarray is a large collection of specific DNA oligonucleotides spotted in a defined pattern on a microscope slide. What is the most useful experiment that can be done with such a tool?
A. Predicting the presence of specific metabolites in a cell
B. Comparing newly synthesized nuclear RNA with cytoplasmic RNA to locate introns
C. Comparing RNA produced under two different physiological conditions to understand patterns of gene expression
D. Comparing proteins produced under two different physiological conditions to understand their function

6. The KDEL sequence, found on luminal proteins of the ER, is responsible for :
A. Translocation of proteins into the ER lumen
B. Insertion of proteins into the membrane of the ER
C. Retrieval of ER luminal proteins from the Golgi
D. Recognition by signal peptidase of the signal sequence

7. Methylation at restriction sites in Bacterial genomes is prevented :
A. By its own endonucleases
B. Immune mechanism
C. Nuclease resistant genome
D. Are not much effective on bacterial genome

8. A C_3 mustard plant was grown at 300 ppm of CO_2 in 14 h light and 10 h dark cycles, it was transferred to 1000 ppm CO_2. This will lead to (other environmental parameters remaining identical) :
A. Increased Photosynthesis
B. Decreased Photosynthesis
C. Increase in Respiration
D. No Change

9. Freshly broken chromosome ends are sticky and tend to fuse, however ends of intact chromosomes are stable. Their stability is due to presence of :
A. Centromeres
B. Telomeres
C. Special membrane around chromosomes
D. Kinetochores

10. The structure formed where two adjacent membrane are thickened with disc shaped adhesive material in between and tonofibrils radiating out from adhesive region is :
A. Gap junction
B. Tight junctions
C. Desmosomes
D. Plasmodesmata

11. Spectrin of erythrocytes and cytochrome *c* of mitochondria, which can be easily dissociated by high ionic strength and metal ion chelating agent are example of :
A. Extrinsic Protein

B. Intrinsic Protein
C. Tunnel Protein
D. Cytoplasmic Protein

12. RER is found abundantly in goblet cells, pancreatic cells and liver cells is mainly engaged in :
A. Glycosylation of protein
B. Folding and secondary structure formation
C. Production of secretory and cytosolic protein
D. Production and excretion of protein

13. Why are individuals with an extra chromosome 21, which causes Down syndrome, more numerous than individuals with an extra chromosome 3 or chromosome 16?
A. There are probably more genes on chromosome 21 than on the others.
B. Chromosome 21 is a sex chromosome and 3 and 16 are not.
C. Down syndrome is not more common, just more serious.
D. Extra copies of the other chromosomes are probably fatal.

14. Humans have 23 pairs of chromosomes, while our closest relatives, chimpanzees, have 24. Chromosome studies indicate that at some point early in human evolution, two chromosomes simultaneously broke into a large portion and a small portion. The large parts combined to form a large chromosome, and the small parts combined to form a much smaller chromosome (which was subsequently lost). This important chromosomal change could best be described as :
A. Nondisjunction followed by deletion
B. Translocation followed by deletion
C. Duplication followed by deletion
D. Translocation followed by inversion

15. The cell wall of micro-organisms is coated with certain plasma protein promoting the attachment of microbe to phagocytes, only then they can be phagocytosised. The coat protein are called as :
A. Globins B. Opsonins
C. Ovulbumins D. Phagosonins

16. After the damage of body tissues, blood vessel is dilated where damage has occurred, due to which permeability of blood vessel also increases. Vasodilation is caused by :
A. Histamine B. Kinin
C. Prostaglandin D. All of these

17. Cancers cell are monoclonal, are characterized by uncontrolled growth, invasion of other tissues and dissemination to other tissues. The phenomenon of invasion to other tissues is termed as :
A. Angiobiogenesis
B. Metastasis
C. Diapedesis
D. Transformation

18. A major protease secreted by cancer cells acts on plasminogen and converts it into plasmin. Plasmin is proteolytic enzyme that dissolves blood clots and also removes exposed protein groups at cell surface. If the plasminogen is removed from the medium, then :
A. The morphology of cancer cells returns to normal
B. The cancer cell will show more exponential growth
C. Cancer cell will die
D. No change will be seen

19. One of the major higher molecular weight glycoprotein component which can be easily isolated from normal cultured fibroblast by mild treatment of urea, also occur at "foot prints" that moving culture cells leaves, is totally absent in cancerous cell is :
A. Fibronectin B. Albumin
C. Ferritin D. Transferin

20. When the human genome draft sequence was released, which was least expected?
A. The large amount of repetitive DNA
B. The size of total genome
C. The size of individual chromosomes
D. The small number of protein coding genes

21. In a Sephadex gel filteration column, a mixture of albumin, lysozyme and thymidine was loaded. In what sequence these will be eluted from the column?

A. Albumin > Lysozyme > Thymidine
B. Lysozyme > Thymidine > Albumin
C. Thymidine > Albumin > Lysozyme
D. Thymidine > Lysozyme > Albumin

22. Animal viruses cannot be seen under phase contrast microscope because :

A. They do not have a defined morphology
B. They are too small to be seen under microscope
C. They do not have any envelop that make them transparent under microscope
D. Since they have no colour, they are not visible

23. You have homogenized plant tissue and would like to separate chloroplast from nuclei. Which of the following methods would be most suitable?

A. PAGE
B. Equilibrium density gradient centrifugation on CsCl gradients.
C. Differential centrifugation using sucrose gradients.
D. Gel filtration.

24. Suppression mutation results in restoration of wild type phenotype. The suppression of mutant phenotype is usually brought about :

A. By misreading of mutant codon and incorporation of a correct amino acid
B. By insertion of another copy of gene
C. By revertion of mutation to wild type
D. Any deletion of mutant gene

25. Leber hereditary optic neuropathy is an inherited condition what causes a loss of central vision resulting from a mutation in mitochondrial DNA. What is the probability of the children of a man with mutations in two genes and a woman with normal mitochondrial DNA inheriting this disorder as carriers?

A. 0 per cent B. 100 per cent
C. 50 per cent D. 66.67 per cent

26. A contig from the genome sequence of *Plasmodium falciparum,* with a single start codon when translated, was found to have high similarity with the enzyme dihydrofolate reductase and thymidylate synthase. Which of following statement is correct?

A. It contains domains present in both proteins, but neither in two enzymes
B. It is single bi-functional protein
C. Both the protein share a common domain
D. The protein is unrelated to both the enzymes. There is a problem with the similarity search program used for the study

27. *m*RNA can be isolated by passing cell lysate through a column of oligo (dT)-Cellulose. The method is example of :

A. Partition chromatography
B. Ion-exchange chromatography
C. Affinity chromatography
D. Adsorption chromatography

28. In many situations it has been found that the nucleotide sequences of two highly homologous proteins are different. It may be due to use of :

A. Different amino acids
B. Non-overlapping genes
C. Different strands of DNA for encoding gene
D. Synonymous codons due to degeneracy

29. If the number of heterozygous pairs involved in a particular cross is three, the phenotype ratio obtained in their F2 generation will be :

A. 9:3:3:1
B. 27:9:9:9:6:6:3:3:1
C. 27:9:9:9:3:3:3:1
D. 1:2:1:2:4:2:1:2:1

30. The genomic size of *E. coli* is 4.6×10^6 bp, if a 6 base pair cutter is utilized to obtain restriction fragments, then what will be total number of fragments obtained :

A. 1.12×10^3 B. 7.66×10^5
C. 25×10^4 D. 3.83×10^3

31. Genetic variations, in form of multiple alleles of many genes, exist in most of natural population. Such genetic differences between individuals are called DNA polymorphism. Such a DNA polymorphism is used as :
A. DNA markers
B. DNA probes
C. *c*-DNA
D. Heteroduplex DNA

32. The best DNA markers utilized to differentiate different human individuals (to establish paternity and divergence) are :
A. Single Nucleotide Polymorphism
B. Rest Fragment Length Polymorphism
C. Random Amplified Polymorphic DNA
D. Simple tandem repeats polymorphism

33. The enzyme coded by allele IA transferase which adds N-acetyl glucasmine and blood group is designated as A, similarly IB encodes transferase which adds galactose to precursor sugar, similarly IO encodes transferase which adds :
A. No sugar
B. N-acetyl glucasmine
C. Galactose
D. Both galactose and N-acetyl glucsamine

34. Which of the following is true of heatshock genes?
A. They are only found in prokaryotes
B. They encode for proteins such as progesterone and estrogen
C. They are only found in eukaryotes
D. They are known to respond to stressful conditions

35. Dioxin, produced as a byproduct of various industrial chemical processes, is suspected of causing cancer and birth defects in animals and humans. It apparently acts by entering cells and binding to proteins, altering the pattern of gene expression. The proteins affected by dioxin are probably :
A. Water-soluble proteins
B. DNA polymerase
C. Transcription factors
D. Enhancers

36. In humans, the hormone testosterone enters cells and binds to specific proteins, which in turn bind to specific sites on the cells DNA. These proteins probably act to :
A. Help RNA polymerase transcribe certain genes
B. Alter the pattern of DNA splicing
C. Inhibit transcription.
D. Unwind the DNA for gene transcription

37. It is possible for a cell to make proteins that last for months; haemoglobin in red blood cells is a good example. However, many proteins are not this longlasting. They may be degraded in days or even hours. Why do cells make proteins with such short lifetimes if it is possible to make them last longer?
A. Most proteins are used only once
B. Most cells in the body live only a few days
C. Cells lack the raw materials to make most of the proteins they need
D. Only cancer cells, which can keep dividing, contain long-lasting proteins

38. Imagine an error occurring during DNA replication in a cell, so that where there is supposed to be a T in one of the genes there is instead a G. What effect will this probably have on the cell?
A. Each of its kinds of protein will contain an incorrect amino acid
B. An amino acid will be missing from each of its kinds of protein
C. One of its kinds of protein might contain an incorrect amino acid
D. The amino acid sequence of one of its kinds of protein will be completely changed

39. All your cells contain proto-oncogenes, which can change into cancer causing genes. Why do cells possess such potential time bombs?
A. Viruses infect cells with proto-oncogenes
B. Proto-oncogenes are genetic junk and have no known function

C. Proto-oncogenes are unavoidable environmental carcinogenes
D. Cells produce proto-oncogenes as a by-product of mitosis

40. A cell biologist found that two different proteins with largely different structures were translated from two different *m*RNAs. These *m*RNAs, however, were transcribed from the same gene in the cell nucleus. Which mechanism below could best account for this?
A. Different systems of DNA unpacking could result in two different *m*RNAs
B. A mutation might have altered the gene
C. Exons from the same gene could be spliced in different ways to make different *m*RNAs
D. The two *m*RNAs could be transcribed from different operons

41. A particular __________ carry the information for making a particular polypeptide, but __________ can be used to make any polypeptide.
A. Gene and ribosome; a *t*RNA and an *m*RNA
B. Gene and *m*RNA; a ribosome and a *t*RNA
C. Ribosome and *m*RNA; a gene and a *t*RNA
D. Gene and *t*RNA; a ribosome and an *m*RNA

42. The nucleotide sequence of a DNA codon is GTA. A messenger RNA molecule with a complementary codon is transcribed from the DNA. In the process of protein synthesis, a transfer RNA pairs with the *m*RNA codon. What is the nucleotide sequence of the *t*RNA anticodon?
A. CAT B. GUA
C. CAU D. GTA

43. The nucleotide sequence of a DNA codon is ACT. A messenger RNA molecule with a complementary codon is transcribed from the DNA. In the process of protein synthesis, a transfer RNA pairs with the *m*RNA codon. What is the nucleotide sequence of the *t*RNA anticodon?
A. TGA B. UGA
C. TGU D. ACU

44. A sequence of pictures of polypeptide synthesis shows a ribosome holding two transfer RNAs. One *t*RNA has a polypeptide chain attached to it; the other *t*RNA has a single amino acid attached to it. What does the next picture show?
A. The polypeptide chain moves over and bonds to the single amino acid
B. The amino acid moves over and bonds to the polypeptide chain
C. The *t*RNA with the polypeptide chain leaves the ribosome
D. A third *t*RNA with an amino acid joins the pair on the ribosome

45. Which of the following would be most likely to lead to cancer?
A. Multiplication of a proto-oncogene and inactivation of a tumor-suppressor gene
B. Hyperactivity of a proto-oncogene and activation of a tumor-suppressor gene
C. Failure of a proto-oncogene to produce a protein and multiplication of a tumor suppressor gene
D. The failure of both a proto-oncogene and a tumor-suppressor gene to produce proteins

46. Which of the following statements about repetitive DNA is NOT true?
A. Repetitive DNA is associated with the centromeres and telomeres in higher eukaryotes
B. Repetitive DNA is restricted to nontranscribed regions of the genome
C. Repetitive DNA sequences are often found in tandem clusters throughout the genome
D. Repetitive DNA was first detected because of its rapid reassociation kinetics

47. The ability of a cell to migrate on a substrate involves all of the following EXCEPT :
    A. Formation and breakage of focal adhesions
    B. Assembly of an actin meshwork at the leading edge
    C. Connexin proteins
    D. *Arp*2/3 complex proteins

48. The glyoxylate cycle is found in plants and bacteria but not in animals. The lack of this cycle in animals results in the inability to :
    A. Synthesize oxaloacetate from isocitrate
    B. Synthesize glutamate from malate
    C. Perform gluconeogenesis from amino acids
    D. Perform gluconeogenesis from fatty acids

49. Mitosis and meiosis accomplish segregation of the replicated DNA to two or more daughter cells. Which of the following is characteristic of both mitosis and meiosis?
    A. Chromosomes attach to spindle fibers composed of actin
    B. The resulting cells are diploid ($2n$)
    C. The resulting cells are haploid ($1n$)
    D. Spindle fibers attach to chromosomes at their kinetochores

50. Treatment of root tip meristem cells with the microtubule inhibitor colchicine results in all of the following EXCEPT :
    A. Induction of polyploidy
    B. Prevention of cytokinesis
    C. Inhibition of mitotic spindle assembly
    D. Cessation of DNA replication

ANSWERS

1	2	3	4	5	6	7	8	9	10
A	D	B	C	C	C	A	A	B	A
11	**12**	**13**	**14**	**15**	**16**	**17**	**18**	**19**	**20**
A	D	D	B	B	D	B	D	A	D
21	**22**	**23**	**24**	**25**	**26**	**27**	**28**	**29**	**30**
A	B	C	A	A	B	C	D	C	D
31	**32**	**33**	**34**	**35**	**36**	**37**	**38**	**39**	**40**
A	D	A	D	C	A	B	C	B	C
41	**42**	**43**	**44**	**45**	**46**	**47**	**48**	**49**	**50**
B	B	D	B	A	B	D	D	D	D

●●●

CHAPTER

14

Transgenic Plants

Introduction

Transgenic plants are plants that have been genetically engineered, a breeding approach that uses recombinant DNA techniques to create plants with new characteristics. They are identified as a class of genetically modified organism (GMO).

- The first genetically modified crop plant was produced in 1982, an antibiotic-resistant tobacco plant. The first field trials occurred in France and the USA in 1986, when tobacco plants were engineered for herbicide resistance.
- The People's Republic of China was the first country to allow commercialized transgenic plants, introducing a virus-resistant tobacco in 1992, which was withdrawn in 1997.
- The first genetically modified crop approved for sale in the U.S., in 1994, was the *FlavrSavr* tomato. It had a longer shelf life, because it took longer to soften after ripening.

Role of Transgenic Plants in Crop Improvement

Transgenic plants or genetically modified plants (GMP) have both basic and applied uses which are briefly summarised below:

1. Transgenes will be important in increasing the efficiency of crop production systems. For instance tragenic plants resistant to herbicides, insects, viruses and other biotic and abiotic stress have already been produced.
2. Transgenic breeding is an effective means of inducing male sterility in crop plants. *e.g.,* Barnase and barstar systems in *Brassica napus.*
3. Transgenic plants which are stable for food processing have also been produced *e.g.,* Bruise resistant and delay ripening in tomato.
4. Several gene transfers have been aimed in improving the produce quality. *e.g.,* Protein and lipid quality: improved quality may be achieved by either suppression of or over production of endogenous genes.
5. Transgenic plants are aimed to produce novel biochemicals like interferon, insulin, immunoglobin etc. useful biopolymers like polyhydroxy-butyrate which are not produced by normal plants. These compounds are extracted from plants can be used as pharmaceutical (or) industrial substrates.
6. The utilization of transgenic plants as bioreacters (or) factories for the production of special chemical and pharmaceuticals is known as Molecular forming.
7. Transgenic plants have been produced that express a gene encoding antigenic protein from a pathogen. Use of transgenic plants as vaccines for immunization against pathogens fast emerging as an important objective.

8. Transgenic plants have proved to be extremely valuable tools in studies on plant molecular biology. Regulation of gene action, identification of regulatory, promoters sequences etc.

Limitations

1. Transgenic plants sometimes exhibit instable performance for character under consideration.
2. The transformants had the undesirable side effects of transgenes.
3. The position and integration of foreign gene in host genome effects the expression of transgene in the transformant.
4. Inability to transfer polygenic traits.
5. Transgenic breeding is a very expensive method of crop improvement and requires high technical skills.
6. Transgenic cells are recovered at a very low frequency in cell culture, more over regeneration of transgenic cells into whole plants is also difficult and time consuming task.
7. Transgenic breeding acts against natural evolution.
8. There are chances of developing new weed species through transgenic breeding.
9. Sometimes the foreign genes have adverse effects on genome of the recipient parent, in such cases it may give rise useless gene combinants which may become a problem.

Current Global Status of Transgenic Crops

In recent years, development of transgenic crops expressing a variety of novel traits such as insect resistance, disease resistance, herbicide tolerance, hybrid production, improved oil quality etc. have led to large scale cultivation of transgenic crops.

- In 2013, GM crops were planted in 27 countries; 19 were developing countries and 8 were developed countries. 2013 was the second year in which developing countries grew a majority (54 per cent) of the total GM harvest.

Country	*2013-GM Planted Area (million hectares)*	*Biotech Crops*
USA	70.1	Maize, Soybean, Cotton, Canola, Sugarbeet, Alfalfa, Papaya, Squash
Brazil	40.3	Soybean, Maize, Cotton
Argentina	24.4	Soybean, Maize, Cotton
India	11.0	Cotton
Canada	10.8	Canola, Maize, Soybean, Sugarbeet
Total	**175.2**	–

- According to the 2013 International Service for the Acquisition of Agri-biotech Applications (ISAAA), a total of 36 countries (35 + EU-28) have granted regulatory approvals for biotech crops for food and/or feed use and for environmental release or planting.
- Japan has the largest number (198), followed by the U.S.A. (165, not including "stacked" events), Canada (146), Mexico (131), South Korea (103), Australia (93), New Zealand (83), European Union (71 including approvals that have expired or under renewal process), Philippines (68), Taiwan (65), Colombia (59), China (55) and South Africa (52).
- Maize has the largest number (130 events in 27 countries), followed by cotton (49 events in 22 countries), potato (31 events in 10 countries), canola (30 events in 12 countries) and soybean (27 events in 26 countries).
- In November 2014, the USDA approved a GM potato that prevents bruising.
- In February 2015, Arctic Apples were approved by the USDA, becoming the first genetically modified apple approved for US sale. Gene silencing is used to reduce

the expression of polyphenol oxidase (PPO), thus preventing enzymatic browning of the exposed fruit after it has been sliced open.

Transgenic Research Status in India

Bt cotton was the first crop approved for commercial cultivation in India. The Bt cotton (Bollgard) was approved in 2002.

- In the year 2002, first transgenic crop planting of Bt cotton was cultivated in about 50,000 ha in India.
- Presently, about 11 million hectare (2013) area is estimated to be under approved Bt cotton hybrids.
- But in February 2010, the government imposed a moratorium on the commercial cultivation of Bt brinjal EE-I, the first biotech food crop, due to concerns aired by green activists.
- The institutions under the ICAR, DST, DBT and CSIR have established a large number of facilities where most advanced transgenic research are being done.

Multiple Choice Questions

1. The first transgenic plants expressing engineered foreign genes were tobacco plants produced by the use of :
   A. *Agrobacterium tumefaciens*
   B. *Bacillus thuringiensis*
   C. *Arabidopsis thaliana*
   D. *Streptomyces hygroscopicus*
2. For developing herbicide resistance in transgenic plants which of the following approach/(es) is/are used?
   A. Target molecule is made insensitive to herbicide
   B. Target protein is overproduced
   C. A pathway should be introduced that detoxify the herbicide
   D. All of the above
3. Resistance to glyphosphate in transgenic petunia plants has been developed by the transfer of :
   A. Gene for EPSPS (5-enol-pyruvyl shikimat 3 phosphate synthase)
   B. Gene for ALS (acetolactate synthase)
   C. Gene for GS (glutamine synthase)
   D. Any of the above
4. Resistance to sulphonyl urea compounds in transgenic tobacco plant has been developed by the transfer of :
   A. Gene for EPSPS (5-enol-pyruvyl shikimat 3 phosphate synthase)
   B. Gene for ALS (acetolactate synthase)
   C. Gene for GS (glutamine synthase)
   D. Any of the above
5. Resistance to L-phosphinothrium in tobacco plants is developed by the transfer of :
   A. Gene for EPSPS (5-enol-pyruvyl shikimat 3 phosphate synthase)
   B. Gene for ALS (acetolactate synthase)
   C. Gene for GS (glutamine synthase)
   D. Any of the above
6. Which of the following gene is transferred to plants that detoxify the herbicide atrazine?
   A. Nitrilase
   B. Glutathione S-transferase (GST)
   C. Phosphinothrium acetyl transferase
   D. All of these
7. Which of the following gene detoxify herbicide bronoxynil?
   A. Nitrilase
   B. Glutathione S-transferase (GST)
   C. Phosphinothricin acetyl transferase
   D. All of these
8. Which of the following gene detoxifies herbicide phosphinothricin?
   A. Nitrilase
   B. Glutathione S-transferase (GST)
   C. Phosphinothricin acetyl transferase
   D. All of these

9. Fusion of karyoplast with the enucleated cell is achieved in presence of :
A. Cytochalasin B
B. Polyethylene glycol
C. Both (A) and (B)
D. Alcohol

10. Transfection refers to which of the following?
A. Synthesis of *m*RNA from DNA template
B. Synthesis of protein based on *m*RNA sequence
C. Introduction of foreign gene in to a cell
D. The process by which a cell become malignant

11. Which of the following is/are the method of transfection for making transgenic animals?
A. Transfer or whole nuclei
B. Transfer of whole individual chromosomes or fragment
C. Transfer of DNA
D. All of the above

12. For the transfer of whole individual chromosomes, they are isolated from the cells at :
A. Prophase B. Metaphase
C. Telophase D. Anaphase

13. Chromosomes may be isolated from metaphase cells by :
A. Hypertonic lysis
B. Hypotonic lysis
C. Either (A) or (B)
D. Isotonic lysis

14. DNA is microinfected into the fertilized egg :
A. After the fusion of male and female nuclei
B. Before the fusion of male and female nuclei
C. At the time of fusion of male and female nuclei
D. Any time, it can be infected

15. DNA microinjection into the egg has been used to produce which of the following transgenic animals?
A. Mice B. Chicken
C. Pigs D. All of these

16. Animal pharming can be defined as :
A. Growing animals for farming
B. Programming animals to produce novel products
C. Generating transgenic animals for farming
D. None of the above

17. Which protein has been produced generating a transgenic sheep that is used for replacement therapy for individuals at risk from emphysema?
A. Plasminogen activator (*t*PA)
B. α-anti trypsin (AAT)
C. Casein
D. Amyloid precursor proteins

18. Transgenic goats have been used to produce which of the following protein that is used for dissolving blood clots?
A. Amyloid precursor protein
B. α-anti trypsin (AAT)
C. Casein
D. The process by which a cell become malignant

19. Transgenic goats produce a variant of human tissue type plasminogen activator protein in :
A. Blood B. Urine
C. Milk D. Muscles

20. In transgenic fish, the genes are introduced by :
A. Microinjection in fish
B. Viruses
C. Transfer of whole nuclei
D. All of these

21. Which of the following gene have been introduced into the transgenic fish?
A. Human or rat gene for growth hormone
B. Chicken gene for delta crystalline protein
C. *E. coli* gene for β-galactosidase
D. All of the above

22. DNA into fish is injected into :
A. Pronuclei B. Cytoplasm
C. Both (A) and (B) D. None of these

23. Integrated nopaline T-DNA occurs as :
A. Single segment B. Two segments
C. Three segments D. Four segments

24. Integrated octopine T-DNA occurs as :
A. Single segment B. Two segments
C. Three segments D. Four segments

25. The left segment of octopine T-DNA (TL) is necessary for :
A. Enzymes for agropine biosynthesis
B. Tumour formation
C. Conjugative transfer
D. All of these

26. The right segment of octopine T-DNA (TR) is necessary for :
A. Enzymes for agropine biosynthesis
B. Tumour formation
C. Conjugative transfer
D. All of these

27. On Ti-plasmid T-region or T-DNA is flanked by a direct repeat of :
A. 12 bp B. 20 bp
C. 25 bp D. 30 bp

28. The direct repeats flanking the T-DNA of *Agrobacterium tumefaciens* are known as :
A. Cos site
B. Flanking sequences
C. Border sequences
D. Transfer sequences

29. Which of the following is not true about the direct repeats flanking T-DNA?
A. They are conserved between nopaline and octopine Ti-plasmids
B. These repeats are transferred intact to the plant genome
C. These are important in integration mechanism
D. All of the above

30. T-DNA transfer and processing into plant genome requires products of which of the following genes?
A. *Vir* A, B B. *Vir* G, C
C. *Vir* D, E D. All of these

31. *Vir* genes required for the T-DNA transfer and processing are located :
A. On the T-DNA
B. Outside the T-DNA region
C. On the plant genome
D. None of these

32. Which of the following plant signal molecules regulate the expression of *vir* B, C, D and E in case of tobacco?
A. Acetosyringone
B. A-hydroxy syringone
C. Both A and B
D. None of these

33. Which of the following genes are constitutively expressed and control the plant induced activation of other *vir* genes?
A. *Vir* A and *vir* G B. *Vir* C and *vir* D
C. *Vir* B and *vir* E D. *Vir* A and *vir* B

34. In response to the activating signal molecule, an endonuclease is produced that causes nicks in the T-DNA. It is encoded by :
A. *Vir* A B. *Vir* B
C. *Vir* C D. *Vir* D

35. Because of large size of Ti-plasmid, intermediate vectors (IV) are developed in which T-DNA has been subcloned into :
A. PBR 322 based plasmid vector
B. PRK 2013
C. PRN 3
D. All of these

36. Intermediate vectors containing T-DNA are conjugation deficient. Thus conjugation is mediated in presence of which of the following plasmid?
A. PRK 2013 B. PRN 3
C. Either A or B D. None of these

37. Which of the following is not true about the helper plasmids?
A. These can replicate in *Agrobacterium*
B. These help in the mediating conjugation of intermediate vectors
C. These can't replicate in *Agrobacterium*
D. All of the above

38. The transfer of intermediate vectors into Agrobacterium are brought about by :
A. Transformation
B. Biparental mating
C. Triparental mating
D. Transduction

39. Which of the following are used as selection marker for the cells transformed with *Agrobacterium*?
A. Neomycin phosphotransferase
B. Streptomycin phosphotransferase
C. Hygromycin phosphotransferase
D. Any of the above

40. Direct DNA uptake by protoplasts can be stimulated by :
A. Polyethylene glycol (PEG)
B. Decanal
C. Luciferin
D. All of these

41. Which of the following is not true for microinjection technique that involves transfer of DNA into protoplast?
A. It is carried out with the help of micromanipulator
B. The recipient cells are immobilized on artificial support or artificially bound to substrate
C. It employs needle with diameter greater than cell diameter
D. All of the above

42. Microinjection involves :
A. Injection of large amount of DNA
B. Injection of DNA into bigger cells
C. Injection with needle having diameter greater than cell diameter
D. All of the above

43. Microprojectile method of gene transfer in plants involves delivery of DNA :
A. With the help of micromanipulator
B. With the help of bolistics
C. With the help of needles
D. Any of the above

44. Advantage of microprojectile method over microinjection method for gene transfer in plants include :
A. Intact cells are used
B. Method is universal in its application irrespective of all shape, size, type and presence or absence of cell wall
C. Gene can be transferred to many cells simultaneously
D. All of the above

45. Liposomes mediated gene transfer in plants involves :
A. Plasmid DNA enclosed in a lipid bag
B. Fusion of liposomes with protoplast
C. Use of polyethylene glycol (PEG)
D. All of the above

46. In the liposome mediated gene transfer in plants, nucleic acids are :
A. Protected from nuclease digestion
B. Stable in liposomes
C. Both A and B
D. Not stable in liposomes

47. Plant transformation vectors based on Agrobacterium can generally be divided into :
A. Two vectors B. Four vectors
C. Six vectors D. Eight vectors

48. Co-integrating transformation vectors must include a region of homology in :
A. The vector plasmid
B. The Ti-plasmid
C. Between vector plasmid and Ti-plasmid
D. None of these

49. Virulent strains of Agrobacterium contain large Ti-plasmids, which are responsible for the DNA transfer and subsequent disease symptoms. It has been shown that Ti-plasmids contain :
A. One set of sequence necessary for gene transfer
B. Two sets of sequence necessary for gene transfer
C. Three sets of sequence necessary for gene transfer
D. Four sets of sequence necessary for gene transfer

50. Which technique is used to introduce genes into dicots?
A. Electroporation
B. Particle acceleration
C. Microinjection
D. Ti-plasmid infection

ANSWERS

1	2	3	4	5	6	7	8	9	10
A	D	A	B	C	B	A	C	B	C
11	**12**	**13**	**14**	**15**	**16**	**17**	**18**	**19**	**20**
D	B	B	B	D	B	B	C	C	A
21	**22**	**23**	**24**	**25**	**26**	**27**	**28**	**29**	**30**
D	B	A	B	B	A	C	C	B	D
31	**32**	**33**	**34**	**35**	**36**	**37**	**38**	**39**	**40**
B	C	A	D	A	C	A	C	D	A
41	**42**	**43**	**44**	**45**	**46**	**47**	**48**	**49**	**50**
D	D	B	D	D	C	A	C	B	D

●●●

CHAPTER

15

Transgenic Animals

Introduction

A *transgenic animal* contains in its genome, a gene or genes introduced by one or the other technique of transfection. The gene introduced by transfection is called *transgene*. In animals, *transfection* specifies the introduction of a DNA segment, either naked or integrated into a vector, into an animal cell. It may be pointed out that the same phenomenon is known as *transformation* in all other organisms. However, in case of animals, transformation has long been used to described the change of normal, *i.e.,* nontumourous, cells in culture to tumour like cells. Therefore, it because necessary to use the term transfection to avoid confusion.

Transfection may be transient or permanent (stable). In *transient transfection*, the introduced gene(s) is(are) gradually lost from the daughter cells of transfected cells. But in case of *stable transfection*, the introduced gene(s) is(are) retained and expressed in all the cells derived from the transfected cells. Since most of the animal vectors are unstable, *i.e.,* are gradually lost, in the extrachromosomal state, stable transfections are ordinarily due to the integration of introduced gene(s) into the cell genome.

Objectives of Gene Transfer

The various objectives for which transgenic cells/animals are produced may be summarised into the following five categories:

1. Genes have been transferred and expressed into cultured cell lines to obtain the proteins encoded by them.
2. Genetic modification of animals may be aimed at improving their milk, meat, wool, etc. production.
3. Genes have been transferred into animals with a view to obtain a large scale production of the proteins encoded by these genes in the milk, urine or blood of such animals; such animals are often called *bioreactors* and the approach is referred to as *molecular farming* or even as *gene farming*.
4. A special case of gene transfer aims at alleviating or even eliminating the symptoms and consequent miseries of genetic diseases. In this approach, normal and functional copies of the defective gene (that produced the genetic diseases) are introduced into the patient (*gene therapy*).
5. Finally, specific transgenic animals strains or lines are created to fulfil specialized experimental and/or biomedical needs. A good example of such animals are "knock out" mice strains in which specific genes have been replaced or knocked out by their disrupted counterparts through a process of homologous recombination. In simple words, the strategy is as follows. The gene to be knocked out say, *tk* (thymidine kinase) gene, is cloned, and disrupted by insertion

with its coding region of another DNA sequence, *e.g.*, the *neo* gene of *E. coli*. The disrupted gene is used to transfect embryonic stem cells where it could replace the normal *tk* gene present in the genome of ES cells by homologous recombination. The ES cells containing such recombination can be selected by a suitable experimental design.

Proliferation of Axillary Buds

This approach of micropropagation utilizes cultures of either shoot tip or nodal explants on a medium supplemented with usually a cytokinin (*e.g.*, 1-3 mg/l BAP) and often an auxin. The cytokinin stimulates pre-existing shoot buds present in the explants (apical meristems in shoot tips, and axillary buds in nodal explants) to develop into shoots. Each leaf on such shoots has an axillary bud. In many species, *e.g.*, strawberry, these axillary branching in a culture reaches the maximum. The individual shoots are then excised and subcultured onto fresh medium to initiate a new cycle of multiplication by axillary branching. In most plant species, each explant would produce 5-6 shoots in 4-5 weeks; this would give upto 5^{10}-6^{12} plants in one year from a single initial explant provided all the shoots produced could be cultured and there was 100% survival.

In some species, *e.g.*, blueberry, *Dalbergia sisso*, etc., the axillary buds present on growing shoots do not develop into new shoots irrespective of the GR combination or culture condition used. In such cases, each preexisting shoot bud present in an explant develops in a single shoot. Each such shoot is excised and cut into pieces to obtain nodal explants of appropriate size which are subcultured to initiate a new cycle of micropropagation. In such cases, the rate of multiplication will depend on the number of nodal explants obtained from each shoot. This approach may give a multiplication rate of 3-4 in about 6 weeks.

Micropropagation of woody trees presents several problems, *e.g.*, poor growth *in vitro*, deterioration on subculturing, difficulties in rooting, browing of medium, fungi and bacteria present within plant tissues (endophytic) causing contamination, etc. Often juvenile or seedling explants of woody species are quite responsive, but explants from mature trees are problematic especially in terms of rooting. Therefore, explants from elite mature/old trees are obtained as follows: some branches of a tree are cut off, and the new shoots arising on the remaining portions of such branches are used when they are about 1 year old. Alternatively, suckers, copices etc. may be used. In some cases, *e.g.*, *Feijoa*, *Eucalyptus* etc., serial subculturing *in vitro* improves rooting and brings about rejuvenation of shoots.

Maintenance of Isolates/Strains

Industrial microbiology continuously uses specific micro-organism isolates/strains as research, assay, development and production cultures. These strains are highly valuable and must be preserved over long periods without any genetic and, as a result, phenotypic changes. A microbial strain discovered to produce a useful product is called *research culture* since it is now used for studies on product isolation and identification, strain improvement, etc. The micro-organisms used for assays of the biological activities of microbial products as well as for various screening techniques are known as *assay cultures*. A research culture producing a metabolite of sufficient commercial interest is used for developing the large scale fermentation procedures; it is now termed as *development culture*. These cultures become *production cultures* if their product is good enough to be marketable.

The various cultures can be maintained for reasonably prolonged period using one of the following approaches.

1. *Low temperature storage* at 2-6°C on agar slants or in liquid medium; useful for relatively shorter periods (2-6 months) for working stock cultures.
2. *Storage as frozen cultures* at –20 to –100°C (both agar slants and liquid cultures).
3. Storage as lyophilized cells under vacuum at low temperatures (5°C or even –20 to –70°C). Lyophilization removes the free

water from cells and spores. Protective agents like skimmed milk or sucrose reduce the detrimental effects of lyophilization. Lyophilization is easily done by using automated freeze-drying machines in which the cells are first frozen and then their free water is withdrawn under high vacuum.

4. Storage of vegetative *cells/spores in liquid nitrogen* (–196°C) or in the vapour phase of liquid nitrogen (–167°C).

   Preservation of lyophilized cells in the most widely used method, but storage in liquid nitrogen is also as effective and widely used. These methods can theoretically maintain the micro-organisms indefinitely.

5. *Air dried at room temperature on sterile loam sand or on other natural substrates* like maize seed, rice bran, etc., and stored at room temperature or in a refrigerator. Bacterial cultures may remain viable upto 70-80 years. It is more suited for organisms that form spores.

6. Storage in glycerine stabs (0.85 ml of cell suspension mixed with 0.15 ml of sterile glycerol and stored at –70 or –75°C). This is commonly used for recombinant bacteria.

Procedure 1 for culture maintenance is used for working stock cultures, while those listed from 2 to 6 are used for the primary stock cultures. *Working stock cultures* are maintained in a vigorous and uncontained condition and are used quite frequently. They must be checked frequently for characteristic features and contamination. In contrast, *primary stock cultures* are kept in long-term storage for later use; they are used to produce new working stock culture as per need. A *stock culture* may be simply defined as a culture which serves as a source of inoculum.

Multiple Choice Questions

1. The techniques for gene transfer and transgenic animal production were developed using which of the following animals as models?

A. Mice B. Monkey
C. Rabbit D. Sheep

2. Anti-freeze protein promoter was isolated from which of the following?

A. Snake B. Frog
C. Fish D. None of these

3. Mammalian artificial chromosomes and human artificial chromosomes are linear vectors that contain which of the following?

A. Transcriptionally active chromosomal domains
B. Centromeres and telomeres
C. Origin of replication recognition sequences
D. All of these

4. The bacterial XGPRT gene serves as dominant selectable marker on a culture medium containing which of the following?

A. Mycophenolic acid
B. Aminoprotein
C. Xanthine and adenine
D. All of these

5. Which of the following are correct?

I. DEAE-dextran is water soluble and polycationic.
II. F-Virosomes deliver the DNA directly into the cytoplasm via fusion with cell plasma membrane.
III. Many cell lines do not like the calcium phosphate precipitate adhering to their substrate.
IV. In case of fish embryos, microinjection and electroporation are routinely employed.

Options:

A. I, II, III, IV B. II, III, IV
C. I, III, IV D. I, II, III

6. Which of the following is not correct?

A. The enzyme thymidine kinase belongs to the salvage pathway.

B. The F-protein of Sendai virus envelope mediates the fusion of viral envelope with cell plasma membrane
C. Lipofection is the method of choice for transfection of mammalian cells *in vitro*
D. Virosomes contain viral genome

7. Which of the following is not correct about SV40 plasmid vectors?
A. Plasmid vectors are generally used for stable transfection
B. They integrate into the host genone with a frequency of 10^{-5} to 10^{-3}.
C. Plasmid vectors have been mainly used to investigate sequences involved in transcriptional and post transcriptional regulation
D. pRSV series of vectors are derivatives of pSV vectors

8. *Xenopus* oocytes can be used for expression of which of the following?
A. *m*RNAs obtained by *in vitro* transcription of eukaryotic genes using T7 RNA polymerase.
B. Eukaryotic *m*RNAs isolated from various tissues
C. Genomic eukaryotic DNA microinjected into the nucleus
D. All of the above

9. In mouse cell lines, plasmid vectors having which of the following replicons yield stable episomal transformation?
A. SV40 replicons
B. BPV replicons
C. Murine-polyomavirus
D. EBV replicons

10. Which of the following polypeptide sequences are used as tags to facilitate recombinant protein purification?
A. Glutathione-*s*-transferase
B. Maltose binding protein
C. *myc* epitope
D. All of these

11. The chromatography column used to separate recombinant protein having polyhistidine tags has which of the following?
A. Divalent nickel B. Ca^{2+}
C. K^{+} D. Both (A) and (B)

12. Which of the following features are correct about *cre/lox* system?
I. The most commonly used recombinase is *cre,* isolated from *E. coli* bacteriophage P_1.
II. *Cre* recognizes a specific 34-bp site in DNA known as *lox P* and upon recognition, catalyzes a site specific recombination between two *lox P* sites.
III. Expression of the *cre* recombinase gene can be controlled by the use of a tissue specific promoter.
IV. Tissue-specific expression (of *cre* recombinase) allows tissue-specific deletion of the DNA sequence flanked by *lox P*.

Options
A. I, III, IV B. II, III, IV
C. I, II, III, IV D. I, II

13. SV40 transducing vectors must have which of the following features?
A. SV40 origin including the surrounding region containing the transcriptional regulatory signals
B. Genes encoding large-T, VP1, VP2, and VP3
C. Total size (including, that of the DNA insert) between 3,900 bp and 5,300 bp for packaging into virions
D. All of these

14. Cells take in calcium phosphate-DNA precipitate particles by which of the following?
A. Pincocytosis B. Phagocytosis
C. Both (A) and (B) D. Lipofection

15. All the transfection techniques are applicable to cultured animal cells, but microinjection is ordinarily not used due to which of the following?
A. Limited number of cells can be handled
B. Tediousness of the technique
C. Availability of less demanding alternative techniques
D. All of the above

16. Mouse is the most preferred mammal for studies/research on gene transfer due to which of the following?

I. Production and maintenance of ES cell lines.

II. Short oestrous cycle and gestation period.

III. Convenient *in vitro* fertilization and production of several offspring per pregnancy.

IV. Short generation time and successful culture of embryos *in vitro* at least for a period of time.

Options:

A. I, II, III, IV B. I, III, IV
C. III, IV D. II, III, IV

17. If inducible gene expression is driven by metallothionein promoter, transcription is induced by which of the following metal ions?

A. Cadmium B. Zinc
C. Iron D. Both A and B

18. Which of the following is often called molecular glue?

A. DNA ligase B. Synthetase
C. DNA polymerase I D. RNA polymerase

19. The human genome project was officially completed in which of the following years?

A. 2003 B. 2008
C. 2006 D. 2009

20. The site where protein binds a DNA molecule is determined by which of the following techniques?

A. FISH
B. DNA Fingerprinting
C. Gel retardation assay
D. DNA Footprinting

21. In Polymerase Chain Reaction (PCR), primer length affects the temperature used for which of the following?

A. Annealing
B. Storage of PCR Product
C. Primer Extension
D. Denaturation

22. The human genome is estimated to contain how many genes?

A. 45,000 B. 55,000
C. 25,000 D. 35,000

23. In case of single cell proteins, high content of which of the following is problematic?

A. Lipids B. Nucleic acids
C. Proteins D. Carbohydrates

24. The term 'Clonal selection' signifies. Which of the following in response to an antigen?

A. T cell proliferation
B. B cell proliferation
C. B/T cell maturation
D. Both A and B

25. Which of the following generates more number of different proteins from the same number of genes?

A. Alternative splicing B. RNA editing
C. Autosplicing D. *Trans* splicing

26. Which of the following mapping populations is immortal?

A. Recombinant inbred lines
B. F_2 population
C. Near isogenic lines
D. Both A and C

27. The term 'wound response' refers to the production of which of the following by the plant cells at wound sites?

A. Acetosyringone B. Lectins
C. Protein *Vir* G D. Protein

28. In hyper sensitive response, the *avr* gene of pathogen is responsible for generating, which of the following?

A. Repressor B. Elicitor
C. Toxin D. Receptor

29. Which of the following is a transgene?

A. A human gene transferred into *E. coli*
B. A human gene used for gene therapy
C. Both (A) and (B)
D. An *Aegilops* sp. gene transferred into *Triticum* sp. by backcrossing.

30. A biotechnological production process may use which of the following?

A. A microbe
B. An animal/a plant cell
C. An enzyme
D. Any of the above

31. SI nuclease is used for, which of the following?
A. Production of blunt ends from protruding ends
B. DNA footprinting
C. DNA sequencing
D. DNA fingerprinting

32. Which of the following techniques does not use antigen, antibody interaction for screening of expression libraries?
A. FACS B. ELISA
C. Complementation D. None of these

33. The animals having specific genes obtained from outside is known as :
A. Transgenic animals
B. Transgenic plants
C. Transgenic microbes
D. None of these

34. The microbes used in industry for producing different biochemical and various functions is called :
A. Transgenic plants
B. Transgenic microbes
C. Transgenic animals
D. None of these

35. A decline in the level of transgene-expression with the passage of generation of the transgenic line is called :
A. Transgene flow
B. Gene splicing
C. Transgene silencing
D. Transgene instability

36. The enzymatic attachment of one gene, or part of a gene, to another is called :
A. Gene splicing
B. Trangene flow
C. Transgene instability
D. None of these

37. Which of the following conditions is likely to interfere with the transfer of genetic material by conjugation in bacteria?
A. Treatment of the recipient cells with cycloheximide
B. Treatment of the mating cell pairs by application of strong shearing forces
C. Pre-treatment of the recipient cells with DNase
D. Pre-treatment of the recipient cells by application of strong shearing forces.

38. A DNA used in DNA cloning, usually made by reverse transcriptase, complementary to a given *m*RNA is called :
A. DNA supercoiling
B. DNA replicase system
C. *c*DNA library
D. Complimentary DNA

39. Which polypeptide sequences are used as tags to facilitate recombinant protein purification?
A. Maltose-binding protein
B. *myc* epitope
C. Glutathione-S-transfication
D. All of these

40. Which techniques is used to obtain information on the relative levels of *m*RNA present in the tissues of different transgenic individuals?
A. *In-situ* hybridization
B. Nuclease protein assay
C. Reverse transciption PCR
D. All of these

41. Human disease caused by mutation in mitochondrial genome :
A. Are inherited from the mother
B. Are inherited from the father
C. Are inherited from both mother and father
D. Do not exist because the mutation is always complemented by the normal

42. RNAi suffers from which limitations?
A. Longer dsRNA molecules trigger antiviral
B. RNAi based on exogenous siRNA is rather short-lived, *e.g.*, for 4-6 days only
C. Exogenous siRNA must be delivered into the cells, which is not highly efficient
D. All of these

43. Which is obtained by Southern hybridization?
A. Confirmation of integration of transgene
B. Number of sites of transgene integration per genome

C. Presence of *m*RNA/Protein produced by transgene
D. Confirmation of integration of transgene

44. Which of the following is not used as a genetic marker?
A. SINE
B. VNTR
C. RFLP
D. Single nucleotide polymorphisms (SNPs)

45. Which of the following is not true about the human Y chromosome (in male)?
A. It is packaged in nucleosomes
B. During mitosis, prior to metaphase, it contains only one chromatid
C. Its genes are known as holandric genes
D. It is a single linear DNA molecule

46. In bacteria, which enzyme binds single-stranded DNA, denatures double-stranded DNA and matches the single stranded DNA with complementary denatured DNA?
A. Uvr ABc B. RecA
C. Uvr D D. RecBcD

47. Galatosemia is a recessive single gene genetic disorder, caused due to the mutation in any one of the three genes involved in galactose catabolism. A family consists of 10 normal children with both parents suffering from galactosemia. This is most likely because of :
A. Reversion B. Suppression
C. Complementation D. Epistasis

48. Which is based on homologous recombination?
A. Gene replacement
B. Targetted gene transfer
C. Gene disruption
D. All of these

49. According to classical genetic, which of the following statements is true?
A. Gene on sex chromosomes segregate with the same pattern as autosomal genes
B. Genes on different autosomes segregate independently
C. Recessive alleles are detected by the phenotype of the F_1 generation
D. The closer two genes are, the more frequently they combine

50. Temperature sensitive mutant for a specific gene does not allow growth at the restrictive temperature by :
A. Degrading the protein as soon as it is expressed
B. Inhibiting translation
C. Expressing an unfolded form of the specific gene
D. Inhibiting the expression of the specific gene

ANSWERS

1	2	3	4	5	6	7	8	9	10
A	C	D	D	A	D	A	D	B	D
11	**12**	**13**	**14**	**15**	**16**	**17**	**18**	**19**	**20**
A	C	D	B	D	A	D	A	A	D
21	**22**	**23**	**24**	**25**	**26**	**27**	**28**	**29**	**30**
A	C	B	D	A	C	A	B	C	D
31	**32**	**33**	**34**	**35**	**36**	**37**	**38**	**39**	**40**
A	C	A	B	C	A	B	D	D	D
41	**42**	**43**	**44**	**45**	**46**	**47**	**48**	**49**	**50**
A	D	C	A	B	B	C	D	B	D

●●●

CHAPTER

16

Plant Biotechnology

Introduction

Plant biotechnology is concerned with developing ways to improve the production of plants in order to supply the world's needs for food, fibre and fuel. In addition, plants provide us with many pharmaceuticals and industrial compounds.

Role of biotechnology in crop improvement

1. **Crop improvement through plant tissue culture**
   - *Anther/ ovule/ pollen culture :* To obtain haploid plants and there by development of dihaploids on chromosome doubling
   - *Micropropagation :* Mass production of plants
   - *Meristem culture :* Virus free plants
   - *Production of secondary metabolites : e.g.;* Caffine from *Coffea arabica* and Nicotine from *Nicotiana rustica.*
   - *Somaclonal variation : e.g.;* 'ono' variety of sugarcane
   - *Embryo culture :* To overcome pre and post fertilisation barriers
   - *Ovary culture :* To obtain haploid plants from egg cell
   - *Endosperm culture :* To obtain triploid plants
2. **Use of protoplasm fusion and somatic hybrids:**
   - Transfer of CMS from *N. tabacum* to *N. sylvestris*
   - Transfer of diseases resistance from *Solanum bravidens* to *S. tuberosum.*
3. **Genome mapping and MAS**
   - MAS for QTLs and other agronomic character
   - MAS in prediction of heterosis and identification of heterotic pool.
   - For locating, identification and isolation of gene.
   - Plant variety protection
   - Gene pyramid and QTL introgression.
4. **Genetic engineering and transgenic plant**
   - *Bollgaurd :* Cotton variety against American bollworm
   - *Flavr savr :* Tomato improved for quality
   - *Endless summer :* Tomato variety
   - *Male sterility :* Barnase - Barstar in Brassica
5. **Edible vaccines**

Plant Organ Culture

(A) Anther Culture

Though *in vitro* culture of anthers was first undertaken by Shimakura (1934), it were Guha & Maheshwari (1964) of Delhi University who first successfully obtained numerous embryo-like structures called *embryoids* and subsequently the plantlets from mature anthers of *Datura innoxia.* These two workers observed the embryoids developed from the inside of the anthers at that time but in 1966 they confirmed that the origin of embryoids has been the regeneration of plants from anthers (androgenesis) as well as from excised pollen grains has been of much interest. The intact anthers containing the developing pollen have been cultured in numerous laboratories of the world and this has resulted in the development of embryoids directly from pollen grains. These embryoids have been induced to develop into whole plants which are true haploids (with reference to higher plants, haploids represents the gametic chromosomes constitution). (N.B. seeds of most flowering plants are diploids). It is evident from a recent review that most species capable of haploid production proceed via *anther culture or androgenesis.*

There is a long list of species in which haploids have been produced by anther culture technique. Some of these are *Datura metal*, *Nicotiana tabacum*, *N. sylvestris*, *N. alata*, *N. rustica*, *N. glutinosa*, *Solanum nigrum*, *Lycopersicon esculentum*, *Petunia hybrida* and *Capsicum annum*.

(a) Technique

For successful anther and excised pollen culture, it is essential to excise the flower buds at the correct time, the correct time is the stage of the first mitotic division of the uninucleate microspores tetrad. Buds are surface sterilized with alcohol, chlorine-water, sodium hypochloride or ethyl mercuric chloride. Now, the anthers are excised and transferred onto nutrient medium. Different nutrient media used for anther culture are Murashige-Scoog media, Nitsch medium and certain other modifications–1% agar is added to the medium to provide solid base. The transferred anthers, in the beginning, show slight swelling followed by initiation of cell division inside the massive connective tissue. This latter show initiations of light green crumbled callus. Simultaneously, the pollen sac swells up and ruptures along its upper margin lengthwise exhibiting a mass of pale-green pollen-callus. The callus arising from pollen shows slow growth as compared to connective callus. Now, the nutrient medium is supplemented with high concentration of kinetin for multiplication of callus, keeping the auxin concentration quite low. However, when formation of root and shoot is desired, the ratio of kinetin and auxin concentration is reserved in the medium.

(b) Applications

(*i*) Contrary to the fact that mutations are difficult to detect in higher plant because they are usually recessive and do not express themselves in the presence of their dominant alleles on the homologous chromosomes, the mutations induced in haploids can be easily detected because they have only a single set of genes and, consequently there is no interference by their dominant alleles.

(*ii*) Haploids provide a source of haploid tissue. Suspension of haploid cells can be obtained from haploid tissue. These haploid cells can be used to carry out new genetic researches on higher plants by applying the techniques used for micro-organisms.

(*iii*) Haploids can be easily treated with colchicine to obtain homozygous diploids in a single generation. It is much advantageous to the traditional plant breeding methods because it involves selfing for many generations in many years to obtain homozygous diploids when conventional method of plant breeding is applied.

(B) Ovary Culture

(a) Technique

LaRue (1942) was probably the first who successfully obtained *in vitro* rooting of the pedicel and a limited growth of ovaries. Nitsch (1951) developed this technique by growing excised ovaries of some plants (*e.g., Cucumis anguria*, *Nicotiana tabacum*, *Lycopersicon esculentum*, *Phaseolus vulgaris*) *in vitro* under aspetic conditions, the excised ovaries from pollinated flowers of *Cucumis anguria* and *Lycopersicon esculentum* viable seeds. But, however, the fruits so developed were comparatively smaller than those naturally developed. Maheshwari and Lal (1958) excised the ovaries from flowers of *Iberris amara* one day after pollination, grew them on a simple medium containing mineral salts and sugar supplemented with B-Vitamins and obtained normal healthy fruits matching the naturally developed ones.

(*i*) *Fruits developments :* The knowledge of fruit physiology is valuable in improving the quality of fruits. The technique of excised ovary culture may allow one to study the effect of chemicals directly on ovaries. Contrary to it, when the entire plant is employed with chemicals to study their effects on ovaries, it remains questionable whether the chemical used act in its original form or its originality is somewhat altered in the plant system before it reaches the ovary.

(*ii*) *Parthenogenetic Haploid Production :* Hess and Wagner ((1974) successfully obtained parthenogenetic haploids of *Mimulus luteus cv. Tigrirus grandiflorus* by pollinating their exposed ovaries in *in vitro* ovary culture studies. Similar results of producing parthenogenetic haploids have been obtained in the cultures of unfertilized ovaries of *Hordeum vulgare*, *Nicotiana tabacum* and *Triticum aestivum*.

(C) Nucellus Culture

(a) Technique

Nucellar tissue excised from post pollinated carpels, when the adventive embryos (the embryos arising from the maternal sporophytic tissue, outside the embryosac, are called *adventive embryos*), are already in torpedo-shaped stage, and cultured on case in hydrolysate supplemented with White's medium, proliferate and result in the development of callus. Numerous outgrowths, referred to as pseudobulbils, differentiate from the callus. The pseudobulbils develop into embryoids which germinate and finally, give rise to seedlings. It has been found for the first-time by Button and Bormann (1971) that the excised nucellar tissue from unfertilized ovule (*i.e.,* the case where there are no adventive embryos present), when grown *in vitro* under suitable conditions, develop embryoids. When the embryoids are excised and planted individually on a medium supplemented with GA_3 they sometime form plantlets. These plantlets have been successfully transplanted to soil by Button and Borman.

(b) Applications

The nucellar tissue culture may prove of great significance in horticulture. The seedlings obtained from embryoids are of parental type, as obtained through vegetative propagation by cuttings. However, nucellar seedlings of *Citrus* furnish better clones of orchard stock than cuttings as a result of the following reasons :

(*i*) The nucellar seedlings, have a tap root, and, therefore develop a better root system than do the cuttings. The latter have only a small lateral root system.

(*ii*) The nucellar seedlings generally restore the vigour which is lost after repeated propagation by cuttings.

(*iii*) The nucellar embryos prove to be disease free. The nucellar polyembryony has been used as a significant practical approach to develop virus-free clones of polyembryonate *Citrus* varieties.

Protoplast Culture

Protoplasts (Torrey and Landgren, 1977, have defined higher plant protoplast as "cells with their walls stopped off and removed from the proximity of their neighbouring cells) represent the plant cells minus their cell wall (*i.e.,* the naked cells) which are removed by either a mechanical or an enzymatic device. They are unusual because of being limited only by plasma membrane which is fully exposed and is the only barrier between external environment and the interior of the living cell.

Protoplasts represents the finest single cell system and offer exciting possibilities in the fields of somatic cell genetics and crop improvements. Since the isolated protoplasts grown in culture often perform better than single whole cells, they may serve as an excellent starting material for cell cloning and development of mutant lines. Besides being useful for somatic hybridization technique, higher plant protoplasts can introduce in them foreign DNA, cell organelles, bacteria, or virus particles through their exposed plasma membrane. These unique properties of protoplasts, combined with the totipotent nature of plant cells, have opened up an entirely new area of fundamental and applied research in experimental biology and somatic cell genetics.

The use of protoplasts for the introduction of foreign germplasm, somatic hybridization, and for studies of hormone action, membrane structure and function, and cell wall synthesis are well documented for model plant systems. These aspects of protoplast applications are described in application section.

Somatic Hybrids and Cybrids

Somatic hybridization is production of cells, tissues or organisms by fusion of non gametic nuclei. Phenomenon may be induced under laboratory conditions in cells that normally never fuse together and is used as tool in plant breeding or genetics. It may occur naturally, especially in fungi.

Somatic hybridization, often called *somatic fusion*, represents a novel technique which involves the fusion of somatic (body) cells, eliminating sex altogether, and results in the production of new hybrids. It is now well established that plant protoplasts/animal cells can be induced to fuse, either chemically or physically (electrically), and the fusion products, after suitable selection, cultured to result in somatic hybrids. The technique of somatic hybridization is being used extensively in many areas of fundamental and applied researches in plants and animals.

Following table lists intraspecific and interspecific somatic hybrids produced in *Nicotiana*

Hybrid ***Intraspecific***	***Reference***
N. debneyi + N. debneyi	Scowcroft and Laskin-1981
N. sylvestris + N. sylvestris	White & Vasil, 1979
N. tabacum + N. tabacum	Melchens and Labib, 1974
Interspecific	***Reference***
N. glauca + N. longsdorffti	Carlson *et al.*, 1972
N. sylvestris + N. knightiana	Maliga *et al.*, 1977
N. tabacum + N. glauca	Evans *et al.*, 1980
N. tabacum + N. glutinosa	Uchimiya, 1986
N. tabacum + N. knightiana	Maliga *et al.*, 1981
N. tabacum + N. nesophila	Evans *et al.*, 1983
N. tabacum + N. otophora	Sidorov *et al.*, 1981
N. tabacum + N. plumbaginifolia	Sidorov *et al.*, 1981
N. tabacum + N. repanda	Nagao, 1982
N. tabacum + N. rustica	Nagao, 1978
N. tabacum + N. stocktonii	Evans *et al.*, 1981
N. tabacum + N. sylvestris	Zelcer *et al.*, 1978
	Medgyergy *et al.*, 1980
	Evans *et al.*, 1983.

Disease Resistance

Though the intraspecific and interspecific protoplast fusion technique available with *Nicotiana* have not been exploited for crop improvement as yet, the potential agricultural usefulness of a somatic hybrid in the area of disease resistance has been examined by Evans *et al.*, (1981).

Stress Tolerance

One of the main objectives of somatic hybridization in plants is the transfer of genes conferring stress tolerance from a related wild species into modern high yielding crop varieties. It has been recognized that *'salt tolerance'* in *Lycopersicum cheesmanii* and *'cold tolerance'* in *L. hirusutum* would be useful if integrated into cultivated tomatoes. An effort in this direction has recently been initiated in the studies on somatic hybridization between *Lycopersicum esculentum* and *L. pennellii* (O'Connell and Hanson, 1985).

Transfer of Cytoplasmic Traits

Protoplast fusion technique could be useful in transferring cytoplasmic traits via transferring cytoplasmic genes from one species to other. Rick (1984) has studied the development of *'cytoplasmic male sterility'* via interspecific hybridization as a significant development in tobacco breeding. He has suggested that by employing this technique it might be possible to find out a workable-system of male sterility for the production of hybrid tobacco cultivars. Pelletier *et al.* (1983) have succeeded in transferring cytoplasmic male sterility into rapeseed (*Brassica napus*). *Brassica napus* plants have been regenerated after fusion between protoplasts bearing cytoplasms of different genera into modern high yielding crop varieties. It has been recognized that salt tolerance' in *Lycopersicum cheesmanii* and 'cold tolerance' in *L. hirusutum* would be useful if integrated into cultivated tomatoes. An effort in this direction has recently been initiated in the studies on somatic hybridization between *Lycopersicum esculentum* and *L. pennellii* (O' Connell and Hanson, 1985).

Transfer of Cytoplasmic Traits

Protoplant fusion technique could be useful in transfering cytoplasmic traits via transferring cytoplasmic genes from one species to other. Rick (1984) has studied the development of **'cytoplasmic male sterility'** via interspecific hybridization as a significant development in tobacco breeding. He has suggested that by employing this technique it might be possible to find out a workable system of male sterility for the production of hybrid tobacco cultivars. Pelletier *et al.*, (1983) have succeeded in transferring cytoplasmic male sterility into rapeseed (*Brassica napus*). *Brassica napus* plants have been regenerated after fusion between protoplasm bearing cytoplasms of different genera.

Multiple Choice Questions

1. Which one of the following scientists produced golden rice?
A. M.S. Swaminathan
B. Norman Borlaug
C. Ingo Potrykhus
D. Barbara McClintock

2. Eco RI is an enzyme that cuts :
A. DNA B. *t*RNA
C. *m*RNA D. Histones

3. Which one of the following is the first genetically modified crop?
A. Golden rice B. Bt cotton
C. Maize D. Flavr Savr tomato

4. Yeast artificial chromosome (YAC) vectors contain selectable markers. Loss of which marker at the cloning site distinguishes the religated YACs from the original vector marker?
A. TRPl B. SUP4
C. URA3 D. CEN

5. Which of the following methods of plant transformation can be used to introduce a gene into chloroplast genome?
A. *Agrobacterium*-mediated transformation
B. Particle delivery system
C. Permeabilization
D. Electroporation

6. A researcher has isolated a restriction endonuclease that cleaves at only one specific 10 base pair site.
P. Would this enzyme be useful in protecting cells from viral infections, given that a typical viral genome is 5×10^4 base pairs long?
Q. Restriction endonucleases are slow enzymes with turnover number of $1s^{-1}$. Suppose the isolated endonuclease was faster with turnover numbers similar to those for carbonic anhydrase (10^6 s^{-1}), would this increased rate be beneficial to host cells, assuming that the fast enzymes have similar levels of specificity?
The correct combination of answer is :
A. P : No; Q : Yes B. P : No; Q : No
A. P : Yes; Q : No D. P : Yes; Q : Yes

7. Which one of the following chemicals is used for inducing roots?
A. Kinetin
B. Indole Acetic Acid
C. Gibberellin
D. Abscisic Acid

8. The enzyme that joins the DNA strand during construction of recombinant DNA is :

A. Polymerase B. Lipase

C. Ligase D. DNA gyrse

9. The technique used to amplify specific DNA sequence is :

A. RFLP B. RAPD

C. PCR D. SNPs

10. Bollgard cotton contains endotoxin gene from :

A. *Bacillus subtilis*

B. *Bacillus cereus*

C. *Bacillus thuringensis*

D. *Agrobacterium tumefaciens*

11. Golden rice developed by the presence of :

A. β-carotene B. Vitamin-D

C. Biotin D. Vitamin-C

12. Hybridoma technology is useful in the production of :

A. Single cell protein

B. Anticancer agents

C. Monoclonal antibodies

D. Biopesticides

13. Vaccines produced by using plant as an expression system is known as :

A. Edible vaccines

B. Oral vaccines

C. Planti bodies

D. DNA vaccines

14. Blue rose is created through the genetic manipulation of :

A. Photosynthesis

B. Amino acid synthesis

C. Primary metabolic pathway

D. Flavonoid pathway

15. The vector used for genetic transformation of plants is :

A. M13 Vector B. T-plasmid

C. Ti-plasmid D. Cosmid

16. *Cry* genes are present in :

A. *Bacillus subtilis*

B. *Bacillus megasporim*

C. *Bacillus aminoliquefaciens*

D. *Bacillus thuringensis*

17. Positional cloning refers to :

A. Using a selection procedure to clone a cDNA

B. Cloning a portion of a gene using PCR

C. Isolating a gene from a specific tissue in which it is being expressed

D. Mapping a gene to a chromosomal region and then identifying and cloning a genomic copy of the gene from the region

18. Most stable vector to insert eukaryotic DNA of 140kb size is :

A. BAC B. YAC

C. C-cosmid D. Plasmid

19. First transformation experiment was carried out using :

A. *Escherichia coli*

B. *Salmonella typhi*

C. *Streptococcus pneumoniae*

D. *Bacillus subtilis*

20. Plasmids containing replication origin sequences of 2 different hosts are :

A. Amplification vectors

B. Shuttle vectors

C. Expression vectors

D. Integrating vectors

21. Which bacterium is called the "guinea-pig" for all *r*DNA manipulations?

A. *E. coli*

B. *Bacillus thuringensis*

C. *Agrobacterium tumefaciens*

D. *Bacillus subtilis*

22. Which of the following genes was engineered in the "Flavr Savr" transgenic tomato variety?

A. 1-amino cyclopropane 1-carboxylic acid synthase

B. 1-amino cyclopropane 1-carboxylic acid oxidase

C. Expansin

D. Polygalacturonase

23. Match the following :

I. Bt cotton	1. Improved shelf life
II. Flavr Savr tomato	2. Improved nutritional quality

III. Round up rady Sorghum — 3. Pest resistance
IV. Golden yellow rice — 4. Herbicide tolerance

	I	II	III	IV
A.	4	2	1	3
B.	3	1	4	2
C.	4	2	3	1
D.	3	2	1	4

24. Bt cotton contains :
A. Dry gene
B. Cry gene
C. Nif gene
D. AmpR gene

25. Given below are two statements.
Assertion (A) : Foreign genes can be cloned into plasmid cloning vectors.
Reason (R) : Cloning vectors can be used for expressing the foreign genes in bacterial systems.
Which of these statements is correct?
A. Both (A) and (R) are correct
B. Both (A) and (R) are wrong
C. (A) is correct, (R) is wrong
D. (A) is wrong, (R) is correct

26. For sustained expression of a transgene in the successive generation of a cell line in culture, the ideal gene transfer can be obtained using :
A. Lentiviral vector
B. Adenoviral vector
C. Plasmid DNA containing the transgene
D. Only transgenic DNA

27. During transgenesis, the location of the genes and their number integrated into the genome of the transgenic animal are random. It is often necessary to determine the copy number of genes and their tissue-specific transcription. The following are the possible methods used for the determination.
(*a*) Polymerase Chain Reaction (PCR)
(*b*) Southern blot hybridization
(*c*) Reverse Transcriptase PCR
(*d*) Western Blot
Choose the correct set of combinations.
A. (*a*) and (*b*)
B. (*b*) and (*c*)
C. (*b*) and (*d*)
D. (*a*) and (*d*)

28. Triticale, the first man made cereal is an example of :
A. Artificial allopolyploidy
B. Artificial autopolyploidy
C. Man made crossing
D. Artificial evolution

29. In order to clone an eukaryotic gene in pBR322 plasmid vector, the desired DNA fragment was produced by PstI cleavage and incubated with PstI digested pRR322 (PstI cleavage site lies within the ampicillin resistant gene) and ligated. Mixture of ligated cells were used to transform *E. coli* and plasmid containing bacteria were selected by their growth in tetracycline containing medium. Which type of plasmid/s will be found?
A. Circular pBR322 plasmid containing the target gene and resistant to only tetracycline.
B. Circular pBR322 plasmid containing the target gene and resistant to tetracycline only and recircularised pBR322 plasmid resistant to both ampicillin and tetracycline.
C. Circular pBR322 plasmid containing the target gene and resistant to only tetracycline, recircularised pBR322 resistant to both ampicillin and tetracycline and concatemerized pBR322 resistant to both ampicillin and tetracycline.
D. Circular pBR322 plasmid containing the target gene and resistant to both ampicillin and tetracycline.

30. Choose the correct sequence of events in a next generation sequencing technology-based whole genome sequencing project.
A. DNA extraction → shearing → library preparation → sequencing → assembly → finishing → annotation → submission to Genbank.
B. DNA extraction → library preparation → sequencing → assembly → annotation → finishing → submission to Genbank.

C. DNA extraction → shearing → adapter ligation → library amplification → sequencing → assembly → finishing → annotation → submission to Genbank.
D. DNA extraction → adapter ligation → library amplification → shearing → sequencing → finishing → annotation → submission to Genbank.

31. Among existing technologics, which of the following vector systems would you prefer to use for generating a library for 140 kb eukaryotic genomic DNA fragments, while giving due consideration to size as well as stability of the insect?
A. Phage
B. Cosmid
C. Bacterial artificial chromosome (BAC)
D. Yeast artificial chromosome (YAC)

32. Which of the following food crops has recently been genetically engineered to obtain edible vaccine to develop immunity against hepatitis B?
A. Banana B. Maize
C. Potato D. Tomato

33. Rice can grow in water submerged soil because the plants :
A. Have spongy roots
B. Can oxidize the micro-environment around their roots and thus able to absorb oxygen
C. Are adapted to enable roots to ward off toxic products
D. All of the above

34. Gypsum is beneficial for :
A. Phosphorus enriched soil
B. Water logged soil
C. Alkaline soil
D. Saline soil

35. The volume of O_2 liberated in photosynthesis has which of the following ratio to CO_2?
A. 1 : 1 B. 2 : 1
C. 1 : 2 D. 3 : 1

36. The major source of sugar in the world is :
A. Watermelon B. Beetroot
C. Sugarcane D. Dates

37. The study of interaction between living organism of environment is called :
A. Ecosystem B. Phytogeography
C. Ecology D. Photosociology

38. Auxanometer is used for measuring :
A. Respiratory activity
B. Photosynthetic activity
C. Growth activity
D. Osmotic pressure

39. The colour of flower is due to the presence of :
A. Chlorophyll
B. Xanthophyll
C. Florigen
D. Chromoplast or anthocyanin

40. In angiosperm, the endosperm is :
A. Haploid B. Diploid
C. Triploid D. None of these

41. Neurospora is used as an experimental tool in :
A. Ecology B. Cytology
C. Genetics D. Physiology

42. Lichen involves two organisms as :
A. Virus and bacteria
B. Algae and bacteria
C. Algae and fungi
D. Fungi and mosses

43. The proteins that forms the walls of the microtubules are :
A. Actin B. Tubulin
C. Pectin D. Hydroxyproline

44. Archaebacteria differ from the true bacteria as they have :
A. Different cell membrane lipids
B. Cell wall composition
C. Ribosomal RNA structure
D. Growth not inhibited by antibiotics

45. Which group of land plants is most restricted to moist environments?
A. Lycophyta
B. Sphenophyta
C. Bryophyta
D. Angiosperms

46. Which group of plants has the greatest diversity (*i.e.*, the most species) living today?
A. Bryophyta B. Lycophyta
C. Gymnosperms D. Angiosperms

47. What single feature is probably most responsible for the success of angiosperms?
A. Seeds B. Fruit
C. Broad leaves D. Flowers

48. A very common mutualistic, symbiotic relationship between a fungus and the roots of a plant is :
A. Lichen B. Mycorrhizal
C. Ascomycete D. Basidiomycete

49. What are the names of the two genders of haploid yeast cells?
A. A and Alpha B. M and F
C. Y (A) and Y (B) D. None of these

50. Mycorrhizae are symbiotic associations between :
A. Algae and fungi
B. Root and fungi
C. Bacteria and root
D. Bacteria and fungi

ANSWERS

1	2	3	4	5	6	7	8	9	10
A	D	B	B	B	A	C	C	C	A
11	**12**	**13**	**14**	**15**	**16**	**17**	**18**	**19**	**20**
C	A	D	C	D	D	B	C	C	C
21	**22**	**23**	**24**	**25**	**26**	**27**	**28**	**29**	**30**
D	B	B	B	A	B	D	D	C	C
31	**32**	**33**	**34**	**35**	**36**	**37**	**38**	**39**	**40**
B	C	D	C	A	B	C	C	D	C
41	**42**	**43**	**44**	**45**	**46**	**47**	**48**	**49**	**50**
C	C	B	B	C	D	D	B	A	B

CHAPTER

17

Animal Biotechnology

Introduction

Animal biotechnology is defined as the application of biotechnology to the processing or production of materials by animals to provide goods and services.

The art and science of producing—genetically engineered animals have advanced very rapidly in the past few years. It now is possible to generate animals with useful novel properties for dairy, meat, or fibre production, for environmental control of waste production, for biomedical purposes or other human consumption, or that are nearly identical copies of animal choosen for useful traits.

Engineering of Animals for Human Health Purposes

Genetic engineering has the potential to produce domestic animals that can be used for biomedical purposes. Such uses can be divided into three major categories:

(*i*) Live cells, tissues and organs for Xenotransplantation.
(*ii*) Biopharmaceuticals for animal or human use.
(*iii*) Raw materials for processing into other useful end products.

Animal provide a number of products we use in every day life *viz.* milk, leather, wool, eggs and meat.

Roles of Animals in Biotechnology

(*i*) Animals are involved in lab experiments.
(*ii*) Without the use of animals humans might be in danger.

Animals Model

(*i*) Four models apply for the use of animals.
(*ii*) Living animals.
(*iii*) Living animal tissues or systems.
(*iv*) Non-living systems.
(*v*) Computer and mathematical approaches.

Living Animals

(*i*) These animals are living and usually have no threat to their well being.
(*ii*) Such animals may be known as laboratory or scientific animals.
(*iii*) Agricultural research often uses experimental groups of animals.

Living Animal Tissues or Systems

(*i*) Animal tissues can be cultured in a lab.
(*ii*) This saves the use of animals as well as the expense of feeding, housing and cleaning up after the animals.

Non-Living Systems

(*i*) Involves using non-living mechanical models that reflect animal activity.
(*ii*) These often relate to skeletal movement and locomotion.
(*iii*) Artificial replacement parts, such as hip-joints can be studies using non-living systems.

Computer and Mathematical Approaches

(*i*) Computer simulations with virtual reality and other uses help in biotechnology.

(*ii*) Computer modelling may be done with a propose biotechnology practice before it is tested with animals.

(*iii*) A laboratory animal is an animal used for laboratory or research purpose.

(*iv*) Good tending is needed to assume that the research is properly carried out.

(*v*) Without research involving animals, scientists would not have a base.

Animal Species

(*i*) Most animals are subjects of study at one time or another.

(*ii*) Most common species used in research laboratories are mice, rats, and hamsters.

(*iii*) A primate is an animal with thumb and forefinger opposition.

(*iv*) A non-human primate is similar to humans but is not a human.

Animal Reproduction

(*i*) It is process by which offspring are produced.

(*ii*) Sexual reproduction is the union of a sperm and egg to ultimately produce a new individual.

(*iii*) The union of sperm and egg is called fertilization.

(*iv*) Semen – the fluid produced by the male reproductive organs, contains sperm.

(*v*) Copulation – is the sexual uniting of animals so the male can ejaculate semen near the eggs in the reproductive tract of the female.

Reproductive Development

(*i*) Puberty is the time at which animal is capable of reproduction.

(*ii*) Fertilization – union of sperm and egg results in the formation of a zygote.

(*iii*) Gestation – is the period of pregnancy in animals.

(*iv*) Parturition – is the birthing process.

(*v*) Lactation – is the secretion of milk by the mammary glands.

(*vi*) Dry time is a time when mammals are not lactating.

The Estrous cycle

(*i*) Estrous is a time when a female is fertile and receptive to a male.

(*ii*) The estrous cycle is the time between the estrous.

(*iii*) The three periods in the estrous cycle that follow estrous are metestrus, diestrus and proestrus.

Artificial Insemination (AI)

(*i*) Artificial insemination is the transfer if collected semen to a recipient female. It is used with sheep, beef cattle, turkeys, and swine.

(*ii*) Generally female estrous cycles are regulated with hormone injections.

(*iii*) Sperm are collected from males by artificially promoting ejaculation.

Gender Preselection

(*i*) Gender preselection is choosing the gender of a potential offspring.

(*ii*) Gender preselection may be done by sorting sperm.

(*iii*) The DNA content of sperm varies slightly based on the gender of the offspring that would result from conception.

Embryo Transfer

(*i*) Embryo transfer is the harvesting of fertilized Ova from a donor and implanting them into a recipient.

(*ii*) The harvested embryos are transferred to a recipient.

(*iii*) Non-surgical transfer involves flushing the embryos from the uterine horn.

Transgenic Animals and Products

(*i*) Pigs have been used to manufacture human haemoglobin.

(*ii*) A Xenograft is the practice of grafting an organ or a tissue from one species into another.

(*iii*) Cattle – are used to control disease such as mastitis in dairy cows.

(*iv*) Mice – transgenic mice have been used in several ways.

The different types of vectors used for gene transfer in animals

	Vector	Derived from	Features
1.	**SV40 vectors**		
	(*a*) Early region replacement vectors	Replacement of large-T gene of SV40	1. Produce virions which infect host cells 2. Transient gene expression 3. Mammalian cells are hosts of SV40
	(*b*) Late region replacement vectors	Replacement of VP1, VP2, and VP3 genes of SV40, *e.g.*, SVGT-5	
	(*c*) Plasmid vectors	Origin of replication and large-T gene of SV40	—
	(*d*) Shuttle plasmid vectors	Plasmid vector [Item 1(c)] plus *p*BR322 origin and amp^r gene, *e.g.*, *p*SV2, *p*SV3, etc.	—
		Rous sarcoma virus promoter in Place of SV40 early promoter, *e.g.*, *p*RSV.	Strong expression of the maker gene
	(*e*) Passive transfecting vectors	SV40 transcription regulatory and polydenylation sequences, plus *p*BR322 origin and amp^r gene	Shuttle vectors; used for gene integration of mammalian cells
2.	**BPV vectors**	Bovine papilloma virus "Transforming region" + *p*BR322.	Shuttle vector; often *p*BR322 sequence deleted prior to transfection, plasmid-like vector
3.	**Retrovirus vectors**	*p*BR322 + retrovirus sequences	Shuttle vector; integrates as provirus into mammalian genome; produces virions
4.	**Polyoma vinis vectors**	Polyoma virus origin and early region + *p*BR322 sequences	Similar to SV40 vectors; mouse cells used as host.
5.	**Vaccinia virus**	DNA insert placed within the thymidine kinase gene of virus by a process of recombination.	Promising as live vaccines; DNA insert is a pathogen gene encoding an antigen.
6.	**P element vectors**	*Drosophila* transposable element P, minimum of 31 bp inverted repeat borders and the neighbouring regions, plus and *E. coli* vector, *e.g.*, *p*UC8; DNA insert of upto 40 kb placed within the two borders.	Gene transfer in *Drosophila*; a helper P element is needed to provide the transposes necessary for transposition or insertion of the recombinant P vector into the *Drosophila genome*.
7.	**Baculovirus vectors**	Nuclear polyhydroma virus (NPV) polyhedrin gene replaced by DNA insert; *e.g.*, AcNPV (*Autographia california* nuclear polyhedroma virus) and BmNPV (*Bombyx mori* nuclear polyhedroma virus) vectors.	Produce virions; expression vector for production of transgenic proteins in silk worm larvae (BmNPV vectors) and in *Spodoptera frugiperda* larvae or cultured cells (AcNPV vectors).

A List of Products (Recombinant Proteins) Generated using Recombinant DNA Technology

(The genes are expressed in animal cell cultures, and their products are either already in therapeutic use or in advanced stages of development).

Product (protein)	*Gene expressed in (host)*	*Application*
Currently in Therapeutic Use		
Human growth hormone (HGH)	Mouse mammary tumour cell line	Dwarfism
Tissue plasminogen activator (tPA)	Chinese hamster ovary (CHO) cell line	Thrombolysis
Erythropoietin	Chinese hamster ovary (CHO) cell line	Anaemia
Blood clotting factor VII	CHO cell line	Haemophilia
In Advanced Stages of Development		
Antithrombin VIII	CHO cell line	Anticoagulant
Blood factor VIII	BHK-21 cell line	Haemophilia
Blood factor IX	CHO cell line	Hemophilia-R
Interleukin-2 (Il-2)	—	Cancer therapy
Human recombinant antibodies	CHO cell line and Mouse myeloma lines	Infections, various cancers
Hepatitis-B surface antigen (HBSAg)	CHO cell line and C-127 cell line	As vaccine against Hepatitis-B
Interferons (IFN)	—	Tumours
Plasminogen activator (including chimeric and modified)	CHO cell line and *Spodoptera frugiperda*	Thrombolysis
CD4 Protein	Mammalian cell line	AIDS vaccine
gp160	Mammalian cell line	AIDS vaccine
*p*24 protein	Mammalian cell line	AIDS vaccine
Interleukin-1 receptor antagonist	Mammalian cell line	Severe sepsis
Atrial natriuretic	Mammalian cell line	Kidney failutre
Ciliary neurostropic factor	Mammalian cell line	Amyotrophic sclerosis
DNAse	Mammalian cell line	Cystic fibrosis

A List of Reporter or Marker Genes used in Gene Transfer in Animal Cells

Reporter gene	*Source*	*Remarks*
Scorable reporter genes		
Chloramphenical acetyl transferase (CAT)	*E. coli* transposon Tn9	Used in fish, mammalian cells

Reporter gene	Source	Remarks
β-galactosidase	*E. coli*	Used in fish
Luciferase (*lux*)	Bacteria	Used in fish
Selectable reporter genes		
Thymidine kinase (TK)	Herpes simplex virus (HSV)	Used in TK cells; HSV gene promoter is weak in mammalian cells
Dihydrofolate reductase (DHFR)	1. Chinese hamster ovary (CHO) cell line 'A29'	Resistance to methotrexate due to over production of DHFR; DHFR gene is amplified during integration.
	2. Bacteria	DHFR intrinsically resistant to methotrexate
CAD protein gene	Syrian hamster gene cloned in *E. coli*	Overproduction of CAD protein confers resistance to N-phospho-acetyl-L-aspartate (PALA). CAD gene is also amplified during integration.
Xanthine-guanine phosphoribosyl transferase (XGPRT)	*E. coli*	Resistance to mycophenolic acid
Neomycin phospho transferase	Bacterial transposons Tn5 and Tn601	Resistance to antibiotic G418

White Blood Cells, Types and Their Function

Cell Type	Function
A. Phagocytes	
1. Neutrophil	It participates in early stage of defence against micro-organisms.
2. Monocyte	It arrives at site after neutrophils transfer into macrophages; Engulf's foreign materials, present antigen to lymphocytes, stimulates lymphocyte proliferation.
3. Eosinophil	It responds to allergies and parasitic infections.
B. Lymphocytes	
1. B-cells	When they are activated by foreign molecules, produce plasma and memory cells.
2. Plasma cells	They secrete antibodies.
3. Memory cells	They respond to antigens during secondary response.
4. Natural killer cells	They directly destroy virus infected and cancerous cells.
5. Cytotoxic T-cells	They destroy virus-infected and cancerous cells.
6. Helper T-cells	They stimulate B-cell and killer T-cell proliferation.
7. Suppressor T-cells	They slow down immune response.
C. Basophils	They release histamine in inflammatory response.

Multiple Choice Questions

1. Polyuria is found in case of :
A. Fever B. Shock
C. Diabetes mellitus D. All of the above

2. Red urine is found in :
A. Haemoglobin (blood, free)
B. Myoglobin
C. Porphyrian
D. All of the above

3. Protein test in urine is done by :
A. Sulphosalicylic acid test
B. Heller's ring test
C. Sulphur test
D. Both A and B

4. Gmelin test is done for :
A. Bile pigment B. Protein
C. Glucose D. Blood

5. Wintrobe tube is primarily used for the estimation of :
A. DLC B. Hb
C. TLC D. PCV

6. Faecal examination is indicated in .... conditions.
A. Eosinophilia B. Anaemia
C. Icterus D. All of these

7. The larvae are separated from faecal culture by ______ method.
A. Zeil-Neelson B. Baerman
C. Gorcotts D. None of above

8. Canine has more ______ than lymphocytes.
A. Monocytes B. Eosinophils
C. Neutrophils D. Basophils

9. pH of normal milk is :
A. 6.5 B. 7.5
C. 5.5 D. 8.5

10. Tissue should be collected directly into ______ for histopathology.
A. Water B. Ice
C. Fixative D. None of these

11. Haemoglobulineria occurs in :
A. Leptospirosis B. Pasteurellosis
C. Tuberculosis D. Brucellosis

12. Type II diabetes mellitus is :
A. Juvinile diabetes
B. Adult on-set diabetes
C. Genetic diabetes
D. Gestational diabetes

13. In pre-renal azotemia kidney will be :
A. Highly affected
B. Moderate affected
C. Normal in function
D. None of these

14. $CuSO_4$ is used in the staining of :
A. Cell membrane B. Capsule
C. Spore D. Flagella

15. BUN and Creatinine is measured for the diagnosis of abnormal function of :
A. Liver B. Pancreas
C. Kidney D. Heart

16. Antidote of nitrate, nitrite and chlorate is :
A. Methylene blue B. BAL
C. Deferoxamine D. Tannic acid

17. The process of exclusion in closely related disease is known as :
A. Test therapy diagnosis
B. Clinical diagnosis
C. Laboratory diagnosis
D. Differential diagnosis

18. Increased TLC is referred to as ______.
A. Leucopenia B. Leucocytosis
C. Leukemia D. Lymphocytosis

19. The TLC count 6000/cu mm of blood will be______ per liter blood.
A. 6×10^9 B. 6×10^6
C. 6×10^3 D. 6×10^{12}

20. If no parasitic ova are detected in direct smear method saturate the sample should be declared ______ .
A. Negative
B. Positive
C. For re-examination by flotation
D. None of the above

21. To preserve the faecal sample, ______ is added as preservative.
A. 10 per cent ethanol
B. 10 per cent formalin
C. 10 per cent methanol
D. 10 per cent acetone

22. Teats should be washed with ______before collection of milk samples for the laboratory examination.
A. 70 per cent ethanol
B. 70 per cent formalin
C. Per cent acetone
D. None of the above

23. Washing solution in ELISA is ______ .
A. PBS B. Tris buffer
C. PBS-Tween D. PBS-conjugate

24. Strychnine can be detected in :
A. Stomach content B. Liver
C. Urine D. All of the above

25. Whole blood minus both the cells and clotting factor is k/a :
A. PCV B. Plasma
C. Serum D. Hb

26. Advantage of anticoagulant is :
A. Least effect on size and haemolysis of RBC
B. Used for blood transfusion
C. For routine haematology
D. All of the above

27. In dog blood is collected from :
A. Jugular vein B. Ear vein
C. Cephalic vein D. Wing vein

28. Which of the protein is involved in Alzheimer's disease?
A. Amylin B. A4
C. Alzhemin D. Fibrin

29. Glucagon harmone is secreted by :
A. Liver B. Kidney
C. Pancreas D. Spleen

30. Elivated serum conc. of triglyceride is due to :
A. Hypothyroidism B. D.M
C. Anaemia D. Both A and B

31. Role of Ca in body is :
A. Skeleton formation
B. Enzyme activation
C. Blood coagulation
D. All of the above

32. In pre-renal azotemia kidney will be :
A. Highly affected
B. Moderate affected
C. Normal in function
D. Least affected

33. Sweetish or fruity odour in urine indicate presence of :
A. Glucose B. Ketone bodies
C. Blood D. Urea

34. In urine sample Sulphosalicylic test is used for detection of :
A. Glucose B. Protein
C. Bile salt D. Urea

35. Majority of the parasites/ectoparasite are :
A. Leeches and ticks
B. Monogenetic trematodes
C. Mangemites
D. Arthopodes

36. Macrocytes are typically seen in ______ deficiency.
A. Folic acid B. Vitamin B_{12}
C. Both A and B D. None of these

37. Morbid changes are :
A. Alterations found in tissue at necropsy
B. Alterations found in tissue at biopsy
C. Alterations found in tissue as symptom
D. None of the above

38. P.M. examination in vet. medicine known as :
A. Necropsy B. Autopsy
C. Biopsy D. Any of the above

39. Main components of blood clot is :
A. Fibrin B. R.B.C.
C. Platelet D. None of these

40. Presence of cast of urine is known as :
A. Bilirubinuria B. Globulinuria
C. Proteinuria D. Cylindruria

41. EDTA is an antidote for which of the following :
   A. Sodium secobarbital
   B. Asprin
   C. Phosphorus
   D. Lead

42. Techoic acid is present in the cell wall of :
   A. Gram positive
   B. Gram negative
   C. Both (A) and (B)
   D. None of the above

43. Bacteria which remain red coloured after ZN staining are :
   A. Acid fast bacteria
   B. Partial acid fast bacteria
   C. Non acid fast bacteria
   D. Gram variable bacteria

44. In *staphylococcus* infection partial haemolysis is on the basis of :
   A. Coagulation test
   B. Dick test
   C. CAMP test
   D. None of the above

45. Medium which is suitable for the growth of a specific organism is termed as :
   A. Selective media
   B. Minimal media
   C. Complete media
   D. None of the above

46. *Taenia saginata* is pathogenic to :
   A. Dog B. Cat
   C. Man D. Pig

47. Normally eggs are operculated in :
   A. Trematode
   B. Roundworm
   C. Both (A) and (B)
   D. None of the above

48. *Fasciola hepatica* present in :
   A. Liver
   B. Caecum
   C. Small intestine
   D. None of the above

49. Erythrocytes arranged in a roll is called :
   A. Rouleaux B. Ovalocyte
   C. Band cell D. Shift to left

50. Thickness of blood smear depends upon :
   A. Size of drop
   B. Angle of spreader
   C. Both A and B
   D. None of the above

ANSWERS

1	2	3	4	5	6	7	8	9	10
C	D	D	A	D	D	B	C	A	C
11	**12**	**13**	**14**	**15**	**16**	**17**	**18**	**19**	**20**
A	B	C	B	C	A	D	B	A	C
21	**22**	**23**	**24**	**25**	**26**	**27**	**28**	**29**	**30**
B	A	C	D	C	D	C	B	C	D
31	**32**	**33**	**34**	**35**	**36**	**37**	**38**	**39**	**40**
D	C	B	B	D	C	A	A	A	C
41	**42**	**43**	**44**	**45**	**46**	**47**	**48**	**49**	**50**
D	A	A	C	A	C	A	A	A	B

●●●

CHAPTER

18

Food Biotechnology

Introduction

Microorganisms are associated in a variety of ways with all of the foods we eat. Microorganisms may influence the quality, availability and quantity of our food. Food can serve as a medium for growth of microorganisms and this growth may cause the food to undergo decomposition and spoilage. Some microorganisms, if allowed to grow in certain food products, produce toxin substances that result in food poisoning when the food is ingested. Still other microorganisms are used in the preparation and preservation of food products.

Factors that determining microbial growth in foods

A variety of factors such as the pH, moisture content, oxidation-reduction potential, nutrient etc. influence microbial activity in foods. For example, each organism has an optimum pH for growth. In general, yeast and fungi are more acid tolerant than bacteria. Most yeasts favour a pH around 4-4.5, while most bacteria favour a pH around 7.0. The pH of foods varies as some may be neutral while others may be acidic. Acidic foods are not readily spoiled by bacteria. Microorganisms have an absolute demand for water and the optimum level of moisture required for growth varies with the organism. Nutrients in food, their kind and proportions determine the type of organism that will grow. Microorganisms vary in their ability to use nutrients. The presence of easily utilizable nutrients will encourage faster growth and quicker spoilage.

Microbes important in foods

Some molds are important in food preparation as molds are involved in ripening of cheese and preparation of oriental foods. Yeasts are both useful as well as problem organism and this depends upon the food. For example, the production of wine is dependent on growth and activity of the yeast *Saccharomyces cerevisae;* while wine can be oxidized to CO_2 and water by wild yeast.

Fermented foods

There are many useful applications of microorganism in food industry.

Various food products prepared by use of microbes

Food	*Raw material used*	*Microbes involved*
Curd	Milk	*Lactobacillus, Streptococcus.*
Ripening of cheese	Milk	*Lactobacillus, Propionibacterium*
Bread	Flour	*Saccharomyces*
Sauerkraut	Cabbage	*Lactobacillus, Leuconostoc, Aerobacter.*
Wine	Fruit juices	*Saccharomyces*
Vinegar	Alcohol	*Acetobacter*
Dosa & Idli	Rice urd daal	*Saccharomyces*
Soysauce	Soyabean	*Aspergillus, Lactobacillus, Saccharomyces*

Single Cell Protein (SCP)

Bacteria, yeasts and algae, produced in massive quantities are attractive sources of food for animals as well as humans. These microorganisms can be cultivated on industrial wastes or by-products as nutrient and yield as large cell crop that is rich in protein.

They are the dried cells of microorganisms such as algae, certain bacteria, yeast and molds and that are grown in large scale culture systems for use as protein for human or animal consumption. The products also contain other nutrients like fat, vitamin and minerals.

Microbial flora of fresh food

The inner tissues of healthy plants and animals are free of microbial activity. However, the surface of raw vegetables and meats are contaminated with a variety of microorganisms.

(*i*) The carcass of a healthy animal slaughtered for meat and held in refrigerated room is likely to have only nominal surface contamination while the inner tissue are sterile. Each new surface of meat, resulting from a new cut, adds more microorganism to the exposed tissue. Common species of bacteria occurring on fresh meats are *Pseudomonas, Staphylococcus, Micrococcus, Enterococcus* and *coliforms.* The low temperature at which fresh meats are held, favours the growth of psychrophilic microbes.

(*ii*) The microorganisms found in milk are *Streptococcus lactis, Streptococcus cremoris* and certain other *Lactobacilli.* The principle change is lactose fermentation forming lactic acid that develop sour flavour.

Microbial Spoilage of Food and Food Poisoning

Given a chance to grow, the organism will produce changes in appearance, flavour, odour and other qualities of foods. This degradation process may be described as follows:

(a) Putrefaction

Protein food + Proteolytic M.O. → Amino acids + amines + ammonia + hydrogen sulphide

(b) Fermentation

Carbohydrates rich food + microbes → Acids + Alcohols + gases or methane and carbon dioxide

(c) Rancidity

Fattyacids + lipolytic microorganisms → Smaller fatty acids + glycerol + CO_2

Common types of spoilage and microbes

Food	*Types of spoilage*	*Microbes involved*
Bread	Moldy	*Rhizopus, Penicillium, Aspergillus.*
Fruits and vegetables	Ropy sour	*Bacillus subtilis, Lactobacillus.*
Fruits and vegetables	Soft rot, gray mold rot and black mold rot	*Rhizopus, Erwinia, Aspergillus, Botrytis*
Meat and Fish	Putrefaction	*Bacillus, Alkaligens, Clostridium, Pseudomonas.*
Concentrated juices	Off flavour	*Lactobacillus*
Ropiness in milk	Off flavour	*Bacillus, Alkaligens, Clostridium, Streptococcus*
Canned foods	Putrefaction	*Clostridium, Bacillus, Micrococcus*

Various foods spoiled by microbes

Food	*Types of spoilage*	*Microbes involved*
Fresh fruit and vegetables	Soft rot, Gray mold rot, Black mold rot	*Rhizopus, Erwinia, Botrytis, Aspergillus niger*

Food	*Types of spoilage*	*Microbes involved*
Pickles, Sauerkraut	Film Yeasts	*Rhodotorula*
Fresh meat	Putrefaction	*Clostridium, Proteus*
Cured meat	Moldy	*vulgaris, Pseudomonas, Aspergillus, Rhizopus*
Fish	Discolouration	*Pseudomonas*
Eggs	Green rot	*Pseudomonas fluorescens*
	Black rots	*Alcaligenes*
Concentrated orange juice	Off flavour	*Lactobacillus, Acetobacter*
Poultry	Slime, odor	*Alcaligenes, Pseudomonas*

Preservation of foods

Fermented dairy products

In the dairy industry, fermented milks are produced by inoculating pasteurized milk with a known culture of microorganism, sometimes referred to as **starter culture.** Starter cultures include bacteria like *Streptococcus, Thermophilus* and *Lactobacillus bulgaricus.*

A. Milk $\xrightarrow{\textit{Streptococcus lactis or Lactobacillus lactis}}$ Curd

B. Milk $\xrightarrow{\textit{Mechanical method}}$ Churning cream $\xrightarrow{\textit{Streptococcus lactis}}$ Butter

Leuconostoc citrovorum add for flavours

C. Milk *Streptococcus lactis* Fermented *Lactobacillus bulgaris* Yoghurt milk

D. Milk *Streptococcus lactis* curd *Lactobacillus lactis*

Cheese *Streptococcus lactis*

The toxin production in foods and food poisoning.

The term food poisoning refers to the illness caused by the presence of a bacterial toxin formed in food or illness caused by the entrance of bacteria into the body through ingestion of contaminated food. Various toxins are as follows:

(*a*) Endotoxin is that toxin which is retained within the cells and liberated on bacterial cell lysis. *e.g., E. coli, Salmonella, Schigella, Bacillus cereus.*

(*b*) Exotoxin is that which is excreted into the medium or surroundings where bacteria are growing *e.g., Clostridium botulinum.*

(*c*) Mycotoxins are produced by fungi. Its maximum occurrence is in grain. Various examples of mycotoxins are
- Patulin produced by *Penicilline patulum*
- Penicillic acid produced by *Penicillium puberulum*
- Ergot in Bajra produced by *Claviceps purpurea* causes abortion in animals. It has medicinal value also.

(*d*) Aflatoxins are produced by *Aspergillus flavus.*

Common Diseases and Their Symptoms caused by Food Poisoning

Disease	*Infection*	*Causative organism*	*Symptoms*
Salmonellosis	+	*Salmonella*	Diarrhoea, abdominal pain, vomiting, fever, dehydration.
Traveller's sickness	+	*E. coli*	Diarrhoea, fever, dehydration.
Schigellosis	+	*Schigella*	Diarrhoea, headache, vomiting, dehydration.
Gastroenteritis	+	*Bacillus cereus*	Diarrhoea, abdominal pain, vomiting, nausea, dehydration.

Disease	*Infection*	*Causative organism*	*Symptoms*
Botulism	Toxin production	*Clostridium botulinum*	Diarrhoea, headache, dizziness, difficulty in speaking, progressive weakness, ultimately cardiac and respiratory failure.
Food poisoning	+	*Staphylococcus aureus*	Collapse, vomiting, nausea and dehydration.

Milk

The number of bacteria present at the time of milking has been reported to range between several hundreds to several thousands per milliliter. From the time the milk leaves the udder until it is dispensed into containers, everything with which it comes in contact is a potential source of contamination.

The various practices used for food preservation are:

- Aseptic handling
- High temperature
- Low temperature
- Dehydration
- Osmotic pressure
- Chemicals
- Radiation

Aseptic handling

Food items undergo considerable handling prior to being processed by some specific method of preservation such as canning, freezing or dehydration.

High temperatures

High temperature is one of the safest and most reliable methods of food preservation. Heat is widely used to destroy organism in food products in cans, jars and other types of containers that restrict the entrance of microorganisms after processing. The various procedures are as follows:

- **Steam under pressure** – Steam kills all vegetative cells and spores.
- **Canning** – The temperatures used for canning foods ranges from 100°C for high acid foods to 121°C for low acid foods.
- **Pasteurization of Milk** – Heating milk or milk product to at least 145° F and holding it continuously at or above this temperature for at least 30 minutes or to at least 161°F for 15 minutes.
- **Sterilization** – Milk sterilization techniques have been developed which expose milk to ultra high temperatures for very short periods of time for example 300°F (148.9°C).

Preservation of food

The various practices utilized for food preservation may be summarized as follows:

1. Aseptic handling
2. Heat
   (*a*) Boiling
   (*b*) Steam under pressure
   (*c*) Pasteurization.
3. Low temperatures
   (*a*) Refrigeration
   (*b*) Freezing
4. Dehydration
5. Osmotic and pressure
   (*a*) In concentrated sugar
   (*b*) With brine
6. Chemicals
   (*a*) Organic acids
   (*b*) Substances developing during processing (smoking)
   (*c*) Substances contributed by microbial fermentation (acids)
7. Radiation
   (*a*) Ultra violet
   (*b*) Ionizing radiations.

All methods of food preservation are based upon following principles

(*i*) Prevention or removal of contamination.
(*ii*) Inhibition of microbial growth and metabolism (microbistatic action).
(*iii*) Killing of microbes (microbicidal action).

Multiple Choice Questions

1. The direct DNA uptake by the protoplast is stimulated by the chemical :
A. Cellulose B. Sucrose
C. Sodium chloride D. PEG

2. The chemical used as an inducer of the vir operon is :
A. Hygromycin B. Fanamycin
C. Acetosyringene D. Kanamycin

3. Which of the following is not an insecticidal protein gene?
A. Sunflower seed albumin
B. Lectins
C. Protease-inhibitor
D. α-amylase inhibitor

4. In Agrobacterium mediated plant transformation which one of the following is used to check the Agrobacterium :
A. Basta B. PPt
C. Timetin D. Kanamycin

5. Most commonly used strategy to develops transgenic plant against any viral class ease is to engineer a :
A. Liposome mediated transfer of viral RNA
B. Antisense RNA for viral genome
C. Coat or capsid protein gene
D. Direct DNA transfer of viral gene

6. The vir region is organized into six operons *viz,* vir A, vir B, vir C, vir D, vir E and out of these which four operons are required for virulence.
A. A, B, C, D B. A, B, D, G
C. C, D, E, G D. None of these

7. The large dark body (or boodles) in the nucleus is the nucleolus, which participates in the formation of :
A. Ribosome B. RNA
C. Protein D. DNA

8. Which property is followed by reducing sugar
A. Do exhibit mutarotation
B. They are in hemi acetal
C. Do form osazones with phenyl hydrazine
D. All of these

9. The combination of genes from two different cells is called :
A. Linkage B. Recombination
C. Ligation D. None of these

10. Nutritionally deficient mutant organisms are :
A. Auxotrophs B. Prototrophs
C. Deficient mutant D. All of them

11. Which DNA is having HMC in the place of cytosine?
A. *E. coli* B. P22
C. Yeast D. T4 phages

12. Which linkage is breaks in the cell wall of microorganism ruptured?
A. β–1-6 linkage B. α–1-4 linkage
C. β–1-4 linkage D. All of them

13. Plasmid replication occurs by which enzyme?
A. Polymerase - I
B. Polymerase A
C. Polymerase - III
D. None of these

14. In nature competence occurs is the cells due to uptake of :
A. Food B. Proteins
C. Biomolecular D. DNA

15. Improvement in the activity and use fullness of existing enzyme in :
A. Enzyme engineering
B. Agricultural Engineering
C. Genetic Engineering
D. All of these

16. Oligosaccharides contain ________ units of monosaccharides.
A. 2-10 B. 10-20
C. 30-40 D. None of these

17. The major techniques employed for enzyme purification includes :
A. Fractional ppt by change in pH
B. Fractional denaturation by heating
C. Fractional PPT by organic solvent
D. All of these

18. The term chromatography was coined by :
A. Tswett
B. Watson and Crick
C. Emil Fischer
D. None of these

19. Normally which carrier gas is used in gas chromatography :
A. Nitrogen B. Helium
C. Argon D. All of these

20. The most widely used technique for analysis and characterization of protein and nucleic acid :
A. Paper electrophoresis
B. Starch electrophoresis
C. Polyacrylamide gel electrophoresis
D. None of these

21. This technique is based on examining the paramagnetic characteristics of molecule :
A. RSE
B. NMR
C. Mass spectrometry
D. None of these

22. International unit for enzyme activity recommended by enzyme commission is :
A. Katal B. Cat
C. Ketel D. Ketal

23. The molecular weight and native double stranded DNA molecule in the smallest DNA virus is :
A. 2 billion B. 1 million
C. 80 billion D. 3 million

24. The process during which the information in a sequence of DNA in used to construct an *m*RNA molecule :
A. Transcription
B. Translation
C. Transduction
D. None of these

25. Which on the following can be used to transfect DNA into mammalian cell P. Liposomes Q. Cholesterol R. $CaCl_2$ + HEPES butter S. $MgCl_2$:
A. Only P B. P and R
C. P and Q D. P and S

26. The enzyme that is used to make the first strand *c*DNA from *m*RNA :
A. Reverse transcriptase
B. Restriction Ends nucleases
C. T4 DNA ligase
D. None of these

27. Choose the right combination of components required to set up a polymerase chain reaction from the following :
A. Template DNA, Two primers d NTP's and DNA ligase
B. Template DNA, Two primers NTP's and DNA ligase
C. Template RNA, Two primers NTP's and DNA Polymerase
D. Template RNA, Two primers d NTP's and DNA Polymerase

28. Which of the following pair of vitamins participates as cofactors in the cellular oxidation reduction reaction?
A. Pyridoxine and folate
B. Riboflavin and cobalamine
C. Ascorbate and nicotinamide
D. Thiamine and pantothenate

29. Most accurate method to determine the molecular weight of a given polypeptide is :
A. Gel Percolation Chromatography
B. Analytical Ultracentrifugation
C. SDS - PAGE
D. MALDI - T of Mass Spectrometry

30. The quantity of bacteriophages in a given sample is best give as :
A. Colony Forming Unit (CFU)
B. Plague Forming Unit (PFU)
C. Minimum Inhibitory concentration (MIC)
D. Lethal Dose 50% (LD50)

31. The back bone of peptidoglycan present in the bacterial cell wall is :
A. N-acetylgalactosamine - N-acetyl muramic acid
B. N-acetylgalactosamine - N-acetyl uramic acid
C. N-acetylgalactosamine - N-acetyl muramic acid
D. None of these

32. Which one of the following movements of phospholids catalysed by the phospholipid trans locaters in the lateral cell membrane :
A. Lateral B. Flip-flop
C. Flexion D. Rotation

33. Papain is derived from ....
A. Papaya
B. Pomegranate
C. *Aspergillus niger*
D. *Panicillium roquefortii*

34. Which enzyme converts co-casein of the milk to the co-para caseinate which has card like appearance?
A. Renin B. Cellulase
C. Pectinase D. Amylase

35. Manufacture of soft center candies and confectionary which enzyme is important?
A. Sucrose B. Papain
C. Pectinase D. Amylase

36. Which techniques of immobilization is base upon occlusion of biocatalyst within the constraining structure tight enough to prevent the leakage of biocatalyst?
A. Entrapment
B. Covalent binding
C. Adsorption
D. Membrane confinement

37. Immobilization may lead to loss of enzyme activity due to which?
A. Denaturation of enzyme
B. Precipitation of enzyme
C. Degradation of enzyme
D. Involvement of active site in immobilization

38. Which of the proteolytic enzyme is used in meat tenderization?
A. α-amylase B. Pectinase
C. Cellulase D. Papain

39. Which enzyme hydrolyse α-1-4 glycosidic bond from non reducing end stares?
A. Sweetase
B. α-amylase
C. Glucanase
D. Trypsin

40. In textile industries which enzyme is used for the dry-cleaning?
A. Lipase B. Protease
C. Amylase D. All of the above

41. Ascorbic acid into dehydroascorbic acid :
A. Acidic condition
B. Alkaline condition
C. Neutral condition
D. None of the above

42. Which is the carbohydrate does not easily digested by human digestive system?
A. Starch B. Glucose
C. Sucrose D. Maltose

43. _________ discovered and helical structure of protein molecules.
A. Paul Ehrlich B. Linus Pauling
C. Jules Bordet D. Paul Partier

44. _________ are specific antibody binding sites on the surface of antigen also antigenic determinates.
A. Paratopes B. Epitopes
C. Clone D. None of these

45. _________ organisms in guinea pigs causes diphtheria.
A. *Mycobacterium tuberculosis*
B. *Corynebacteria*
C. *Clostridium tetani*
D. *Streptococcus cremoris*

46. The immunity acquired the lifetime of an individual is known as _________ .
A. Innate immunity
B. Adaptive immunity
C. Non-specific immunity
D. Specific immunity

47. _________ is the founder of animal cell and tissue culture.
A. Ross Harrison B. Jolly
C. Carrel D. Roux

48. _________ cell which are phenotypically modified during *in vitro* to become anchor independents.
A. NK-cell B. Adherent cells
C. Non-adherent cell D. Transferred cell

49. The go phase of the animal cell cycle can occurs :

A. Just before the G1 phase
B. Just before the mitotic (m) phase
C. Late in the G1 phase
D. During the G2 phase

50. What is the key enzyme is Glycogen synthesis?

A. Glycogen synthase of phosphorylase
B. Phosphorylase
C. Phosphokinase
D. Glycogen synthase

ANSWERS

1	2	3	4	5	6	7	8	9	10
D	C	A	C	C	B	B	D	B	A
11	**12**	**13**	**14**	**15**	**16**	**17**	**18**	**19**	**20**
D	C	C	D	A	A	D	A	D	C
21	**22**	**23**	**24**	**25**	**26**	**27**	**28**	**29**	**30**
B	A	B	A	B	A	D	B	C	B
31	**32**	**33**	**34**	**35**	**36**	**37**	**38**	**39**	**40**
C	B	A	A	A	A	D	D	B	D
41	**42**	**43**	**44**	**45**	**46**	**47**	**48**	**49**	**50**
A	A	B	B	B	B	B	D	C	A

CHAPTER

19

Environmental Biotechnology

Introduction

Environmental biotechnology is science that is applied to and used to study the natural environment. The International society for Environmental Biotechnology defines environmental biotechnology as "the development, use and regulation of biological systems for remediation of contaminated environments (land, air, water) and for environment-friendly processes.

Environmental Biotechnology deals with preservation and restoration of the natural resources and will be one of the most valuable tools for improving the quality of life for future generations.

Issues in Environmental Biotechnology

Pollution

"Any undesirable, unpleasant and harmful change in the environment due to industrial or other activities of man."

The objective of environmental education and awareness is to enlighten public, specially students, about the importance of protection and conservation of our environment and the need to restrain public activities which lead to indiscriminate discharge/release of pollutants into environment.

Major pollutants and their sources are summarized in the following table:

S. No.	*Major pollutant*	*Major sources of pollutants*
1.	CO	Auto exhaust, transportation, paper industries, petroleum, steel and iron industries, solid waste disposals.
2.	CO_2	Industrial processes and fuel combustion.
3.	SO_2	Combustion of coal, metallurgical operations, combustion of oil, paper and manufacture of H_2SO_4.
4.	H_2S	Natural gas plants, petroleum-refining and coke evens.
5.	NO_2, NO	Explosive industries, fertilizer industries and manufacture of HNO_3.
6.	NH_3	Natural biological processes and industrial processes.
7.	Particulates	Metallurgical operation like smelting and mixing, incomplete combustion in industrial processes and involcanic eruptions.
8.	Hydrocarbons	Chemical processes, rubber-manufacture, solid waste disposal, transportation and petroleum refineries.

Major sources of atmospheric pollution are :

(*i*) Combustion of fossil fuels in homes, factories, thermal plants, automobiles, aircrafts, railways etc.

(*ii*) Mining and processing

(*iii*) Chemical industries

(*iv*) Processing industries etc.

(*a*) **Particular Matter :** It consists of soot, flyash, dusts of various types, fur, hair, spores, pollen grains etc.

(*b*) **Carbon Monoxide :** It is produced due to incomplete combustion and naturally by plants as well as animals. 70% emissions are from automobiles. Carbon monoxide combines with haemoglobin, produces carboxyhaemoglobin or COHb.

(*c*) **Carbon Dioxide :** It is a green house gas the concentration of which is constantly rising (0.033 from 0.029 and expected to be doubled by 2020). In excess it causes headache and nausea.

(*d*) **H_2S :** It is a product of putrefaction, treatment of sulphur containing over, refineries, chemical plants and bituminous fuels. It causes mottled *chlorosis* and defoliation in plants, decolourises paints, produces eye irritation, throat irritation and nausea.

(*e*) **Hydrocarbons :** They are produced naturally (*e.g.,* marsh gas) as well as due to incomplete combustion. Hydrocarbons are carcinogenic, cause irritation or eyes and mucous membrane and bronchial constriction.

(*f*) **Sulphur Dioxide :** It is produced during combustion of fossil fuels and smelting of sulphur containing ores. Sulphur dioxide produces *smog*. Maximum SO_2 pollution is found in Kolkata. It causes membrane damage, destruction of chlorophyll (change to phaeophytin), necrosis and water soaked appearance. Lichens are most sensitive to SO_2 pollution.

(*g*) **Nitrogen Oxides :** They are formed electro-photo-chemically in air, burning of fossils fuels and denitrifying bacteria. Maximum NO_2 pollution is recorded from Baroda. Nitrogen oxides causes *necrosis*, *defoliation*, dieback and death of plants. Like SO_2, they corrode metals and deteriorate paints as well as various articles. At less than 0.1 ppm, nitrogen oxides impair sensation of smell. The oxides produce eye irritation, lung oedema, dilation of arteries, injury to liver and kidneys. Carcinogenic, form secondary pollutants.

(*h*) **Fluorides :** They are given out during refining of minerals (*e.g.,* aluminium) as also from ground water. Fluorides cause *fluorosis*. In plants there is chlorosis and necrosis of leaf tips and leaf margins, followed by abscission. In animals, fluorides bring about abnormal calcification of bones and teeth (making them weak), frequent diarrhoea and swelling of knee bones. In humans, there is mottling of teeth, weak bones, boat-shaped posture, knocking knees, gastrointestinal and neuromuscular disorders.

(*i*) **Chlorofluorocarbons/Chlorofluoromethane/Freon/Aerosols :** They are chemicals used as refrigerants, propellants and solid plastic foams. The chemicals are release as aerosols by jets flying at high altitudes. Alongwith nitrogen oxides, chlorofluorocarbons react with ozone of ozonosphere and depletes the same. Hole in the ozone layer over Antarctic region has widened from 129 to 133 dobson units in 1994 alone. This can increase the amount of ultraviolet radiations reaching the earth.

(*j*) **Photochemical Oxidants :** They are secondary pollutants (ozone, peroxyacyl nitrates, aldehydes and phenols) produced due to unsaturated hydrocarbons.

(*k*) **Other Air Pollutants :** Mercury (burning of coal, smelting), methyl isocyanate (pesticide manufacture), phosgene (pesticide manufacture, dye industry), ammonia (fertilizer, dye and lacquer industries) and lead (automobile exhausts) are pollutants added to atmosphere though they may not reside in them (*e.g.,* lead in soil). *Bhopal gas tragedy* (Dec. 1984) was due to release of phosgene and methyl isocynate (MIC).

(l) Automobile Exhausts : Automobiles burn petroleum inefficiently causing 80% of air pollution and 75% of noise pollution in urban areas. They release hydrocarbons (13.7%), carbon monoxide (77.2%), nitrogen oxides (7.7%), sulphur oxides, ammonia, aldehydes and lead (90% of total lead poisoning).

(m) Smog : Smog is opaque or dark fog having condensed water vapours, dust, smoke and gases (SO_2, H_2S, NO_2 etc.). It causes silvering/glazing and necrosis in plants, allergies and asthma/bronchitis in humans. Some famous smogs of the world are Los Angels smog (1946), London smog (1952), Tokyo (Rome, New York, Sydney, 1870).

(n) Pollen/Spores : They cause allergic reactions or hay fever in certain persons.

Control of Air Pollution

(A) Automobiles

(*i*) Two stroke engines fitted in two wheelers (waste fuel 20-30%) be changed to either four-stroke engines or fitted with catalytic converters specially designed for them. A low cost-catalytic converter has been developed at BHU.

(*ii*) Leaded petrol be replaced with unleaded one and diesel with low sulphur diesel.

(*iii*) Tune-ups (for high air-fuel ratio) and catalytic converters (for oxidising $CO \rightarrow CO_2$ and reducing $NO \rightarrow N_2$) be fitted in automobiles.

(*iv*) Periodic checkup of pollution control for all vehicles.

(B) Bagasse and Rice Husk

Bagasse should not be used as fuel. Rice Husk should be first converted into briquetts.

(C) Fly Ash

About 38% fly ash is produced by coal based thermal plants. It should be removed through wet method and used in building material.

(D) Industrial Pollution

(Control measures)

(*i*) *Tall Chimneys :* They disperse smoke more throughly.

(*ii*) *Gravity Setting Chambers :* Particles larger than 50 µm settle down.

(*iii*) *Wet Scrubbers :* A fine spray of water or alkaline fluid is used to remove soluble gases and particles.

(*iv*) *Bag Filters :* Porous bags of teflon or polyester filterout particulate matter.

(*v*) *Cyclone Collectors :* They cause settling down of particulate matter through centrifugation.

(*vi*) *Electrostatic Precipitators :* (ESPs). They are electrically charged plates or electrodes which remove most of the particles present in chimney exhausts.

(E) Vegetation

A broad strip of vegetation along road and around industrial areas reduces particulate pollution. Vegetation can also metabolise toxic gases like CO (*e.g., Ficus, Coleus, Daucos, Phaseolus*), and nitrogen oxides (*e.g., Vitis, Pyrus, Robinia, Rhamnus*).

Acid Rain

Acid rain is the term coined by Robert Augus (1872). It is rainfall and other forms of precipitation with a pH of less than 5. The most acidic rain has occured over west Virginia, USA with a pH of 1.5. It is caused by large scale emission of acidic gases into atmosphere from thermal power plants, industries and automobiles. The common ones are *sulphur dioxide, nitrogen oxides* and *hydrogen chloride*. Sulphur dioxide and nitrogen oxides are changed in the atmosphere into sulphuric acid and nitric acid by combining with oxygen and water.

Acid rain damages plants by direct effect on foliage and growing points-chlorosis, necrosis, defoliation, dieback. It causes leaching of essential minerals of soil. Toxic minerals left in the soil further kill the plants. 50% of natural forests have been destroyed by acid rain in Germany, Sweden, north east USA, Romania, Poland etc. Acid rain has also ruined fresh water reservoirs of most industrialised countries, *e.g.*, 80% in Norway, 25% in Sweden, 20% in USA. Acidity dissolves toxic metals like Hg, Pb, Zn, Al. Both acidity and toxic metals kill all types of aquatic life except some algae and fungi. Acid rain corrodes metals, marble, painted surfaces, slate, stone, etc. The phenomenon is called *stone leprosy*.

Ozone Depletion

Ozone layer or shield is present in the stratosphere. It protects the earth from short-wave ultraviolet rays (below 300 nm) by changing the same into infrared rays. A large hole has appeared in ozone shield over Antarctica (first detected by Farman, 1982), and a smaller one over North Pole. Size of the hole varies with the seasons. Thinning of ozone shield has also been reported elsewhere (*e.g.*, 8% between 30°-50°N). Depletion of ozone layer allows harmful UV-radiations to reach earth. It is the major cause of *skin cancer, cataract, dimming of eye-sight*, decrease in immune system and increased susceptibility to herpes.

Thinning of ozone shield is being caused by a number of pollutants like chlorofluorocarbons (14% of total depletion), nitrogen oxides (3.5% depletion), sulphur dioxide, halogen, carbon tetrachloride, being released by jets flying in the stratosphere and rockets being fired into space. Others are persistent in the troposphere and gradually pass into stratosphere. Maximum ozone depleting potential or ODP is of chlorofluorocarbon due to release of chlorine by it. A single chlorine atom converts 1 lakh molecules of ozone into oxygen. Consequently, chlorofluorocarbons *(CFCs)* are being replaced by hydrofluorocarbons *(HFCs)* and hydrochlorofluorocarbons *(HClFCs)*. Carbon tetrachloride, halogen and methyl chloroform also deplete ozone by a similar method. Nitric oxide (NO) and other gases released by jets directly react with ozone to form oxygen.

Water Pollution

It is the degradation of quality of water due to addition of substances (inorganic, organic, biological, radiological) or factors (*e.g.*, heat).

(A) Organic Wastes/Sewage

It is waste water having food residue, animal and human excreta, detergents, discharges from commercial and industrial establishments.

(*i*) Raw sewage contains a number of pathogens (dysentery, diarrhoea etc.) coliforms and enterococi.

(*ii*) It stimulates the activity of several decomposer organisms collectively called *'sewage fungus'*. The property of becoming decomposed through microbial activity is known as *'putrescibility'*. Degree of impurity of water due to organic matter is measured in terms of B.O.D. Biochemical Oxygen Demand or B.O.D. is the oxygen in milligrams required for five days in one litre of water at 20°C for the microorganisms to metabolise organic waste. Low pollution–below 1500 mg/l, medium pollution—1500-4000 mg/l and high organic pollution–above 4000 mg/l of B.O.D.

(*iii*) Water has a brown colouration and an unpleasant odour due to formation of secondary pollutants like CH_4, NH_3, H_2S.

(*iv*) Scum and sludge (H_2S + metallic ions)

(*v*) There may be algal blooms and eutrophication. The latter is due to phosphates present in detergent.

(*vi*) Animals may die due to depletion of oxygen by sewage fungus.

(*a*) *Primary Treatment :* This consists of shredding, churning, floatation, screening and sedimentation.

(*b*) *Secondary Treatment :* The organic matter is decomposed through sewage fungus by either **trickling filter method** (passing through thick bed of gravel) or **activated sludge method** (aeration/oxidation tanks). The cleared water is now chlorinated to kill pathogens. The water can be passed into fields as manured water or into water courses where it may cause eutrophication.

(*c*) *Tertiary Treatment :* Where water is to be recycled, the chlorinated waste water is mixed with alum, ferric chloride and lime for precipitation of salts. Zirconium is also a good precipitant. Another technique is *reverse osmosis*.

(B) Fertilizers

Part of fertilizers added to crop fields are passed down to water bodies during rains through surface run-off. Presence of extra nutrients brings about dense growth of plant and animal life. The phenomenon is called *'eutrophication'*. Eutrophication leads to organic loading, depletion of oxygen, death of animals and fouling of water.

(C) Pesticides

Pesticides spread over crops also pass into water bodies due to surface run-off. In excess they cause immediate and mass scale deaths of aquatic animals. Persistent-pesticides (*e.g.*, organochlorine or chlorinated hydrocarbons like DDT) pass into food chain and increase in amount per unit weight of organisms with the rise in trophic level due to their accumulation in fat. The phenomenon is called *biomagnification/bioconcentration/biological amplification*. *e.g.*, 0.01-0.05 parts per billion in water, 10 parts per billion or 0.001 ppm in phytoplankton, 0.04 ppm in zooplankton, 0.4 ppm in clams, 3-8 ppm in fish and 18-75 ppm in water birds like Sea-Gulls. Higher amounts of pesticides cause cerebral haemorrhage, softening of brain, liver cirrhosis, hypertension, thinning of egg shell, malfunctioning of sex hormones. Population of Bald Eagle has declined due to it.

(D) Silt

Silt from soil erosion makes water muddy and unfit for plant growth. Silt may clog gills of fishes.

(E) Thermal Pollution

Hot effluents and hot water (*e.g.*, thermal plants/atomic reactors) bring about rise in water temperature. Warm water contains less oxygen, has lower rate of putrescibility resulting in increased organic loading, replacement of green algae by blue-green ones. Many animals fail to reproduce *e.g.*, Trout, Salmon.

El Nino is warm ocean *Peru current* that recurs after every 5-8 years. It kills marine animals and plankton over large tracts along Peru and Ecqudor coasts.

(F) Oil Pollution

Oil spills and refinery discharges cause oil pollution. It reduces oxygenation, inhibits plankton growth and kills animals.

(G) Industrial Effluents

Major water pollutants and their sources are summarised in the following table:

S. No.	*Name of the pollutant*	*Sources*
1.	Microorganism and organic wastes	Domestic sewage, animal and human excreta, animal manures, decaying animals and plants, detergents, and wastes from food processing industries.
2.	Toxic heavy metals	Chemical factories and other industries
3.	Plant nutrients	Fertilizers (chemicals)
4.	Heat	Industrial units, generally water used for cooling and which is discharged. (as hot water)
5.	Radioactive wastes	Mining uranium containing minerals and accidental release from nuclear plants.
6.	Sediments	Organic compounds, insoluble particles of soil and rock from mining construction activities, soil erosion and agricultural activites.
7.	Pesticides	Used for killing insects, weeds and fungi.

Soil Pollution

It is alteration in soil caused by removal or addition of substances and factors which decreases its productivity, quality of plants and ground water.

Negative soil pollution is reduction in soil productivity due to erosion and over use. *Positive soil pollution* is reduction in soil productivity due to addition of undesirable substances (*e.g.,* pesticides, fertilizers, industrial wastes, air pollutants washed down by rain, faulty sanitation). *Landscape/third pollution* is converting fertilie land into barren one by dumping wastes (*e.g.,* ash, sludge, garbage, rubbish, industrial wastes, broken cans, bottles etc.) over it.

(A) Pesticides

They include insecticides, fungicides algicides, rodenticides and weedicides. Pesticides are generally broad spectrum and function as biocides.

(*i*) *Organochlorides/chlorinated hydrocarbons:* The important ones are DDT (dichlorodiphenyl trichloroethane), BHC (Benzene hexachloride), aldrin, dieldrin and endrin. They are persistent, fat soluble and show biomagnification whence they prove harmful to higher trophic level organisms. Hence their use is being restricted.

(*ii*) *Organopesticides :* Degradable, but toxic to workers *e.g.,* malathion, parathion, carbamates.

(*iii*) *Inorganic pesticides :* They contain arsenic and sulphur. Persistent. Hence, their use is highly restricted.

(*iv*) *Weedicides :* Often persistent and harmful.

(B) Fertilizers

Excessive use causes soil deterioration through decrease of natural microflora. Leading down causes pollution of underground water (third poison). Salts entering crop plants in excess may prove harmful For example, nitrate rich leaves, fruits and water produce nitrite in alimentary canal that enters bloods, combines with haemoglobin forming *methaemoglobin* and reducing oxygen transport. It may prove fatal in infants. Organic farming involves used of biofertilisers, manures, pesticides of organic origin, biological control and resistant varieties.

(C) Industrial Effluents/Wastes

They contain a number of toxic substances including cyanides, chromates, acids, alkalis and metals like mercury, copper, zinc, lead, cadmium etc.

(D) Mine Dust

It destroys vegetation and produces many deformities in animals and human beings.

(E) Soil Salination

It is increase in salt concentration of soil making the soil *halomorphic*. Salination is often accompanied by white incrustation over the surface. It makes the soil barren. Only halophytes can grow over it. Soil salination is caused by the following factors :

(*i*) Parent-rock.

(*ii*) Poor drainage and elevated water table.

(*iii*) Salt rich ground/canal water.

(*iv*) Excessive fertilizers.

(*v*) Salt blown from rocks/sea.

(*vi*) Plants that absorb ions.

Noise Pollution

It is a physical form of pollution that affects the receiver directly. Noise pollution is due to release of unwanted sound generally with the value of 80 dB and above. Frequency of sound is measured in Hz (Hertz) while unit of sound in dB (deci Bel).

Range of human hearing is 20 Hz to 2000 Hz. Infrasonic vibrations (below 20 Hz) can be felt because some body parts resonate at this frequency. Ultrasonic vibrations (above 20,000 Hz) are used in imaging, cleaning, drilling cutting, welding and sealing. Moderate conservative produces 60 dB sound, loud conversation 70 dB, plying of trunk/bus 90 dB, jet aeroplane–150 dB, rocket 180 dB. Noise becomes uncomfortable above 100 dB. Sound is produced by loud speakers/public address system, transport vehicles, defence equipment, industries, domestic gudgets and outdoor equipment.

Effects of Noise : Noise brings about

(*i*) Damage to ear drum and impairment of hearing (a ten year exposure to 80 dB impairs hearing by 15 dB)

(*ii*) Interference in conversation and hearing.

(*iii*) Emotional disturbance, development of anxiety and stress (first effect).

(*iv*) Damage to eye sight, colour perception, night vision, etc.

(*v*) Hypertension, decreased heart output and gastric problems.
(*vi*) Startle reaction.

Control Measures

(*i*) Delimination of acoustic soning
(*ii*) Use of cotton pluges or ear muffs in occupational exposure
(*iii*) Development of quieter machines.
(*iv*) Restricted use of loud speakers.
(*v*) Acoustic furnishing and low voice radio/TV.
(*vi*) Regulation of noise on road.
(*vii*) Green muffler.

Biodiversity, Global Warming, *El Nino* and Green House Effect

Biodiversity is the number of species of different organisms present in an area. Thus, biodiversity includes all plants, animals and microorganisms of an area. Earth is endowed with an immensely rich variety of life forms which roughly consists of 30,000 green plants, 8,00,000 insects, 23,000 fishes, 2,000 birds, 6,500 reptiles, 4,100 mammals and few thousand microbes. Thus, *biodiversity* means variation and abundance of species and their habitats. According to the *World Resources Institute*, 'biodiversity' is variety of the world's organisms including their genetic diversity and the assemblage they form.

Conservation of Biodiversity

There is currently an enormous amount of general concern over the threat to biological diversity. Extinction of species may be natural or man made. However, the cause of concern is the rate at which species are becoming extinct today due to human interference. Some of the other causes for the loss of biodiversity are :

(*i*) Degradation and destruction of habitats due to human needs.
(*ii*) Husting and over-exploitation mainly for commercial purposes.
(*iii*) Introduction of exotic species.
(*iv*) Pollution and poisoning stress on ecosystem.
(*v*) Global warming and green house effects.
(*vi*) Improper use of agro-chemicals and pesticides etc.

The loss of biodiversity is not only an ethical tragedy, but also a great social, economic and cultural one. Without biological diversity use should perish. Hence, efforts are being made to save biodiversity. Conservation of biodiversity can be divided into two categories : *In-situ conservation* and *Ex-situ conservation.*

(*a*) *In-situ conservation* : This is the conservation of biodiversity through their maintenance within natural or even human-made ecosystems in which they occur. This type includes a system of protected areas of different categories, managed with different objectives to bring benefit to the society. National Parks, Sanctuaries, Nature Reserves, Natural Monuments, Cultural Landscape, Biosphere Reserves, etc. belong to this type of conservation.

(*b*) *Ex-situ conservation :* This is conservation outside their habitat by perpetuating sample populations in genetic resource centres, zoos, botanical gardens, culture collections, etc. or in the form of gene pools, and gamete storage for fish, germplasm, banks for seeds, pollen, semen, ova, cells, etc. Plants are more readily maintained than animals. In this type of conservation, seed banks, botanical gardens, pollen storage, tissue culture and genetic engineering have been playing important role.

Climate Change

Back in the 1960s and 1970s, after some year of careful observations of CO_2 in atmosphere, it became evident that its concentration was increasing. In this context, earlier findings which indicated a possibility of climate change due to so called 'green house effect' of carbon dioxide and other radioactive gases assumed great significance. The increase of concentration of CO_2 and subsequently, observed increase of concentration of methane and other greenhouse gases were attributed to a considerable extent to human activities, in particular, to emissions of green house gases into the atmosphere from the burning of fossil fuels. It was recognized that global climate can be modified by inadvertent humans action. We can study the global climate change under the following three sub-headings (*i*) Global Warming, (*ii*) Green House Effect, (*iii*) Ozone Depletion.

Global Warming

Global warming has often been described as one of the most serious environmental problems ever to confront humanity as this problem is closely linked to the process of development and economic growth itself.

While the pattern of future warming is very much open to debate, it is indisputable that the surface of the earth has warmed, on an average, 0.3 to 0.6°C since the late 19th century. The ten warmest years in the last 130 years, have all occurred in the 1980s, and 1990s. And within this ten years the three warmest years were in 1990s.

Global climate models show that due to a temperature rise, warmer ocean release more energy into the atmosphere, leading to more intense and violent storms. Illustrative of this is the 1991 cyclone of Bangladesh and more recent is Odisha cyclone of 1999.

The frequency and intensity of *El Nino* have been increasing. We have faced a strong *El Nino* is 1997-98, which, in turn, was responsible for forest fires in Indonesia and Sumatra. The seasonal rain which could behave doused the fire did not arrive due to *El Nino.* All this supports the theory that what is altering the Earth's climate is the accumulation in the atmosphere of green house gases generated by human activities.

Scientists continue to be worried by the phenomenon of global warming. Not withstanding the highly exaggerated predictions by environmentalists, scientists are veering to the view that the world is indeed facing a grave threat from the effects of global warming scientists have recently discovered that the Quelccaya ice cap in the South American Andes—the hemisphere's largest glacier is melting.

There are other precurious indicators too : half the glacier ice on the European Alps has dwindled over the past 100 years. Antarctica is experiencing its hottest temperatures in over 4,000 years. In October 1998, an iceberg, 7,125 sq. km in area, separated from the Ronniee ice shelf, the second largest ice cap in Antarctica. The Arctic ice, too is now one third thinner than it was in 1976. Meanwhile, the Inter-governmental Panel on Climate Change (PCC) in its scientific consensus has warned that global mean temperature will rise by 1 degree = 3.5 degree in the 21st century statistical evidence also speak for themselves. Thirteen of the 14 hottest years is history of the planet have occurred since 1980 and six of the eight months of 1998 were the warmest since 1866, with July being the hottest single month ever recorded.

El Nino

El Nino literally means Child Christ in Spanish. It takes its name from the fact that it arrives usually during season of christmas. This weather phenomena which occurs in cycles, resulting in a reversal of wind flows. Instead of blowing from South America towards Asia across the Pacific, *El Nino* results in the reverse flow from Asia towards South America. So instead of bringing in rains to Asia, it takes away the moist air and dumps unwanted rains across the Pacific, causing floods. For Asia, it leaves large tracts like Indonesia without rains and creates drought in some South Pacific Islands or even Australia. It occurs because of the warming up (which is generally cold during winter) of the coastal waters of the eastern pacific in South America, off the cost Peru. This warming spread westward into the Central Pacific. The stronger the *El Nino,* the more intense the changes in the weather. Due to warm waters, a low pressure is created over southern pacific ocean (which usually has a high pressure) and to balance it a high pressure develops over Indonesia (which usually has a low pressure over it). As a result, the reversal of wind flow occurs. The 1997 *El Nino* was worse for Asia, particularly, Indonesia. Papua New Guinea faced severe droughts. Forest fires in Indonesia, worsened by the drought and dry weather are supposed to be caused by *El Nino.*

Green House Effect

The green house effect is a fundamental property of the Earth's atmosphere. The atmospheric green house effect warms the Earth surface by reflecting back down to it a portion of the solar heat energy, the Earth surface sends skyward. The atmosphere is transparent to the Sun's visible light (short-wave), which is absorbed by the Earth's surface. The Earth then slowly expels this stored heat in the form of invisible infrared radiation (long wave radiation). But, the atmosphere acts something like a one-way mirror, reflecting some but not all of this radiated heat energy. It comes back down to earth to warm its surface further.

The chief greenhouse gases are water vapour, CO_2, methane, nitrous oxide and various complex human-made compounds. They are entirely responsible for reflecting downward heat that the Earth's surface would otherwise send into space. The gases that makes rest of the atmosphere, nitrogen (78%), oxygen (21%) and Argon (0.9%) are transparent to heat as well as to light and contribute nothing to the green house effect. Without this effect, the Earth's surface on an average would be 33 degrees celsius colder, yielding an ice-covered planet inhospitable to life as we know it. Having some green house gases in the air, thus, makes life possible. But if their concentrations exceed critical levels, conditions could become less favourable for human beings and other species.

Today, however, concentration of most of the greenhouse gases are rising due to the growth of a relatively new influence on the atmosphere-people, with their modern consumer life styles and their large numbers. Human activities have come to rival natural processes as a source of heat trapping gases without producing comparable sinks to aid in the gases removal from the atmosphere. As a result of this imbalance, greenhouse gases are accumulating in the atmosphere.

(A) The Green House Gases

Although there are dozens of greenhouse gases, four of them are most important, which are closely linked to common human activities. These are as follows :

Green House Gases	Green House Effect
Carbon dioxide	55%
Methane	15%
CFCs	24%
Nitrous oxide	6%

Carbon dioxide is responsible for more than half the human contribution to the greenhouse effect, CO_2 concentrations have increased from 280 parts per million at the dawn of the industrial revolution to around 355 ppm today. Each year, human beings shift 6 billion metric tonnes of fossil carbon (coal, petroleum, natural gas) from the Earth to the air in the form of 22 billion tonnes of CO_2. The destruction of forests and the degradation of soils adds an estimated 5-9 billion tonnes of CO_2 to the atmosphere. About half of CO_2 released by human activities is quickly absorbed by the oceans and plants and half of the global emission remains, in the atmosphere indefinitely, contributing to greenhouse effect. Methane (CH_4) is released as a result of the combustion of carbon and microbial decays in the absence of oxygen. This occurs in the cultivation of wetland rice, the burning of plant material, land fills and the digestive systems of cattle and termites. Methane is also emitted from coal mines and the production and distribution networks of natural gas. Methane is 50 times more powerful in trapping heat than CO_2. Human activities have more than doubled its concentration from 0.7 ppm (parts per million) in pre-industrial times to more than 1.7 ppm today. Because it is removed from the atmosphere relatively rapidly by chemical reactions in the air, emissions, reductions of only about 15 to 20 per cent would be required to stabilize methane at its current high concentrations.

Nitrous oxide (N_2O) is produced by a variety of biological processes in soils and water. Although its concentration increase of 8 per cent since pre-industrial times (from 288 to 310 parts per billion) is because of human influence, soil cultivation, biomass burning and fossil fuel combustion—all play some role in nitrous oxide production. Bringing new land into cultivation may be largest source of the gas. Two recently identified major sources are the production of nylon and nitrogen fertilizers. It is estimated that a 70 to 80 per cent reduction in human emissions of N_2O would be needed in order to stabilize concentrations at their present high level.

(a) *Chloro Fluoro Carbons (CFCs) :* Industrial gases known as CFCs are powerful heat trappers. But, the net impact of CFCs on climate change may be natural because they also contribute to the depletion of upper atmospheric ozone layer. Production of CFCs is being gradually phased out under the Montreal Protocol (1987), but they are still abundant in the atmosphere. Some of the compounds being developed to replace them are also greenhouse gases.

(b) *Ozone (O_3) :* It is a powerful greenhouse gas in upper atmosphere. Ozone protects living beings from the sun's potentially harmful UV-β radiation. Near the Earth's surface, however ozone is a human-made pollutant produced by the interaction of sunlight and industrial and vehicle waste gases that damages living tissues of animals and some plants.

(c) *Water Vapour :* It is the most abundant greenhouse gas. Its concentrations are regulated by overall atmospheric temperature and pressure, not human emissions. However, water vapour plays a major role in human-induced climate change through what is called a 'feed-back loop'. If increasing concentrations of CO_2 and other gases raise global temperatures that will draw more water vapour by evaporation and evapotranspiration into the atmosphere, complifying the warming.

(B) Impact of Green House Effect

It is estimated that due to the greenhouse effect, the Earth's temperature will rise. On a worldwide average, temperatures would rise by 1 to 3.5°C by the year 2100. If CO_2 concentration doubles, global sea level could rise from 8 to 29 centimeters by the year 2030 and by as much as one metre by the end of 21st century. A rise of one metre will permanently inundate thousands of square miles of low-lying coastal land.

Ozone Depletion and Ozone Hole : The sun emits radiation over a broad range of wavelengths, to which humans eye respond in the region from approximately 400-700 nm. The range can be divided into three categories.

UV–A (ultra violet) 320-400 nm : Not absorbed by ozone.

UV–B (286-320 nm) : Partially absorbed by ozone.

UV–C (200-280 nm) : Completely absorbed by ozone.

Ozone produced is destroyed at a wide range of altitudes in the atmosphere. About 90% of the ozone is present in the stratosphere. The maximum concentration (about 0.5 ppm) occurs between the altitude of 20 to 35 km and the layer at this level is called *ozone layer*. The presence of ozone is an essential necessity for life on Earth. Stratosphere ozone layer absorbs dangerous (harmful) UV-B rays of the sun and thus protects the earth's surface from these high energy radiation.

Over the last few decades O_3 layer is thinning out because of man made pollutants which catalyse the dissociation of O_3 at a very fast rate. Major pollutants responsible for depletion of ozone are chlorofluorocarbons (CFCs), nitrogen oxides, hydrocarbons and oxides of chlorine and bromine.

Application of Biofilters

Industry	*Application*	*% Efficiency*
Gelatin production	Elimination of odour	70-93
Cocoa and Chocolate processing	Elimination of odour	99
Fishmeal factory	Elimination of odour	50-90

Industry	Application	% Efficiency
Tobacco industry	Elimination of nicotine	95
Waste-water treatment	Elimination of odour, H_2S, acetone	90-95
Flavour and fragrance	Removal of odour, H_2S, acetone	98
Paint production	Organic solvent	75
Pharmaceuticals	Aromatic, aliphatic, chlorinated compounds	80
Photofilm production	Organic solvent	75
Food processing	Elimination of oil, odour	93
Ceramics processing	Ethanol	98
Metal foundry	Benzene	80

Air Pollutants Degradation for Various Source Industries

Industry	Pollutants	Systems/Micro-organisms	Technology
Food processing, Sewage treatment	Mercaptans, alkyl sulphides and H_2S		Obayashi Corp. and Hitachi Zosen Corp.
Coal industry	NO_x from flue gas	*Pseudomonas denitrificans*	Idaho National Engineering Laboratory
	SO_2 from flue gas	Sulphate reducing bacteria, *Thiobacillus ferroxidans*	Paques BV and Hoogovens Technical Services Energy and Environment BV
	Trichloroethylene	*Pseudomonas* or genetically engineered *E. coli*	Envirogen, Inc.
Automotive parts	Styrene and VOC's	Naturally occurring bacteria	
Polystyrene plant	Styrene	Naturally occurring bacteria	Envirogen, Inc.
Mining, chemical and petrochemical	H_2S, SO_2	*Chilobacillus ferroxidans*	Dowa Mining Company
Agricultural and animal wastes	Sulphites, H_2S mercaptans, low mol. weight fatty acids, etc.	*Streptomyces griseus, Streptomyces antibioticus, Thermoactinomyces* spp.	
Livestock production	NH_3		
Animal rendering	Furans, sulphur compounds, H_2S, thiophene, NH_3 etc.	Biofilters	Germany
Resin, pesticides, petrochemical, pharmaceutical and textile	Phenol derivatives	*Candida tropicalis, Pseudomonas putida, Fusarium flocci-ferium, Trichosporon cutaneum*	

Biosorption Parameters

Organism used	*Metals removed*	*System*	*Reported efficiency and application*
Pseudomonas fluorescence	Lead, zinc	Immobilised on PVC and packed in columns	Used in Hungarian chemical company
Pseudomonas aerugenosa	Plutonium, uranium	Immobilised on plasma treated polypropylene	75-80% removal
Citrobacter spp.	Cadmium, lead, copper, uranium	Immobilised in polyacrylamide gel columns	80-90% removal, long use possible
Bacillus subtilis	Lead, zinc, copper, nickel, cadmium, mercury, silver	Fixed bed reactor packed with granules of micro-organisms	Metal recovery agent of (Advanced Mineral Technology Inc.)
Pseudomonas, Micrococcus, Saccharomyces	Gold, palladium,		98% removal from industrial and mining waste-water (Miyazaki Medical College)
Sphaerotilus natans and *Leptothrix*	Iron, copper, magnesium cobalt, cadmium		Not commercially exploited
Streptomyces viridochromogenes	Uranium	Immobilisation on polyacrylamide gel columns	80-100% removal
Rhizopus arrhizus (obtained as waste from fermentation industry)	Uranium, thorium	Packed and fixed reactors	170 mg per gram dry weight of biomass, uranium accumulated is 18% of dry weight of biomass
Rhizopus arrhizus	Cadmium		30-130 mg per gram of dry weight of biomass
Aspergillus niger	Silver		Used in Austria
Penicillium chrysogenum	Zinc		
Claviceps paspali heat killed biomass of fungi	Gold, copper, zinc, platinum		17 kg fungus captures 1 gram of platinum, 2 patents in USA
Rhizomucor mehei		Alkali treated biomass in column	
Mucor mucedo, Trichoderma viride	Copper	Immobilised on china clay	5.94-127.5 mcg per ml
Aspergillus oryzae	Cadmium	Immobilised on reticulated foam	99% removal
Penicillium chrysogenum mixed with urea-formaldehyde polycondenstate	Uranium, radium		'Biosorbent M' product in semi-commercial use
Chlorella vulgaris	Gold, silver, mercury, zinc, copper, uranium	Immobilised in polyacrylamide gel and packed in columns	From sea water
Dead algae		Trapped in silica matrix	90% uptake, 'Algasorb' product of Biorecovery System Company, New Mexico and BB Sorbex Canada

Multiple Choice Questions

1. Bangalore process - a system of composting was initiated by :
A. Hutchinson, H.B and Richards, E.H.
B. Fowler and Ridge
C. Howard and Ward
D. Dr. C.N. Acharya

2. Who discovered *Azotobacter,* a free living bacterium that can fix atmospheric nitrogen and improves the soil fertility?
A. H. Hellriegel and H. Wilfarth
B. William Beijerinck
C. T J Burrill
D. Chamberland

3. Indore process - a system of composting was initiated by :
A. Hutchinson, H.B and Richards, E.H.
B. Fowler and Ridge
C. Howard and Ward
D. Dr. C.N. Acharya

4. Who first isolated nitrifying bacteria responsible for production of nitrate from ammonia?
A. Carnaham
B. Winogradsky
C. Del Tredici
D. Dobereiner

5. During Bio-geochemical cycle, some amount of elemental carbon was utilized by the microorganisms. The phenomenon is called as :
A. Dissimilation B. Immobilization
C. Decomposition D. Neutralization

6. The ruminants are a group of ________ mammals.
A. Carnivorous B. Herbivorous
C. Omnivorous D. None of these

7. Bioluminescence property is common to several species of ________.
A. *Vibrio* and *Shigella*
B. *Photobacterium* and *Vibrio*
C. *Photobacterium* and *Shigella*
D. Only *Vibrio*

8. ________ is the obligatory interaction.
A. Mutualism
B. Commensalis
C. Cooperation
D. Ammensalism

9. ________ device is used for the enumeration of bacteria in air.
A. Lemon Sampler
B. Andersons air sampler
C. Capillary impinge
D. All of these

10. ________ are used frequently as indices of faecal pollution in water and food.
A. Only *E. coli*
B. *E. coli* and *Cl. perfringens*
C. *Cl. perfringens* and *Strept. faecalis*
D. *E. coli* and *Strept. faecalis*

11. ________ test/tests are used in the routine analysis of coliform.
A. Presumptive test
B. Confirmed test
C. Completed test
D. All of these

12. ________ medium is used in the presumptive test of *E. coli.*
A. Lauryl sulphate tryptone broth
B. Mac Conkeys broth
C. Sodium azide glucose broth
D. Yeast extract mannitol broth

13. For enumeration of coliform organism ________ method is used.
A. IMViC reaction
B. MPN
C. Standard plate count
D. None of these

14. In laboratory many sea bacteria grow at 300°C and 15000 Ib/inch these are known as ________.
A. Barophilic
B. Obligatory barophilic
C. Psychrophilic
D. Mesophilic

15. The organism whose natural habitat is terrestrial and which are able to grow in media without sea water yet able to tolerate varying degrees of salinity are known as________.
A. Indigenous organism
B. Transient organism
C. Halotolerant organism
D. None of these

16. The size, weight, moisture content and ________ of air born particles are of importance in consideration of methods for disinfecting air.
A. Shape
B. Opacity to UV light
C. Both A and B
D. None of these

17. The coliform group includes all the ________ *Bacilli.*
A. Aerobic and facultative anaerobic
B. Gram negative
C. Non-spore forming
D. All of these

18. ________ test is designate to enumerate total viable pollution.
A. Presumptive test
B. Completed test
C. Standard plate count
D. Confirmed test

19. ________ inhibit virtually all bacteria except faecal *Streptococci.*
A. Ethyl violet and Azide
B. Glucose
C. Lauryl sulphate
D. Tryptose

20. *E. coli* and ________are normally referred to as fecal and non-fecal contaminants of water respectively.
A. *Cl. perfringens*
B. *Enterobacter aerogenes*
C. *Bacillus* spp.
D. *Vibrio cholerae*

21. Archaebacteria are found in environments as bogs :
A. Oceans depths B. Salt brines
C. Hot acid springs D. All of these

22. ________ are early colonizers of newly exposed material in wet situations.
A. Algae B. Fungi
C. Bacteria D. Viruses

23. Strict anaerobic phase in biogas production is :
A. Carboxylic phase
B. Hydrolytic phase
C. Acetogenic phase
D. Methanogenic phase

24. Bacterial cells that can take up DNA from the environment are said to be :
A. Transformants B. Conjugates
C. Competent D. None of these

25. Which among the following is entomo-pathogenic bacteria?
A. *Bacillus thuringiensis*
B. *Bacillus cereus*
C. Both A and B
D. None of these

26. Elimination of pathogenic or deleterious organisms, insect larvae, intestinal parasites, and weed seeds is called as :
A. Sanitization B. Quarantine
C. Both A and B D. None of these

27. In the process, trees indirectly remediate by controlling groundwater movement.
A. Rhizofiltration
B. Phyto-volatilization
C. Phyto-degradation
D. Hydraulic Control

28. Micro cyclic rust is caused by :
A. *Puccinia helianthus*
B. *Urmyces appendiculatus*
C. *Puccinia recondite*
D. Both A and B

29. Which of the following produces Galls?
A. *Agrobacterium tumefeciens*
B. *Ustolago maydis*
C. Both A and B
D. None of these

30. The antiviral compound produced in response to viral infection by eukaryotic :
A. Organism B. Antivironin
C. Interferon D. Plonteferon

31. In which virus, RNA directly serves as the *m*RNA?

A. Rhabdo virus B. Influenza virus
C. TMV D. Orthomixo virus

32. The group of bacteria (Carboxysome) that fix the CO_2, will follow one of these pathways :

A. Ribulose monophosphate pathway
B. Calvin cycle
C. Ribulose biphosphate pathway
D. Both A and B

33. _______ is amount of dissolved oxygen consumed in five days by biological processes breaking down organic matter.

A. BOD B. COD
C. DO D. None of these

34. ________ is continuous culture device usually controlled by the concentration of limiting nutrient and dilution rate.

A. Chemotaxis
B. Chemostat
C. Chlamydospore
D. None of these

35. The biochemical mechanism of nitrate (NO_3^-) respiration or dissimilative metabolism of :

A. Nitrate has been well studied in microorganism like
B. *E. coli*
C. *Paracoccus denitrificans*
D. *Pseudomonas stutzeri*
E. All of these

36. Role of bacteria in carbon cycle is :

A. Photosynthesis
B. Chemosynthesis
C. Breakdown of organic compounds
D. Assimilation of nitrogen compounds

37. In the process, plants actually metabolize and destroy contaminants within plant tissues.

A. Rhizofiltration
B. Phyto-volatilization
C. Phyto-degradation
D. Hydraulic control

38. Removal of Cl as Cl^- from an organic compound by reducing the carbon atom from C-Cl to C-H is called as :

A. Replication
B. Reductive dechlorination
C. Reduction potential
D. Reduction

39. Ability of an organism to colonize the rhizosphere is :

A. Rhizomorph
B. Rhizoplane
C. Rhizosphere competence
D. None of these

40. Which of the Organisms action results in the loss of nutrients from soil?

A. *Thiobacillus denitrificans*
B. *Thiobacillus thiooxidans*
C. *Azospirillum*
D. *Azotobacter*

41. Which of the Organisms action results in the loss of nutrients from soil?

A. *Thiobacillus thiooxidans*
B. *Azospirillum*
C. *Azotobacter*
D. *Desulfovibrio* species

42. Organisms involved in the decomposition of crop residues includes :

A. *Cellulomonas* species
B. Cellulolytic bacteria
C. Fungi
D. All of these

43. Organisms that promote above-ground and/ or below-ground plant growth :

A. *Pseudomonas* spp.
B. *Bacillus* spp.
C. *Streptomyces* spp.
D. All of these

44. Tillage-induced shifts in the fungi:bacteria ratio influences :

A. The rate of organic matter decomposition
B. Nutrient availability
C. Both (A) and (B)
D. None of these

45. Management practices affect soil biota by :
A. Direct effects on the populations
B. Changes in the soil habitat
C. Influences on above-ground productivity and community composition
D. All of these

46. Some of the agronomic practices that impact on populations of biota include :
A. Stubble retention
B. Tillage
C. Application of herbicides
D. All of these

47. Some of the agronomic practices that impact on populations of biota include :
A. Insecticide and fungicide application
B. Application of manures or fertilisers
C. Application of chemical amendments or waste products
D. All of these

48. Which of the following is a Microbial pesticides that fight pests through a variety of ways?
A. Bacteria B. Fungus
C. Virus D. Protozoa
E. All of these

49. Active SOM includes :
A. Living (2-12 per cent)
B. Non-living (8-24 per cent)
C. Humus (70-90 per cent)
D. None of these

50. Active SOM includes :
A. Viruses B. Bacteria
C. Actinomycetes D. All of these

ANSWERS

1	2	3	4	5	6	7	8	9	10
D	B	C	B	B	B	B	A	B	A
11	**12**	**13**	**14**	**15**	**16**	**17**	**18**	**19**	**20**
D	A	B	B	C	A	D	C	A	B
21	**22**	**23**	**24**	**25**	**26**	**27**	**28**	**29**	**30**
D	B	D	C	C	A	D	A	C	C
31	**32**	**33**	**34**	**35**	**36**	**37**	**38**	**39**	**40**
B	A	A	B	E	C	C	B	C	A
41	**42**	**43**	**44**	**45**	**46**	**47**	**48**	**49**	**50**
D	D	D	C	D	D	D	E	A	D

CHAPTER

20

Medical Biotechnology

Introduction

Medical biotechnology is the use of living cells and cell materials to research and produce pharmaceutical and diagnostic products that treat and prevent human diseases.

Examples of discoveries in the field of medical biotechnology include insulin and growth hormone. Both discoveries were the result of research studies related to deoxyribonucleic acid (DNA).

Biotechnology has made a huge difference in human health care and has now enabled scientists to develop products which can give quicker and more accurate tests, therapies that have a lot less side effects and vaccines which are safer than ever before.

Diagnosing and Biotechnology

Medical conditions and diseases are now being detected more accurately and quickly due to the advancement of biotechnology based tools. e.g., Home pregnancy kit, certain types of cancer such as ovarian and prostrate cancer now rely on biotechnology-based tests by taking a simple blood test and thus eliminating the need for expensive and invasive surgery.

Vaccine

A vaccine is a biological preparation the improves immunity to a particular disease. A vaccine typically contains an agent that–resembles a disease-causing micro-organism, and is often made from weakened or killed forms of the microbe or its toxins. The agent stimulates the body's immune system to recognise the agent as foreign, destroy it, and 'remember' it, so that the immune system can more easily recognise and destroy any of these microorganisms that it later encounters.

Types of Vaccines

There are several types of vaccines currently in use:

Killed : Some vaccine contain killed, but previously virulent, micro-organisms that have been destroyed with chemical or heat.

e.g., Influenza vaccine, Cholera vaccine, Polio vaccine, Hepatitis vaccine, Rabies vaccine etc.

Attenuated : Some vaccines contain live, attenuated microorganisms.

e.g., Yellow fever, measles, rubella and mumps and the bacterial disease typhoid.

Toxoid : Toxoid vaccines are made from inactivated toxic compounds that cause illness rather than the microorganism.

e.g., Toxoid-based vaccines include tetanus and diphtheria.

Subunit: Protein subunit–rather than introducing an inactivated or attenuated micro-organism to an immune system, a fragment of it can create an immune response.

e.g., Vaccine against Hepatitis B virus that is composed of only the surface proteins of the virus.

Conjugate : Certain bacteria have polysaccharide outer coats that are poorly immunogenic. By linking these outer coats to proteins, the immune system can be led to recognise the polysaccharide as if it were a protein antigen. e.g., *Haemophilus influenzae* type B vaccine.

Live vaccines : Live vaccines tend to be more immunogenic than non-living antigens because they can proliferate and elicit strong innate and adaptive responses. e.g., Attenuated measles virus, polio virus, rubella virus and *S. typhi* strain Ty 21a.

Some Selected Valuable Products from Micro-organisms : These Represent Potential Targets for Directed Screening

Product	*Activity*	*Source (producing microorganism)*
Avermectin	Antihelmintic	*Streptomyces avermitilis*
Bestatin	Antitumour	*S. olivoreticuli*
1,3-Diphenethylurea	Antidepressant	*Streptomyces* sp.
Detoxin	Detoxicant	*S. caespitosus*
Herbicidin	Herbicide	*S. sagamonensis*
Cyclosporin	Immunosuppressor	*Trichoderma polysporum*
Piercidin	Insecticide	*S. mobaraensis*
Tetranactin	Miticide	*S. aureus*
Gibberellic acid	Plant growth regulator	*Gibberella fujikuroi*
Slaframine	Salivation inducer	*Rhizoctonia leguminicola*
Azalomycin	Antiprotozoal	*S. hygroscopicus*
Aabomycin A	Antiviral	*S. hygroscopicus*

Some Selected Applications of Enzymes Used in Treatment of Diseases

Enzyme	*Reaction catalyzed*	*Used for*
α-Amylase	Starch hydrolysis	Digestive disorders
Asparaginase	Asparagine → aspartate	Leukaemia
Collagenase	Collagen hydrolysis	Skin ulcers
Ficin	Protein hydrolysis	Deworming
Glutaminase	L-Glutamine → L-glutamate	Leukaemia
Lysozyme	Bacterial cell wall hydrolysis	Used as antibiotic
Lipase	Lipid hydrolysis	Digestive disorders
Papain	Protein hydrolysis	Deworming
Proteases	Protein hydrolysis	Digestive disorders
Rhodonase	Formation of cyanide sulphide	Cyanide poisoning
Ribonuclease	RNA digestion	As antiviral agent
β-Lactamase	Penicillin → penicilloate	Penicillin allergy
Streptokinase	Plasminogen → plasmin	Blood clots
Uricase	Urate → allantoin	Gout
Urokinase	Plasminogen → plasmin	Blood clots

Some Selected Antibiotics Produced Commercially

Antibiotic	*Produced by*	*Activity*	*Chemical nature*
Amphotericin B	*Streptomyces nodosus*	Antifungal	Polyene
Bacitracin	*Bacillus subtilis*	$Gram^+$ bacteria	Peptide
Cephalosporin C	*Cephalosporium acremonium*	$Gram^+$ and $^-$ bacteria	Peptide
Cycloheximide	*S. griseus*	Antifungal	–
Fungimycin	*S. coelicolor*	Antifungal	Polyene
Gentamycin	*Micromonospora purpurea*	$Gram^+$ bacteria	Aminoglycoside
Gramicidin	*Bacillus brevis*	$Gram^+$ bacteria	Peptide
Griseofulvin	*Penicillium griseofulvum*	$Gram^+$ and $^-$ bacteria	Spirolactone
Kanamycin	*S. kanamyceticus*	$Gram^+$ and $^-$ bacteria, and mycobacteria	Aminoglycoside
Neomycin	*S. fradiae*	$Gram^+$ and $^-$ bacteria	Aminoglycoside
Pimaricin	*S. natalensis*	Antitumour	Polyene
Penicillin G	*P. chrysogenum*	$Gram^+$ bacteria	Peptide
Polymixin B	*Bacillus polymyxa*	Antifungal	Peptide
Streptomycin	*S. griseus*	$Gram^+$ and $^-$ bacteria, and mycobacteria	Aminoglycoside
Tetracyclines	*Streptomyces* sp.	$Gram^+$ and $^-$ bacteria	Tetracycline
Trichomycin	*S. hachijoensis*	Antifungal	Polyene

Some Antibiotics and the Mode of Their Action

Antibiotic	*Mode of Action*
Amino glycoside Manamycin Neomycin Streptomycin	Inhibit protein synthesis by interacting with 30S or 50S subunit of ribosome.
β-Lactams Ampicillin Carbenicillin Penicillin	Inhibit synthesis of bacterial cell wall.
Trimethoprim and sulphonamides	Inhibit synthesis of tetrahydrofolate.
Chloramphenicol	Inhibits protein synthesis by acting on 50S subunit of ribosome.
Tetracyclines	Inhibits protein synthesis by interacting with 30S subunit of ribosome.
Glycopeptides Vancomycin	Interferes with synthesis of bacterial cell wall.

Antibiotic	*Mode of Action*
Macrolides and lincosamides Erythromycin Lincomycin	Inhibit protein synthesis by interacting with 50S subunit of ribosome.
Polymixins Polymixin B Polymixin E	Attach to cell membrane and modify ion flux, resulting in cell lysis.
Quinolones Nalidic acid Norfloxin Ofloxacin	Interferes with replication of DNA by inhibition of DNA gyrase.
Rifampicin	Interferes with formation of mRNA by binding to RNA polymerase.

Some Tumour Associated Antigens being Evaluated as Targets for Cancer Therapy using Monoclonal Antibodies

Tumour Type	*Tumour antigen*	*Target antigen*
	Limphoid cell-surface markers	
T-cell leukaemia/lymphoma	T-cell marker	CD5
B-cell lymphoma	B-cell marker	CD20
B-cell lymphoma	Anti-idiotype	Immunoglobulin
	Nonlymphoid tissue markers	
Colon cancer	Carcinoembryonic antigen (CEA)	Glycoprotein
Neurorectodermal tumours	Gangliosides like GD2 and GD3	Glycolipids associated with neutral tissues
Breast cancer	MUC1	Glycoprotein
Some lung, head, neck and breast-tumour	Epidermal growth factor receptor (EGFR)	EGF-binding cell surface protein
Breast and ovarian tumours	HER2 (an EGF like receptor)	Cell surface EGF binding protein

Some Diseases/Syndromes that have been Mapped onto Specific sites on Human Chromosomes

Chromosome	*Disease/Syndrome*
1	Ductal breast cancer, Charcot-Marie Tooth Neuropathy 2a, Chediak-Higashi syndrome, Gaucher disease, Usher syndrome type 2.
2	Tibial muscular dystrophy, Alstrom syndrome, Autosomal recessive deafness 9, Limb Girdle muscular dystrophy 2b, Holoprosencephaly 2.
3	*Alkaptonuria,* Acute Myeloid leukaemia, von Hippel-Lindau syndrome, Cornelia de Lange syndrome.

Chromosome	Disease/Syndrome
4	*Huntington disease,* Wolf Hirschhorn syndrome, Wolfram syndrome, Juvenile Periodontitis.
5	Spinal muscular atrophy, *Cri-du-chat,* Coloractal cancer, Diastrophic dysplasia, Teacher Collins syndrome.
6	Hemochromatosis, Lafora myoclonus epilepsy, Celiac disease, Spinocerebellar atrophy 1.
7	*Cystic fibrosis,* Type 1 Williams-Beuren syndrome, Pendred syndrome, Split hand/foot malformation, Pallister-Hall syndrome.
8	Werner syndrome, Nijmegen-breakage syndrome, Lipoprotein lipase: Human, Cohen syndrome, Langer-Giedion syndrome.
9	Torsion Dystonia 1, Fanconi anemia type C, Dysautonomia, Friedreich ataxia, Nail-Patella syndrome.
10	Wolman disease, Cowden disease, Jackson-Weiss syndrome, Hermansky-Pudlak syndrome.
11	Long QT syndrome 1, Wilms tumour type 1, Ataxia-telangiectasia, Beckwith-Wiedemann syndrome, Bardet-Biedl syndrome 1.
12	*Phenylketonuria,* Darier disease, Familial periodic fever, Dentatorubro-pallidoluysian atrophy, Noonan syndrome.
13	Wilson disease, Stargardt disease-2, Retinoblastoma, Rieger syndrome type 2.
14	Graves disease, Machato-Joseph disease, Alzheimer disease, Spastic paraplegia 3A.
15	*Tay-Sachs disease,* Amytrophic lateral sclerosis-5, Bloom's syndrome, Prader-Willi syndrome.
16	Fanconi anemia type A, Familial Mediterranean fever, Inflammatory bowel disease 1, Polycystic kidney disease adult 1.
17	*Breast cancer 1,* Canavan disease, TP53 tumour suppressor gene, Miller-Dieker, Lissencephaly, Neurofibromatosis type 1.
18	Colorectal cancer, Niemann-pick disease type C, Tourette syndrome, Holoprosencephaly 4.
19	Diamond-Blackfan anemia, *Acute T-cell leukaemia,* Myotonic dystrophy.
20	Corneal dystrophy, polymorphous SCID due to ADA deficiency, Alagille syndrome.
21	*Down syndrome,* Usher syndrome type 1E, Holoprosencephaly 1.
22	Cat eye syndrome, Ewing sarcoma, DiGeorge syndrome, Velocardiofacial syndrome, Neurofibromatosis type 2.
X	Duchenne muscular dystrophy, Adrenoleukodystrophy, Lowe syndrome, Norrie disease, Rett syndrome.
Y	Gonadal dysgenesis, *Prostrate cancer,* Mobius syndrome, *Acute myeloid-leukaemia,* Adenocarcinoma.

Source: Genome Database, cancer genome anatomy project website and NCBI's genome resource site.

Multiple Choice Questions

1. Application of gene therapy involves which of the following basic developments in genetics, molecular biology and biotechnology?

I. Identification of the genes that play the key role in the development of genetic disorder.
II. Determination of the role of its product in health and disease.
III. Isolation and cloning of the gene.
IV. Development of an approach for gene therapy.

Options:

A. I, II, III, IV
B. II, III, IV
C. I, III, IV
D. I, II, IV

2. Recombinant insulin is produced in which of the following?

A. Continuous cell lines
B. Yeast
C. *E. coli*
D. Both A and C

3. Which of the following is regarded as the third revolution in vaccines?

A. Recombinant polypeptide vaccines
B. DNA vaccines
C. Recombinant protein vaccines
D. All of these

4. The antigen presenting cell :

A. May be a T cell
B. Does not produce cytokines which influence the adaptive response
C. May be a dendritic cell in the skin
D. Matures upon antigenic stimulation and becomes a plasma cell

5. Mucosal immunity provides most of its protection by blocking :

A. Blocking penetration of undigested food products into the mucosal tissues
B. Microbial receptors specific for colonization
C. The complete cascade
D. Both (A) and (B)

6. SDS-PAGE is used in one of the following immunological techniques :

A. Western Blotting
B. Immunoelectrophoresis
C. Immunoprecipitation
D. Immunodiffusion

7. The live vaccines are available against the following viruses, except :

A. Influenza
B. Measles
C. Rabies
D. None of these

8. Treatment with bacterial endotoxins results blast formation in :

A. T cells
B. Monocytes
C. B cells
D. None of these

9. Which of the following is correct?

A. T cell receptors are generated by random assortment of gene cassettes
B. All T cells are $CD4^+$
C. MHC class II molecules are generated by random assortment of gene cassettes
D. An epitope can have multiple antigens

10. .... is a genetic disease characterized by a total or partial inability to synthesize globulins.

A. Apitososis
B. A gamma globulinemia
C. Sickle cell anemia
D. None of these

11. In AIDS, Kaposis sarcoma may respond to :

A. Azathioprine
B. Interleukin-2 infusion
C. Alpha interferon
D. None of these

12. Lymphogranuloma venerum (LGV) is a sexually transmitted disease is caused by :

A. *Chlamydia trachomatis*
B. *Chlamydia psittasi*
C. *Chlamydia pneumonias*
D. *Chlamydia opthalmia*

13. Bacilli Calmette Guerin (BCG) contains the avirulent strains of :

A. A typical mycobacteria
B. Human tubercle bacilli
C. Avian tubercle bacilli
D. Bovine tubercle bacilli

14. Neurological complications following rabies vaccines is common with :
A. Chick embryo vaccine
B. Semple vaccine
C. BPL vaccine
D. HDCS vaccine

15. The following characters are true about Neisseria gonorrhoea except :
A. Oxidase positive organism
B. Gram-negative aerobic bacteria
C. Non-motile diplococci
D. Airborne infection

16. The antitumour antibiotics act by inhibiting :
A. The DNA structure and function
B. Cell membrane synthesis
C. Cell wall synthesis
D. RNA synthesis

17. Clear-zones formation around antibiotic disc is due to :
A. Growth of the bacterium surrounding the disc
B. The destruction of paper disc
C. Lysis the bacterial cells surrounding the disc (antibiotic)
D. None of these

18. The smallest amount of chemotherapeutic agents required to inhibit the growth of the organism *in vitro* is known as :
A. Thermal death point (TDP)
B. MIC (Minimum Inhibitory Concentration)
C. Death rate
D. None of these

19. The best test for the susceptibility of a micro-organism to antibiotics and other chemotherapeutic agents :
A. Tube-dilution test
B. Paper-disc test
C. Both A and B
D. None of these

20. Diagnostic test for the identification of primary syphilis :
A. Kahn's test
B. Dark ground microscopic examination
C. VDRL test
D. *Treponema pallidum* immobilization test

21. AIDS disease is caused by a virus which belongs to :
A. Retro virus group
B. Rhabdo virus group
C. Adeno virus group
D. Hepatitis virus group

22. Very effective, less time consuming and at a time so many samples can be detected by :
A. CFT B. Neutralization
C. Agglutination D. ELISA

23. Anti-tuberculosis drug, isonicotinic acid hydrazide (INH) leads to the deficiency of Vitamin ....
A. Pyridoxine (B_6)
B. Cobalamin (B_{12})
C. Vitamin D
D. Vitamin C

24. Haptoglobin binds and prevents the excretion of the compound ....
A. Heme B. Iron
C. Haemoglobin D. Calcium

25. The disorders characterized by decreased synthesis or total absence of globin chains of haemoglobin is collectively known as ....
A. Thalassemia B. Thick cells
C. Thalamus D. Thin cells

26. Cholesterol synthesis is controlled by feedback inhibition of the enzyme ....
A. GGGCoA reductase
B. HHGCoA reductase
C. MMGCoA reductase
D. HMGCoA reductase

27. Hypouricemia is due to deficiency of the enzyme ....
A. Pantothine oxidase
B. Lycopene oxidase
C. Xanthine oxidase
D. None of these

28. The copper containing protein involved for the conversion of ferrous iron (Fe^{++}) to ferric iron (Fe^{+++}) in the plasma ....

A. Neuroplasmin B. Ceruplasmin
C. Protoplasmic D. None of these

29. The hormone controlling water excretion via kidney is ....

A. Diuretic hormone
B. Retarding hormone
C. Antidiuretic hormone
D. Promoting hormone

30. The endocrine organ most predominantly associated with BMR is ....

A. Gall bladder
B. Thyroid gland
C. Tubular gland
D. Testosterone gland

31. The biochemical parameter often used as an index for monitoring the recovery from Kwashiorkor is ....

A. Plasma glutelin
B. Plasma globulin
C. Plasma prolamin
D. Plasma abbumin

32. The EBV-hybridoma technique :

A. Gives much higher hybridization frequencies
B. Immortalizes the donor B-cells
C. Facilitates the proliferation of antigen specific B-cells
D. All of these

33. Preliminary clinical results with a humanized antibody against the interleukin-2 receptor have suggested the :

A. Poor recognition of immunoglobulin, Ig constant region
B. Presence of HAMA response
C. Absence of human immune response against murine proteins (HAMA) response
D. All of these

34. The Cross linkage of antigens by antibodies is known as :

A. Agglutination
B. A cross reaction
C. Complement fixation
D. All of these

35. A cytokine that stimulates the activity of B and T cells is :

A. Lymphotoxin B. Interleukin-2
C. Interleukin-1 D. All of these

36. Live vaccine is :

A. Low dose of the infectious bacteria administered as prophylactic
B. A low dose of toxin that is produced by the bacterium
C. A sample of cells from a patient who recently recovered from the disease
D. A low dose of bacterial strain in a modified form which retains immunogenicity but is not pathogenic

37. Endotoxins are :

I. Found in gram positive cells
II. Lipopoly saccharides
III. Found in gram negative cells
IV. Lipoproteins
V. Found in mycoplasma

A. I, II and V
B. III, IV and V
C. I and II
D. II and III

38. The pathogenesis associated with *Vibrio cholerae* infection depends on the colonization of the small intestine by the organism and secretion of an enterotoxin. Which of the following statement related to pathogenicity of cholera is correct?

A. CTX bacteriophage carries the gene of cholera toxin and other virulence factors
B. Choleragen toxin alone cannot reproduce the symptoms of cholera in the absence of *Vibrio cholerae*
C. *Vibrio cholerae* secretes the enderotoxin choleragen
D. Choleragen consists of A (active) and B (binding) subunit.

39. Which of the following statement about HIV is correct?

A. HIV is an enveloped retro virus

B. HIV infects T-cells that carry the CD4-antigen on their surface
C. HIV contains two copies of plus single stranded RNA genome
D. HIV-1 contains only three genes namely *gag, pol* and *env*.

40. HIV a retrovirus contains :
A. Two copies of dsRNA
B. Two copies of positive sense ssRNA
C. One ssRNA
D. Two copies of negative sense ssRNA

41. Which of the following is not related to HIV?
A. HIV genome contains one copy of each gene
B. Long terminal repeat contains the promoter for HIV transcription
C. HIV belongs to lentiviridae subform
D. gp 120 is the membrane protein that HIV transcription

42. The ability of a virus to produce a disease can result from a variety of mechanisms. Which of the following is not a likely option?
A. Cytopathic effect in infected cells
B. Malignant transformation of infected cells
C. Secretion of exotoxin that activates adenylyl cyclase
D. Immune response to virus-induced antigens on the surface of infected cells.

43. Choose the incorrect statement :
A. Mutations are expressed more frequently in bacteria because they are monoploid
B. Rickettsia is a gram-negative and obligate intracellular parasitic bacteria
C. A microorganism which required a growth factor in order to grow is called an auxotroph
D. Cyanobacteria is an example of photo-autotroph

44. Sulpha drugs are effective antibiotics which inhibits the formation of folic acid by acting as a :
A. Non-competitive inhibitor
B. Uncompetitive inhibitor
C. Competitive inhibitor
D. Irreversible inhibitor

45. Mycoplasma are not inhibited by penicillin because they :
A. Are gram-positive
B. Are gram-negative
C. Produce penicillinase
D. Do not have a cell wall

46. Penicillin and lysozyme prevent synthesis and cause lysis, respectively, of cell walls of :
A. *Micrococcus lysodeikticus*
B. *Escherichia coli*
C. *Saccharomyces cerevisiae*
D. *Methanobacterium barkeri*

47. Two antibiotics have different sites of action in a bacterial cell. The frequency of occurrence of resistance to these antibiotics used singly are 10^{-5} and 10^{-6}, respectively. When the antibiotics are used in combination the frequency of occurrence of resistance to both antibiotics are :
A. 10^{-5} B. 10^{-11}
C. 10^{-30} D. 10^{-6}

48. Penicillin-resistant bacteria produce-lactamase which inactivates penicillin by :
A. Hydrolysing the C-N bonds
B. Oxidising S atom
C. Esterifying the Carboxyl group
D. Hydrolysing the side chain

49. Endochitinase can be used as an :
A. Antiviral
B. Antibacterial
C. Antifungal
D. Antinematodal

50. Which of the following is incorrect?
A. Bacterial chromosome rarely contains introns
B. Rickettsia is a gram-negative obligate intracellular parasite
C. Endospore formation is common in genera *Bacillus* and *Clostridium*
D. Antibiotic rifampicin inhibits the cross linking in peptidoglycan cell wall of bacteria

ANSWERS

1	2	3	4	5	6	7	8	9	10
A	D	B	A	D	B	C	A	A	B
11	**12**	**13**	**14**	**15**	**16**	**17**	**18**	**19**	**20**
C	A	D	B	D	A	C	B	C	B
21	**22**	**23**	**24**	**25**	**26**	**27**	**28**	**29**	**30**
A	D	A	C	A	D	C	B	C	B
31	**32**	**33**	**34**	**35**	**36**	**37**	**38**	**39**	**40**
D	D	C	A	B	D	D	B	D	B
41	**42**	**43**	**44**	**45**	**46**	**47**	**48**	**49**	**50**
A	C	D	B	D	A	B	A	C	D

●●●

CHAPTER

21

Industrial Biotechnology

Introduction

Biotechnology has applications in four major industrial areas, including health care (medical), crop production and agriculture, non-food (industrial) uses of crops and other products (e.g., biodegradable plastics, vegetable oil, biofuels) and environmental uses.

For example, one application of biotechnology is the direct use of organisms for the manufacture of organic products (examples include beer and milk products). Another example is using naturally present bacteria by the mining industry in bioleaching. Biotechnology is also used to produce biological weapons.

White Biotechnology

It is also known as industrial biotechnology, is biotechnology applied to industrial processes.

An example is the designing of an organism to produce a useful chemical. Another example is the using of enzymes as industrial catalysts to either produce valuable chemicals or destroy hazardous/ polluting chemicals. White biotechnology tends to consume less in resources than traditional processes used to produce industrial goods.

Applications of Industrial Biotechnology

(*i*) Industrial biotechnology has produced enzymes for use in our daily lives and for the manufacturing sector. For instance, meat tenderizer is an enzyme and some contact lens cleaning food contain enzymes to remove sticky protein deposits.

(*ii*) It working with nature to maximize and optimize existing biochemical pathways that can be used in manufacturing.

(*iii*) The industrial biotechnology revolution rides on a series of related developments in the field of study of detailed information derived from the cell *i.e.,* genomics, proteomics and bioinformatics.

(*iv*) It uses renewable raw materials and is one of most promising, innovative approaches towards lowering greenhouse gas emissions.

(*v*) It uses in metabolic production.

(*vi*) Industrial biotechnology has produced fermented foods.

(*vii*) It uses for methane production.

(*viii*) It uses in production of biocontrol agents, and bio-based fuel and energy.

(*ix*) It is also used for the recovery of metals. *e.g.,* Leaching of copper as copper sulphate from ores, Desulphurization of coals by *Thiobacillus* spp.

Industrial Biotechnology Products

	Products	*Source*
1.	Production of useful compounds such as ethanol, lactic acid, glycerine, citric acid, gluconic acid, acetone etc.	Produced by micro-organisms mainly bacteria, from less useful substrate.

	Products	*Source*
2.	Production of antibiotics such as *penicillin, streptomycin, erythromycin, mitomycin, cycloheximide* etc.	Produced by fungi, bacteria and actino mycetes as secondary metabolites.
3.	Transformation of less useful and cheaper compounds into more useful and valuable ones such as steroid hormones from sterols, sorbose from sorbitol etc.	Generally by micro-organisms or immobilized enzymes as secondary metabolites.
4.	Production of enzymes such as α-amylase, proteases, lipases etc.	From fungi, bacteria etc. for use in detergent, textile, leather, dairy industries and in medicines.
5.	Single cell proteins from bacteria, yeasts, fungi or algae for human food and animal feed as a supplements.	Single cell proteins are the total microbial biomass freed from toxins, contaminants, if any.
6.	Production of immunotoxins by joining a natural toxin with a specific antibody.	These destroy specific cell types, may provide a potent treatment for cancer.
7.	Immobilization of enzymes for their repeated industrial application.	More attractive than the use of whole micro-organisms.
8.	Mineral extraction through leaching from low grade ores such as copper, uranium etc.	Due to action of microbes, mainly bacteria.
9.	Fuel mainly ethanol, sometimes biogas production from cheaf, less useful and abundant substrates such as sugarcane biogas, wood etc.	Produced through fermentation by micro-organisms cowdung based biogas being popularized in India.
10.	Protein/enzyme engineering to change the primary structure of existing proteins/enzymes to make them more efficient, change their substrate specificity as success with T4 lysozyme, trypsin, subtilism, lactate dehydrogenase etc.	Extensive use of computers for generating models of protein molecules. It is hoped to change RuBisCO so as to minimize its affinity for O_2.

List of Some Enzymes used for Commercial Activities

Enzymes	*Source*	*Application*	*Use in the form*
α - Amylase	1. *Bacillus licheniformis* 2. *B. amyloliquefaciens*	Hydrolysis of starch to dextrans	Extracellular, soluble
Glucoamylase	*Aspergillus niger*	Dextran hydrolysis to glucose	Extracellular, soluble
Xylose (Glucose) isomerase	*Bacillus coagulans**	Pure glucose to equilibrium mixture of glucose + fructose	Immobilized whole cells
Alkaline protease	*B. licheniformis, B. subtilis*	Detergents (protein digestion)	Extracellular, soluble
Neutral protease	*B. amyloliquefaciens, B. subtilis*	Protein digestion in brewing substances	Extracellular, soluble
Acid protease, *e.g.,* rennet	*A. niger***	Milk coagulation and cheese flavour enhancement	Extracellular, soluble

Enzymes	Source	Application	Use in the form
Pectinase	*A. niger, B. subtilis*	Pectin hydrolysis in fruit	Extracellular, soluble
Cellulase	*A. niger, Trichoderma reesei*	Cellulose hydrolysis, in fruit juices	Extracellular, soluble
Lipase	*Rhizopus species*	Detergents, lipid hydrolysis	Extracellular, soluble
Lactase	*A. niger*	Milk lactose hydrolysis to glucose + galactose	Immobilized on glass beads

* Other source organisms: *Streptomyces olivaceus, Microbacterium arborescens, Actinoplanes missouriensis.*

** Other source organisms: *Mucor miehei, Mucor pusillus, Endothica parasitica.*

+ Other source organisms: *Candida lipolytica, Geotrichum candidum.*

Enzymes Widely used in the Production of Fruit Juices, Beer and Distilled Alcoholic Drinks

Product	Substrate	Problem	Enzyme(s)	Action/purpose
Fruit Juices, wines	Pectins	Cloudiness	A mixture of: (*i*) Polygalacturonase (*ii*) pectin esterase, (*iii*) pectin lyase, and (*iv*) hemicellulases	Reduced viscosity, no cloudiness increased juice yield, enhanced flavour, shorter fermentation time (wines)
Beer	1. Starch	Nonfermentable	Barley malt + bacterial and fungal α-amylases, β-glucanase	Saccharification; stopped by boiling the 'wort' when 75% starch is saccharified.
	2. Protein	–	Proteases (neutral)	Hydrolysis of protein and increased fermentation rate (later); especially to obtain high gravity beer.
	3. Cellulose and barley β-glucans	–	Cellulases	Used to hydrolyse cellulose and barley β-glucans, especially when wheat is added as adjunct*
	4. Protein (in association with tannin)	Chill-haze on cooling the beer	Papain	Used in post-fermentation stages
	5. Starch	More total solids in beer	Glucoamylase of fungal α-amylase	Higher degree of saccharification, reduced alcohol and total solids; to produce 'light' beers.
	6. Dextrins**	Nonfermentable; more total solids in beer	Genetically engineered yeast *(S. cerevisiae)*	Utilizes dextrins to produce alcohols; to produce light beers.
Distilled alcoholic drinks	Starch	Nonfermentable	More heat stable bacterial α-amylases	Saccharification of starch present in the substrate

* An adjunct is a starch containing material added to barley malt in order to increase the amount of fermentable sugars.

** Up to 25% of the carbohydrates in original fermentation liquid.

List of Enzymes used in Food Processing (other than cheese)

A.	**Proteases**		
1.	Endogenous proteases	Meat ('hanging' of meat)	Tenderisation of meat, flavour development.
2.	Subtilisin	Soy protein	Partial hydrolysis; increases whipping expansion, emulsifying capacity; hydrolysate may be added to cured meats.
3.	Neutral or alkaline proteases	Animal and fish bones (contain about 5% meat)	Mashed bone incubated at 60°C for 4 hours; meat slurry used in canned meats and soups.
4.	Subtilisin	Red blood cells (RBCs)	RBC haemolysate subjected to hydrolysis; haeme molecules precipitate and are removed; purified hydrolysate spray-dried and used in cured meats, sausages, luncheon meats, etc.
5.	Papain (in active form)	Meat (tenderisation)	Injected into the jugular vein shortly before slaughter; after slaughter, papain is activated and tenderises the meat; only 2-5 ppm (of body weight) enzyme injected.
6.	Heat labile fungal protease	Dough from high gluten wheat varieties	Hydrolysis of gluten; makes dough suitable for biscuit, pie-pastry making.
B.	**Glucose oxidase**	D-glucose	D-glucose oxidized to gluconic acid, O_2 utilized, H_2O_2 produced.
C.	**Catalase**	H_2O_2	Degrades H_2O_2 into water and O_2; used in combination with glucose oxidase to remove glucose and/or O_2 from foods, drinks, etc.
D.	**Amylases (α- and β-), glucoamylase, pullulanase, invertase, glucose isomerase (immobilized)**	Starch, sucrose, D-glucose	Production of glucose, maltose and high fructose syrups.

Multiple Choice Questions

1. Which of the following Ascomycetes are used for riboflavin production?
A. *Eremnothecium ashbyii*
B. *Ashbya gossypii*
C. *Candida guilliermondia*
D. Both A and B

2. The major organism used in the microbial production of citric acid is :
A. *Penicillin notatum*
B. *Rhizopus nigrificins*
C. *Aspergillus niger*
D. *Lactobacillus delbrueckii*

3. Which of the following organism is utilized for the production of D-L-alanine?
A. *Microbacterium ammoniaphilum*
B. *Brevibacterium flavum*
C. *Arthrobacter paraffineus*
D. *C. glutamicum*

4. The aerated submerged bacterial fermentation to produce vitamin B_{12} employs strains of :
A. *Propionibacterium*
B. *Pseudomonas*
C. *Acetobacter*
D. Both (A) and (B)

5. The rapid process of microbial oxidation developed to produce acetic acid from ethanol at Schutzenbach, Germany is also known as :
A. Packed generator B. Trickle method
C. Orleans process D. Both A and B

6. Which of the following micro-organism utilizes lactose as a carbon source in production of lactic acid from whey media?
A. *Lactobacillus bulgaricus*
B. *Lactobacillus delbrueckii*
C. *Lactobacillus pentosus*
D. *Leuconostoc mesenteroides*

7. The organism which utilizes the pentose of sulphite waste liquor for lactic acid production is :
A. *Streptococcus lactis*
B. *Lactobacillus bulgaricus*
C. *Lactobacillus casei*
D. *Lactobacillus pentosus*

8. The production of acetic acid from ethanol is
A. An anaerobic process
B. A aerobic process
C. None of these
D. Combination of both

9. Which of the following agents involve biosensors linking with electronic circuit?
A. Microorganisms
B. Microbially derived enzymes
C. Either microorganisms or microbially derived enzymes
D. Neither microorganisms nor microbially derived enzymes

10. The maximum concentration of ethanol that is achieved by successive distillation of fermentation broth in continuous stills is :
A. 60 per cent B. 95 per cent
C. 70 per cent D. 99 per cent

11. Which of the following microorganism produces only L(+) - lactic acid without further racemization?
A. *Leuconostoc mesenteroides*
B. *Lactobacillus delbrueckii*
C. *Rhizopus oryzae*
D. *Lactobacillus bulgaricus*

12. Naringinase, an enzyme commonly used for removal of bitter taste from citrus juice is obtained from :
A. *A. niger* B. *A. flavus*
C. *A. oryzae* D. *S. cerevisiae*

13. Protease, which is used for flavouring of sake and haze removal in sake is produced by :
A. *A. oryzae* B. *A. flavus*
C. *B. cereus* D. *A. niger*

14. The organism *B. brevis* can be used commercially for the production of :
A. Gramicidin A
B. Tyrocidin
C. Tyrothricin (bacitracin tyrocidin complex)
D. All of these

15. The early assay employed for measuring penicillin activity were based upon the quantity required to prevent growth of S. aureus over :
A. 20 mm diameter zone using the cylinder plate method of assay
B. 24 mm diameter zone using the cylinder plate method of assay
C. 200 mm diameter zone using the cylinder plate method of assay
D. None of these

16. Protease, which is used for flavouring of sake and haze removal in sake is produced by :
A. *A. oryzae* B. *A. flavus*
C. *B. cereus* D. *A. niger*

17. A major organic acid produced by a microbial process used in foods is :
A. Sulphuric acid B. Oxalic acid
C. Citric acid D. Uric acid

18. Microbially produced streptokinase is used to
A. Remove spots from laundry products

B. Dissolve clots for patients with coronary blockages
C. Clarify fruit juice
D. Breakdown starch molecules into smaller sugars

19. Which of the following organism is utilized to obtain commercial L-lysine by direct method?
A. *Micrococcus glutamicus*
B. Mutated strain of *Micrococcus glutamicus*
C. *Brevibacterium aminogenes*
D. Mutated strain of *Brevibacterium aminogenes*

20. "Vinegar mother" in Orleans process is :
A. Inoculum of acetobacter bacteria used
B. Considerable slime and film growth of Acetobacter bacteria on the surface of the alcoholic broth
C. Cellulosic strands produced by Acetobacter bacteria to hold cells together
D. None of these

21. Commercial scale production of amino acids is typically carried out using :
A. Regulatory mutants to cause over-production of biochemical intermediates
B. Creation of an intentional increase in membrane permeability to increase release of the amino acid
C. Both A and B
D. None of these

22. The beet molasses is pretreated with ________ to reduce great quantity of trace metals.
A. Ferrocyanide or ferricyanide
B. Ethylenediamine tetrachloro acetic acid
C. Hydrochloric acid
D. Sulphuric acid

23. In *Corynebacterium glutamicum,* tryptophan biosynthesis is regulated by the activity of :
A. DAHP synthetase
B. Anthranilate synthase
C. Both A and B
D. Phosphoenol pyruvate

24. Which of the following is true about α-ionone, non-ionic detergent used in α-carotene fermentation medium?
A. It is a precursor of a-carotene fermentation process
B. It is regarded as "steering" factor associated with bringing about the presence and activity of the enzymes responsible for α-carotene formation
C. It is an inducer of α-carotene fermentation process
D. None of these

25. Invertase has its application in confectionaries to prevent crystallization of sugar. It is obtained from :
A. *A. oryzae*
B. *A. niger*
C. *S. cerevisae*
D. *B. subtilis*

26. Which one of the following is called as the 'Brewers yeast'?
A. *Saccharomyees ludwigi*
B. *Saccharomyces cerevisial*
C. *Saccharomyces boulardii*
D. *Saccharomyces pastorianys*

27. Which of the following species is used for vitamin B_{12} production by fermentation process?
A. *Streptomyces griseus*
B. *Streptomyces olivaceus*
C. *Bacillus megaterium*
D. All of these

28. Cephalosporin an antibiotic closely related to penicillin in structure and antibiotic activity are produced by species of :
A. *Cephalosporium*
B. *Cephalospodium*
C. *Pseudomonas*
D. *Bacillus*

29. The production of alcohol from lactose of whey is generally carried out by :
A. *Saccharomyces cerevisiae*
B. *Candida pseudotropicalis*
C. *Candida utilis*
D. None of these

30. Which of the following is not a principal media for industrial alcohol production by fermentation?
A. Black strap molasses
B. Sulphite waste liquor
C. Wood waste
D. Glucose

31. The precursor used for production of vitamin B_{12} by *Streptomyces olivaceus* is :
A. Cobalt chloride B. Cobalt sulphate
C. Calcium chloride D. Calcium sulphate

32. Bioremediation uses :
A. Microorganisms' natural capacities to break materials down
B. Added oxygen, water, and nutrients to increase rates of degradation
C. Organisms such as *Pseudomonas* and *Bacillus*
D. All of these

33. Industrial microbiologists use "tricks" to increase the amount of chosen end product; this may include :
A. Manipulating the growth environment
B. Using small scale containers to concentrate production
C. Selecting microbial strains that lack a feedback system
D. Both A and C

34. The production of substances in industrial microbiology occur in the sequence :
A. Fermentation, downstream processing, removal of waste, inoculation
B. Inoculation, downstream processing, fermentation, removal of waste inoculation
C. Fermentation, downstream processing, removal of waste
D. Removal of waste, inoculation, fermentation, downstream processing

35. Amino acids are used as food additives for which of the following reasons?
A. As natural antibiotics
B. As natural growth inhibitors
C. For nutritive purposes as antioxidants
D. As food preservatives

36. Commercial sterilization of processed canned foods uses a 12D treatment. This means that if there were 1012 endospores in a can of food, after treatment there would theoretically be:
A. 12 endospores B. 100 endospores
C. 0 endospore D. 1 endospore

37. All of the following can occur in canned foods stored at high temperatures except :
A. Flat sour spoilage may occur
B. The food can undergo mesophilic-spoilage
C. Thermophilic species of *Clostridium* may cause spoilage
D. *Bacillus stearothermophilus* spores may cause spoilage

38. All of the following are true of irradiation methods used for food preservation and sterilization except :
A. It makes foods safer for immuno compromised patients
B. It prolongs the shelf-life of fruits and vegetables
C. The food becomes radioactive
D. It eliminates pathogens present in meats and poultry

39. Several cans of food held in an automobile trunk during warm weather for several days are bulging at the ends. What type of spoilage is this?
A. Bacterial soft rot
B. Flat sour spoilage
C. Thermophilic anaerobic spoilage
D. Hydrogen sulphide spoilage

40. Which of the following microbial products and their uses are mismatched?
A. Phenylalanine and aspartic acid: aspartame
B. Glutamic acid : monosodium glutamate
C. Citric acid : sweetening
D. Glucose isomerase : fructose production agent

41. Industrially important microorganism are grown in large vessel containing nutrient media. This vessel called as ________ .
A. Baffles B. Sparger
C. Fermenter D. Impeller

42. Methods used to get immobilized enzymes :
A. Adsorption B. Encapsulation
C. Covalent bonding D. All of these

43. Raw-material used for the production of alcohol is :
A. Molasses
B. Starch
C. Sulphite waste water
D. All of these

44. Industrial Production of Vitamin-B_{12} is from
A. *Propionibacterium* sps.
B. *Pseudomonas* sps.
C. Both A and B
D. None of these

45. *Clostridium acetobutylicum* is used for the production of :
A. Acetone-Butanol B. Ethanol
C. Vitamin-B_{12} D. None of these

46. In the citric acid production, the pH to be maintained in the fermenter is :
A. 7.0 B. 5.0 to 6.0
C. 8.0 to 9.0 D. 1.0 to 6.0

47. The required temperature for the production of citric acid is :
A. 10ºC-80ºC B. 30ºC-50ºC
C. 20ºC-50ºC D. 25ºC-30ºC

48. Which of the statements is correct about single cell protein?
(*i*) *Spirulina* is good source of SCP.
(*ii*) High nucleic acid content of SCP results in kidney stone formation in humans.
(*iii*) They are dried cells of microorganisms such as algae, fungi, yeast, molds and bacteria.
A. Only (*i*) and (*ii*) are correct
B. Only (*i*) and (*iii*) are correct
C. Only (*ii*) and (*iii*) are correct
D. Only (*i*), (*ii*) and (*iii*) are correct

49. Which of the following are used as biopesticides and biofertilizers?
(*i*) NPV and *Mesorhizobium*
(*ii*) *Bacillus papillae* and *Azotobacter*
(*iii*) *Beauveria bassiana* and phosphate solubilizing bacteria
A. Only (*i*) and (*ii*)
B. Only (*i*) and (*iii*)
C. Only (*ii*) and (*iii*)
D. Only (*i*), (*ii*) and (*iii*)

50. In the process of composting, which of the statement is correct :
(*i*) It is a stabilization process carried out at the thermophilic temperature for 3-4 weeks.
(*ii*) The final product is free of pathogens
(*iii*) The product is used to enhance soil fertility and crop productivity
A. Only (*i*) and (*ii*)
B. Only (*i*) and (*iii*)
C. Only (*ii*) and (*iii*)
D. Only (*i*), (*ii*) and (*iii*)

ANSWERS

1	2	3	4	5	6	7	8	9	10
D	C	A	D	D	A	D	B	B	B
11	**12**	**13**	**14**	**15**	**16**	**17**	**18**	**19**	**20**
C	A	C	D	B	B	B	B	B	B
21	**22**	**23**	**24**	**25**	**26**	**27**	**28**	**29**	**30**
C	A	C	B	C	B	D	B	B	D
31	**32**	**33**	**34**	**35**	**36**	**37**	**38**	**39**	**40**
A	D	D	C	C	A	B	B	C	C
41	**42**	**43**	**44**	**45**	**46**	**47**	**48**	**49**	**50**
C	D	D	C	B	B	B	D	D	D

●●●

CHAPTER

22

Immunology

Introduction

It is the branch of biomedical science that deals with the defence mechanisms including all physical, chemical and biological properties of the organism that help it to combat its susceptibility to foreign organisms, material etc.

Immune Response

"A protective or defense reaction against foreign substances. These foreign substances, known as antigens, release from the surface of the lymphocytes, the signals for cellular or humoral immune response."

The immune defense mechanism of the body involves the action of white blood cells, or leukocytes. Leukocytes include **neutrophils, eosinophils, basophils and monocytes,** all of which are phagocytic, as well as two types of lymphocytes (T-cells & B-cells), which are not phagocytics but are critical to the specific immune response and the humoral response.

T-cells

A type of lymphocyte involved in cell mediated immune responses and interactions with β-cells, includes both cytotoxic T-cells and helper T-cells, also called T-lymphocytes. After their origin in the bone marrow, T-cells migrate to the thymus (hence the designation "T"), a gland just above heart. There they develop the ability to identify microorganisms and virus by the antigens exposed on their surfaces. Tens of millions of different T-cells are made, each specialized in the recognition of one particular antigen. No invader can escape being recognized by at least a few T-cells. There are four principal kinds of T-cells.

(A) *Helper T-cells (T_H)* : Commander of the immune response, detects infection and sounds the alarm, initiating both T-cell & B-cell responses.

(B) *Inducer T-cells* : Not involved in the immediate response to infection, mediates the maturation of other T-cells in the thymus *i.e.,* oversee the development of T-cells in thymus.

(C) *Cytotoxic T-cells (T_C)* : (cells-poisoning) Detects and kill infected body cells, recruited by helper T-cells.

(D) *Suppressor T-cells* : They terminate the immune response. Dampen the activity of T and B-cells, scaling back the defense after the infection has been checked.

B-cells

Unlike T-cells, B-cells do not travel to thymus, they complete their maturation in bone marrow (B-cells are so named because they were originally characterized in a region of chickens called the bursa). From the bone marrow, B-cells are released to circulate in the blood and lymph. Individual B-cells, like T-cells, are specialized to recognize particular foreign antigen. When a B-cell encounters the antigen to which it is targeted, it begins to divide rapidly, and its progency differentiate into plasma cells and memory cells.

Plasma cells

Biochemical factors devote to the production of antibodies (proteins) directed against specific foreign substance. Antibodies produced stick like flags to that antigen wherever it occurs in the body, making any cell bearing the antigen for destruction. The immunity that Pasteur observed resulted from, such antibodies and from the continued presence of the B-cells that produced them.

The humoral immune response, carried out by B-cells, protects the body from bacteria and other invading cells, by labelling these cells for destruction.

The B-cell lymphocytes that carry out this response produce and secrete antibodies that circulate in the blood plasma, lymph and other extracellular fluids, hence the term HUMORAL (relating to fluid) is used in describing their immune responses. In response to antigen exposure, a B-cell divides to produce plasma cells that serve as short lived antibody factories, and to produce long-lived memory cells.

Mast cells

Initiator of the inflammatory response, which aids the arrival of leukocytes, at a site of infection, secrete histamine and are important in allergic responses.

Monocyte

Precursor of macrophage.

Macrophage

The body's first cellular line of defense, also serves as antigen-presenting cell to B & T-cells and engulfs antibody covered cells.

Killer cells

They recognize and kill infected body cells, natural killer cells (NK) detect and kill cells infected by a broad range of invaders, killer (K) cells attack only antibody coated cells.

(Above mentioned are various type of cells of the immune system *i.e.,* (A-I)).

Active Immunity through Clonal Selection

B- and T-cells have receptors on their cell surfaces that recognize and bind to specific antigens. When a particular antigen enters the body, it must, by chance, encounter the specific lymphocyte with the appropriate receptor in order to provoke an immune response. The first time a pathogen invades the body, there are only a few B- or T-cells that may have the receptors that can recognize the invader's antigens. *Binding of the antigen to its receptor on the lymphocyte surface, however stimulates cell division and produces a clone (a population of genetically identical cells).* This process is known as *clonal selection.* In this first encounter, there are only a few cells that can mount an immune response and the response is relatively weak. This is called *Primary Immune Response.*

If the primary immune response involve B-cells, some become plasma cells that secrete antibodies and some become *memory cells.* Because a clone of memory cells specific for that antigen develops after the primary response, the immune response to a second infection by the same pathogen is swifter and stronger. The next time the body is invaded by the same pathogen, the immune system is ready. As a result of the first infection, there is now a large clone of lymphocytes that can recognize that pathogen. This more effective response, elicited by subsequent exposures to an antigen, is called a *Secondary Immune Response.*

Memory cells can survive for several decades, which is why people rarely contract measles a second time after they have had it once. Memory cells are also the reason that vaccinations are effective. The viruses causing childhood diseases have surface antigens that change little from year to year, so the same antibody is effective for decades. Other diseases, such as influenza are caused by viruses whose protein coat-sepcifying genes mature rapidly. This rapid genetic change causes new strains to appear every year or so that are not recognized by memory cells of previous infections.

Antibodies

Antibodies are proteins in a class called *Immunoglobulins* (abbreviated Ig), which is further divided into five classes based on the structures and functions of the antibodies. The different immunoglobulin subclasses are as follows :

(A) *Ig M* (μ, *H chain*) : This is the first type of antibody to be secreted during the primary response and they serve as receptors on the lymphocyte surface. These antibodies also promote agglutination reactions (causing antigen containing particles to stick together or agglutinate). Does not cross placenta.

(B) *Ig G* (γ, *H chain*) : This is the major form of antibody secreted in a secondary response and the major one in the blood plasma. Fixes complement, which enhances bacterial killing. Neutralizes bacterial toxin & virus. Crosses placenta. Fixes complement.

(C) *Ig D* (δ, *H chain*) : These serve as receptors for antigens on B-cells surface, also found in serum.

(D) *Ig A* (α, *H chain*) : This is the form of antibody in external secretions, such as saliva & mother's milk. Provide protection of nursing infants whose immune system is not developed fully. Does not fix complement.

(E) *Ig E* (ε, *H chain*) : This form of antibodies promotes the release of histamine and other agents that produce allergic symptoms, such as those of hay fever. They bind to *mast cells*. Does not fix complement.

Each antibody molecule consists of two identical short polypeptides, called *light chains* and two indentical long polypeptides called *heavy chain*. The four chains in antibody molecule are held together by disulphide (-S-S-) bonds, forming a Y shaped molecule.

Specificity of antibodies for antigens resides in the two arms of the antibody, which have a variable amino acid sequence. The amino acid sequence of the polypeptides in the stems of the Y is a constant within a given class of immunoglobulins. Most of the sequence variation between antibodies of different specificity is found in the terminal half of each arm. Here, six small *hypervariable segments* (three in the light chain and three in the heavy chain form a cleft that acts as the binding site for the antigen. Both arms always have exactly the same cleft and so bind to the same antigen. But antibodies may differ in the stem portions of the antibody molecule. The stem is formed by "constant" regions of the heavy chains. In mammals there are five different classes of heavy chains that form five classes of immunoglobulins. They include Ig M, Ig G, Ig A, Ig D & Ig E. Ig E antibodies bind to mast cells. The heavy-chain stems of the Ig E antibody molecules insert into receptors on the mast cell plasma membrane, in effect creating B-receptors on the mast cell surface. When these cells encounter the specific antigen recognized by the arms of the antibody, they initiate the inflammatory response by releasing histamine. The resulting vasodilation and increased capillary permeability enable lymphocytes, macrophages, and complement proteins to more easily reach the site where the mast cell encountered the antigen.

Each B-cell has, on its surface, about 100,000 Ig M or Ig D receptors. Unlike the receptor on T-cells which bind only to antigens presented by certain cells. B-receptors can bind to free antigens. This provokes a primary response in which antibodies of the Ig M class are secreted and also stimulates cell division and clonal expansion. Upon subsequent exposure, the plasma cells secrete large amount of antibodies that are generally Ig G class. Although plasma cells live only a few days, they produce a vast number of antibodies. In fact antibodies constitute about 20% by weight of the total protein in blood plasma. Production of Ig G antibodies peakes after about three weeks.

When Ig M (and to a lesser extent Ig G) antibodies bind to antigens on a cell, they cause the aggregation of complement proteins. These proteins form a pore that pierces the plasma membrane of the infected cell, allowing water to enter and causing the cell to burst. In contrast, when Ig G antibodies bind to antigens on a cell, they serve as markers that stimulate phagocytosis by macrophages. Since certain complement proteins attract phagocytic cells, activation of complement is generally accompanied by increased phagocytosis. Antibodies don't kill invading pathogens directly, rather, they cause destruction of the pathogens by activating the complement system and by targetting the pathogen for attack by phagocytic cells.

All immunoglobulin molecules consists of two indentical light chains (23 KDa) and two identical heavy chains (H) (53-73 KDa), held together as a tetramer (L_2H_2) by disulphide bonds. Each chain can be divided conceptually into specific domains or regions that have structural and functional significance. The half of the light chain (L) toward the carbonyl terminal is referred to as *constant region* (C_L); while the amino terminal half is the *variable region* of light chain (V_L). Approximately one quarter of the heavy chain (H) at the amino terminal is referred to as its variable region (V_H), and the other three quarters of the heavy chain are referred to as the constant regions $\left(C_H{}^1, C_H{}^2, C_H{}^3\right)$ of that H chain. The portion of immunoglobulin molecule that binds the specific antigen is formed by the amino terminal portion (variable regions) of both H & L chains *i.e.,* V_H & V_L domains. The domains of the protein chain consist of two sheets of antiparallel distinct stretches of amino acids that bind antigen. Antigen binding fragment is termed Fab and crystallising fragment as Fc, which is responsible for functions of immunoglobulins other than direct binding of antigens. In Ig bacteriaG, two Fab regions are there, so it binds two molecules of antigen and are termed **divalent**. The site on the antigen to which an antibody binds is termed an *antigenic determinant* or **Epitope**. The region between $C_H{}^1$ & $C_H{}^2$ is called *Hinge region* which confers flexibility and allows both Fab arms to move independently, thus helping them to bind to antigenic site that may be variable distances apart (*e.g.,* on bacterial surfaces). Fc and hinge regions differ in the different classes of antibodies. Fab includes CH_2CH_3. There are two general types of light chains, Kappa (κ) and Lambda (λ), which can be distinguished on the basis of structural differences in their C_L regions. A given immunoglobulin molecule always contain two κ and λ light chains, never a mixture of κ & λ. In humans, κ chains are more frequent than λ chains in immunoglobulin molecules.

Five classes of H chains have been found in humans; distinguished by differences in their C_H regions γ, α, μ, δ, ε. The μ and ε chains each have 4 C_H domains rather than the usual three. The type of H chains determines the class of immunoglobulin and thus its effector functions. No two variable regions are identical.

Immunological Tolerance

A mature animal's immune system normally does not respond to that animal's own tissue. This acceptance of self cells is known as *Immunological Tolerance.* The immune system of an embryo is able to respond to both foreign and self molecules but it loses the ability to respond to self molecules as its development proceeds.

Immunological tolerance sometimes break-down, causing either B or T-cells (or both) to recognize their own tissue antigens. This loss of immune tolerance results in autoimmune disease like *Myasthenia gravis* in which individuals produce antibodies directed against acetylcholine receptors on their own skeletal muscle cells, causing paralysis.

Monoclonal Antibodies

These antibodies are produced by cells that arise from successive divisions of a single B-cell, and hence all of the antibodies target a single antigenic determinant site. Such antibodies are used for a variety of medical application, including pregnancy testing.

Role of MHC Proteins in T-cell Function

MHC proteins serve as 'self labels', identifying leukocytes and every other body cell as belonging to a specific individual. There are two classes of MHC proteins. MHC-I is present on every nucleated cell of the body & MHC-II is found only on macrophages, B-cells and a subtype of T-cell called CD^{4+} T-cells. These three cell types work together in one form of immune response and their MHC-II markers permit them to recognize one another.

Cytokines

Each cytokine has many different actions, named *Interleukin* followed by a number, to indicate a cytokine whose amino acid sequence has been determined is now used.

Interleukin-1 → Secreted by macrophages and can activate the T-cell system.

Interleukin-4 → B-cell stimulating factor → Is secreted by T-cells and is required for the proliferation and clone development of B-cells.

Interleukin-2 → Released by helper T-cells and among its effects is that it is required for the activation of cytotoxic T lymphocytes.

Autoimmune Diseases

In case autoimmune diseases and allergies, the immune system is the cause of the problem, not the cure. These are produced by failure of the immune system to recognize and tolerate self antigens. This failure results in the activation of autoreactive T-cells and the production of autoantibodies by B-cells, causing inflammation and organ damage. There are over 40 known or suspected such diseases that affect 5 to 7% of the population.

Autoimmune diseases can result from a variety of mechanisms. The self antigen may normally be hidden from the immune system so that the immune system treats it as foreign if exposure later occurs. Protein normally trapped in thyroid follicles triggers autoimmune destruction of thyroid (Hashimoto's thyroiditis). Immune suppression is generally accomplished with corticosteroids (including hydrocortisone) and by nonsteroidal anti-inflammatory drugs, including aspirin.

Allergy (hypersensitivity)

Refers to particular types of abnormal immune responses to antigens, which are called *allergens.* Two major forms of allergy are :

(*i*) *Immediate hypersensitivity :* This is due to an abnormal B-cell response to an allergen that produces symptoms within seconds or minutes. *e.g.*, conjunctivitis, rhinitis (runnynose), asthma.

(*ii*) *Delayed hypersensitivity :* This is an abnormal T-cell response that produces symptoms within about 48 hours after exposure to an allergen.

Multiple Choice Questions

1. The primary purpose of the adaptive immune system is to :

A. kill tumor cells
B. cure allergic reactions
C. protect from disease upon re-infection with a specific pathogen
D. block all pathogens from entering the body

2. In human B cells and T cells are matured in :

A. Bone marrow and thymus respectively
B. Lymph nodes and spleen respectively
C. Bursa and thymus respectively
D. None of these

3. In monoclonal antibody technology, tumor cells that can replicate endlessly are fused with mammalian cells that produce an antibody. The result of this fusion cell is a :

A. Lymphoblast B. Hybridoma
C. Myeloma D. Natural killer cell

4. Neutrophils provide infection from a variety of microbes and are arguably the most important WBC's eliminating :

A. Non-viral infections
B. Viral infections
C. Fungal infections
D. Parasitic infections

5. Lack of thymus leads to increased infection in humans such a condition is called :

A. Nude mice
B. Thymectomy
C. Di George syndrome
D. None of these

6. The antibiotic penicillin is a small molecule that does not induce antibody formation. However, penicillin binds to serum proteins and forms a complex that in some people induces antibody formation resulting in an allergic reaction. Penicillin is therefore :
   A. An antigen
   B. An immunogen
   C. A hapten
   D. None of these

7. In addition to adjuvants, generation of anti-hapten antibodies will require :
   A. Injection of the mixture of hapten and lipopoly saccharide
   B. Injection of the hapten into mice
   C. Injection of a mixture of hapten and protein
   D. Injection of the hapten covalently coupled with protein

8. The immunoglobulin that results in histamine release is :
   A. Ig E B. Ig D
   C. Ig G D. Ig M

9. Which is the predominant antibody in saliva is?
   A. Ig D B. Ig A
   C. Ig G D. Ig M

10. Type I hypersensitivity is known as :
    A. Contact hypersensitivity
    B. Agglutination
    C. The transfusion reaction
    D. Anaphylaxis

11. An autoimmune disease of humans usually involving anti-nuclear antibodies :
    A. SLE
    B. Rheumatic fever
    C. Myasthemia gravis
    D. Sclerosis

12. Allergy symptoms are produced when antigen binds to Ig E on FcR on :
    A. Neutrophils
    B. Macrophages
    C. Th_1 cells
    D. Mast cells

13. The form of diabetes curable by insulin injection is :
    A. Diabetes rugosa
    B. Insulin dependent diabetes mellitus
    C. Insulin independent diabetes mellitus
    D. None of these

14. Tc Cells are important in controlling :
    A. Allergy B. Autoimmunity
    C. Viral infection D. All of these

15. T cells are the source of :
    A. Interleukin B. Interferon
    C. Lymphotoxin D. All of these

16. Some cross reactions with monoclonal antibodies (MAbs) can occur. Unexpected cross reactions occur more frequently with :
    A. Ig MAbs B. Ig A
    C. Ig E D. Ig G

17. Antibodies are :
    A. Proteins B. Nucleic Acid
    C. Glycoproteins D. Carbohydrates

18. Antibodies consists of :
    A. 2 light chains and 2 heavy chains arranged in a Y-shaped configuration
    B. A light chain and 2 heavy chains arranged in a Y-shaped configuration
    C. 2 light chains and a heavy chains arranged in a Y-shaped configuration
    D. All of these

19. Which is the largest Ig?
    A. Ig M B. Ig G
    C. Ig A D. Ig E

20. Ig that mediates allergic reaction is :
    A. Ig M B. Ig A
    C. Ig C D. Ig E

21. Antigens can be :
    A. Carbohydrates B. Nucleic acids
    C. Proteins D. All of these

22. Skin prevents pathogen entry :
    A. Dead cells of the skin outer layer prevents pathogen entry
    B. Skin secretes oil by sebacous gland that makes surface acidic

C. Skin secretes sweat by sudoriferous gland that makes surface acidic
D. All of these

23. Stomach clearout pathogens by :
A. Secreting HCl
B. Secreting digestive enzymes
C. Both (A) and (B)
D. None of these

24. Vaginal bacterial symbionts like lactobacilli prevents pathogen by :
A. Producing lactic acid thereby reducing pH.
B. Secreting toxins
C. Secreting antibiotics
D. None of these

25. Bodies internal defence or second line of defence include :
A. Phagocytes B. Interferons
C. Fever D. All of these

26. Kuffer cells are macrophages found on :
A. Bone B. Nephrons
C. Liver D. Lungs

27. Tissue damaged caused by wound on invading pathogenic organisms induces a complex sequence of events collectively known as :
A. Phagocytosis
B. Inflammation
C. Opsonisation
D. None of these

28. Internal second line of defence involves all except :
A. Antibodies
B. Natural killer cells
C. Interferons
D. Complement system

29. Antiviral glycoproteins released by living cells in response to viral attack and induce a viral resistant state to neighbouring cells is called as :
A. Natural killer cells
B. Phagocytes
C. Complement system
D. Interferons

30. Humoral immunity is mediated by :
A. Antibody mediated immunity
B. Non-specific immune response
C. Antigen mediated immunity
D. All of these

31. Infection of anti-venom against snake bite is an example of :
A. Active immunity
B. Passive immunity
C. Phagocytic immunity
D. Non-specific immunity

32. The inappropriate response of immune system towards a relatively harmless antigen causing harm to the host is referred as :
A. Hypersensitivity
B. Immunodeficiency
C. Tolerance
D. Auto-immune diseases

33. Rheumatoid arthritis mostly occur in individual carrying :
A. HLA-DR4 gene (HLA-human leucocyte antigen)
B. HLA-DR1 gene
C. HLA-DR3 gene
D. All of these

34. In severe combined immune deficiency (SCID), the patients are deficient in :
A. B-cells B. T-cells
C. Both A and B D. Ig A

35. HIV attacks :
A. T-helper cells B. B-cells
C. T-cytotoxic cells D. Macrophages

36. Severe Combined Immune Deficiency (SCID) can occur due to absence of an enzyme :
A. Adenosine deaminase
B. Guanosine deaminase
C. Phosphorylase
D. Thymidine deaminase

37. The transfer of tissue between individuals of different species is known as :
A. Autograft
B. Xenograft
C. Syngeneric graft
D. Allograft

38. The major molecules responsible for rejection of transplant is :
A. B-cells
B. Antibodies
C. T-cells
D. MHC (Major Histocompatibility Complex) molecule

39. The process of introduction of weakened pathogen into human body is called :
A. Vaccination B. Attenuation
C. Immunization D. None of these

40. The concept of vaccination was first developed by :
A. Louis Pasteur B. Edward Jenner
C. Carl Landsteiner D. Joseph Miester

41. Which of the following statements are true regarding transplantation?
A. The compatibility of MHC proteins of donor and recipient will determine the success of transplantation
B. MHCs are just like fingerprints and all nucleated cells possess, this fingerprint
C. The compatibility of MHC/HLA proteins of donor and recipient will be high if they are genetically closely related and may lead to successful transplantation
D. All of these

42. The genes of HLA proteins are clustered in the major histocompatibility complex located :
A. On the short arm of chromosome 6
B. On the long arm of chromosome 6
C. On the short arm of chromosome 10
D. On the long arm of chromosome 10

43. Which of the following is a combined vaccine?
A. Hib vaccine B. Var vaccine
C. DPT vaccine D. Hepatitis B vaccine

44. Which of the following is a polysaccharide vaccine?
A. Rabies vaccine B. Hib vaccine
C. Anthrax vaccine D. Hapatitis A vaccine

45. Which of the following is a combined vaccine?
A. Hepatitis B vaccine
B. Var vaccine
C. Hib vaccine
D. DPT vaccine

46. The first recombinant antigen vaccine approved for human use is :
A. Hepatitis B vaccine
B. DPT vaccine
C. Hib vaccine
D. Var vaccine

47. What is the characteristic of early stages of local inflammation?
A. Fever
B. Anaphylactic shock
C. Release of histamine
D. Attack by cytotoxic T cells

48. Live vaccine is :
A. Low dose of infectious bacteria administered as prophylactic
B. A dose of bacterial strain in a modified form which retains immunogenicity but is not pathogenic
C. A low dose of toxin that is produced by the bacterium
D. A sample of cells from a patient who recently recovered from the disease

49. Which one of the following character is not found in natural killer cells?
A. Ability to lyse a variety of tumor cells and virally infected cells without prior sensitization
B. Inability to rearrange TCR (Antigen Specific Receptor) genes
C. Presence of two cell surface molecules, CD 16 and CD 56
D. Much smaller than T and B lymphocytes

50. A fundamental difference between the antigen receptor on the B Cells (BCR) and on T cells (TCR) is their :
A. Different requirements for antigen presentation
B. Function following antigen binding
C. Heterogeneity from one lymphocyte to the next
D. Heterogeneity on each lymphocyte

ANSWERS

1	2	3	4	5	6	7	8	9	10
C	A	B	A	C	C	D	A	B	D
11	**12**	**13**	**14**	**15**	**16**	**17**	**18**	**19**	**20**
A	D	B	D	C	A	C	A	A	D
21	**22**	**23**	**24**	**25**	**26**	**27**	**28**	**29**	**30**
D	D	A	A	D	C	B	A	D	A
31	**32**	**33**	**34**	**35**	**36**	**37**	**38**	**39**	**40**
D	A	A	C	A	A	B	D	A	B
41	**42**	**43**	**44**	**45**	**46**	**47**	**48**	**49**	**50**
D	A	C	B	D	A	C	B	D	A

●●●

CHAPTER

23

Bioethics and Biosafety

Biosafety or the prevention of potential risks from products derived from genetic engineering has emerged as an important scientific and political concern worldwide.

- The Cartagena protocol on biosafety, 2003, negotiated under the United Nations Convention on Biological Diversity (CBD) is the first legally binding international agreement that purports to safeguard biodiversity concerns by regulating GMOs intended for use in agriculture.
- The Cartagena protocol on Biosafety to the convention on Biological Diversity states that the protocol aims to ensure an adequate level of protection in the field of the safe transfer, handling and use of living modified organisms resulting from modern biotechnology that may have adverse effect on the conservation and sustainable use of biological diversity, taking into account the risks to human health and specifically focusing on transboundary movements.
- In India, the Ministry of Environment and Forest (MoEF) promulgated in December, 1989 the rules and procedures for the manufacture, import, use, research and release of GMOs as well as products made from such organisms; this was done under the provisions of the Environment Protection Act, 1986 (EPA).
- In addition, The Indian Recombinant DNA safety Guidelines and Regulations have been prepared (initially in 1990; revised in 1994) by and are available on request from Recombinant DNA Advisory Committee (RDAC), Department of Biotechnology (DBT), New Delhi.
- The DBT implements the research and development experiments utilizing GMOs and recombinant DNA products, while MoEF implements the large scale commercial use of these.
- The Indian regulatory structure comprises of six committees to regulate GMOs.
  1. *Recombinant DNA Advisory Committee (RDAC)*
  2. *Institutional Biosafety Committee (IBSC)*
  3. *Review Committee on Genetic Manipulation (RCGM)*
  4. *Genetic Engineering Approval Committee (GEAC)*
  5. *State Biotechnology Coordination Committee (SBCC)*
  6. *District Level Committee (DLC)*

The salient features of these guidelines and regulations are as follows (Ghosh, 1997).

1. Every organization involved in research and development using recombinant DNA technology is required to set up an Institutional Biosafety Committee (IBC), which has a DBT nominee.
   - IBC is the nodal point for interaction of the organization with the Government.

- It is the competent authority to which experiments that are likely to have bid-hazard potential and genetic manipulation of plants under containment have to be reported.
- It also authorizes interstate exchange of etiologic agents, diagnostic specimens, etc. IBC will provide half-yearly reports on the on-going projects to Review Committee for Genetic Manipulation (RCGM).

2. The DBT has a RCGM, which reviews all the approvals of ongoing projects on GMOs and several other issues related to recombinant DNA research and develop-ment. Based on the recommendations of RCGM, trial permits are issued by DBT.
3. Each state has, in addition, a State Biotechnology Coordination Committee, and a District Level Committee, which are also involved, along with RCGM, in the inspection and monitoring of the experiments at the field sites.
4. The MoEF has an inter-ministerial committee, called The Genetic Engineering Approval Committee (GEAC), which has subject specialists as members, and is the competent authority to decide on the large-scale use of GMOs.
   - As in the case of GEAC, state and district level committees also participate, along with GEAC, in the watch on the field applications of GMOs.
   - GEAC would also involve the user Ministries and the other regulatory authorities, *e.g.* Directorate General of Health Services in the case of drugs and pharmaceuticals, Ministry of Agriculture for transgenic plants and other organisms related to agriculture, etc.
5. The guidelines recognize four levels of risk in case of experiments with microorganisms, based on pathogenicity of the microorganisms, local prevalence of the concerned disease and of epidemic causing strains in India.
6. Experiments with microorganisms, plants and animals are grouped into the following three categories:
   (*i*) Exempt category (for self-cloning experiments),
   (*ii*) Category requiring intimation of initiation to competent authority (*e.g.* experiments involving non-pathogenic DNA vector systems), and
   (*iii*) Category requiring review and approval by the competent authority (*e.g.* cloning of genes for toxins, antibiotic resistance, etc.).
7. Four different biosafety levels are recognised and containment facilities for each level are recommended for necessary safeguards.
8. Physical containment envisages to limit the spread of dangerous microorganisms by good laboratory practices, safety equipment and laboratory design and facilities.
9. Biological containment consists of the use of vectors and hosts in such a way so that it can limit the infectivity of vector to specific hosts and, control host-vector survival in the environment.
10. Experiments with plants are to be done under two types of special environment that may be achieved by using glass-house containment.
   (*i*) Glass-house containment A is used when no plant pathogens are used, and where non-pathogenic vector systems and regeneration from single cells are used.
   (*ii*) Glass-house containment B is recommended for experiments involving genetically manipulated plant pathogens, including plant viruses, and the growth of plants regenerated from cells transformed by genetically manipulated pathogen vector system, which still contain the pathogen.

11. Application for recognition of research facility to carry out genetic manipulation should be made to the Department of Environment before the commencement of work.
12. All products obtained using recombinant DNA technology shall be subject to the general regulations normally applicable for such products.
13. The controlled release of GMOs should be done under appropriate containment facilities to ensure safety and to prevent unwanted release in the environment.
14. Prerelease tests of GMOs in agriculture should include elucidation of requirements for vegetative growth and persistence and stability in small plots and experimental fields.
15. Planned field experiments with transgenic plants are permitted only after a step-wise (laboratory to growth chamber and green-house) evaluation, either in India or else-where, to generate data on the following:
    - Promoter sequence,
    - Target gene sequence,
    - Regulatory mechanism utilized in the expression cassette,
    - Diagram of the expression Cassette to describe fully the marker genes used,
    - Cell lines used for shuttling and amplification of the cassette and several other details of the procedure, and the consequences thereof, used for production of the transgenic plants.

Bioprospecting

Bioprospecting, also known as biodiversity prospecting is the exploration of biological material for commercially valuable genetic and biochemical properties.

- It involves mainly systemic search for and development of new sources of chemical compounds, genes, microorganisms, macro-organisms and other valuable products from nature *i.e.*, looking for ways to commercialize biodiversity.
- Biodiversity refers to all living things, including plants, animals, insects and marine life.
- When biodiversity or related knowledge is collected without permission from the owners of these resources and then patented, it is known as biopiracy.
- The main aim of bioprospecting is to find new resources and products from nature that can be used by humans.
- Bioprospecting plays a dominant role in discovering leads for drug development, since existing/known compounds for developing drugs for human use are limited.
- Bioprospecting can be divided into three phases: collecting, analysis and commercia-lization.
- The most significant change and the starting point of international regulation and legislation took place at the Rio Earth Summit in 1992, when participating countries discussed bioprospecting and signed the Convention of Biological Diversity (CBD).
- The CBD recognizes the need for conservation, sustainable use and equitable benefit sharing as cornerstones, and for the first time, acknowledged the sovereign rights of countries over their natural resources.

Multiple Choice Questions

1. *E. coli* model organism database is :

A. ColGene
B. EcoGene
C. EcoBase
D. EcoSeq

2. Which is the comprehensive database for the study of human genetics and molecular biology?

A. PSD B. STAG
C. OMIM D. PDB

3. In the early, 1970s which of the following were considered as the possible potential health hazards of cloning recombinant DNA molecules?
   A. Hybrid organism may escape from the laboratory with unpredictable consequences
   B. Molecular biologist may not be well versed with the laboratory practices needed to handle serious human and animal pathogens
   C. Hybrid organism could be created with biological activities of an unpredictable nature
   D. All of these

4. Which of the following is not correct about NIH guidelines?
   A. In 1977, NIH prepared an Environmental Impact Statement
   B. The first NIH guidelines were prepared in 1975
   C. NIH guidelines were more liberal than the recommendations of the Asilomar conference
   D. The NIH guidelines were prepared in 1975.

5. For biosafety considerations, modern biotechnology includes which of the following?
   I. Recombinant DNA techniques
   II. Direct injection of nucleic acids into cells or organells
   III. Fusion of cells beyond the taxonomic family that overcomes natural, physiological, reproductive or recombination barrirss.

   Options:
   A. I, III　　B. I, II
   C. I, II, III　　D. II, III

6. Which of the following terms is not used as synonym of planned introduction?
   A. Contained use
   B. Deliberate release
   C. Field Testing
   D. None of these

7. The Biosafety guidelines are developed to ensure an adequate level of protection during which of the following activities of modern biotechnology?
   A. Use of GMOs　　B. Safe transfer
   C. Handling　　D. All of these

8. The chief objectives of risk assessment during laboratory research are to decide about which of the following of the proposed research?
   A. Physical containment
   B. Biological containment
   C. Laboratory containment
   D. All of these

9. In which of the following risk groups, agents are not associated with disease in healthy humans?
   A. Risk group 1
   B. Risk group 2
   C. Risk group 3
   D. Risk group 4

10. Which of the following statements is correct about comprehensive risk assessment?
    A. The level of containment appropriate for an experiment is decided on the basis of comprehensive risk assessment
    B. It is done after the initial risk assessment
    C. The containment level for the experiment may, in the end, be comparable to that of the risk group (RG) of the agent or it may be raised or lowered as a result of the comprehensive risk assessment
    D. All of these

11. In case of *E. coli,* the host for any HV1 system is which of the following?
    A. Strain K–12
    B. A derivative of strain K–12
    C. Either (A) or (B)
    D. Strain Chi-177L

12. The vector under host-vector 1 system includes which of the following?
    A. Variants of bacteriophages
    B. Derivative of non-conjugative plasmids
    C. Non-conjugative plasmids
    D. All of these

13. Non-infectious illness like allergic asthma and dermatitis are associated with which of the following production processes?
A. Single cell protein
B. Antibiotic
C. Brewing
D. All of these

14. Which of the following non-infectious illnesses is associated with endotoxin production?
A. Asthma
B. Dermatitis
C. Flu symptoms
D. All of these

15. Which of the following hazards is associated with operation of bioreactor fermentation?
A. Aerosols from reactor
B. Effluent contamination of gases
C. Spillage
D. All of these

16. Which of the following non-infectious illnesses is associated with production of steroids?
A. Hypertension
B. Increased body weight
C. Feminization of males
D. All of these

17. In case of experiments with animals, which of the following procedures is followed at all biosafety levels?
A. The animals are confined to security fenced areas
B. Animal carcass is disposed off to avoid its use as food
C. A permanent record is maintained of the experimental use and disposal of each animal
D. All of these

18. According to biosafety guidelines of India, physical containment envisages to limit the spread of dangerous microorganisms by which of the following?
A. Safety equipment
B. Laboratory practices
C. Laboratory design and facilities
D. All of these

19. Crop yields can be stabilized by introducing which of the following genes in transgenic varieties?
A. Genes for better tolerance/resistance to biotic and/or abiotic stresses
B. Genes for improved nutritional quality
C. Genes for herbicide resistance
D. All of these

20. The rate of gene-flow is affected by which of the following factors?
A. Mating system
B. Mode of seed dispersal
C. Mode of pollination
D. All of these

21. The potential risks from novel DNA sequences used during transgenic development may be summed up as which of the following?
A. Protein products of transgenes or products generated by them may be toxic
B. The transgene product may be allergic
C. Antibiotic resistance genes may be taken up by human intestinal microflora, which, as a consequence, become resistant to these antibiotics
D. All of these

22. Which is correct about wheat germ agglutinin?
A. It is effective against insects at concentration of 10 mg/kg body weight
B. It produces cytotoxic effects in rats at a lower concentration (7 mg/kg)
C. Safe food has it only at 0.3 mg/kg
D. All of these

23. Which of the following is not correct about bromoxynil?
A. It is non-biodegradable herbicide
B. Bromoxynil is absorbed by the skin
C. It causes birth defects in laboratory animals
D. It is toxic to fish and may cause cancer in humans

24. For minimising pollen-mediated gene-flow from transgenic varieties of *Solanum tuberosum,* how many meters of isolation distance is recommended?
A. 25 B. 40
C. 20 D. 60

25. The principal purpose of transgenic plant containment is to avoid the unintentional transmission of a recombinant DNA containing plant genomes including which of the following?
A. Nuclear hereditary material
B. Organellar hereditary material
C. Release of recombinant DNA-derived organisms associated with plants
D. Any of the above

26. "Toxic Shock Syndrome" is caused by the toxin of :
A. *Vibrio cholerae*
B. *Streptococcus pyogenes*
C. *Candia*
D. *Staphylococcus aureus*

27. Seed sterility can be reduced by using which of the following techniques?
I. Terminator technology
II. Transgenic mitigation
III. Recoverable block of function (RBF)
IV. Apomixis
Options:
A. I, II B. I, II, III
C. I, II, IV D. I, II, III, IV

28. Most of the recombinant microorganism released in the environment consisted of only a simple inactivation of genes by which of the following?
A. Genetic tagging of bacteria with specific markers
B. Transposon insertion
C. Deletion of genes by molecular genetic procedure *in vitro*
D. Any of the above

29. The α-amylase inhibitor from *Phaseolus vulgaris* (Rajma) caused allergy when expressed in which of the following by Australian scientists?
A. Peas B. Tomato
C. Maize D. Rice

30. The assessment of risk during laboratory research is usually done in which of the following steps?
A. Initial risk assessment
B. Comprehensive risk assessment
C. Advanced risk assessment
D. Both A and B

31. In case of which of the following products, an empirical approach based on data obtained through experimentation about the source of potential hazards has proved satisfactory?
A. Drugs B. Vaccines
C. Pesticides D. All of these

32. In risk assessment, a lack of scientific knowledge or scientific consensus should not necessarily be interpreted as indicating which of the following?
A. Absence of risk
B. A particular level of risk
C. Acceptable risk
D. Any of the above

33. Physical containment relates to which of the followings?
A. Laboratory design
B. Special procedures
C. Containment equipment
D. All of these

34. A transgenic variety of which of the following crop had to be withdrawn from cultivation following deaths due to allergy?
A. Soyabean B. Tomato
C. Maize D. *Brassica napus*

35. Which of the following is not correct?
A. Transgenes integrated into plasmids may be transferred to other bacteria
B. Auxotrophic mutants can survive in nature
C. Rec A^- strains lack recombination
D. Transposon vectors lacking transposase gene are not able to transpose to other bacteria

36. Transgenic plants could serve as vectors for transgene dissemination through which of the following?
A. Pollen
B. DNA released from debris of transgenic plants
C. Seed
D. All of these

37. Which of the following is not correct about Biosafety Guidelines in India?
A. Canada become the first country to begin commercial use of virus resistant transgenic tobacco and tomato in the early 1990
B. Fields trials, in India, using transgenic plants began in 1995
C. Initially, the Indian Recombinant DNA safety Guidelines and Regulations were prepared in 1990 by Recombinant DNA Advisory Committee, Deptt. of Biotechnology, New Delhi
D. The Indian Recombinant DNA Safety Guidelines and were revised in 1994

38. Which of the following is considered as primary physical containment?
A. Laboratory design
B. Containment equipment
C. Special procedures
D. Both A and B

39. It is feared that transgenic plants, may lead to which of the following?
A. Degrade the environment
B. May themselves become weeds
C. Transfer of their transgenic features to weeds
D. All of these

40. Which of the following is known to be a food allergen?
I. Wheat germ agglutinin (WGA)
II. Soyabean kunitz trypsin inhibitor
III. Trypsin inhibitor
IV. Peanut lectins
Options:
A. I, II, III, IV B. I, II, III
C. I, III, IV D. II, IV

41. The NIH, USA guidelines specify the practices for constructing and handling :
A. Recombinant DNA molecules
B. Organisms and viruses containing recombinant DNA molecules
C. Both (A) and (B)
D. None of these

42. Transgenic crops are perceived by many to pose which of the following threats to the environment?
I. They have adverse impact on biodiversity
II. Insect resistant transgenic crops may affect non target species.
III. They may encourage use of agro-chemicals.
IV. They may after the composition of rhizosphere and phyllosphere microflora
Options:
A. I, II, III, IV B. II, III, IV
C. I, II, III D. I, II, IV

43. In *Agrobacterium,* the *vir* operons are subject to which of the following mechanisms of regulation?
A. Inducible negative
B. Inducible positive
C. Repressible negative
D. Repressible positive

44. Which of the following is correct about yeast *Ty* elements?
A. It is a retroposon
B. It is a composite transposon
C. It produce two *m*RNAs
D. It produces infective particles

45. Synaptonemal complex is associated with which of the following?
A. Paired meiotic chromosomes
B. Polytene chromosomes
C. Lampbrush chromosomes
D. Mitotic chromosomes

46. In *E. coli,* which of the following is not required for transcription elongation phase?
A. β-subunit
B. β′-subunit
C. σ-Factor
D. Both (A) and (B)

47. Choose the right combination of components required to set up a polymerase chain reaction from the following :

A. Template DNA, two primers d NTP's and DNA ligase
B. Template DNA, two primers NTP's and DNA ligase
C. Template RNA, two primers NTP's and DNA polymerase
D. Template RNA, two primers d NTP's and DNA polymerase

48. Which of the following involves preventing the transmission of diseases in the laboratory?

A. Biohazard B. Biosecurity
C. Biosafety D. Bioinformatics

49. Which of the following is the nodal center for Indian biosafety framework?

A. Department of Science and Technology
B. Ministry of Environment and Forests
C. ICAR-Indian Agricultural Research Institute
D. Department of Biotechnology

50. Reverse isolation would be appropriate for :

A. A patient who has had minor surgery
B. A patient with glaucoma
C. A patient with leukemia
D. A patient with tuberculosis

ANSWERS

1	2	3	4	5	6	7	8	9	10
B	C	D	C	C	A	D	D	A	D
11	**12**	**13**	**14**	**15**	**16**	**17**	**18**	**19**	**20**
C	D	A	C	D	D	D	B	A	D
21	**22**	**23**	**24**	**25**	**26**	**27**	**28**	**29**	**30**
D	D	A	C	D	D	A	D	A	D
31	**32**	**33**	**34**	**35**	**36**	**37**	**38**	**39**	**40**
D	D	C	A	B	D	A	D	D	A
41	**42**	**43**	**44**	**45**	**46**	**47**	**48**	**49**	**50**
C	A	B	A	A	B	D	C	D	A

Your Space